# Laser Optofluidics in Fighting Multiple Drug Resistance

**Edited by**

**Mihail Lucian Pascu**
*Laser Department, National Institute for Laser, Plasma and Radiation Physics, Romania*

**Bentham Science Publishers Ltd.**
Executive Suite Y - 2
PO Box 7917, Saif Zone
Sharjah, U.A.E.
Email: subscriptions@benthamscience.org

BENTHAM SCIENCE

# CONTENTS

# FOREWORD

The book has an inciting title, "*Laser Optofluidics in Fighting Multiple Drug Resistance*" and is dedicated to a subject of high interest that is a challenge for the biomedical specialists as well as chemists, physicists, public health experts and even outer space applicants: fighting multiple drug resistance acquired by bacteria and tumours in normal and/or extreme conditions.

The editor and the invited authors propose two action lines, each of them implying pluridisciplinary experiments and data interpretation:

1. Exposure of selected non-antibiotic medicines at UV pulsed laser beams to modify their chemical structure and generate photoproducts with enhanced properties in fighting multiple drug resistance. At the origin, the parent compounds (mainly phenothiazines, quinazolines and hydandoin derivatives) do not have significant effects on bacteria or tumour tissues, but after being exposed to laser radiation in water solutions they generate photoproducts with individual or synergistic effects on biological targets. The book shows most recent results in the action of exposed chlorpromazine and thioridazine on Gram-positive and Gram-negative bacteria and their antibacterial and antibiofilm enhanced activity. Complementary, a report about clinically used methotrexate exposed to continuous wave (UV-Vis) optical radiation emitted by lamps and then utilised on eye pseudotumours evidenced that the mixture of photoproducts has in some cases anti-inflammatory effects higher than the parent compound. Another clinically used cytostatic, 5–Fluorouracil, exposed to UV pulsed nitrogen laser beams evidenced the same effects as methotrexate.
2. The "simple" identification of the obtained photoproducts constitutes a complex problem since the photochemistry of the processes is quite complicated; consequently, many procedures are utilised with this purpose and the obtained results are described in detail in the book. One speaks about laser spectroscopy (fluorescence), mass spectrometry, thin layer chromatography, UV–Vis and FTIR absorption spectroscopy, microfluidics (surface tension, contact angles, wetting properties) measurements and many others. This is correlated with the rigorous description of microvolumetric droplets as vectors to transport medicines to targets by applying microfluidics methods and procedures. Particular attention is devoted to description of the interaction – unresonant and resonant – between a laser beam and a single droplet which is of atmost interest in biomedical applications since it allows to fast modify the content of a microdroplet and to send parts of it on the target.

In presenting results, the editor took care that the book provides more interdisciplinary and multidisciplinary information about: the laser systems used to modify pendant droplets and bulk solutions, the properties of laser beams with emphasis on those of them which are essential in reported applications, the behaviour of droplets containing medicines exposed to laser radiation in terrestrial gravity and hypergravity conditions, the micro-spectroscopy specific methods to explore droplets' content.

The hypergravity experiments and results are groundbreaking since they show that microdroplets of chlorpromazine water solutions have better wetting properties when exposed long time to laser beams compared to water at different gravity levels. In general, unexposed/exposed medicine droplets to laser radiation have better wetting ability for cotton as well as activated charcoal target surfaces.

More, new results are shown about the way in which very small concentrations of photoproducts may be seized after exposure of compounds at laser beams, such as an antibiotic like vancomycin. The developed method is based on a microfluidic approach which allows measurement of surface tension at the interface between a gas bubble and the laser exposed solution. It may be also interpreted in terms of cleaning procedure of water from pollutants found at very low concentrations.

This E-book is very well organised in a series of condensed chapters that are illustrated with high quality figures. The text is accessible and easy to read.

The approach of the multiple drug resistance combat in the manner proposed by the editor is ground-breaking and the book opens promising perspectives in solving this threatening issue nowadays in flexible, rapid and adjustable ways.

**Jean Pierre Delville**
CNRS/University of Bordeaux,
Laboratoire Ondes et Matière d'Aquitaine,
351 Cours de la Libération,
F-33405 Talence,
France

# PREFACE

This book is proposed as a synthesis of inter- and multi-disciplinary results about new and somewhat unconventional methods and means to fight multiple drug resistance acquired by microorganisms and tumours. Essentially, two are the main directions along which the book text is elaborated:

(i) Modification of non-antibiotic medicines by exposing them at un-coherent, or laser optical radiation so that from an initial compound that is not efficient in combating bacteria or tumours one obtains photoproducts which alone or by synergetic action receive bactericide or, possibly, tumouricide properties. Examples are given regarding several classes of medicines, pointing out particular compounds amongst cytostatics, phenothiazines, quinazolines and hydantoins and showing how the generated photoproducts may be identified by methods belonging to spectroscopy, microfluidics, optofluidics, chromatography and mass spectrometry.

(ii) Developing new vectors to transport medicines to targets based on optics and micro-spectroscopy methods. These vectors are microvolumetric droplets of water solutions that contain parent medicines and have two roles. First, is to allow fast modification of their content by exposing them to pulsed laser beams so that photoproducts with bactericide properties are generated *via* resonant interaction between one beam and one single droplet. The second, is splitting the droplet by its unresonant interaction with another laser beam having suitable properties, so that nano-droplets and micro-droplets are generated which contain the photoproducts and propagate at supersonic or, respectively, subsonic speeds towards the target.

The target readers of the book are medical doctors, physicists, optofluidics and microfluidics specialists, photochemists, biologists, laser spectroscopists as well as specialists in a broad area of domains ranging from delivery methods of medicines using different sorts of fabrics, to the use of multifunctional medicines in outer space missions, after passing hypergravity conditions. A particular target group is constituted by students experimenting in laser spectroscopy, biology, biomedicine, photochemistry, biophotonics and microfluidics since the book provides new and innovative information about behavior of liquid drops, foams, emulsions and bubbles at interaction with laser radiation and the possible applications of the results in the former mentioned fields.

The book is conceived not only as a coherent synthesis of new results, but also as a source of novel ideas, yet untreated, that are proposed to the readers as working variants in future research. This approach would allow, among others, a fast and flexible reaction in the fight against naturally or accidentally occurring multiresistant microorganisms and tumours, with fast enough results to allow a rapid deal with environment unexpected changes.

A particular interest is devoted to the use of laser spectroscopy and related methods for making available multifunctional drugs that may be applied for treatment of humans or for the decontamination of modules during space flights, in the conditions in which confined small spaces are used in isolation regime for long time intervals, as happens in interplanetary missions. Another subject of interest is the micro-lasers or micro-lasing droplets that emit in free space around them and may be used in a large area of biomedical and technological applications.

The editor would like to thank:

- Dr. Tatiana Tozar for valuable assistance in placing the book text in the printing house template and for detailed checking of figures, tables, list of contributors and abbreviations.
- Dr. Andra Dinache for final overall critical reading of the text book.
- Dr. Viorel Nastasa for assistance in internet connections with the printing house.
- The Laser Spectroscopy Group of the National Institute for Laser, Plasma and Radiation Physics in Bucharest, for the team work that made possible harvesting together this book.

**Mihail Lucian Pascu**
Laser Department,
National Institute for Laser,
Plasma and Radiation Physics,
Romania

# List of Contributors


| | |
|---|---|
| **S. B. Aidarova** | Kazakh National Technical University, Almaty, Kazakhstan |
| **I. R. Andrei** | National Institute for Laser, Plasma and Radiation Physics, Magurele, Ilfov, Romania |
| **M. Boni** | National Institute for Laser, Plasma and Radiation Physics, Magurele, Ilfov, Romania<br>Faculty of Physics, University of Bucharest, Magurele, Ilfov, Romania |
| **M. C. Chifiriuc** | Faculty of Biology, University of Bucharest, Bucharest, Romania<br>Research Institute of the University of Bucharest, Bucharest, Romania |
| **M. Costache** | Faculty of Biology, University of Bucharest, Bucharest, Romania |
| **F. Cotorobai** | National Institute of Materials Physics, Magurele, Ilfov, Romania |
| **A. Dinache** | National Institute for Laser, Plasma and Radiation Physics, Magurele, Ilfov, Romania |
| **A. Dowson** | European Space Agency, European Space Research and Technology Centre, TEC-MMG, Noordwijk, The Netherlands |
| **L. Frunza** | National Institute of Materials Physics, Magurele, Ilfov, Romania |
| **S. Frunza** | National Institute of Materials Physics, Magurele, Ilfov, Romania |
| **C. P. Ganea** | National Institute of Materials Physics, Magurele, Ilfov, Romania |
| **G. Gochev** | Max Planck Institute of Colloids and Interfaces, Potsdam, Germany<br>Institute of Physical Chemistry, Bulgarian Academy of Sciences, Sofia, Bulgaria |
| **X. W. Hu** | Max Planck Institute of Colloids and Interfaces, Potsdam, Germany<br>State Key Laboratory of Multiphase Flow in Power Engineering, Xi'an Jiaotong University, Xi'an, P.R. China |
| **D. Izbassarov** | Department of Mechanical Engineering, Koc University, Rumelifeneri Yolu, Sariyer, Istanbul, Turkey |
| **A. Javadi** | Max Planck Institute of Colloids and Interfaces, Potsdam, Germany<br>Chemical Engineering Department, University of Tehran, Tehran, Iran |
| **T. Kairaliyeva** | Max Planck Institute of Colloids and Interfaces, Potsdam, Germany<br>Kazakh National Technical University, Almaty, Kazakhstan |
| **M. Karbaschi** | Max Planck Institute of Colloids and Interfaces, Potsdam, Germany |
| **J. Krägel** | Max Planck Institute of Colloids and Interfaces, Potsdam, Germany<br>Sinterface Technologies, Berlin, Germany |
| **J. J. W. A. van Loon** | Dutch Experiment Support Center, Department of Oral and Maxillofacial Surgery / Oral Pathology, VU University Medical Center & Academic Centre for Dentistry Amsterdam, Amsterdam, The Netherlands<br>European Space Agency, European Space Research and Technology Centre, TEC-MMG, Noordwijk, The Netherlands |
| **A. V. Makievski** | Sinterface Technologies, Berlin, Germany |
| **R. Miller** | Max Planck Institute of Colloids and Interfaces, Potsdam, Germany |

| | |
|---|---|
| **M. Muradoglu** | Department of Mechanical Engineering, Koc University, Rumelifeneri Yolu, Sariyer, Istanbul, Turkey |
| **V. Nastasa** | National Institute for Laser, Plasma and Radiation Physics, Magurele, Ilfov, Romania |
| **M. L. Pascu** | National Institute for Laser, Plasma and Radiation Physics, Magurele, Ilfov, Romania<br>Faculty of Physics, University of Bucharest, Magurele, Ilfov, Romania |
| **R. Pirvulescu** | Emergency University Hospital, Ophthalmology Clinic, University of Medicine and Pharmacy "Carol Davila", Bucharest, Romania |
| **M. Popa** | Faculty of Biology, University of Bucharest, Bucharest, Romania<br>Research Institute of the University of Bucharest, Bucharest, Romania |
| **A. Popa-Cherecheanu** | Emergency University Hospital, Ophthalmology Clinic, University of Medicine and Pharmacy "Carol Davila", Bucharest, Romania |
| **M. T. Rahni** | Max Planck Institute of Colloids and Interfaces, Potsdam, Germany<br>Sharif University, Tehran, Iran |
| **M. O. Romanitan** | Södersjukhuset AB, Internmedicin kliniken, Sektion för Neurologi, Stockholm, Sweden |
| **A. A. Sharipova** | Max Planck Institute of Colloids and Interfaces, Potsdam, Germany<br>Kazakh National Technical University, Almaty, Kazakhstan |
| **S. Simion** | National Institute for Laser, Plasma and Radiation Physics, Magurele, Ilfov, Romania |
| **Á. Simon** | National Institute for Laser, Plasma and Radiation Physics, Magurele, Ilfov, Romania<br>Faculty of Physics, University of Bucharest, Magurele, Ilfov, Romania |
| **A. Smarandache** | National Institute for Laser, Plasma and Radiation Physics, Magurele, Ilfov, Romania |
| **A. Staicu** | National Institute for Laser, Plasma and Radiation Physics, Magurele, Ilfov, Romania |
| **A. Stoicu** | National Institute for Laser, Plasma and Radiation Physics, Magurele, Ilfov, Romania |
| **A. Tleuova** | Max Planck Institute of Colloids and Interfaces, Potsdam, Germany<br>Kazakh National Technical University, Almaty, Kazakhstan |
| **T. Tozar** | National Institute for Laser, Plasma and Radiation Physics, Magurele, Ilfov, Romania |
| **V. Ulaganathan** | Max Planck Institute of Colloids and Interfaces, Potsdam, Germany |
| **J. Y. Won** | Max Planck Institute of Colloids and Interfaces, Potsdam, Germany |
| **I. Zgura** | National Institute of Materials Physics, Magurele, Ilfov, Romania |

**CHAPTER 1**

# Introduction

**Mihail Lucian Pascu**[1,2,*]

[1] *National Institute for Laser, Plasma and Radiation Physics, Magurele, Ilfov, Romania*

[2] *Faculty of Physics, University of Bucharest, Magurele, Ilfov, Romania*

This book is triggered by the current progress in two emerging fields: (i) fighting multiple drug resistance acquired by microorganisms (in particular bacteria) by laser/optical means and methods and (ii) developing new transport vectors to deliver medicines to targets. Each of them is related to a – respectively - larger, distinct and self-consisting domain that undergoes also a current accelerated expansion.

The use of laser radiations to produce new substances from parent compounds is part of photochemistry which deals with generation of ultrapure products by interaction of molecules with photons of laser radiation. This is a chemistry in which chemical reactions are controlled instantaneously by the interaction of parent chemicals with laser beams, to yield new (photo) products, instead of using with the same purpose thermal/pyrotechnical procedures. The interaction produces either an inner modification of molecules structure which makes them more chemically active, or a break-up of molecules in radicals which interact with surrounding environment and/or between themselves and lead to new products. In both cases resulting materials are ultrapure due to the selective interaction of laser radiation with the molecular targets. This kind of procedure becomes very useful in pharmacology because it allows to produce in not too large quantities new and ultrapure medicines starting from substances already utilised in treatments.

The vectors considered in this book for possible transportation of medicines to targets are micro- and/or nano-droplets either directly generated by a capillary system or obtained by interaction of a laser beam with one single pendant droplet when the beam is not absorbed by droplets' compounds and the light pressure on it dominates the interaction. The same kind of vectors may be droplets included in aerosols, where a particular distribution of theirs function of diameters/volumes may be produced.

[*] **Corresponding author Mihail Lucian Pascu:** Laser Department, National Institute for Laser, Plasma and Radiation Physics, Magurele, Ilfov, Romania; Tel/Fax: 0040 21 4575739; E-mail: mihai.pascu@inflpr.ro

The multiple drug resistance (MDR) which is also called multi-drug resistance or multiresistance of microorganisms that cause infections or even diseases may be defined as a state or property of a microorganism to resist antimicrobial drugs used either as single drugs or as "cocktails" of drugs. Depending on the kind of microorganisms, the medicines with respect to which MDR is acquired may be antibiotics, antifungal, antiviral or antiparasitic chemicals having functions and molecular structures originally conceived to efficiently eradicate the targets. Examples of most common MDR microorganisms which developed resistance are: Gram-positive (*Staphyloccocus aureus*) and Gram-negative (*Pseudomonas aeruginosa, Escherichia coli*) bacteria, viruses (*HIV-Human immunodeficiency virus*), fungi (*Candida* species, *Scedosporium prolificans, Scedosporium apiospermum*), parasites (*Plasmodium falciparum* and *Plasmodium vivax* producing malaria). Drug resistance was acquired from the very first antibiotic, penicillin (widely used since 1943, but tested before 1940) for which the resistance of *Pneumococcus* was reported in 1965, but the first resistance was reported in 1940, in the testing time interval, exhibited by *Staphylococcus*. Some antibiotics for which resistance of bacteria were shown as well as bacteria and respective years when resistance was mentioned are, as presented in SwitchYard Media [1]: tetracycline (utilised in 1950)/*Shigella* (resistance reported in 1959); erythromycin (introduced in 1953)/*Streptococcus* (1968); methicillin (introduced in 1960)/*Staphylococcus* (1962); gentamicin (introduced in 1967)/*Enterocuccus* (resistance first reported in 1979); vancomycin (first used in 1972)/*Enterocuccus* (1988) and *Staphylococcus* (2002); linezolid (introduced in 2000)/*Staphylococcus* (2001); ceftaroline (introduced in 2010)/*Staphylococcus* (resistance first reported in 2011).

A more complete description and a set of MDR related definitions (such as extensive drug resistance-XDR and pandrug-resistance-PDR) are introduced in Magiorakos *et al.* [2] as a result of a study made by the European Centre for Disease Prevention and Control (ECDC) and the Centers for Disease Control and Prevention (CDC) for creating a standardised terminology introduced to describe acquired resistance profiles in *Staphylococcus aureus*, *Enterococcus spp.*, *Enterobacteriaceae* (other than *Salmonella* and *Shigella*), *Pseudomonas aeruginosa* and *Acinetobacter spp.*. Lists of antimicrobial categories were made using the expertise of the Clinical Laboratory Standards Institute (CLSI)-US, the European Committee on Antimicrobial Susceptibility Testing (EUCAST) and the United States Food and Drug Administration (FDA), as well. MDR was defined as acquired non-susceptibility to at least one agent in three or more antimicrobial categories, XDR as non-susceptibility to at least one agent in all but two or fewer antimicrobial categories (bacterial isolates remain susceptible to only one or two categories) and PDR was defined as non-susceptibility to all agents in all antimicrobial categories.

On the other hand, one may speak about resistance of tumours and, in particular, malignant tumours to treatment with drugs belonging to several categories such as cytostatics, antibiotics, specifically designed (quinazolines, pyridinium) compounds, phenothiazines.

The delivery of medicines to targets is made using several procedures among which the local delivery of relatively large volumes (ml) and the systemic delivery through injections (usually one or more ml) are most common. In this book the introduction of drops and droplets as vectors to transmit the medicines to target tissues is described, taking into account the small volume (μl or less) of the delivered medicine which meets the needs to treat a particular system using the smallest necessary and possible quantity of drugs to avoid toxic and related to toxicity effects (photo-toxicity included). On the other hand, a single droplet in pedant/hanging/suspended position at interaction with laser beams, emitted either in pulsed regime or in continuous wave may be used in technological applications and a more general description of the optical properties of droplets is given in the book.

In general, a drop or droplet is a small quantity of liquid which may have different shapes (cylinder, sphere, ellipsoid or combinations of them in static or dynamic evolution) bounded completely or almost completely by free surfaces defined with respect to surrounding media (gases, liquids and high viscosity media immiscible with the droplet's content and keeping it confined). The droplet may be generated in a hanging position with respect to a capillary in which case it is called pendant droplet or on a surface being called sessile droplet. If instead of generating a drop of liquid in gas environment one generates a small gas volume in a liquid, one may define a bubble as shown in de Gennes *et al.* [3]. One may also speak about a droplet in an emulsion or a bubble in a foam, or one may study droplets of emulsions or foams in pendant position in air or another media, for instance. All these physical entities are studied in microfluidics and constitute basic elements for applications in pollution control, dedicated industrial technologies and even outer space experiments.

The interaction of an optical beam with a droplet, a bubble, a liquid containing a bubble or a droplet of immiscible liquid with it, is treated by a relatively new field, the optofluidics which was coagulated mainly in the last 10-15 years. Optofluidics is a mixture of photonics and microfluidics (which deals with non-solid entities) that brings together light and non-solids to provide possible advanced technologies such as fluid waveguides, deformable lenses, microdroplet lasers, new photochemistry and low toxicity biomedicine (see [4 - 6]).

The coupling of optofluidics with the fight against MDR and its forms is the main idea behind this book which is conceived as a multi- and inter-disciplinary collection of data describing the authors' main results in the field. So, the modification of existing medicines (antibiotics, non-antibiotics, cytostatics) for which MDR was acquired, by exposure to laser radiation to obtain photoproducts efficient against biological targets is reported. The process lasts as long as the sample is exposed to laser beam and the photoproducts are generated in minutes to hours if exposure is performed in bulk (1-2 ml volumes) or in, at most minutes if the irradiation of droplets (microvolumetric liquid entities with less than 10 μl volume) containing solutions of medicines is made.

The modification of molecular structures of parent compounds in liquid samples that interact with laser radiation is based on the absorption of laser beam by molecules in the droplet which is otherwise called resonant interaction because it is based on close values (resonance) of the energy difference between two molecular singlet states and the energy of laser beam photons.

If laser radiation is not absorbed by droplet materials, the interaction (usually called unresonant) generates only light pressure effects and produces mechanical effects on the droplet that may vary from changing its shape and producing vibrations to emission of smaller droplets having dimensions at nanoscale. Droplets at milli-, micro- and nano-scale may be used as vectors to transport medicines (parent compounds or photoproducts) to biological targets.

This is also function of the relation between the strengths/intensities of resonant and unresonant interaction effects of laser beams on one hand and the droplets contents, shapes and volumes on the other.

Fighting MDR acquired by bacteria, viruses, fungi, parasites and tumours can be made by working on the molecular structures of existing medicines which are not efficient anymore in treatments, *via* exposure to optical/laser radiation with the purpose to obtain new photoproducts efficient against targets that resist or are not too sensitive to the corresponding parent compounds. Further, generation of photoreaction products can be combined with droplets as new vectors to transport medicines to targets. In doing this, pendant droplets have the advantages of a minimal interaction with environment as well as with the generating capillary. On the other side, characterisation of compounds generated in droplets by optical means, *i.e.* spectroscopic investigations of the content of droplets that have very small volumes and contain very small quantities of substances to be analysed is another emerging field: micro- and nano-spectroscopy.

Starting from these general considerations, the book is conceived as a coherent collection of information about interdisciplinary and multidisciplinary research in the fast currently developing field devoted to applications of optically processed small volume samples in biomedicine and technology. The small volume samples are produced as droplets in pendant positions in different environment media.

In treating the subject, the description of pendant droplets is made from microfluidics (Chapter 2) and optofluidics (Chapter 3) points of view considering mainly single droplets in different media with which they do not mix, but also droplets containing foams or emulsions. Then, drop profile tensiometry results are shown that characterize liquid interfacial dynamics with emphasis on pendant droplets (Chapter 4) and an overview of dynamics and applications of pendant drops is shown in Chapter 5. To make the connection between the description of droplets and their biomedical applications, in Chapter 6 is presented an up-date about MDR acquired by microorganisms and tumours. Further, laser beam properties are introduced in Chapter 7 given that in the applications of interest here the single droplets interact with laser beams emitted in pulsed regime and many effects are directly related to laser beam shape, energy, time and space characteristics. Usually, the interaction takes place with a single pulse or with a controlled number of pulses. The next two chapters deal with unresonant (Chapter 8) and resonant (Chapter 9) interaction of laser beams described in Chapter 7 with droplets in pedant position. In the results shown here the droplets have microvolumetric dimensions and are hanged in air. The investigations of the effects of a laser beam on a single droplet utilise high speed optical recording, laser induced fluorescence and Raman spectra monitoring. In direct connection with resonant and unresonant interaction in Chapter 10 are described relevant data about surface tension and contact angles of microdroplets containing water solutions of medicines exposed to laser radiation either in bulk solutions or in droplets with the same content as bulk. Since droplets considered in the reported studies may contain solutions of medicines exposed to and modified by optical incoherent beams or laser beams prior to be generated as microdroplets, in Chapter 11 are shown results of the interaction of such beams with medicines water solutions in bulk. The studied medicines are part of the following categories: cytostatics (methotrexate and 5–fluorouracil), phenothiazines that are utilized normally as antipsychotic drugs (chlorpromazine-CPZ, promazine-PZ, thioridazine-TZ, promethazine-PMZ), quinazoline derivatives developed in Marseille for MDR experiments (BG 204, BG 1120, BG 1188), hydantoin derivatives developed in Krakow for MDR applications (SZ-2 is a typical compound). Here, emphasis is placed on the identification of new photoproducts generated from medicines molecules by exposure to optical radiation. The methods/techniques used with this purpose are standard UV-Vis optical absorption, laser induced fluorescence (LIF), phosphorescence based singlet oxygen detection, Raman scattering, Fourier transform infrared spectroscopy (FTIR) and thin layer chromatography (TLC).

In Chapter 12 results on studies regarding the processes that take place when a laser beam interacts with foams or emulsions in order to monitor their properties or to modify their content are shown. The substances considered here are vancomycin, oily vitamin A, polidocanol (aethoxyscklerol) and rhodamine 6G, separated or in combinations. Some of them are considered in interaction with surfactants such as xanthan gum and tween 80. Additionally, colourless and odourless glycerin is also used in experiments.

The next two chapters treat the applications of laser modified medicines in order to combat MDR acquired by microorganisms (Chapter 13) such as Gram-positive, Gram-negative bacteria and fungi as well as tumours (Chapter 14). As for the tumours, some of studied medicines interacted with optical uncoherent radiation emitted by Hg or Xe lamps in cw regime and other were exposed to pulsed laser radiation emitted in UV by an nitrogen pulsed laser or in UV-Vis by a Nd:YAG laser coupled to nonlinear crystals. The tumours were, actually, psuedotumours produced on eye conjunctiva and treated with cytostatics (MTX, 5–FU) quinazoline derivatives (BG 204, BG 1120) and phentothiazine derivatives (CPZ).

A section dedicated to materials that may be used to deliver medicines to superficial tissues, such as skin or to clean infected hydrophobic surfaces, is Chapter 15 which deals with the interaction of medicines exposed to laser beams with fabrics/materials in view of biomedical applications.

Further, Chapter 16 is dedicated to droplets optofluidic properties in extreme conditions such as hypergravity (up to 20 g), having in mind applications in outer space or during trips towards outer space.

Another subject of quite largely foreseen perspective in optics of droplets and its applications is described in Chapter 17 which presents data about lasing properties of optically/laser pumped pendant droplets containing fluorophores such as laser dyes in view of scientific, technological and biomedical applications.

Finally, in Chapter 18 is shortly and synthetically discussed a new branch of spectroscopy which may be defined based on data such as those reported in some of the chapters of the book: droplet based spectroscopy as an alternative to bulky materials spectroscopy, when the drop and bulk materials are the same.

## CONFLICT OF INTEREST

The authors confirm that this chapter content has no conflict of interest.

## ACKNOWLEDGEMENTS

This work has been financed by the National Authority for Research and

Innovation in the frame of Nucleus programme-contract 4N/2016 and the project PN-II-ID-PCE-2011-3-0922.

## REFERENCES

[1] "Timeline of antibiotic resistance produced by SwitchYard for The Food and Environment Reporting Network and Medium", *SwitchYard Media.* Available at: http://switchyardmedia.tumblr.com/post/80782524007/timeline-of-antibiotic-resistance-produced-by

[2] A.P. Magiorakos, A. Srinivasan, R.B. Carey, Y. Carmeli, M.E. Falagas, C.G. Giske, S. Harbarth, J.F. Hindler, G. Kahlmeter, B. Olsson-Liljequist, D.L. Paterson, L.B. Rice, J. Stelling, M.J. Struelens, A. Vatopoulos, J.T. Weber, and D.L. Monnet, "Multidrug-resistant, extensively drug-resistant and pandrug-resistant bacteria: an international expert proposal for interim standard definitions for acquired resistance", *Clin. Microbiol. Infect,* vol. 18, no. 3, pp. 268-281, 2012. [http://dx.doi.org/10.1111/j.1469-0691.2011.03570.x] [PMID: 21793988]

[3] P-G. de Gennes, F. Brochard-Wyart, and D. Quere, *Capillarity and wetting phenomena: drops, bubbles, pearls, waves*Springer Science & Business Media, 2003.

[4] D. Pile, and F. Ligler, "An interdisciplinary approach", *Nat. Photonics,* vol. 5, no. 10, pp. 569-569, 2011. [http://dx.doi.org/10.1038/nphoton.2011.224]

[5] X. Fan, and I.M. White, "Optofluidic microsystems for chemical and biological analysis", *Nat. Photonics,* vol. 5, no. 10, pp. 591-597, 2011. [http://dx.doi.org/10.1038/nphoton.2011.206] [PMID: 22059090]

[6] V. Nastasa, K. Samaras, I.R. Andrei, M.L. Pascu, and T. Karapantsios, "Study of the formation of micro and nano-droplets containing immiscible solutions", *Colloids Surf. Physicochem. Eng. Asp.,* vol. 382, no. 1-3, pp. 246-250, 2011. [http://dx.doi.org/10.1016/j.colsurfa.2011.01.016]

**CHAPTER 2**

# Pendant Droplets – Microfluidic Approach

**Viorel Nastasa**[1,2], **Angela Staicu**[1] and **Mihail Lucian Pascu**[1,3,*]

[1] *National Institute for Laser, Plasma and Radiation Physics, Magurele, Ilfov, Romania*

[2] *ELI-NP, "Horia Hulubei" National Inst. for Physics and Nuclear Eng., Magurele, Ilfov, Romania*

[3] *Faculty of Physics, University of Bucharest, Magurele, Ilfov, Romania*

**Abstract:** This chapter contains basic data about the microfluidic description of pendant droplets. Results are shown regarding the surface/interfacial tension measurements performed on water based solutions following the interaction with laser radiation. A synthesis is introduced of the main methods used to produce simple or complex droplets in different media. A method to evidence surface active products obtained after exposure of medicine solutions to laser radiation is presented. It consists in measuring in real time the dynamic interfacial tension at the interface between air and irradiated solution, when solution is in bulk form. The variation of dynamic interfacial tension is an indicator of the presence of laser produced amphiphilic molecules in solution. These results belong to series of reports dedicated to new methods used to fight multiple drug resistance developed by bacteria by decreasing the concentration of active compounds with bactericide effects. In line with microfluidic approach of droplets with µl volumes, surface tension measurements on DMSO-water mixtures containing a dye are presented.

**Keywords:** Aerosols, Bubbles, Capillarity, Colloidal systems, Contact angle, Dynamic surface tension, Hanging droplet, Hydrophilicity, Hydrophobicity, Immiscible fluids, Pendant droplet, Sessile droplet, Suspended droplet, Vancomycin.

## INTRODUCTION AND GENERALITIES

Droplets are an important component of daily life with multiple applications in various domains. A droplet "component" that plays an essential role in understanding its behaviour, is the surface it "shows" to the environment. The shape of an individual, unperturbed droplet is mainly determined by two forces: surface tension that tends to decrease the surface to volume ratio by giving a

* **Corresponding author Mihail Lucian Pascu**: Laser Department, National Institute for Laser, Plasma and Radiation Physics, Magurele, Ilfov, Romania, Phone/Fax: 0040 214575739, E-mail: mihai.pascu@inflpr.ro

(quasi) spherical shape to the droplet and gravity that tends to deform it. There is an increased interest in using droplets with milli-, micro-, and nano-meter dimensions in various domains (industry, medicine, space science, *etc*.), which created new research domains such as optofluidics in which small volumes behaviour at interaction with light beams are mainly studied.

A particular field within microfluidics and optofluidics is the development of fluid materials able to be confined in a droplet and to produce effects on targets on which these droplets are sent or deposited.

Microfluidic systems are able to provide exact liquid volumes for each droplet. This can be made in an active or a passive mode. Active control uses local forces that allow manipulation of each droplet, individually in the intended direction; it can be obtained using several methods, such as: electrowetting, electrophoresis, electrostatic manipulation, pneumatic pressure and/or thermocapilarity actuation [1 - 5]. In passive control, externally generated flows are used where each flow is modified locally through capillary geometry [6]; most of these devices are focused on droplet generation from continuous flows [7].

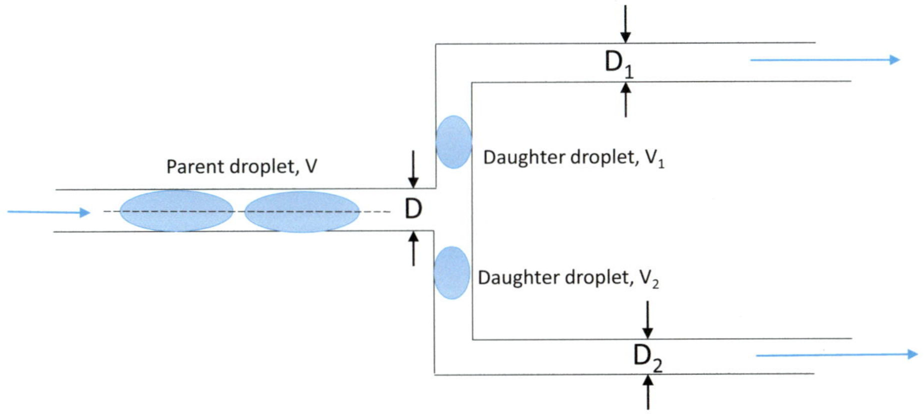


**Fig. (1).** Schematic of a bifurcating junction in a microfluidic device.

Passive microfluidic channels can control the droplet volume through droplet fission, fusion and sorting. The dimensions of channels are used to control the size interval between daughters and parent droplets (Fig. **1**). Microfluidic channel systems for high-throughput bio-chemical analyses, named “micro total analysis systems” (μTAS) or “lab-on-a-chip” are versatile and have multiple biomedical applications due to their capability to control droplet volume with picoliter accuracy (*e.g.* protein screening, crystal growth in mixed droplets, electrophoresis, DNA analysis, cell growth and analysis, emulsification) [3, 5, 7 - 9].

Another method to generate droplets is represented by the use of a pumping system which sends programmed liquid samples through capillaries under computer control. The system can be used to generate droplets of a single liquid component in pendant positions (only one hanging droplet, if needed) or to mix different kinds of liquids with controlled volumes and/or concentrations of ingredients. The characteristics of the pumping system (*e.g.* syringe volume, capillary diameter) are selected depending on the volume of liquid that needs to be generated (Fig. **2a**). By adding another syringe and by changing the simple capillary with double coaxial capillary, this system can also generate layered droplets that are needed for interfacial adsorption measurements (Fig. **2b**).

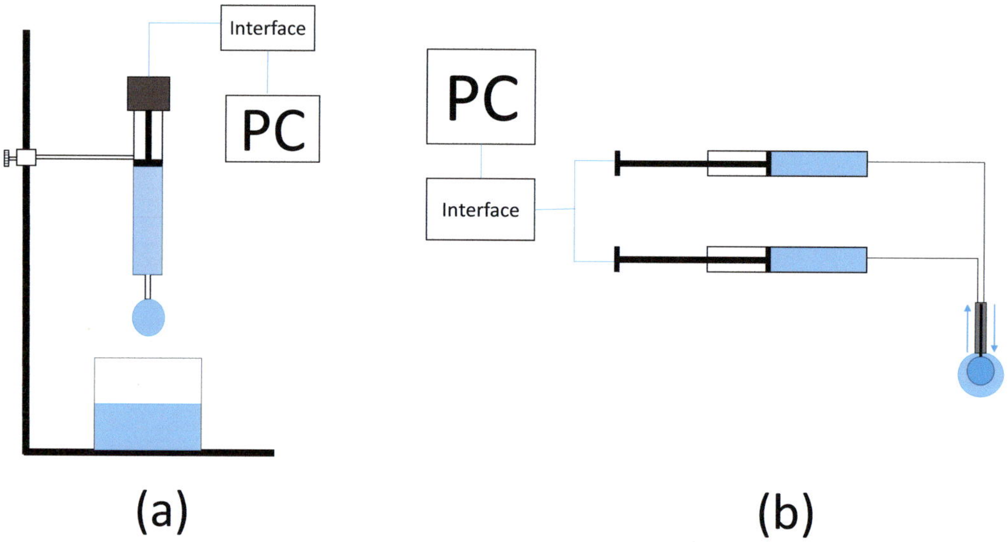


**Fig. (2).** Pumping system for pendant droplets generation [10]. (**a**) generation of single pendant droplets and (**b**) double syringe system for layered droplets generation.

Another way of generating complex pendant droplets is by using previously obtained emulsions. Fig. (**3**) shows a schematic presentation of a simple liquid droplet (a) and a droplet containing an oil in water emulsion (b) associated with the method to generate emulsions (c). The shape of such a droplet is (quasi)spherical and it is due to selected liquid volume (*i.e.* mass) and to liquid density, surface tension and dimensions of the capillary [11]. A more detailed presentation of this type of complex droplets and their applications is presented in Chapters 4 and 10.

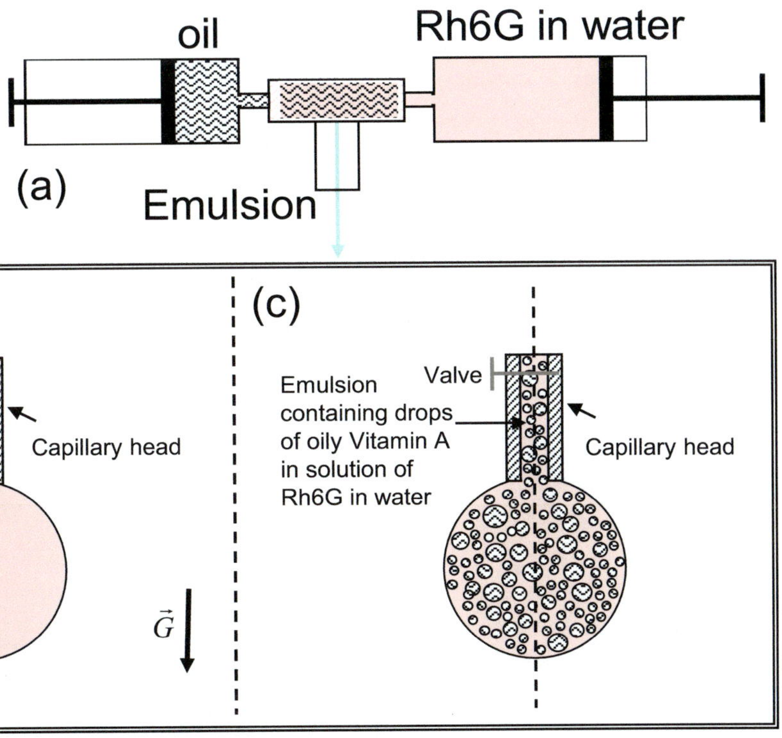


**Fig. (3).** Schematic presentation of the generation of pendant droplets containing emulsions: (**a**) a simple method using a double syringe system to generate emulsions, (**b**) a simple droplet containing a solution of Rhodamine 6G, and (**c**) a pendant droplet containing an emulsion.

Since the properties of micro and nano-droplets are not entirely known based on constituent liquid properties in bulk, their measurement is needed, the most important of them being the surface or interfacial tension that leads to effects observed every day: leaves that float on the surface of a lake, a drop of water climbing to the edge of a surface until its mass is high enough so that the weight defeats the adhesive forces that keep it on that surface and it falls down *etc.* A mechanical model to describe surface tension of a drop can be a balloon filled with air, since air pressure inside it is higher than atmospheric pressure. At the same time, by increasing the surface, the balloon will stretch until it ruptures while a surface of a liquid can be expanded without modifying its surface tension. The water molecules inside an electrically neutral fluid interact to each other with van der Waals forces that cancel reciprocally. At the boundaries, in some cases, there is an inward attraction of molecules due to the difference between the material of the droplet and that of its environment (the molecules cohere stronger to those in their vicinity within droplet's material) which makes it acquire the smallest possible surface area. The interface between two immiscible liquids has a tension as well, called interfacial tension [12]. A schematic of this process is presented in Fig. (**4**). Polar liquids (*e.g.* water) have strong intermolecular bonds and therefore a high value of surface tension. Any variation of the interaction strength (depending on temperature, solution contamination with surfactants *etc.*) will modify the surface tension. In some cases, (Fig. **4a**), the meniscus at the liquid contact with a solid wall may be oriented upward if the cohesion forces between the molecules in the liquid are smaller than those with the wall, or downward if these forces are larger than the cohesion with the walls. In droplet in air, the shape of the droplet is spherical if its weight does not exceed the surface tension of the liquid (Fig. **4b**).

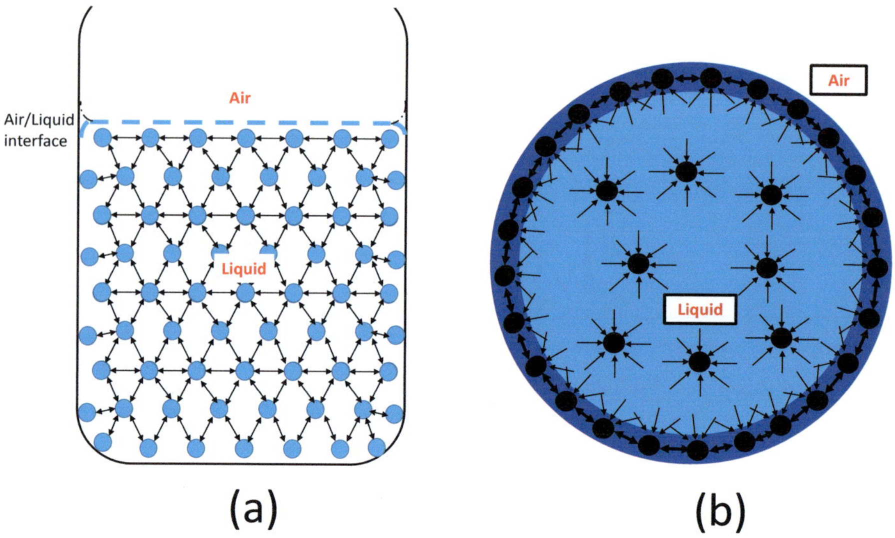


**Fig. (4).** Molecular interactions at the liquid surface in bulk and in a droplet. (**a**) air/liquid interface in bulk solution; (**b**) air/liquid interface in droplet.

There are several methods to measure the surface tension of a liquid, namely: capillary rise, Wilhelmy plate, du Noüy ring, maximum bubble pressure, drop weight or volume, spinning drop, pendant drop and sessile drop [12]. The results presented in this book are obtained with devices that use the pendant drop method to measure the surface tension of simple or layered droplets. A description of such a device is made in more detail in Chapter 4.

Another important characteristic is the contact angle made by liquids at the intersection with a solid surface, which is obtained by applying a tangent line to the contact point along the drop profile and by measuring the angle between the tangent and the solid surface in this point. The contact angle allows the quantitative measurement of a material surface wettability. It highly depends on the surface tension value of the liquid: a small surface tension is associated with high wettability.

Over 200 years ago [13], Thomas Young proposed a theoretical description that considers the thermodynamic equilibrium of a sessile drop on a solid surface under the influence of three phases: liquid, solid and vapor phase. Each interface has a specific interfacial tension $Y_{SL}$ (interfacial tension between liquid and solid), $Y_{SG}$ (surface free energy of the solid) and $Y_{LG}$ (surface tension of the liquid): $Y_{LG} \cos\theta = Y_{SG} - Y_{SL}$ [14, 15]. A more detailed theoretical background is presented in Chapter 10 (Fig. **5**).

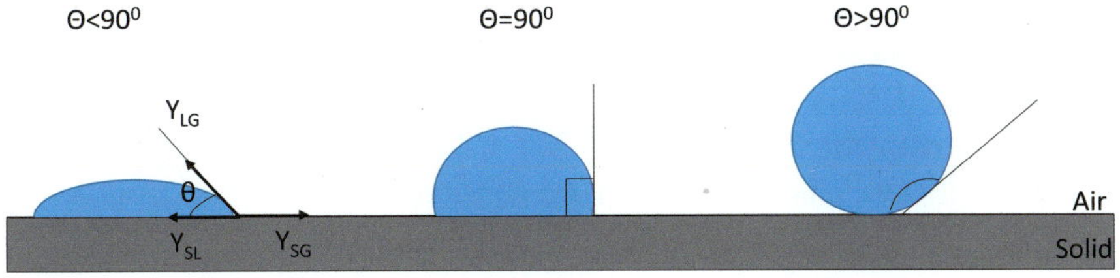


**Fig. (5).** Schematic diagram for contact angles formed by sessile liquid droplet on solid surface [14].

Depending on the contact angle values, a liquid is considered as wetting (angle smaller than 90°) or not wetting (angle between 90° and 180°) liquid. A hydrophilic surface belongs to a material with a special affinity to water, on which the water spreads [16]. Similarly, a hydrophilic molecule, or portion of a molecule is attracted to water or other polar substances; the interaction with water is more thermodynamically favorable than with non-polar or hydrophobic solvents. In contrast, a hydrophobic surface is repelling water (mostly, the attraction is absent) [17] and therefore it has a contact angle with water higher than 90°. Superhydrophobic surfaces have the contact angles higher than 150° for a drop of water on the same kind of substrate. The term "hydrophobic" is often replaced by "lipophilic" although they are not synonyms (some hydrophobic substances such as silicones are not "fat loving"). The hydrophobic interaction is mostly an entropic effect that originates from the break of highly dynamic hydrogen bonds between molecules of liquid water by the nonpolar solute, and leads to formation of a clathrate-like structure around the non-polar molecules. This structure is highly ordered, more than free water molecules since the molecules are arranged with respect to each other to be able to interact as much as possible between them. This leads to a higher entropic state which causes non-polar molecules to clump together with the result of the reduction of surface area exposed to water and the decrease of system's entropy. Thus, the two immiscible phases (one hydrophilic and the other hydrophobic) will be changed so that their interfacial area will reach a minimum. This effect can be evidenced by the so-called phase separation [18,

19]. Both, hydrophilic and hydrophobic classes of materials have applications in areas such as: coating, painting, cleaning, printing, bonding, dispersing, airplane wings building, electronics, power plants and other industrial applications. Superhydrophobic surfaces are used in lab-on-a-chip microfluidic devices.

In this chapter, recent results are shown about microfluidic properties of:
(i) vancomycin in water solutions; (ii) Zn-phthalocyanine in dimethyl sulfoxide; (iii) single wall carbon nanotubes conjugated with 5-mono (4-carboxyphenyl) - 10,15,20-triphenyl porphine in solutions with N, N-dimethylformamide.

## MATERIALS AND METHODS

Vancomycin (VCM) is a hydrochloride powder (Sigma, DE) and its molecular structure is presented in Chapter 12. The experimental set-up used to expose VCM samples in ultrapure water ($2\times10^{-3}$ M) at laser radiation is shown in Fig. (**6**) where are also mentioned the methods used to identify generated photoproducts.

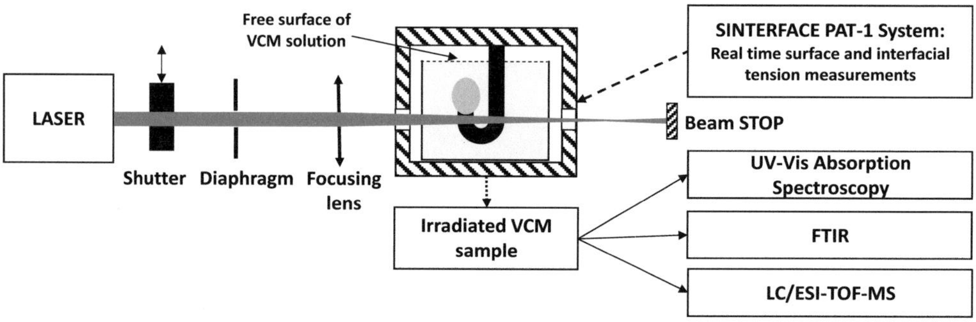


**Fig. (6).** The experimental set-up used to irradiate VCM solutions in ultrapure water. Laser: pulsed Nd:YAG; fourth harmonic beam at 266 nm; pulse repetition rate: 10 pulses per second; full time width at half maximum: 5 ns; beam average energy 7 mJ. SINTERFACE PAT-1: Profile Analysis Tensiometer (Sinterface, DE); liquid volume: 5 ml; air bubble volume: 10 µl [20]†.

Vancomycin is used in clinical applications at concentrations between 0.25 mg/l and 1 mg/l for susceptible strains and 8 mg/l ($5.4\times10^{-6}$ M) for bacteria that are resistant to other treatments. The results presented below use a concentration of $2\times10^{-3}$ M (*i.e.* $2.9\times10^{3}$ mg/l) since the values used in clinical applications are too small to be used for optical/laser measurements with the purpose to monitor solutions' evolution under irradiation conditions or after this stage. At a higher concentration, the modifications induced by laser radiation may be stronger and easier to measure.

For measurements presented here, a volume of 5 ml of VCM solution (bulk) is exposed to UV radiation emitted as FHG by an Nd:YAG laser system (266 nm), within which an air bubble in emerged position was generated through a curved capillary.

A description of the laser systems used for drug irradiation is presented in Chapters 11 and 12. Dynamic interfacial tension is measured with a Profile Analysis Tensiometer (PAT-1) described in Chapter 4, for a time interval ranging from seconds to hours. Laser beam propagation direction is parallel to the capillary, passing at 1 mm distance below its apex. From irradiated solution, smaller volumes may be extracted to be submitted to other types of measurements performed to identify the modifications induced in solution (VCM molecules included) by exposure to laser radiation.

Other measurements were performed on mixtures of two solvents currently used in pharmaceutics: dimethyl sulfoxide (DMSO) and water. The effect of the mixture on a compound which is solvated only in DMSO, Zn-phthalocyanine (ZnPc), was also studied. The DMSO used for these measurements was EMSURE grade (Merck, DE), ZnPc dye was produced by Sigma-Aldrich (purity 97%) and distilled water was home-prepared with Merit Water Still W4000 equipment.

Surface tension and viscosity measurements were performed with PAT-1 system based on pendant droplet profile analysis which makes use of Young-Laplace equation where. The fitted profile of droplet is estimated to calculate the value of surface tension. PAT -1 maintains constant the droplet volume, which eliminates the influence of liquid evaporation on the accuracy of the results.

Another studied immiscible samples were solutions in N, N-dimethylformamide (DMF) of single wall carbon nanotubes (SWCNT) conjugated with a porphyrin type sensitizer 5-mono(4-carboxyphenyl)-10,15,20-triphenyl porphine (TPP). There were analyzed two types of samples: (i) the covalent functionalized SWCNT with TPP (SWCNT-TPP linked) and (ii) a mixture with the same proportion of the concentrations of SWCNT and TPP (SWCNT-TPP mixed) [4].

The chemical compounds and solvents were: amino-functionalised single walled carbon nanotubes (Nanocs Inc. USA) with Ø=0.7-1.2 nm, bundle diameter 20 nm, fiber length 20-50 μm, amino moiety to ends 0.25 to 0.45 mmol/g nanotubes; TPP from Frontier Scientific (USA); N, N'-dicyclohexylcarbodiimide (DCC) from Merck (DE); N, N-dimethylformamide (DMF) from Merck (DE).

Amino functionalised nanotubes were conjugated with TPP porphyrin molecules using a carboxylation method that implies the activation and interaction of a carboxylic moiety with an amino group and has as result an amide bond.

# RESULTS

## Vancomycin Exposed to Laser Radiation - Tensiometric Data

Dynamic interfacial tension (DIT) measurements were performed on a bubble generated in bulk VCM solution exposed to UV pulsed laser radiation, to evidence the presence of surface active molecules in solution obtained as a result of irradiation.

The dynamic surface tension (DST) measurement is a sensitive method to prove the adsorption of amphiphilic molecules at water/air interface.

The measurements were performed for different experimental parameters (*e.g.* position of laser beam in the cuvette, bubble's dimensions resulting from its volume and bubble illumination conditions) before, during and after exposure to laser radiation. A distinct inflection point in DIT data obtained for irradiated samples is evidenced, an example being shown in Fig. (7).

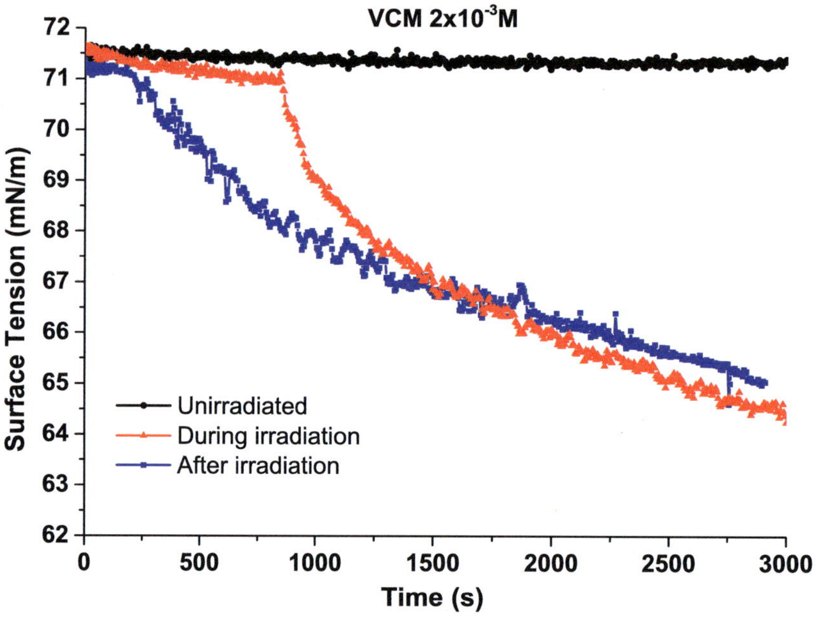


**Fig. (7).** Surface tension measurements of VCM ultrapure water solution at $2\times10^{-3}$ M‡.

The starting value for all water based solutions were close to the values of ultrapure water and for unexposed samples the value remained constant for whole measurement, which shows that VCM molecule is not adsorbed at water/air interface. Irradiated sample shows a small decrease of DIT for the first 800 s and, at around 800 s, a steep decrease followed by a further exponential decrease. This non-regular behaviour indicates the appearance of a surface transition at the break point that can be explained by a phase transition in the adsorption layer [18], which at water/air interface is constituted by transient segments in surface relaxations curves.

After one hour, irradiation was stopped and a new bubble was generated, for which DIT was also measured; the time evolution of it was quasi-exponential without showing the steep decrease at 800 s. During all measurements that involved exposure to laser radiation, DIT behaviour was specific to a process in which a surface active substance is gradually accumulated at surface. The real-time observation of surface active hydrophobic by-products formation is made. An induction time characteristic for the photoreaction process is needed, after which formation of a surface active layer occurs. This time may be influenced by the diffusion time of an adsorbing species having larger molecular weight and being produced in VCM solution. This holds if more species of hydrophobic or partially hydrophobic products are generated by exposure to laser radiation. Some characteristics of the interfacial behaviour can be related to other data reported on interfacial adsorption of slightly soluble or insoluble surfactants. The slope variation observed during real time measurements can be produced by an interfacial phase transition as that which takes place in n-dodecanol adsorbed layers [21, 22]. The ongoing formation of the layer can be accompanied by a process of competitive adsorption and formation of domains or phases in the adsorbed layer [20].

These results are also confirmed by literature reports that mention the coexistence of gaseous and liquid-like 2D phases beyond a critical surface coverage at interface [23]. The fact that after one hour of irradiation, DIT value does not reach a plateau means that the time necessary to reach adsorption equilibrium conditions is longer than this one and a continued exposure to laser beam could lead to generation of further surface active molecules that would be adsorbed at interface. The temperature variation of the sample was also assessed with an infrared camera ThermaCAM E45 (FLIR) and it was very small during the irradiation (~1°C). This temperature variation is not high enough to influence surface tension values, which means that the solution is not significantly heated during irradiation and therefore the measured surface tension effect may be due to production of amphiphilic compounds generated by exposure to UV laser beam.

### DMSO-Water Mixture

The surface tension behaviour function of co-solvent ratio in DMSO-water mixture was also analysed. In the literature, reports are made on microfluidic properties of DMSO-water mixtures showing data about surface tension, viscosity, and enthalpy of mixtures [24]. This kind of mixture has a highly nonlinear behaviour which was assigned to changes of clusterisation state of the component molecules (water and DMSO) [24 - 26].

The surface tension is shown in Fig. (**8**), together with the data reported in [24]. It can be noticed that the values of surface tension for the mixture has a non-linear variation with the mole fraction of the separated components.

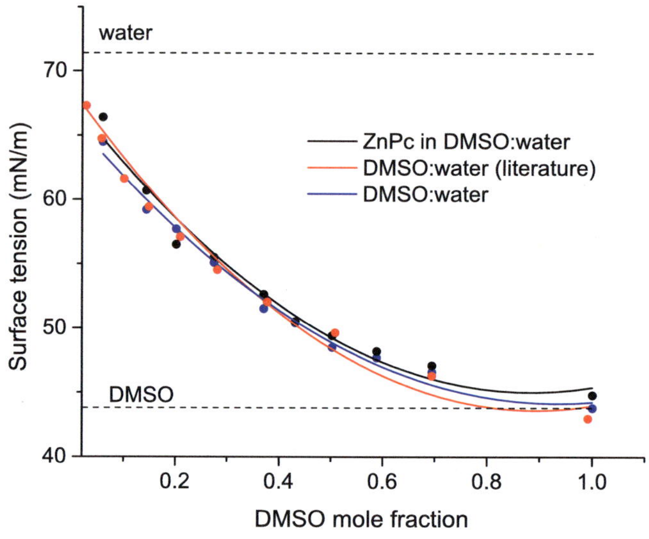


**Fig. (8).** Surface tension of DMSO-water mixture and of ZnPc ($5\times10^{-6}$ M) in DMSO-water mixtures.

The fitting polynomial curve starts from the value of water ending to the one of DMSO. The surface tension values of DMSO and water are illustrated as interrupted lines in Fig. (**8**).

The present experiment, based on the analysis of a pendant droplet shape with PAT-1, consisted in the measurement of the surface tension for two series of DMSO-water samples. First series contained binary mixtures DMSO-water at different mole fractions of DMSO. The other series of samples, free of dye, was composed only of solvent mixtures at the same ratios as in the first series.

## Covalent Functionalized Single Walled Carbon Nanotubes with Porphyrin-Type Photosensitiser

Table **1** shows the equilibrium surface tension values of DMF and DMF with nanotubes (SWCNT-TPP linked and SWCNT-TPP mixed). The surface tension values are slightly influenced by the presence of carbon nanotubes in solutions. The values of ST for SWCNT samples are placed above and below the solvent value and the higher is the one for linked compound.

**Table 1. Surface tension equilibrium values.**

| Solution | ST (mN/m) |
|---|---|
| DMF | 0.3 |
| SWCNT-TPP linked | 38.4±0.2 |
| SWCNT-TPP mixed | 36.2±0.4 |

For the evaluation of the rheological properties of samples, a surface tension measurement was followed by a "surface disturbance-relaxation experiment" similar to that reported in [24] that consists in harmonic perturbations induced to the pendant droplet followed by Fourier analysis of the obtained data. The droplet volume was varied with ±1 $mm^3$ and the used frequencies were 0.005, 0.008, 0.01, 0.02, 0.04, 0.05, 0.08, 0.1, 0.16 and 0.2 Hz. The data obtained after Fourier analysis of harmonic perturbation measurements are in accordance with surface tension results.

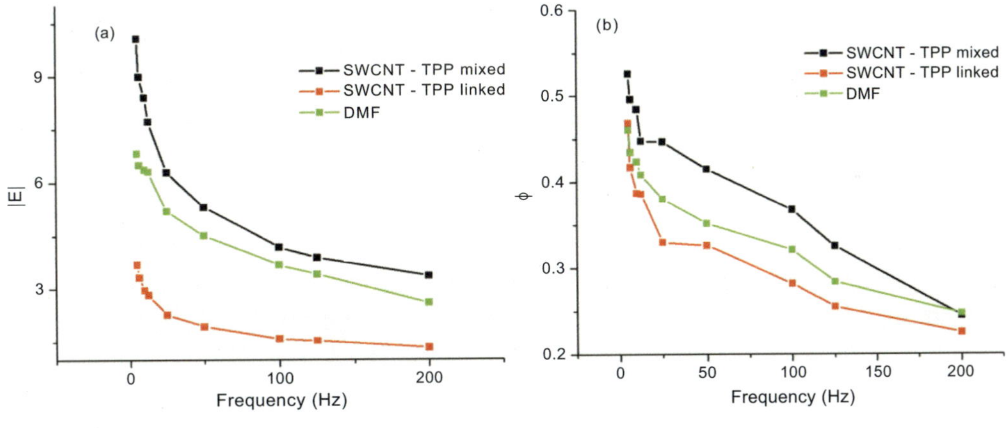


**Fig. (9).** (**a**) Visco-elastic modulus (|E|) and (**b**) the phase lag (Φ) values.

Fig. (**9**) shows: (a) the visco-elastic modulus (|E|) and (b) the phase lag (Φ) values for the studied compounds. These values were calculated from the $E_{im}$ si $E_{re}$ obtained from Fourier analysis of rheological data generated by the induced harmonic perturbations.

$$|E| = \sqrt{E_{re}{}^2 + E_{im}{}^2} = \frac{\Delta\sigma_m}{\Delta A_m/A_0} \quad \textbf{(1)}$$

$$E_{re} = |E| \times cos\phi;\ E_{im} = |E| \times sin\phi \quad \textbf{(2)}$$

The values of $E_{re}$ are bigger than $E_{im}$ which indicates that gas/liquid interface of solutions has a predominantly viscous character. SWCNT-TPP, *mixed*, have the smallest viscoelasticity value, while SWCNT-TPP, *linked*, has the highest values of both |E| and Φ. DMF values are distributed between the values obtained for the nanotubes mixtures. The visco-elastic phase lag (Φ) results present a similar behaviour.

The obtained differences for SWCNT compounds highlight an increase of viscosity and surface tension values due to conjugation with porphyrin molecules.

## CONCLUSIONS

A new method is introduced to evidence the surface active products obtained after exposure of VCM to laser radiation by real time measurements of DIT at air/irradiated solution interface. DIT measurements are a sensitive indicator of surface active products that are generated in the solution exposed to laser radiation. The variation of DIT values are an indicator of presence of laser produced amphiphilic molecules in solution. The reported results may be of interest for biomedical applications because they make possible to evaluate, based of interfacial tension measurements, the adsorption and wetting properties of some amphiphilic compounds originating from VCM after interaction with laser beams. Surface activity modifications may lead to variations of wetting properties of VCM solutions, as well as of speed with which modified VCM and the resulting photoproducts are delivered and penetrate biological tissues/targets.

These results are part of a broader effort made to find new methods for fighting MDR acquired by bacteria through the decrease of the concentration of the active compounds. The use of lasers as sources of electromagnetic radiation in order to generate, out of existing medicines, small amounts of new photoproducts which may be efficient in fighting MDR developed by bacteria is one of the most actual and promising methods in this field.

The surface tension measurements on the $5\times10^{-6}$ M ZnPc in DMSO-water mixtures proved that, whatever the ratio of co-solvents, there is no significant adsorption of the dye molecule at water/air interface of the sample. This is a proof that the dye is uniformly distributed in the volume of the solution.

## NOTES

† Reprinted from A. Dinache, M. Boni, T. Alexandru, E. Radu, A. Stoicu, I. R.Andrei, A. Staicu, L. Liggieri, V. Nastasa, M. L. Pascu, M. Ferrari "Surface

properties of Vancomycin after interaction with laser beams", Colloid Surface A., vol 480, pp. 328-335, 2015.

‡ Reprinted from A. Dinache, M. Boni, T. Alexandru, E. Radu, A. Stoicu, I. R.Andrei, A. Staicu, L. Liggieri, V. Nastasa, M. L. Pascu, M. Ferrari "Surface properties of Vancomycin after interaction with laser beams", Colloid Surface A., vol 480, pp. 328-335, 2015.

## CONFLICT OF INTEREST

The authors confirm that this chapter content has no conflict of interest.

## ACKNOWLEDGEMENTS

This work has been financed by the Romanian National Authority for Research and Innovation in the frame of Nucleus programme-4N/2016 and projects PN-II-ID-PCE-2011-3-0922, 641/2013, PN III-P2-2.1-PED-2016-0420 and COST network MP1106.

## REFERENCES

[1] B. Mathew, A. Alazzam, M. Abutayeh, A. Gawanmeh, and S. Khashan, "Modeling the trajectory of microparticles subjected to dielectrophoresis in a microfluidic device for field flow fractionation", *Chem. Eng. Sci,* vol. 138, pp. 266-280, 2015. [http://dx.doi.org/10.1016/j.ces.2015.07.014]

[2] V. Pratap, N. Moumen, and R.S. Subramanian, "Thermocapillary motion of a liquid drop on a horizontal solid surface", *Langmuir,* vol. 24, no. 9, pp. 5185-5193, 2008. [http://dx.doi.org/10.1021/la7036839] [PMID: 18399689]

[3] M.G. Pollack, A.D. Shenderov, and R.B. Fair, "Electrowetting-based actuation of droplets for integrated microfluidics", *Lab Chip,* vol. 2, no. 2, pp. 96-101, 2002. [http://dx.doi.org/10.1039/b110474h] [PMID: 15100841]

[4] T. Taniguchi, T. Torii, and T. Higuchi, "Chemical reactions in microdroplets by electrostatic manipulation of droplets in liquid media", *Lab Chip,* vol. 2, no. 1, pp. 19-23, 2002. [http://dx.doi.org/10.1039/B108739H] [PMID: 15100855]

[5] Y-C. Tan, J.S. Fisher, A.I. Lee, V. Cristini, and A.P. Lee, "Design of microfluidic channel geometries for the control of droplet volume, chemical concentration, and sorting", *Lab Chip,* vol. 4, no. 4, pp. 292-298, 2004.

[6] M.V. Koch, K.M. VandenBussche, and R.W. Chrisman, *Micro Instrumentation: For High Throughput Experimentation and Process Intensification - a Tool for PAT.* John Wiley & Sons, 2007. [http://dx.doi.org/10.1002/9783527610624]

[7] T. Nisisako, T. Torii, and T. Higuchi, "Droplet formation in a microchannel network", *Lab Chip,* vol. 2, no. 1, pp. 24-26, 2002. [http://dx.doi.org/10.1039/B108740C] [PMID: 15100856]

[8] R.B. Fair, V. Srinivasan, H. Ren, P. Paik, V.K. Pamula, and M.G. Pollack, "Electrowetting-based on-chip sample processing for integrated microfluidics", In: *Electron Devices Meeting, 2003. IEDM '03 Technical Digest. IEEE International*, 2003. [http://dx.doi.org/10.1109/IEDM.2003.1269396]

[9] S. Kaneda, K. Ono, T. Fukuba, T. Nojima, T. Yamamoto, and T. Fujii, "A rapid method for optimizing running temperature of electrophoresis through repetitive on-chip CE operations", *Int. J. Mol. Sci,* vol. 12, no. 7, pp. 4271-4281, 2011.
[http://dx.doi.org/10.3390/ijms12074271] [PMID: 21845077]

[10] R. Miller, and L. Liggieri, *Bubble and Drop Interfaces*. CRC Press, 2011.

[11] M. Boni, V. Nastasa, I.R. Andrei, A. Staicu, and M.L. Pascu, "Enhanced fluorescence emitted by microdroplets containing organic dye emulsions", *Biomicrofluidics,* vol. 9, no. 1, p. 014126, 2015.
[http://dx.doi.org/10.1063/1.4913648] [PMID: 25784965]

[12] "Surface Tension Measurement", Available at: http://www.academia.edu/20783641/Surface-Tension-Measurement

[13] T. Young, "An Essay on the Cohesion of Fluids", *Philos. Trans. R. Soc. Lond,* vol. 95, pp. 65-87, 1805.
[http://dx.doi.org/10.1098/rstl.1805.0005]

[14] Y. Yuan, and T.R. Lee, "Contact Angle and Wetting Properties", In: *Surface Science Techniques,* G. Bracco, B. Holst, Eds., Springer Berlin Heidelberg, 2013, pp. 3-34.
[http://dx.doi.org/10.1007/978-3-642-34243-1_1]

[15] ZismanW.A."Relation of the Equilibrium Contact Angle to Liquid and Solid Constitution", In: *Contact Angle, Wettability, and Adhesion*vol. 43. American Chemical Society, 1964, pp. 1-51.

[16] D.Y. Kwok, and A.W. Neumann, "Contact angle measurement and contact angle interpretation", *Adv. Colloid Interface Sci,* vol. 81, no. 3, pp. 167-249, 1999.
[http://dx.doi.org/10.1016/S0001-8686(98)00087-6]

[17] A.C. Zettlemoyer, "Hydrophobic Interactions", *Adv. Colloid Interface Sci,* vol. 15, no. 2, p. 170, 1981.
[http://dx.doi.org/10.1016/0001-8686(81)80008-5]

[18] E.J. Walsh, "Biochemistry", *Cengage Learn,* vol. 74, no. 2, p. 189, 2012.

[19] T.P. Silverstein, "The Real Reason Why Oil and Water Don't Mix", *J. Chem. Educ,* vol. 75, no. 1, p. 116, 1998.
[http://dx.doi.org/10.1021/ed075p116]

[20] A. Dinache, M. Boni, T. Alexandru, E. Radu, A. Stoicu, I.R. Andrei, A. Staicu, L. Liggieri, V. Nastasa, M.L. Pascu, and M. Ferrari, "Surface properties of Vancomycin after interaction with laser beams", *Colloid Surface A,* vol. 480, pp. 328-335, 2015.
[http://dx.doi.org/10.1016/j.colsurfa.2014.08.023]

[21] M. Ferrari, F. Ravera, L. Liggieri, H. Motschmann, Z. Yi, J. Krägel, and R. Miller, "Adsorption and surface rheology of n-dodecanol at the water/air interface", *J. Colloid Interface Sci,* vol. 272, no. 2, pp. 277-280, 2004.
[http://dx.doi.org/10.1016/j.jcis.2003.10.009] [PMID: 15028487]

[22] R. Palazzolo, F. Ravera, M. Ferrari, and L. Liggieri, "Dynamic Surface Elasticity of Adsorption Layers in the Presence of a Surface Phase Transition from Monomers to Large Aggregates", *Langmuir,* vol. 18, no. 9, pp. 3592-3599, 2002.
[http://dx.doi.org/10.1021/la0156917]

[23] D. Vollhardt, V.B. Fainerman, and G. Emrich, "Dynamic and Equilibrium Surface Pressure of Adsorbed Dodecanol Monolayers at the Air/Water Interface", *J. Phys. Chem. B,* vol. 104, no. 35, pp. 8536-8543, 2000.
[http://dx.doi.org/10.1021/jp001454l]

[24] J. Catalán, C. Díaz, and F. García-Blanco, "Characterization of binary solvent mixtures of DMSO with water and other cosolvents", *J. Org. Chem,* vol. 66, no. 17, pp. 5846-5852, 2001.
[http://dx.doi.org/10.1021/jo010415i] [PMID: 11511261]

[25] D.B. Wong, K.P. Sokolowsky, M.I. El-Barghouthi, E.E. Fenn, C.H. Giammanco, A.L. Sturlaugson, and M.D. Fayer, "Water dynamics in water/DMSO binary mixtures", *J. Phys. Chem. B,* vol. 116, no. 18, pp. 5479-5490, 2012. [http://dx.doi.org/10.1021/jp301967e] [PMID: 22510039]

[26] I.I. Vaisman, and M.L. Berkowitz, "Local structural order and molecular associations in water-DMSO mixtures. Molecular dynamics study", *J. Am. Chem. Soc,* vol. 114, no. 20, pp. 7889-7896, 1992. [http://dx.doi.org/10.1021/ja00046a038]

CHAPTER 3

# Pendant Droplets - Optofluidic Approach

**Mihail Lucian Pascu[1,2,*], Angela Staicu[1] and Mihai Boni[1,2]**

[1] *National Institute for Laser, Plasma and Radiation Physics, Magurele, Ilfov, Romania*

[2] *Faculty of Physics, University of Bucharest, Magurele, Ilfov, Romania*

**Abstract:** This chapter presents a synthesis regarding the optical properties of pendant droplets in view of describing and understanding their interactions with laser and, more general, optical beams. The main methods used for pendant droplets investigation are described, from the optical point of view, at unresonant and resonant interaction with a laser beam focused or sent on it. The interaction is considered in the excitation scheme one laser pulse - one microvolumetric droplet in pendant position in air with typical volumes of the droplet in 1-15 μl range. In the unresonant interaction case the laser beam is not absorbed by droplet's material(s). Beam propagation in droplet is made according to geometrical optics rules when the separation surface between two optical media (air and water, for instance) is spherical. Total reflection of laser beam within droplet at separating surface with respect to the air can be produced and this can give a particular brightness of an illuminated droplet. At resonant interaction, the beam is absorbed by droplet's material(s) and typical phenomena may take place such as laser induced fluorescence (LIF) emission and Raman scattering. These effects are described here. LIF emitted by microdroplets of Rhodamine 6G solution in water is described and results about its amplification in the droplet considered as a spherical optical resonator are shown to the limit of obtaining lasing effects. Raman spectra of dimethyl sulfoxide and ultrapure water microvolumetric droplets are shown and a comparison between Raman beams emitted by microdroplets and bulky samples is made.

**Keywords:** Critical angle, Fluorescence, Hanging droplet, Immiscible liquids, Lasing, Laser induced fluorescence, Light pressure, Light scattering, Optical absorption, Optofluidics, Pendant droplet, Raman scattering, Reflection, Refraction, Refraction index, Rhodamine 6G, Total reflection.

## INTRODUCTION

From the optical point of view, a droplet is an optical body and behaves like one, *i.e.* as an optical medium with a small volume, from micro-liter upward. It may be homogeneous, with isotropic properties, or may have a structured composition if immiscible materials are introduced in it, in which case the properties are not

* **Corresponding author Mihail Lucian Pascu:** Laser Department, National Institute for Laser, Plasma and Radiation Physics, Magurele, Ilfov, Romania; Tel/Fax: 0040 214575739; E-mail: mihai.pascu@inflpr.ro

optically isotropic anymore.

In a liquid droplet one may introduce nanoparticles (such as $TiO_2$-see Chapter 17 and [1]) and then some of the optical properties may be changed with respect to the pure liquid. One may also generate droplets which contain emulsions [2] of two or more normally immiscible substances or even foams. In this case too, optical properties of droplets are those of the respective emulsions or foams. Droplets may be generated in different media and positions, but in all cases their intrinsic optical properties remain the same. Nevertheless, the use of their optical properties differs from case to case because the interaction of droplet with its surrounding media and the transmission of light from media to the droplet and back are affected by the fluidic properties of both droplet and its environment, as well as by the interface between them. In other words, the optical properties of a pendant droplet in air may be exploited in a different way with respect to the same properties of a sessile droplet. If the pendant droplet is hanging in a liquid with which it is immiscible, the interface phenomena when light is crossing it become very important. For a sessile droplet, the surface on which it is standing and the interaction with surrounding materials may regulate the input of light in the droplet as well as the output. The interaction with the surrounding media may also influence the intensity of light transmitted to and from droplets. Another factor to consider with respect to optical effects on droplets is their volume. At volumes larger than 1 µl (droplet diameter about 1.2 mm), optical properties may be treated using the known approach from standard optics and spectroscopy [3]. At microscale, *i.e.* at diameters lower than 1 mm down to nanometer scale the crossing of a droplet by light and the associated phenomena are still issues to study due to, among others, the ratio between light wavelengths and droplets dimensions, to consider only one contributing factor in a droplet's behaviour under light action.

Considering the pendant droplet as a liquid optical body with standard properties, in principle, all optical phenomena specific to an optical passive or active body may be studied on them, such as, but not exclusively: reflection, refraction, absorption, fluorescence, phosphorescence, Raman.

## REFLECTION AND REFRACTION OF LASER BEAM ON A SINGLE PENDANT DROPLET

The pendant droplet may be produced in air or another environment that is immiscible with its liquid materials using a computer controlled droplet generator described in [3 - 6]. A typical example is a water droplet hanging in air on a capillary head that is hydrophobic.

The droplet is shaped under the action of its material weight and surface tension at the interface with air, but its behaviour is also influenced by the viscosity of the constituent liquid or liquid components. Normally, under the action of surface tension in air or interfacial tension in another liquid, the droplet has a spherical shape when its weight is (much) smaller than the surface tension. If the volume is increased, the droplet has an elongated, pear like shape due to the increase of its mass and it may be detached from the capillary. As mentioned above, the process is also related to droplet's material viscosity and its theoretical approach is a complex issue to deal with [7].

In Fig. (**1**) a droplet is depicted which has 1 μl volume, *i.e.* a diameter of 1.2 mm, in which an incident plane parallel laser beam is sent that covers the arcAB with the length L one eighth of the circle full length, *i.e.* 0.47 mm. At the surface of water droplet, the beam is partially reflected and most of it penetrates the droplet and is refracted according to Snell's law being bent towards the normal at the surface or propagating, for 0° incidence angle, along horizontal radius of the droplet. For a droplet with 1.2 mm diameter the time necessary for light to cross the droplet is about 5 ps ($n_2$ water refraction index is considered 1.334 and the speed of light in water at room temperature is taken $225\times10^6$ m/s). In general, function of droplet's volume and material, the lifetime of the photon in the droplet at one single pass may be different from a few picoseconds to some tens of ps. This may become important if the beam is pulsed since the time length of one laser pulse (function of the utilized laser) may vary from sub-picosecond to some nanoseconds or even longer, regardless the pulse repetition rate.

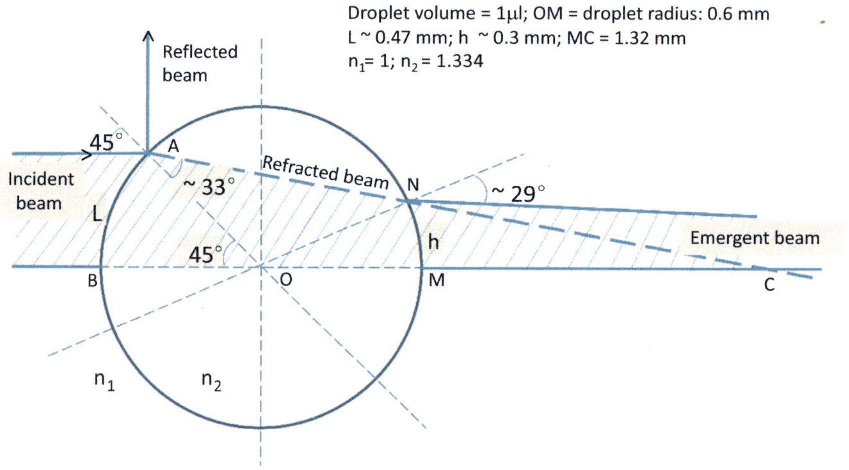


**Fig. (1).** Optical path of a laser ray in a transparent droplet. The input beam is plane parallel and parallel with droplet's horizontal axis; it covers 1/8 of its vertical plane cross section length, between the incidence angle 0° and 45°.

In Fig. (**1**), the light will come out of the droplet between points N and M on the circle and in those points it will be refracted towards larger angles with respect to the normal at the surface in N. The result is that the laser beam will be convergent after the droplet is focused at a distance which is larger than droplet's diameter. Another consequence is that the dimension of the laser spot on the droplet circle at output, h=0.3 mm, is smaller than the spot linear dimension at input, L.

If laser beam is plane parallel and propagates along the horizontal, but it is narrower, say the length of the arcAB=1/16 of the circle length, and is sent above the horizontal symmetry axis of droplet, its behaviour at passing through droplet is the same, but the corresponding propagation values are different. Fig. (**2**) shows the image of the droplet and the laser beam falling on it between the incidence angles 22.5° and 45°, so that L~0.235 mm and h~0.089 mm.

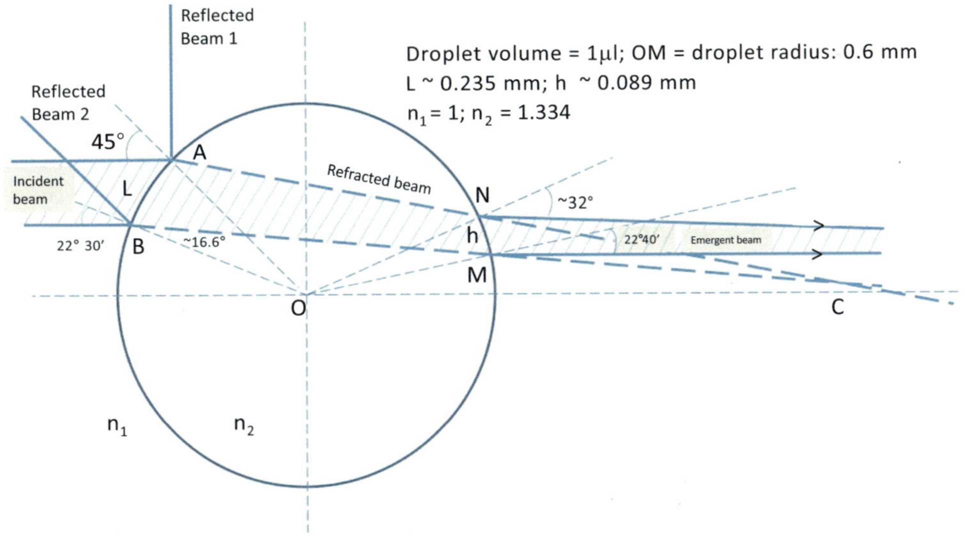


**Fig. (2).** Optical path of a laser ray in transparent droplet. The input beam is plane parallel and parallel with droplet's horizontal but is placed upwards with respect to horizontal symmetry axis and covers 1/16 of its vertical plane section, between the incidence angle 22.5° and 45°.

The beam is also convergent after crossing the droplet but the focus is off-droplet optical axis. Refractions are accompanied by reflections in input-output points on droplet's circumference. In points A and B the reflection is external, in the surrounding air but between points N and M in Figs. (**1-2**) the beam is internally reflected at the interface water-air. So, one may obtain droplet's surface spots of higher luminosity produced by light reflected at lower angles than total internal reflection as shown in Fig. (**3**).

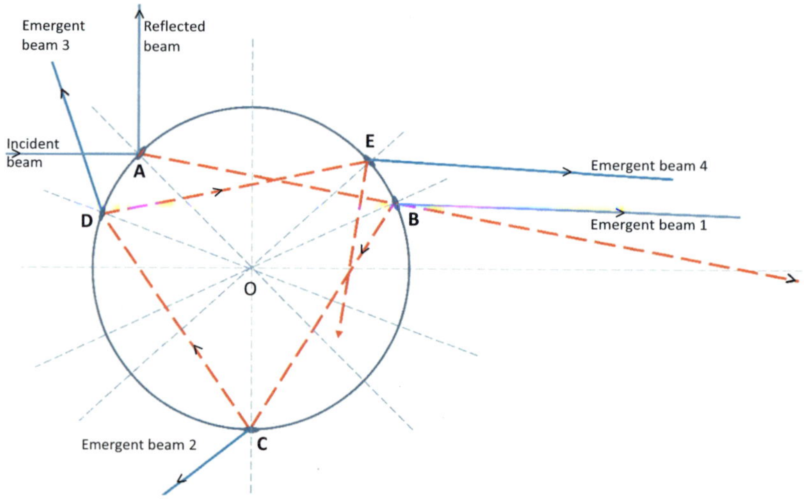


**Fig. (3).** Optical path in a transparent droplet of a narrow cross section/waist, plane parallel laser beam at incidence angles lower than the corresponding total internal reflection angle.

The incident beam in point A covers a spot around it. At input in A, the beam is partially reflected on droplet interface back in the air and refracted so that the refracted beam follows an optical path inside the droplet defined by points B, C, D, E and so on. The spot dimensions around points B, C, D and E show divergence evolution of laser beam inside the droplet and usually are not constant. Each spot is more intense than the rest of surrounding circumference although the intensity is decreasing when the number of reflections is higher due to losses at contact points with droplet interface. Around droplet, refracted beams noted Emergent beam 1, 2 *etc.* in Fig. (**3**) are sent in air.

This is confirmed by experimental measurements made on a 1.2 mm diameter water droplet on which is sent a laser beam of 1 mJ energy, emitted at 532 nm (Fig. **4**). The spot size in A has 90 μm diameter and the angular divergence is around 11°. The spot in B represents the beam waist after the first refraction, the image being recorded with a fast camera (10,000 frames per second) and a Notch filter to cut-off parasite scattered light. In Fig. (**4**) one may see the capillary on which droplet is hanging and one may observe that its material is hydrophobic and the droplet remains in suspended position only due to the favorable relation volume (weight) -surface tension-viscosity. The laser beam has finite cross section and some rays may propagate out of the observed and recorded plane. At the same time, if observing the interaction picture in space, in the experiment, the beam can make a very small angle with respect to vertical picture plane and so, the spots on droplet surface leave the vertical plane and are distributed outside it (Fig. **4**).

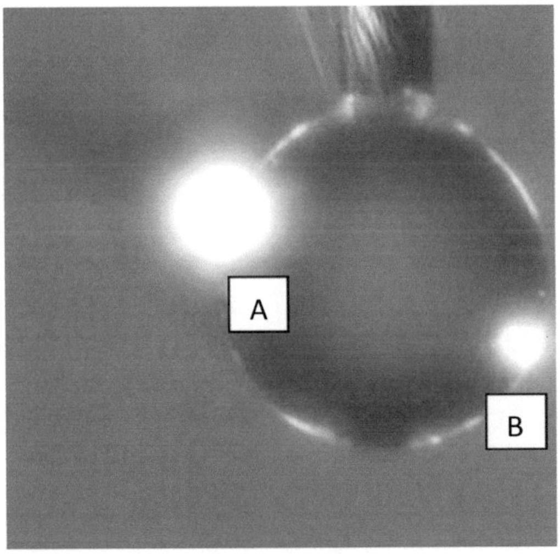


**Fig. (4).** Laser spots on droplet interface when droplet material is not absorbing the beam. Droplet volume is 1 μl. Beam cross section/waist diameter at input in point A, 90 μm and beam input in a random position. Laser beam energy 1 mJ.

Another example is given in Fig. (**5**) where the image taken with a fast camera (10,000 frames/s recording speed) of a droplet crossed by a 0.8 mJ laser beam is shown, when the beam is sent in droplet's equatorial plane and focused on its front surface.

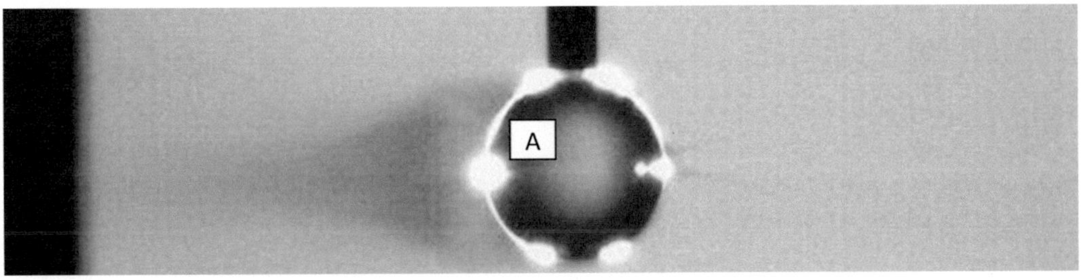


**Fig. (5).** Laser spots on droplet interface with air when droplet material does not absorb the laser radiation. Beam wavelength 532 nm for water droplet study. Beam energy 0.8 mJ. Droplet volume 3.5 μl corresponding to 1.9 mm diameter. Beam waist diameter at input in point A (as in Fig. **4**), 90 μm. Beam propagation at input in droplet's horizontal symmetry plane.

Due to angular divergence which is 11.6° more spots on the interface may be observed; it is possible that the capillary has its own contribution in the multiple reflection processes. The spots are due to successive reflections of the beam at the interface water-air following the scheme depicted in Fig. (**3**). The quite continuous distribution of light along interface outside the spots is due most probably to the totally reflected beams. The fact that the interface is not illuminated on all its length may be due to the slight deformation of droplet during its interaction with laser beam, as well as to the propagation of beams outside observation/recording plane.

A more general case of a beam propagation through droplet is shown in Fig. (**6**) where the incident beam in A has 11.6° convergence angle and the spot diameter is 90 μm. At the output from droplet (points B and C) the spot diameter is 140 μm and the beam is focused off-axis at a distance which is greater than droplet diameter. Fig. (**6**) is simplified by omitting beam's reflections. If total reflection of a laser beam within a droplet is followed-up, the beam incidence angle on droplet has to be computed function of refractive index of droplet material and its surrounding environment. For water droplet, to produce inner reflection of the laser beam within the droplet the incidence angle should be 85.94° (Fig. **7**).

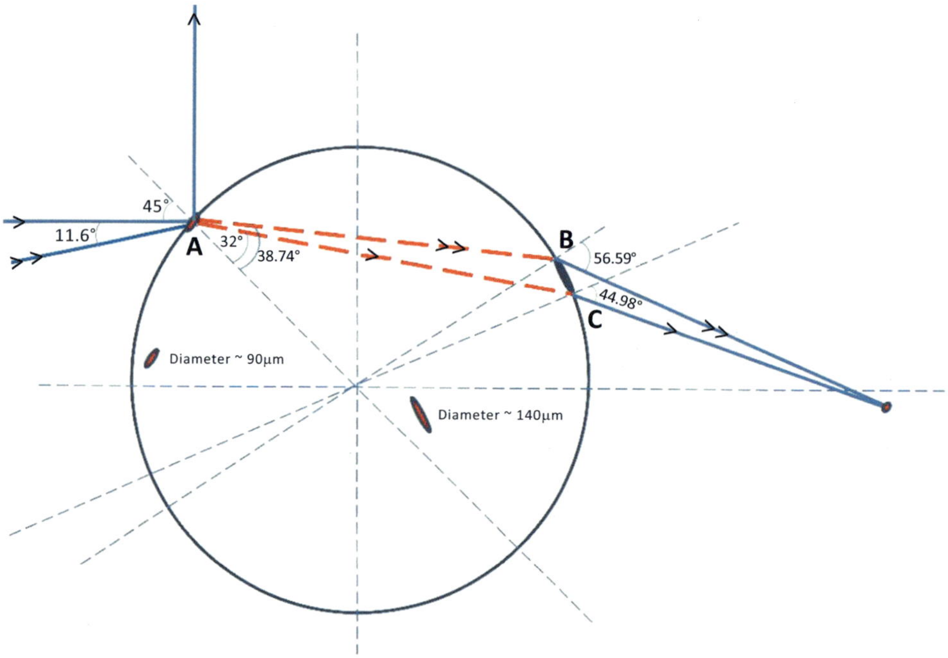


**Fig. (6).** Optical path of a focused laser beam through 1 μl droplet. Input point chosen at 45° from droplet symmetry axis.

This corresponds to the limit angle 48.59° starting from which, at larger angles, the total reflection is produced.

In studying laser beam propagation through a droplet when radiation is not absorbed by droplet's components an important effect is the light pressure of the beam on droplet which is described in Chapter 8 where the unresonant interaction between a laser beam and a single droplet is treated. This effect is always accompanying the light absorption phenomena in the droplet, if any.

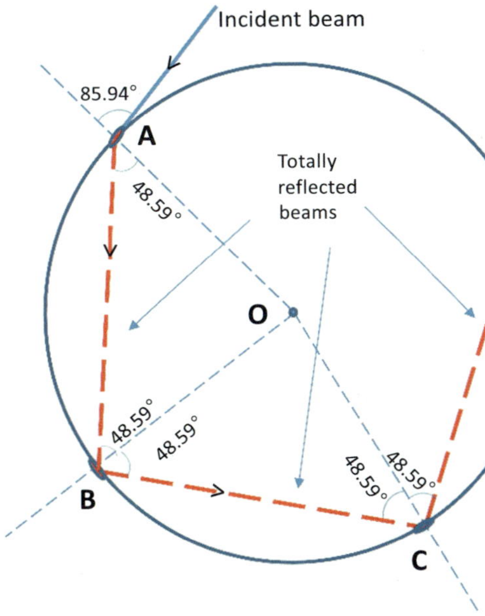


**Fig. (7).** Scheme of the incidence of a laser beam at an angle corresponding to inner total reflection of it within a water droplet.

## SPECTRAL PROPERTIES

### Generalities

An optical beam interaction with a liquid droplet depends on the properties of droplet material(s) such as absorption coefficient(s), refractive index, reflection coefficient etc. If droplet component molecules are chromophores that absorb incident laser radiation, the laser beam-droplet interaction is called resonant and after absorption chromophores emit fluorescence and/or, at longer time and longer wavelength, phosphorescence radiation. Another process at interaction of laser radiation with the droplet is the inelastic scattering on bead molecules otherwise called Raman scattering. The transitions between the energetic levels responsible for the phenomena that take place at interaction of light with droplet molecules are illustrated in Fig. (**8**). Electronic levels are labelled with $S_0$, $S_1$, (singlet electronic levels, respectively, fundamental and first excited electronic single states) and vibrational associated levels are labelled with 0, 1, 2, 3 corresponding to successive vibrational numbers and related to nuclei movements within molecule.

Absorption of light and the consequent fluorescence emission take place between two real electronic levels, whereas the transitions corresponding to Raman scattering imply a virtual level from which the molecules are de-excited to the vibrational levels of the $S_0$ fundamental singlet state.

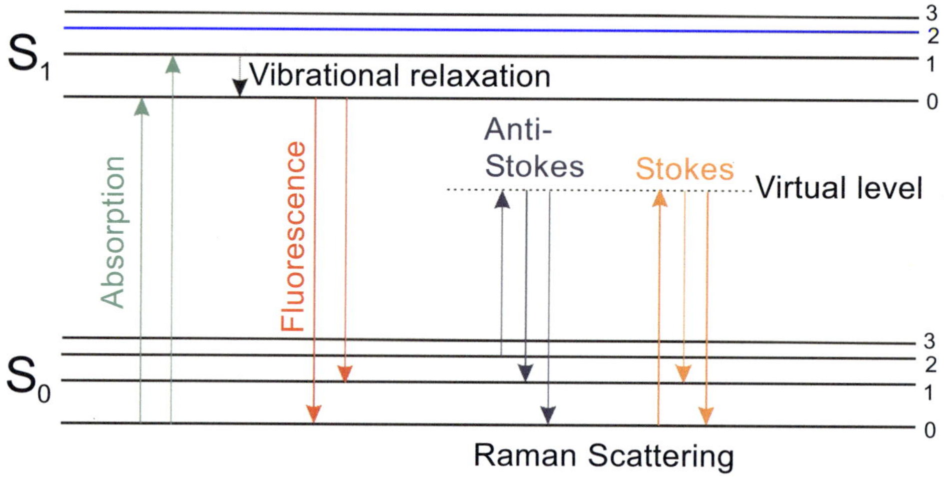


**Fig. (8).** Energy diagram describing transitions of molecule at interaction with light.

The Raman signals leading to Raman spectrum give information about the energy of the vibration levels associated to the fundamental electronic state of molecules. When excitation energy is larger than the scattered signal, a Stokes transition takes place, while the anti-Stokes lines correspond to a scattered signal with energy larger than incident radiation.

## Absorption

When the pumping beam is passing through droplet, its intensity is decreasing according to Beer-Lambert law:

$$I = I_0 e^{-\varepsilon l c} \quad \textbf{(1)}$$

where $I_0$ and $I$ are the intensity of the incident and, respectively, transmitted radiation, $\varepsilon$ is the molar attenuation coefficient, $c$ is the concentration of the absorbent and $l$ is the optical path length of the beam in the droplet. Absorption spectra of Rh6G at concentration in the range $10^{-6}$ M-$10^{-4}$ M are shown in Fig. **(9)** [4]. The absorbed energy contributes to excite molecules of Rh6G from ground singlet state ($S_0$ in Fig. **8**) to the first excited singlet state ($S_1$ in Fig. **8**). From this state a significant number out of the excited molecules could decay on the ground state mainly through emission of fluorescence but other may pass to $S_0$ by other processes as well [8].

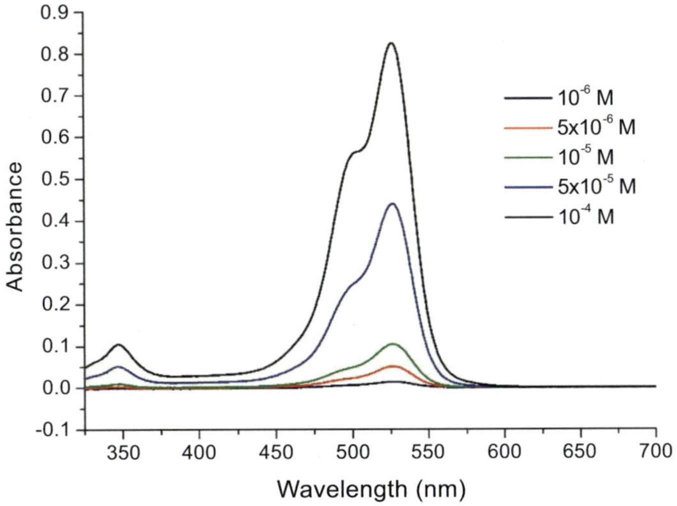


**Fig. (9).** The absorption spectra of solutions of Rh6G in ultrapure water at several concentrations.

With respect to the Beer-Lambert law, after laser beam passes through the droplet, its intensity is attenuated and the transmittance together with penetration depth (pathlength in the sample when *I,* becomes 1/e from $I_0$) are given in Table **1** for different concentrations.

**Table 1. Absorption parameters (attenuation coefficient, penetration depth and transmittance) for several Rh6G molar concentrations in water.**

| Concentration (M) | Attenuation coefficient ($M^{-1}cm^{-1}$) | d (penetration depth) (mm) | T [%] |
|---|---|---|---|
| $10^{-6}$ | 126,400 | 79.11392 | 96.76 |
| $5\times10^{-6}$ | 93,020 | 21.50075 | 88.60 |
| $10^{-5}$ | 95,620 | 10.45806 | 77.98 |
| $5\times10^{-5}$ | 80,308 | 2.49041 | 35.20 |
| $10^{-4}$ | 75,330 | 1.32749 | 14.10 |
| $5\times10^{-4}$ | 49,499 | 0.40405 | 0.16 |
| $10^{-3}$ | 35,868.6 | 0.2788 | 0.008 |

## Emission

At resonant interaction of light with the droplet, *i.e.* when light wavelength corresponds to an absorption band of the molecules contained in droplet, fluorescence or even lasing emission may be produced. The intensity of emitted fluorescence depends as in the case of bulk samples on intrinsic properties of chromophores (fluorescence quantum yield), solvent type, concentration, collections geometry of the optical emission etc. Besides, the properties of radiation emitted by droplets depend on droplet size, shape and structure. The incident beam angle with respect to droplet surface is another parameter that determines the fluorescence intensity and its overall properties. The emitted laser induced fluorescence (LIF) radiation may be trapped inside a water droplet for an inner incident angle larger than 48.59° with respect to droplet surface and in these conditions LIF radiation may be amplified. On the other hand, this angle corresponds to the input angle of the incident beam on water droplet surface of 85.94° when the beam comes from environmental air, and in this case the input beam is totally reflected inside the droplet at incidence on droplet surface. The confined inputted radiation may interact with other dye/chromophore molecules and can be at the origin of amplified stimulated emission by droplet itself. A spherical or even a quasi-spherical droplet acts as a resonant cavity where radiation is amplified and a spectral narrowing of emission band occurs. The resonance frequencies of the cavity can modulate droplet emission and narrow lines structures of it can be observed for emission by microvolumeric droplets when the resolution of detection system is high enough to observe and measure the modal structure. More details about this kind of process are shown in Chapter 17 of this book.

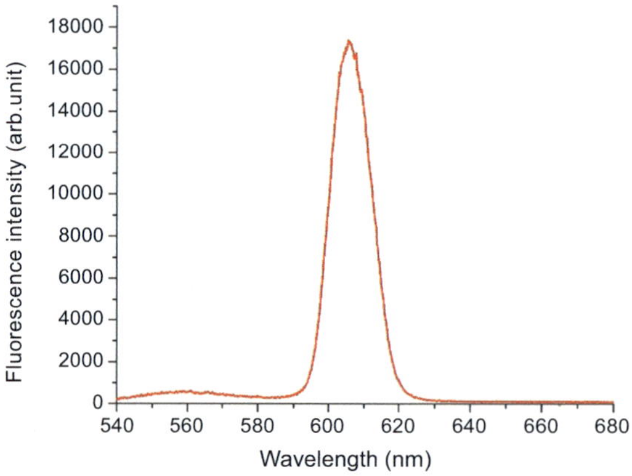


**Fig. (10).** LIF spectrum emitted by an optically pumped 1µl droplet containing a solution of Rh6G in water at $5\times10^{-4}$ M. Laser pulse excitation energy 6 mJ at 532 nm. Pulse repetition rate is 10 pps.

As an example, in Fig. (**10**), the fluorescence spectrum measured for a solution of $5\times10^{-4}$ M Rh6G in ultrapure water in a pendant droplet in air is shown. It can be noticed that the bandwidth of droplet's emission *i.e.* the full width at half maximum (FWHM) is about 15 nm which is much smaller than the typical fluorescence bandwidth for a bulk sample [2 - 4].

In addition, a splitting of the structure in 2 or more bands can be obtained function of excitation energy or droplet content (see Chapter 17 for details).

## Raman Scattering

Raman scattered signal generated at the interaction of laser radiation with a pendant microdroplet was produced and measured for ultrapure water and dimethyl sulfoxide (DMSO) beads generated in open air in normal conditions of temperature, pressure and humidity. Both liquids were of pure solvent grade quality to avoid unwanted LIF that could cover the Raman signal in these experiments and to have a better signal to noise ratio.

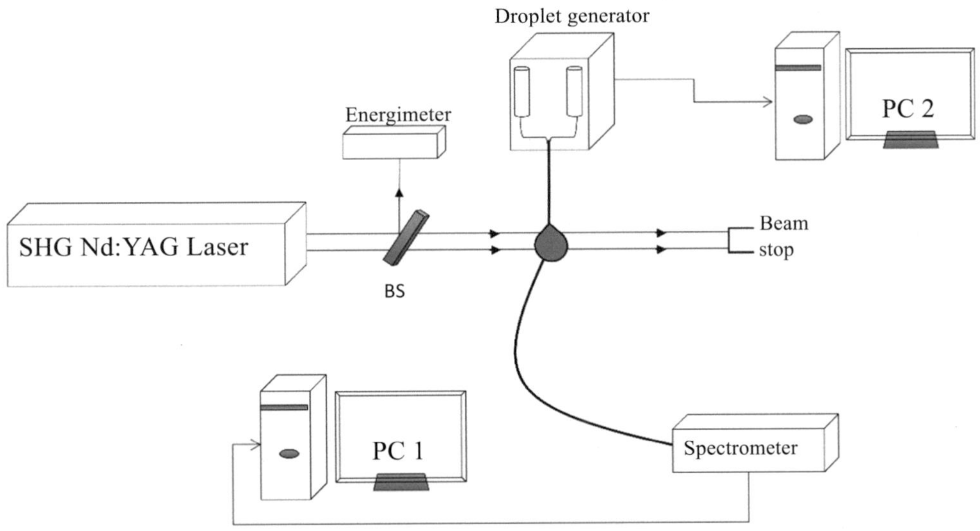


**Fig. (11).** Experimental set-up for Raman measurements.

The laser radiation used to excite Raman signals was the second/third harmonic of a pulsed Nd:YAG laser emitted at 532 nm/355 nm, 10 pulses per second repetition rate, with the pulse time length at half maximum 6 ns. The set-up depicted in Fig. (**11**) allows the collection by an optical fiber (1 mm core) of the Raman emission at $90^0$ with respect to the propagation direction of the laser excitation beam. The Raman spectra for a single droplet were compared with the ones measured for bulk samples (placed in a standard fluorescence quartz cell). The signal is spectrally analysed with a spectrograph (SpectraPro 2750, Acton Research) coupled with an ICCD (PIMAX1024, Princeton Instruments).

In Fig. (**12**), Raman spectra of DMSO obtained for measurements on bulk samples are shown for comparison with the spectra measured on single droplet. The energy of laser pulse was 1 mJ and the beam was tailored to avoid droplet destruction at contact with laser pulse. Specific DMSO vibrational bands observed are listed in Table **2** and are in good correspondence with data given in literature reports [9].

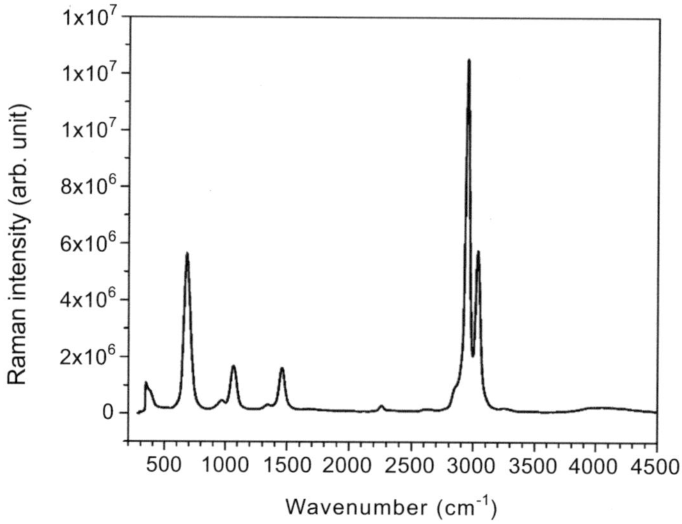


**Fig. (12).** Raman spectrum of DMSO measured for a bulky sample.

**Table 2. Main vibrations of the Raman spectra of DMSO in bulk.**

| Type of vibration | ν ($cm^{-1}$) |
|---|---|
| CSO in-plane rock | 346 $cm^{-1}$ |
| CS anti-symmetric stretch | 684 $cm^{-1}$ |
| Symmetric stretch of monomer | 1,068 $cm^{-1}$ |
| HCH deformation | 1,454 $cm^{-1}$ |
| CH symmetric stretching | 2,945 $cm^{-1}$ |
| CH anti-symmetric stretching | 3,030 $cm^{-1}$ |

For comparison, Raman signal obtained from a DMSO single droplet with 7 μl volume is shown in Fig. (**13**). There is a decrease in intensity and a loss of several spectral lines by comparison with the spectrum registered for DMSO in a bulk sample. This is because the used laser energy for Raman excitation in droplet is limited to a few mJ to avoid mechanical phenomena or breakdown that can be produced at higher energy (see Chapter 8 and [3]). But, on the other hand, the obtained Raman signals on beads strengthens the idea that, although quite weak, Raman signal can be collected even from small samples such as micro-droplets. This can be helpful in investigating various small samples of interest in biomolecular and biomedical studies.

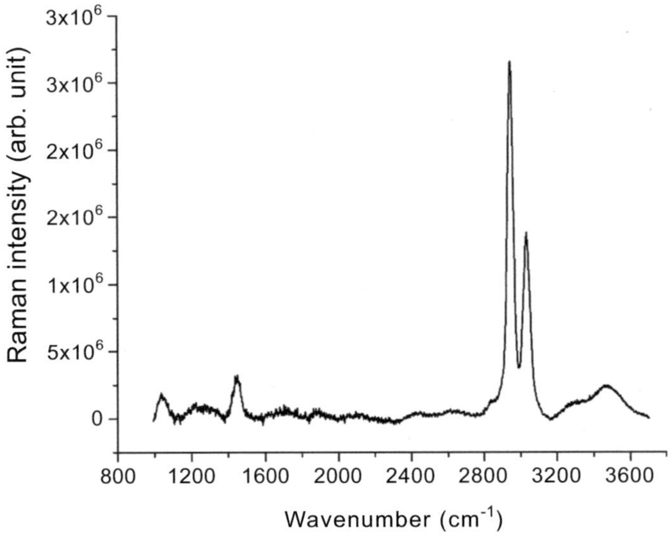


**Fig. (13).** Raman spectrum of DMSO measured on pendant droplets. Bead volume 7 µl. Beam energy 1 mJ.

Raman spectra were also registered for ultrapure water. In Fig. (**14**), the Raman spectra of a bulky sample (placed in a standard spectrophotometric 1 cm thick cuvette) when excited with 355 nm and 532 nm laser radiation are shown. Laser pulse energy was 100 mJ for 532 nm and 60 mJ for 355 nm.

Important information about intra- and inter-molecular interactions in water can be obtained from data measured during Raman experiments. Literature reports present the following Raman bands of water [10, 11]:

- 300-900 $cm^{-1}$–band of intermolecular libration;
- 1,600-1,700 $cm^{-1}$–bending band with maximum at 1,645 $cm^{-1}$;
- 2,000-2,400 $cm^{-1}$–associative band due to overtones of intermolecular vibrations;
- 3,000-3,800 $cm^{-1}$–valence band with maximum at 3,400 $cm^{-1}$;
- 3,900-4,200 $cm^{-1}$–weak intensity band with maximum at 4,000 $cm^{-1}$ due to overtones of intermolecular vibrations and combination frequencies;
- 6,000-7,000 $cm^{-1}$–weak intensity band, an overtone of the valence band.

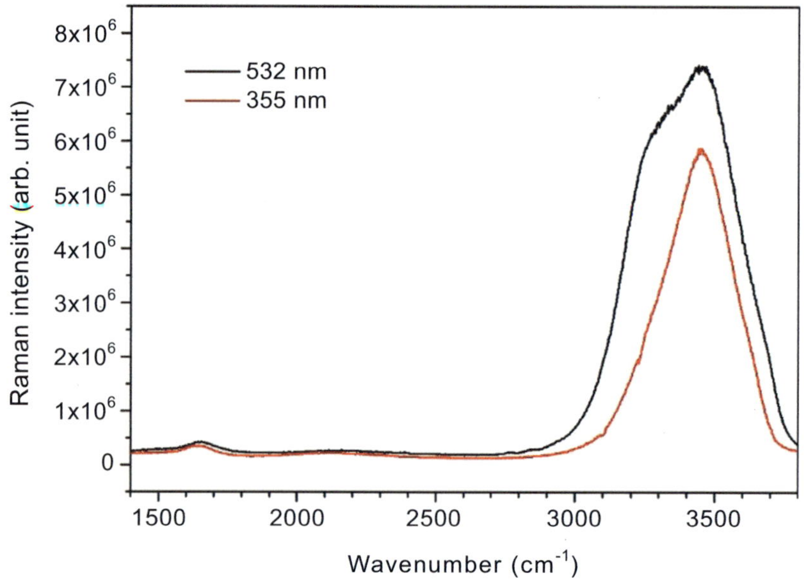


**Fig. (14).** Raman spectra of ultrapure water in bulk sample (volume 2 ml). Pulsed beam energy 100 mJ at 532 nm and 60 mJ at 355 nm.

If Raman spectra measured on bulk samples are compared to spectra measured on micro-droplets containing only ultrapure water, the results are shown in Fig. (**15**) and the common bands are listed in Table **3**. The excitation wavelength was in this case 532 nm and the laser pulse energy was 1 mJ for droplet measurements and 100 mJ for bulk sample. The droplet volume was 7 µl and the bulk sample was introduced in one cm pathlength cuvette.

**Table 3. Main vibrations of Raman spectrum of water for measurements on bulky and droplet samples.**

| Bands | Bending | Associative | Valence |
|---|---|---|---|
| Excited at 532 nm (Bulk) | 1,660 $cm^{-1}$ | 2,184 $cm^{-1}$ | 3,436 $cm^{-1}$ |
| Excited at 532 nm (Droplet) | 1,500 $cm^{-1}$ | 2,361 $cm^{-1}$ | 3,503 $cm^{-1}$ |

It can be noticed that although spectra are quite similar, several narrow bands rise up in the case of droplet samples at the following frequencies: 2,361, 2,800, 2,892, 3,503 $cm^{-1}$. This can be explained by the fact that these vibrations (fundamental or overtones) are enhanced in the spherical cavity formed by the droplet due to the coincidence with some frequency oscillation modes of the optical resonator.

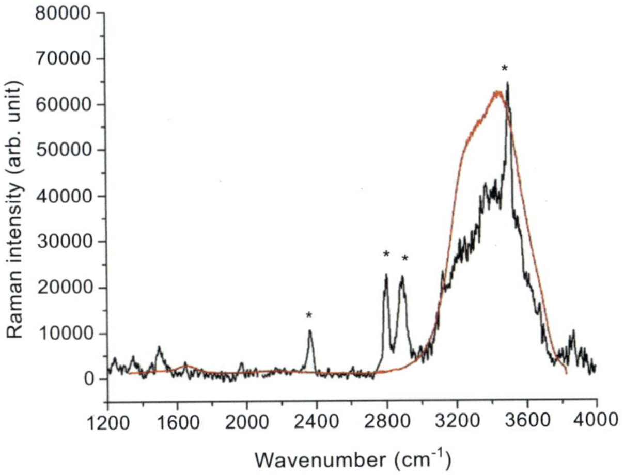


**Fig. (15).** Raman spectra measured on bulk (red curve) and pendant droplet (black curve).

## CONCLUSIONS

Pendant droplets have optical properties that depend on their content and structure. When they interact with a laser beam, several optical phenomena may be produced which take place at the contact of optical radiation with bulky liquids as well. The difference is that due to the small dimensions of droplets (diameters of, at most, some mm) the lifetime of photons in the droplet is of the order of some picoseconds and if the beam remains in the droplet at propagation angles above the total inner reflection, one may reach fast a high number of reflections at droplet's surface which would correspond to a longer lifetime. This could be combined with pulse time length of the pumping laser beam, if pulsed lasers are used.

If unresonant and resonant effects are combined, laser pumping beam power should be chosen so that the droplet does not bear significant mechanical effects after interaction with the beam. This would allow to use LIF and Raman beams in applications.

## CONFLICT OF INTEREST

The authors confirm that this chapter content has no conflict of interest.

## ACKNOWLEDGEMENTS

This work has been financed by the National Authority for Research and Innovation in the frame of Nucleus programme-contract 4N/2016 and the projects PN-II-ID-PCE-2011-3-0922 and 41-018 PALIRT Project.

# REFERENCES

[1] M. Boni, A. Staicu, I.R. Andrei, A. Smarandache, V. Nastasa, M. Comor, Z. Saponjic, and M.L. Pascu, "Studies on laser induced emission of microdroplets containing Rhodamine 6G solutions in water doped with TiO2 nanoparticles", *Colloid Surface A,* vol. 519, p. 238p. 244, 2017. [http://dx.doi.org/10.1016/j.colsurfa.2016.07.033]

[2] M. Boni, V. Nastasa, I.R. Andrei, A. Staicu, and M.L. Pascu, "Enhanced fluorescence emitted by microdroplets containing organic dye emulsions", *Biomicrofluidics,* vol. 9, no. 1, p. 014126, 2015. [http://dx.doi.org/10.1063/1.4913648] [PMID: 25784965]

[3] M. L. Pascu, G. V. Popescu, C. M. Ticos, and I. R. Andrei, "Unresonant interaction of laser beams with microdroplets", *J. Eur. Opt. Soc. Rapid Publ.,* vol. 7, 2012. [http://dx.doi.org/10.2971/jeos.2012.12001]

[4] M. Boni, V. Nastasa, A. Staicu, I.R. Andrei, and M.L. Pascu, "Characterisation of fluorescent pendant droplets", *Rom. Rep. Phys.,* vol. 67, no. 4, pp. 1278-1287, 2015.

[5] M.L. Pascu, I.R. Andrei, M. Ferrari, A. Staicu, A. Smarandache, A. Mahamoud, V. Nastasa, and L. Liggieri, "Laser beams resonant interaction with micro-droplets which have a controlled content", *Colloid Surface A,* vol. 365, no. 1–3, pp. 83-88, 2010. [http://dx.doi.org/10.1016/j.colsurfa.2010.02.036]

[6] A. Smarandache, V. Nastasa, M. Boni, A. Staicu, J. Handzlik, K. Kiec-Kononowicz, L. Amaral, and M-L. Pascu, "Laser beam resonant interaction of new hydantoin derivatives droplets for possible biomedical applications", *Colloid Surface A,* vol. 505, pp. 37-46, 2015. [http://dx.doi.org/10.1016/j.colsurfa.2015.12.002]

[7] M.R. de Saint Vincent, J. Petit, M. Aytouna, J.P. Delville, D. Bonn, and H. Kellay, "Dynamic interfacial tension effects in the rupture of liquid necks", *J. Fluid Mech.,* vol. 692, pp. 499-510, 2012. [http://dx.doi.org/10.1017/jfm.2011.550]

[8] J.R. Lakowicz, Ed., *Principles of Fluorescence Spectroscopy.* Springer US: Boston, MA, 2006. [http://dx.doi.org/10.1007/978-0-387-46312-4]

[9] W.N. Martens, R.L. Frost, J. Kristof, and J. Theo Kloprogge, "Raman spectroscopy of dimethyl sulphoxide and deuterated dimethyl sulphoxide at 298 and 77 K", *J. Raman Spectrosc.,* vol. 33, no. 2, pp. 84-91, 2002. [http://dx.doi.org/10.1002/jrs.827]

[10] S.A. Burikov, T.A. Dolenko, and D.M. Karpov, "The contribution of Fermi resonance to the formation of the Raman valence band of water", *Opt. Spectrosc.,* vol. 109, no. 2, pp. 272-278, 2010. [http://dx.doi.org/10.1134/S0030400X10080205]

[11] S. Burikov, S. Dolenko, T. Dolenko, S. Patsaeva, and V. Yuzhakov, "Decomposition of water Raman stretching band with a combination of optimization methods", *Mol. Phys.,* vol. 108, no. 6, pp. 739-747, 2010. [http://dx.doi.org/10.1080/00268970903567288]

*Laser Optofluidics in Fighting Multiple Drug Resistance*, 2017, 41-74 

# CHAPTER 4

# Profile Analysis Tensiometry for Studies of Liquid Interfacial Dynamics

**Joo Y. Won[1], Vamseekrishna Ulaganathan[1], Ayim Tleuova[1,2], Talmira Kairaliyeva[1,2], Altynay A. Sharipova[1,2], Xiu W. Hu[1,3], Mohsen Karbaschi[1], Georgi Gochev[1,4], Aliyar Javadi[1,5], Mohammad Taeibi Rahni[1,6], Alexander V. Makievski[7], Jürgen Krägel[1,7], Saule B. Aidarova[2]** and **Reinhard Miller[1,*]**

[1] *Max Planck Institute of Colloids and Interfaces, Potsdam, Germany*

[2] *Kazakh National Research Technical University K.I.Satpayev, Almaty, Kazakhstan*

[3] *State Key Laboratory of Multiphase Flow in Power Engineering, Xi'an Jiaotong University, Xi'an, P.R. China*

[4] *Institute of Physical Chemistry, Bulgarian Academy of Sciences, Sofia, Bulgaria*

[5] *Chemical Engineering Department, University of Tehran, Tehran, Iran*

[6] *Sharif University, Tehran, Iran*

[7] *Sinterface Technologies, Berlin, Germany*

**Abstract:** Bubble and drop profile analysis tensiometry is one of the most versatile tools for the characterisation of liquid interfaces. This technique provides the time dependence of surface and interfacial tension in a brought time range. The obtained interfacial tensions are well suited for a quantitative characterisation of adsorption layers of surfactants, proteins, polymers, and of their mixtures at liquid interfaces. Instruments equipped with an accurate dosing system allow experiments under well controlled conditions, such as constant drop/bubble volume or surface area. Also pre-programed changes are feasible, for example drop volume oscillations in a frequency range between 1 mHz up to about 0.1 Hz. This type of experiments allows determination of the dilational visco-elastic properties of liquid interfacial layers. Using a coaxial double capillary, experiments on sequential and simultaneous adsorption routes can be performed. This allows an analysis of the specific aspects of complex formation of *e.g.* proteins with other components or multilayer formation at the interface and determination of their dilational rheology. From a hydrodynamic point of view, the drop profile analysis methodology is limited to equilibrium drop/bubble profiles. Hence, when the size of a drop or bubble is changed too quickly their profiles are no longer Laplacian and the method would deliver wrong results. Therefore, the limits of PAT for a specific system need to be determined experimentally and validated by CFD simulations. Examples of CFD simulations have been presented in order to show how reliable the calculated interfacial tensions are.

[*] **Corresponding author Reinhard Miller:** Max Planck Institute of Colloids and Interfaces, Potsdam, Germany; Tel: +49-331-5679252; E-mail: miller@mpikg.mpg.de

**Mihail Lucian Pascu (Ed.)**

**Keywords:** CFD simulations, Dilational surface viscoelasticity, Drop bulk exchange, Liquid/gas interface, Liquid/liquid interface, Profile analysis tensiometry, Surfactant adsorption.

## INTRODUCTION

The tension $\gamma$ is one of the most easily accessible experimental parameters of liquid interfaces and it is an essential parameter that describes the work that has to be done to create a new interface of a unit area. The concept of the surface tension has been introduced by Segner as early as 1751 [1] which was later advanced by prominent scientists such as Young (1805), Laplace (1806), Poisson (1830), Worthington (1884), Bakker (1928), Brown (1947), Prandtl (1947) and Gauss (1830) [2]. The term surface tension is used for the tension of a liquid surface which is in contact with a gas phase while interfacial tension refers to the tension of the interface between two immiscible liquid phases. For simplicity any surfaces will be named interface during the entire chapter, as long as no specificities of liquid/gas surfaces are concerned.

The properties of an interface can be modified by the addition of surface active molecules which adsorb at the interface due to their amphiphilic character and decrease the interfacial tension. This adsorption effect is described by the Gibbs fundamental equation:

$$\Gamma = -\frac{1}{RT}\frac{d\gamma}{d\ln c} \quad \textbf{(1)}$$

Here $\Gamma$ is the adsorbed amount at the interface, c is the concentration of the adsorbing surface active compound, and $R$ and $T$ are the ideal gas constant and absolute temperature, respectively. Thus, measuring the interfacial tension as a function of surfactant bulk concentration $\gamma(c)$ the adsorbed amount $\Gamma$ can be obtained from the slope of the plot $\gamma(ln\ c)$. A detailed description of the history of adsorption models at liquid interface has recently been summarised [3].

Note, adsorption is a process that requires time. Therefore, measurements of the interfacial tension at a given surfactant concentration c as a function of time $\gamma(t)$ provide information about the change of the adsorption of the surfactant $\Gamma(t)$ at the given interface, assuming that the Gibbs' equation applies also under dynamic conditions. The most famous theory for describing the adsorption kinetics of surfactants at a liquid interface was derived in 1946 by Ward and Tordai [4] and is based on a diffusion controlled adsorption mechanism, *i.e.* surfactant molecules have first to be transported by diffusion close to the interface, from where they can then adsorb directly at the interface:

$$\Gamma(t) = 2\sqrt{\frac{D}{\pi}}\left[c_0\sqrt{t} - \int_0^{\sqrt{t}} c(0,t-\tau)d\sqrt{\tau}\right] \qquad \textbf{(2)}$$

where $D$ is the diffusion coefficient, $t$ is the time, and $c_0$ is the surfactant bulk concentration far from the interface. This non-linear integral equation is not easy to be applied to experimental data and was therefore often the subject of theoretical work in order to allow a comparison of experimental data with this adsorption model [5].

Measurements of dynamic interfacial tensions can be performed by various experimental methods. Some of these methods are particularly dedicated to the short time adsorption process, such as the capillary pressure tensiometry. A special version of this methodology is the maximum bubble pressure tensiometry, which however is only suitable for liquid/gas interfaces. A recently published book was dedicated to various interfacial tension methods, mainly based on drop and bubble interfaces, and discusses many scientific and even technical aspects of these methods [6] so that no details will be presented on how other methods compare with the PAT that is described here extensively.

This chapter, however, is focussed only on the workhorse in modern surface science laboratories, the bubble and drop profile analysis tensiometry. It was first proposed as routine method for measuring the tension of liquid interfaces by Neumann and his group [7]. Here the history of this method is shown, as well as the state of art of its theoretical background and practical implementation, and some examples to demonstrate the potential of its application to various liquid interfaces. Note, however, although PAT is a very suitable method for studies of contact angle and wetting properties of solid surfaces, this topic is not presented here (see [8]).

## DEFINITION AND HISTORY OF PROFILE ANALYSIS TENSIOMETRY

For more than a century the dimensions of large sessile drops have been used to determine the surface or interfacial tension of a liquid. Quincke in 1858 [9] practiced this method by measuring the height of a sessile drop above its equatorial diameter. It was necessary to use very large drop, which in turn was not so easy experimentally, so that some correction factors were proposed to compensate for the finite size of the used drops. Much later, in 1963, Padday [10] applied the drop-height method to determine the spreading coefficient of a liquid drop on a plane solid surface.

The equation of capillarity was first formulated by Laplace in 1804 [11, 12]:

$$\gamma\left[\frac{1}{R_1}+\frac{1}{R_2}\right]=\Delta\rho g(z+z_o) \tag{3}$$

$R_1$ and $R_2$ are the two main radii of curvature of the axisymmetric drop, $\Delta\rho$ is the density difference between the two fluid media, g is the gravity constant and z is the height, while $z_0$ is a reference line, typically located in the apex of the drop. This equation relates the mean curvature of the interface to the pressure difference across the interface.

In 1972, Padday [13] presented the first results of computer calculations for sessile drop profiles and compared them with the tables published by Bashforth and Adams [14]. These tables were the main reference for drop and bubble profiles until then. Padday also compared his calculations with empirical formulae, such as those of Worthington [15] or Porter [16] developed for the determination of a liquid's surface tension from the shape of a sessile drop. The results of Padday's calculations show that all tables and empirical formulae are applicable only in very limited intervals. Also Hartland and Hartley [17] published computed drop profile data to be used for the determination of the interfacial tension of liquid/fluid interfaces.

Among many other computational work to calculate the profile of drops and bubbles, the procedure proposed by Maze and Burnet [18, 19] was finally the most reliable one and is now used in many software packages for measurements of interfacial tension from drop and bubble profiles. They used a non-linear regression for fitting the experimental drop profile by calculated drop profiles, so that no specific set of coordinate points was required for the calculations.

The most important step in the methodology of interfacial tension measurements from liquid profiles, however, was done by Rotenberg, Boruvka and Neumann in 1983 [7]. They developed the so-called Axisymmetric Drop Shape Analysis (ADSA), which for the first time really fitted the Laplacian profiles to a measured drop profile in order to obtain $\gamma$ as the best fit parameter. Hence, one can say that ADSA was the first method to obtain the interfacial tension from the profile of a sessile or pendent drop. The ADSA method was further refined by Cheng *et al.* [20] and is now a routine methodology and belongs to the standard equipment in many laboratories. While in the early years of ADSA a high-end workstation was still required for the data processing, now standard PCs are able to manage the job of image analysis and profile fitting. The PAT described in this chapter works in real-time including the image acquisition and data analysis so that the progress of an experiment can be followed by the experimentalist easily. A good overview of PAT as part of the experimental possibilities for characterizing liquid interfaces and the state of the art of computational fluid dynamics simulations is presented in [5].

## Design of Hardware and Software

After the first experimental set-up has been designed in 1983, many modifications were yet required until the profile analysis tensiometry became a routine tool for interfacial science. Both hardware and software improved significantly with the progress in the available video technique and computer controlled dosing tools. Also the new powerful platforms for programming the required image analysis routines, calculations and fitting algorithms allowed having now user friendly and efficient instruments at hand.

## Experimental Set-up

In Fig.(**1**) the main elements of a PAT instrument are shown. The key elements are a video camera with objective, a homogenous light source, and an accurate dosing system, and all of them are controlled by a computer. The experimental protocols can be quite different however, the easiest protocol comprises a quick formation of a fresh liquid drop or bubble in a liquid, and then of its size control over a preprogramed period of time. The drops or bubbles are typically formed at the tip of a circular capillary, as it is shown in the scheme of Fig. (**1**).

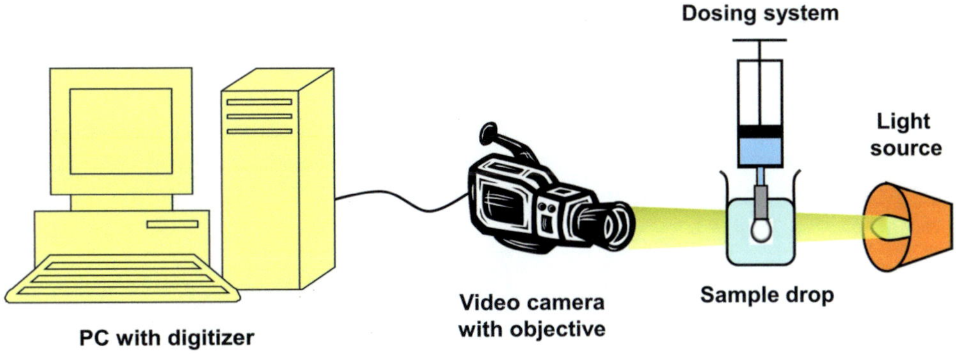


**Fig. (1).** Scheme of a PAT with its main elements.

The high-end PAT instruments have also accurate mechanical tools for a perfect focussing of the video image. In many cases, dedicated software provides various experimental features, for example for the optimisation of the back light and for finding the focus of the image.

The quality of the experimental set-up does of course depend on the quality for setting the optical parameters, and also on the quality of the video camera. The first instruments have been yet built with video cameras having rather irregular pixel sizes so that a pixel-wise correction was required [20]. In modern instruments the resolution of the video chips is much larger and its quality by far higher so that a general determination of the magnification *via* a steel sphere or rod of well-known size is sufficient. High-end PATs are built with cameras having a pixel resolution of 1280×1024 (in contrast to 512×512 of the first video cameras used for ADSA in the nineties). While first PAT instruments were run with extremely expensive workstations, all instruments work now with PCs. There are even instruments on the market with a full-function built-in computer with touch screen now in order to run the instrument completely independent of any external computer.

As mentioned above, the software of all PAT-type instruments provides various possibilities to pre-program the planned experiments. In this way, experimental protocols can contain an initial part with a constant drop size, followed then by drop size oscillations at different frequencies which is required to determine the dilational visco-elasticity of interfacial layers. Accurate dosing systems are required for these drop size manipulations.

A special feature of the PAT-1 and PAT-2 from SINTERFACE Technologies is the concentric double capillary as additional equipment. In Fig. (**2**) the schematic construction of this special capillary is shown. It was invented by Wege *et al.* [21] and became a useful tool for several types of experiments, such as desorption of surfactants and polymers from pre-adsorbed interfacial layers, compensation of evaporation in long term studies, performance of sequential adsorption routes of several adsorbing species, built-up of multilayers on drop surfaces and determination of their mechanical properties, generation of convection for an enhanced adsorption process, mass transfer through membranes [22]. Some examples are presented further below.

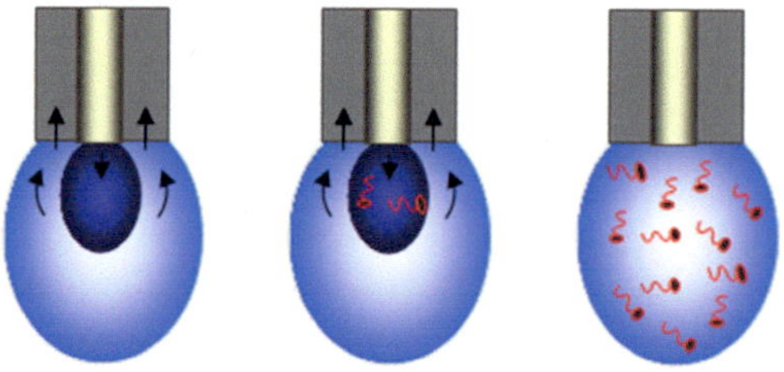

**Fig. (2).** Double capillary for special manipulations of drop formation and composition: left–drop is formed by the matrix liquid, middle–solution is injected *via* the centre capillary while *via* the outer capillary the total drop size is compensated, right–the drop volume is completely exchanged; according to [21].

## Optimisation of the Edge Detection of a Drop/Bubble Profile

For the detection of the profile coordinates, the simplest procedure is a grey level threshold. However, an analysis of the pixels' grey levels along the profile allows determining the coordinates of a bubble or drop with a sub-pixel resolution. In Fig. (**3**) the change of the grey levels is shown along the crossing over the edge of a drop, *i.e.* the value is changing from about 230 (255 is completely white) down to about 10 (0 is completely black).

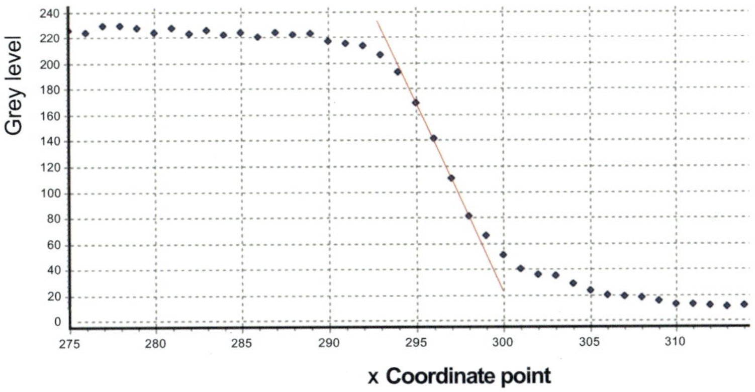


**Fig. (3).** Change of the grey levels across a drop profile.

The drop edge is then defined in the place of the deepest slope of this grey level curve. This point is easily obtained by calculating the first derivative and finding the maximum in this dependency, *i.e.* finding the zero of the second derivative of the curve given in Fig. (**3**).

## Solution of the Laplace Equation to Calculate Drop/Bubble Profiles and the Fitting Routine

For its use in PAT, the Laplace equation, given by equation (3) is typically rewritten into a set of first-order differential equations, based on the geometrical parameters of a liquid axisymmetric profile. These parameters are the arc length s, the normal angle $\phi$, and $x$ and z are the abscissa and ordinate of the profile point (see Fig. **4**):

$$\frac{dx}{ds} = \cos(\phi) \quad \mathbf{(4)}$$

$$\frac{dz}{ds} = \sin(\phi) \quad \mathbf{(5)}$$

$$\frac{d\phi}{ds} = \pm\frac{\beta z}{b^2} + \frac{2}{b} - \frac{\sin(\phi)}{x} \tag{6}$$

The sign in equation (6) is "plus" for sessile drops or captive bubbles and "minus" for pendant drops or emerging bubbles, *b* is the radius of curvature at the bubble apex (0, 0) and β is called the shape factor

$$\beta = \frac{\Delta\rho g b^2}{\gamma} \tag{7}$$

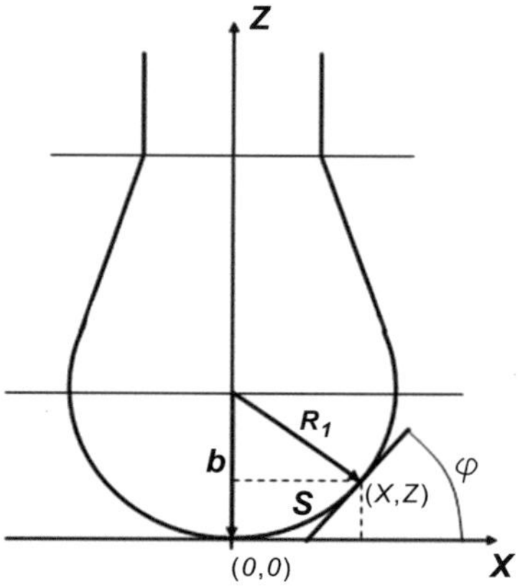


**Fig. (4).** Axisymmetric liquid profile with the definition of all geometric parameters.

The set of equations (4)-(6) can be solved numerically by various methods. Most efficient is the fourth-order Runge-Kutta routine, for which the following initial values at the apex of the drop or bubble at s=0 are used:

$$x(0)=z(0)=\phi(0)=0\ ,\ \frac{d\phi}{ds} = \frac{1}{b} \tag{8}$$

Assumed that Δρ and g are known, the profile obtained from the Laplace equation depends only on the radius of curvature at the apex *b* and on the interfacial tension γ.

Once an efficient code is available to solve the Laplace equation, an efficient strategy is required to fit the calculated profile to the measured one. To evaluate the quality of fitting the experimental profile, a target function is required. The typical measure is the geometric distance between the experimental and calculated profile. This distance d for a profile point ($x_e$,$z_e$) to the calculated ($x_c$,$z_c$) is given by:

$$d = \sqrt{(x_e - x_c)^2 + (z_e - z_c)^2} \quad \textbf{(9)}$$

Various versions have been used, while the one using the normal distance to the profile curve turns out to be the most efficient. As target function, the differences for all measured profile points (few hundreds, depending on the resolution of the video image) are summed up. The three important steps in the PAT methodology are schematically shown in Fig. (**5**): image acquisition, determination of the profile coordinates, fitting the Laplace equation to the experimental profile coordinates.

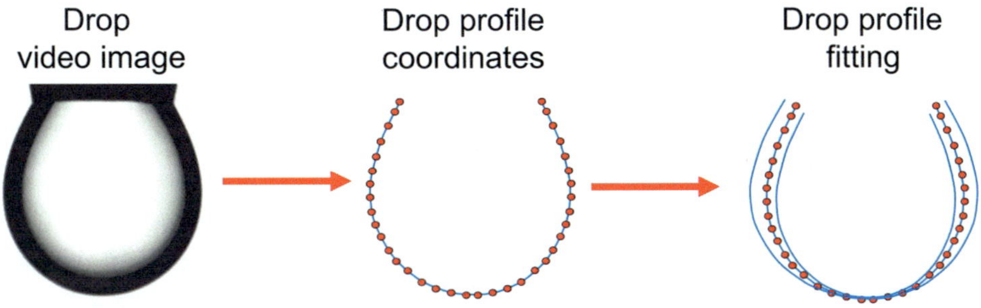


**Fig. (5).** Scheme of the main steps in drop profile analysis tensiometry.

The strategy of changing the value of γ such that its optimum value can be found in the minimum of the target with a least number of calculations and highest accuracy of the parameter γ is another important factor for a fast software of a drop/bubble profile tensiometer. There are simple and more sophisticated strategies (see for example [23] or [24]), while the one proposed by Levenberg [25] and Marquardt [26] turned out to be the most efficient non-linear optimisation method.

## Corrections of Vertical Misalignments and Aspect Ratio

Most cameras and lenses available on the market now are of high quality and do not produce distortions on the video images of drops or bubble. In the past, when the ADSA method was pioneered these distortions caused systematic errors in the interfacial tension results. Therefore, images of drops were taken and each pixel was corrected according to an image of a calibration grid pattern taken at the same position. This process therefore corrected all optical distortion and provided at the same time an accurate calibration [20].

As mentioned above, the quality of electronic and optical elements in PAT set-ups is nowadays very high so that complex corrections of distortions are no more required, however, an accurate calibration of any instruments is yet required. In a paper by Loglio *et al.* [27] it was demonstrated how an error in the aspect ratio of the absolute pixel size leads to wrong results. In particular, a drop size dependence of the determined interfacial tension appears, which is physically unreasonable as long as the drop is of millimeter size.

## APPLICATION OF PAT TO DIFFERENT LIQUID-FLUID INTERFACES

The main part of this book chapter consists of experimental examples in order to demonstrate the capacity of PAT to investigate different types of interfacial layers under various protocols [5]. Of course, the given selection of examples is quite subjective and by far not completed, however, it covers the most popular application of the PAT methodology for the characterisation of liquid interfacial layers. Note again that PAT also includes studies of contact angle, however, about this type of experiments it will not be reported here.

### Dynamic Surface and Interfacial Tensions

The typical application of PAT is the measurement of the dynamic interfacial tension and dilational interfacial visco-elasticity of aqueous solutions of surfactants and polymers. In this and the next paragraph a report is given about studies on the adsorption of ß-lactoglobulin (BLG), a bovine whey protein, at the water/air and water/tetradecane interfaces. For an easier comparison of the data, plots are suitable of the surface pressure $\Pi = \gamma_0 - \gamma$ which is the change of interfacial tension $\gamma$ as compared to the interfacial tension $\gamma_0$ of the surfactant or protein free system.

The Figs. (**6-7**) show the surface pressure $\Pi$ as a function of time $t$ plotted on a logarithmic scale. In certain moments, covering the whole large time interval, intermediate oscillations (three area oscillations each at a frequency of 0.1 Hz) have been performed in order to determine the dilational visco-elasticity as a function of time.

In Fig. (**6**) the results for the solution/air interface and in Fig. (**7**) those for the solution/tetradecane interface are summarized. It is immediately visible that for a given BLG bulk concentration the $\Pi$ values at the water/oil interface are much larger than those at the water/air interface.

For example, at a protein concentration of $10^{-9}$ mol/l the surface pressure at the water/air interface starts to slowly increase after a period of nearly 3 h and reaches values of only 3-4 mN/m after around 22 h whereas at the water/tetradecane interface $\Pi$ starts to increase after few minutes and finally reaches values of around 10-11 mN/m. This is obviously connected to the different environment for the protein molecules at the interface. At the water/air interface the protein molecules adsorb and change their conformation in a complicated and time consuming way in order to find an optimum state in the adsorption layer whereas at the water/oil interface they can protrude with the hydrophobic parts into the oil phase and stay anchored in a limited spectrum of conformations. A more detailed analysis of the data interpretation is given elsewhere [28, 29].

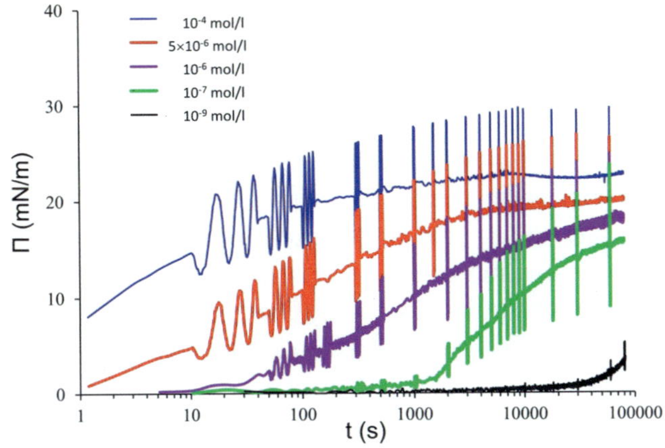


**Fig. (6).** Surface pressure $\Pi$ as a function of the BLG concentration, air bubble in BLG solution at pH 7, $C_{buff}$=10 mM; adapted from [28].

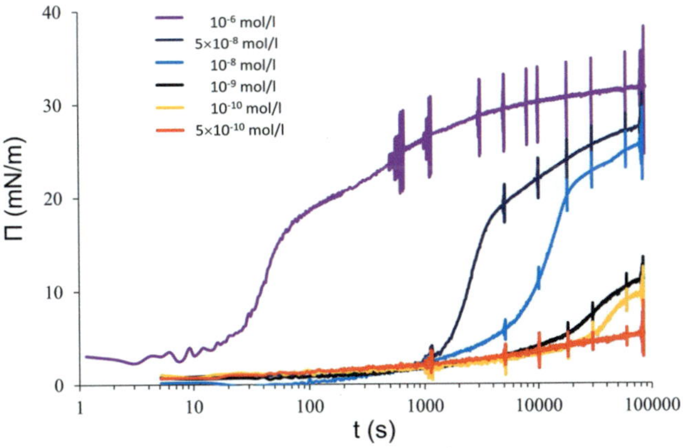


**Fig. (7).** Surface pressure $\Pi$ as a function of the BLG concentration, tetradecane drop in BLG solution at pH 7, $C_{buff}$=10 mM, adapted from [29].

Another important feature is the so-called induction time, *i.e.* the time interval during which the protein molecules adsorb but do not yet remarkably change the interfacial tension. At the water/air interface this induction time is about 10 s for a concentration of $10^{-6}$ mol/l, while it is only of the order of 100 s at the water/oil interface at a much lower BLG concentration of $10^{-8}$ mol/l.

The amplitudes of the interfacial tension oscillations along the dynamic interfacial tension curves are a measure of the dilational interfacial elasticity of the adsorption layers, which are discussed in more detail in the next paragraph.

## Oscillating Drops/Bubbles for 2D Dilational Rheology

The drop and bubble profile analysis tensiometry allows many experimental protocols to be used. This is mainly possible due to the very accurate computer controlled dosing system. Its basic function is to form a drop or bubble and keep its size constant over a certain time, in order to compensate for evaporation or dissolution effects. In addition, this dosing system can also be used to generate various types of drop size changes (and in this way changes of the interfacial area), such as ramp or square pulse expansions and compressions. In this way, the two interfacial layers are compressed or expanded and a relaxation mechanism of the adsorbed molecules set in. Note, for pure liquids all possible interfacial area changes should not lead to a measurable change in the interfacial tension.

In addition to these transient area perturbation (ramps or square pulses) most efficient from a point of view of data analysis are harmonic perturbations. Such sinusoidal drop surface area oscillations can be easily interpreted by a Fourier analysis in order to determine the amplitude of the area and the interfacial oscillations, and the phase shift between these two functions. The amplitude is a measure for the dilational elasticity:

$$E = -\frac{d\gamma}{d\ln\Gamma} = \frac{d\gamma}{d\ln A} \qquad \textbf{(10)}$$

while the phase shift is a measure for the dilational viscosity (*A* is the interfacial area). As discussed in [30], the whole set of equations to describe the visco-elasticity of interfacial layers can be given by complex functions and E can also be defined in the complex domain by:

$$E(i\omega) = E' + iE'' = E_0 \frac{\sqrt{i\omega}}{\sqrt{i\omega} + \sqrt{2\omega_o}} \qquad \textbf{(11)}$$

where E' and E" are the real and imaginary components of the complex visco-elasticity $E(i\omega)$, $E_0$ is the elasticity modulus, ω is the circular frequency $\omega=2\pi f$ ($f$ is the frequency of the oscillation process), and $\omega_0$ is the characteristic frequency defined by the main characteristics of the studied surfactant:

$$\omega_0 = \left[\frac{dc}{d\Gamma}\right]^2 (D/2) \tag{12}$$

For a diffusion controlled relaxation mechanism, Lucassen and van den Tempel [31] derived the relationships for the two components of the complex visco-elasticity:

$$E' = E_0 \frac{1+\sqrt{\omega_0/\omega}}{1+2\sqrt{\omega_0/\omega}+2\omega_0/\omega} \tag{13}$$

$$E'' = E_0 \frac{\sqrt{\omega_0/\omega}}{1+2\sqrt{\omega_0/\omega}+2\omega_0/\omega} \tag{14}$$

This theoretical model is the most frequently applied one to describe the frequency and concentration dependence of the dilational interfacial visco-elasticity of liquid interfaces. For more details, see [5].

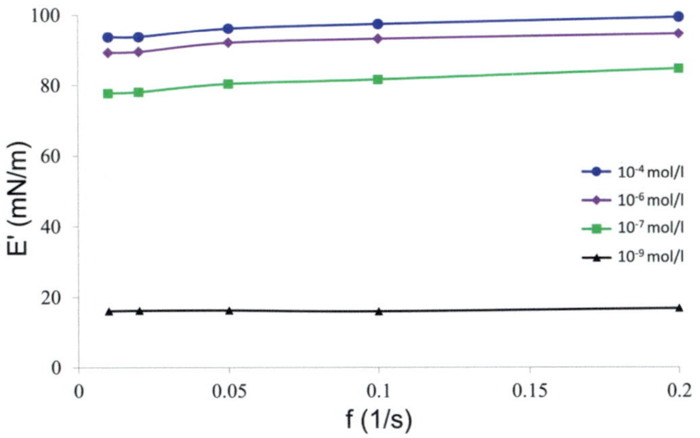


**Fig. (8).** Dilational elasticity as a function of the oscillation frequency at constant area deformation (oscillation amplitude) ΔA/A, air bubble in BLG solution at pH 7, $C_{buff}$=10 mmol/l; adapted from [28].

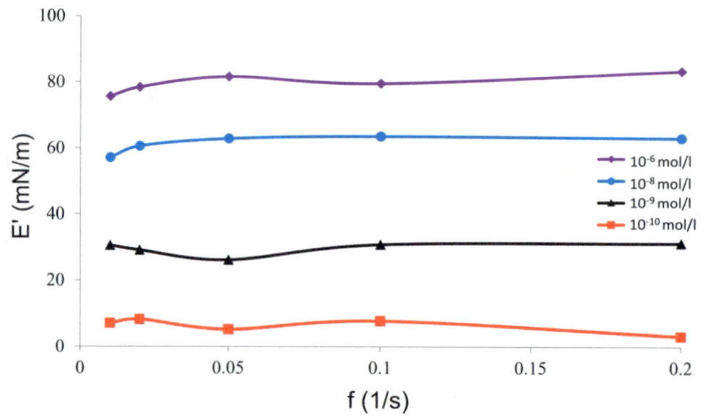


**Fig. (9).** Dilational elasticity E' as a function of the oscillation frequency at constant area oscillation amplitude ΔA/A, tetradecane drop in BLG solution at pH 7, $C_{buff}$=10 mmol/l, adapted from [32].

The data shown in Figs. (**8** and **9**) are the results for measured at the end of the maximum adsorption time of 80.000 s in the frequency range between 0.01 Hz and 0.2 Hz. As one can see, there is small dependence in this frequency range, as expected for adsorbed protein molecules in this concentration range. In contrast to the interfacial pressure (see Figs. **6-7**) the obtained elasticity values for the two studied interfaces are more or less the same.

Even at bulk concentrations as low as $10^{-9}$ mol/l, a concentration range in which classical surfactants do not show any adsorption behaviour, the protein BLG adsorbs strongly and the corresponding adsorption layers exhibit a remarkable dilational interfacial elasticity.

## Experiments at the Solution/Vapour Interface

Although there is much experimental and theoretical work performed at liquid/gas and liquid/liquid interfaces, only few investigations were performed on aqueous surfactant solutions at the interface to organic solvent vapors. The significant effects, for example of alkane vapor, on the adsorption of surfactants from aqueous solutions were reported only very recently [33]. By using a profile analysis tensiometer (PAT) the experimental protocol shown schematically in Fig. (**10**) was applied.

This protocol is based on a drop of water or aqueous surfactant solution formed in air and after a certain time (in most case after 5 min) a certain amount of an alkane (in most experiments 1 ml of hexane) is injected into the closed measuring cell. Due to its volatility, the alkane evaporates and saturates the air atmosphere. The corresponding measured surface tensions for a pure water drop are shown in

Fig. (**11**). The results obtained for several runs show a very good reproducibility. This is remarkable result, re-calling that hexane molecules are not amphiphilic and hence no adsorption due to a certain hydrophobic/hydrophilic balance would be expected.

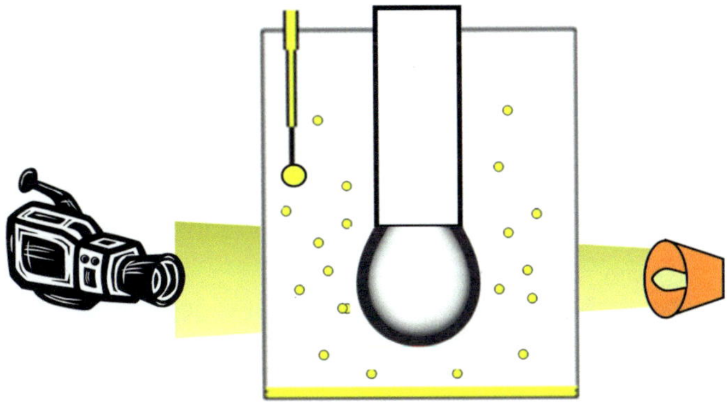

**Fig. (10).** Scheme of Profile Analysis Tensiometry (PAT) experimental setup and protocol for investigation of vapor/liquid interfacial properties, according to [33].

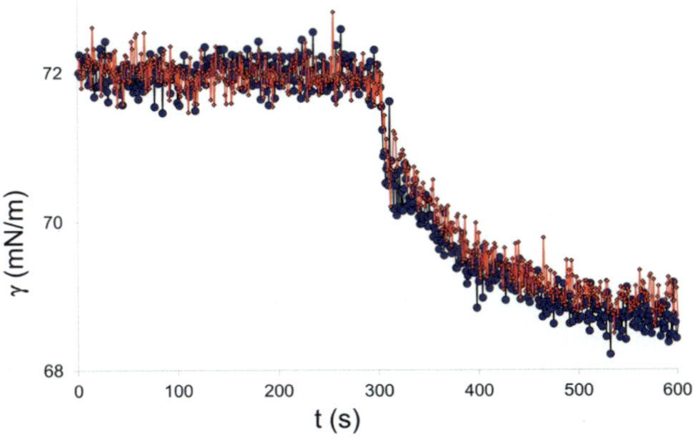


**Fig. (11).** Surface tension of a water drop against air, after 300 s the amount of 1 ml of hexane was injected into the closed cell in order to produce hexane vapor; different symbols refer to different experimental runs; according to [33].

In Fig. (**12**) the results for water and several aqueous SDS (sodium dodecyl sulfate) solutions against air and hexane vapor (hexane was injected into the cell after 300 s) are summarized. Already at very low SDS concentration, the presence of hexane vapor leads to a significant decrease of the surface tension.

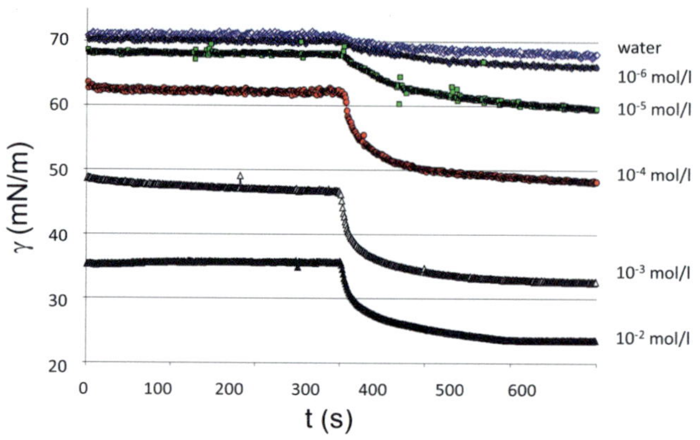


**Fig. (12).** Dynamic surface tension of SDS solutions to air, after 300 s the amount of 1 ml hexane was injected into the closed measuring cell; according to [34].

Even for concentrations above the CMC the presence of hexane vapor leads to a strong decrease in surface tension due to the co-adsorption of surfactant molecules and hexane from the vapor phase [34]. The final tension values around 23 mN/m are much lower than those typically reach by hydrocarbon surfactants.

An overview article on experiments performed for various surfactants and different gases and oil vapors was given recently in [35]. The observed phenomenon can be very important for many applications, such as tertiary oil recovery, when oil wells are flooded by foaming aqueous surfactant solutions.

## PAT with Bulk Exchange for the Characterisation of Multilayers

A double capillary arrangement was used as described in [21]. Using this arrangement, polymer multilayers can be built up, as it is known from the LbL (layer by layer) technique [36]. The scheme is shown in Fig. (**13**). State 1 refers to a drop with a lipid layer serving as the target layer. Then, a solution of a polymer with a charge opposite to the lipid is injected into the drop (step 2). In this way, the polymer forms a second layer (3). Replacing the solution inside the drop by the pure solvent leaves the drop with the lipid target layer and the first polymer layer (4). In a subsequent step a solution with the second polymer, having the opposite charge than the first polymer, is injected into the drop (5), which now forms the second polymer layer (6). Again, the excess polymer in the drop is replaced by the pure solvent (7) before the whole procedure of steps 2 to 7 is repeated. In this way multilayers are formed as described in detail in [37].

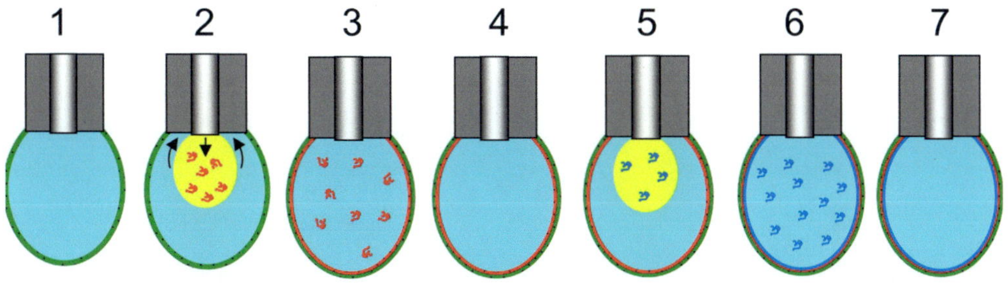


**Fig. (13).** Steps of the process to build polymer multilayers at the surface of a pendent drop, according to the LbL technique.

The measurements proceed during the liquid bulk exchange, which leads to a higher scattering of the surface tension data, however, the results are reliable even during these small distortions. Note, of course the bulk exchange must not be performed at too high liquid flow rates. A blank test with pure water as drop and as injected liquid allows finding out the optimum dosing rate for the given geometry. The scheme of a PAT equipped with such special capillary and two dosing systems is shown in Fig. (**14**).

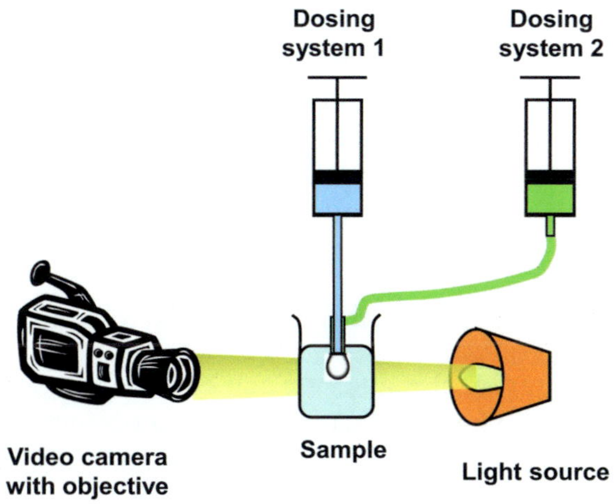


**Fig. (14).** Scheme of PAT with a coaxial double capillary and a double dosing system; according to [20].

The measured surface tension as control of the multilayer built-up process is shown in Fig. (**15**). As a target layer a monolayer of the anionic lipid dimyristoyl phosphatidyl glycerol (DMPG) was spread on the surface of a pendent drop having an optimum size (see details in [37]). Thereafter, alternating polymer layers were added sequentially, starting with the polycation poly(allylamine hydrochloride) (PAH, Mw≈70 kDa) followed by the polyanion poly(styrene sulfonate) sodium salt (PSS, Mw≈70 kDa).

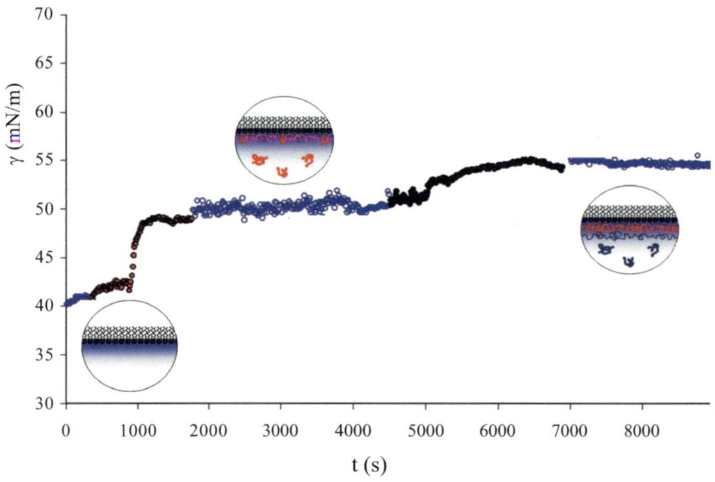


**Fig. (15).** Change in surface tension during the process of multilayer formation at the surface of a drop; the inserts show the progress of the LbL process.

The results shown in Fig. (**15**) show that due to the formation of the first polymer layer the measured surface tension is increased by about 7 mN/m, and it increases again when the second polymer layer is formed. Note, these obtained tensions are only effective values because a drop covered by a multilayer does not necessarily fulfil the conditions of a free liquid interface, for which the classical Laplace equation (equation (3)) has been derived. For a correct determination of the so-called membrane tension the Laplace equations has to be modified, as discussed in detail by Ferri *et al.* [38].

## PAT with Bulk Exchange for Sequential Adsorption Protocols

Proteins are surface active molecules and one can generally say that they adsorb at almost any type of interface. Their replacement from an interface, however, is not so easy and the mechanisms can be quite different. Using surfactant molecules, adsorbing in competition to the proteins at this interface is one of the options. Depending on whether the surfactant is ionic or non-ionic, surfactant molecules co-adsorb at the interface and start modifying the protein. It was found that small amounts of ionic surfactants can even increase the protein's surface activity. With a further increase of the concentration, however, a hydrophobic interaction becomes dominant which results in a hydrophilization of the protein molecules. This and a stronger competition at the interface due to the increasing surfactant concentration lead to a final displacement of the proteins from the interface [22].

The PAT with the extra equipment called coaxial double capillary with double dosing system appears to be a perfect tool to study the replacement of pre-adsorbed proteins from the water/air [22] and water/oil interface [38]. In Fig. (**16**) the protocol of a typical sequential adsorption route is shown. After an adsorption part at constant drop size (I.), the equilibrium state is reached (II.). Then the drop volume is exchanged against the pure buffer solution, *i.e.* all protein molecules will be removed from the drop, except those adsorbed at the drop surface (III.). Then, a respective surfactant solution is injected into the drop (IV.) in order to replace the protein molecules from the liquid interface. Finally, in step V. the surfactant solution in the drop is replaced by a pure buffer solution. In case the surface tension rises up to the value of the pure buffer (close to the interfacial tension of water/air or water/oil, respectively) the pre-adsorbed proteins are replaced from the interface. A difference to these values is a clear sign that proteins are left at the interface. The existence of protein molecules in the interfacial layer can be additionally probed by performing low frequency surface layer oscillations (in Fig. (**16**) this is made at the moments a, b, c and d). In case no protein is present at the interface, the resulting interfacial tension oscillations should be extremely small or not existing.

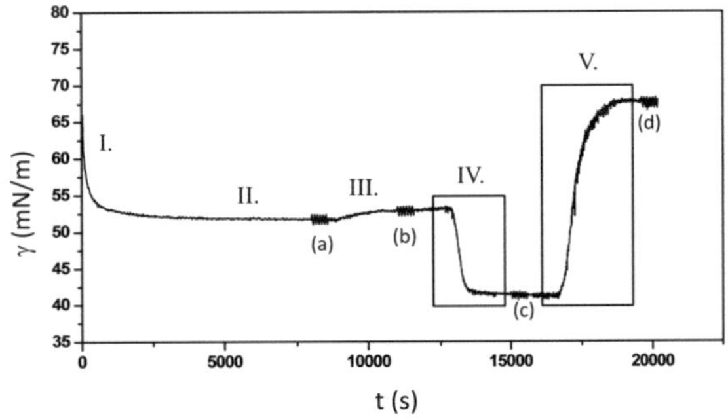


**Fig. (16).** Experimental protocol for a sequential adsorption of a protein solution with subsequent adsorption of a surfactant solution; I–adsorption of the protein solution, II–equilibration of the with some drop oscillations at the end, III–first drop volume exchange, *i.e.* replacement of the protein solution by the pure buffer solution with some drop oscillations at the end, IV–second drop volume exchange, *i.e.* injection of the surfactant solution with some drop oscillations at the end, V–third drop volume exchange of the surfactant solution by the pure buffer solution with some drop oscillations at the end; according to [22].

As an alternative experimental protocol a so-called simultaneous adsorption from a mixed protein/surfactant solution has been performed. In these experiments the two components are first mixed and the interface is then formed from such a mixed solution *via* the outer capillary. After the adsorption equilibrium is established, the droplet can again be rinsed with the buffer solution by injecting buffer solution into the drop, such as done in step V.

As shown in [39], the isotherms of ß-casein (BCS)/surfactant mixtures for both routes of adsorption layer formation suggest that the composition of the layers is similar. This is true for the water/air as well as water/oil interface, respectively. The washing-out experiments for BCS/$C_{12}$DMPO mixtures proved, however (see Fig. **17**), that after the sequential adsorption, the protocol in which the complexes are formed only in the adsorption layer, the interfacial tension reaches much higher values, *i.e.* the proteins are much better displaced from the interface by the competing surfactants, as compared to the protein displacement from a layer formed *via* simultaneous adsorption from a mixed solution.

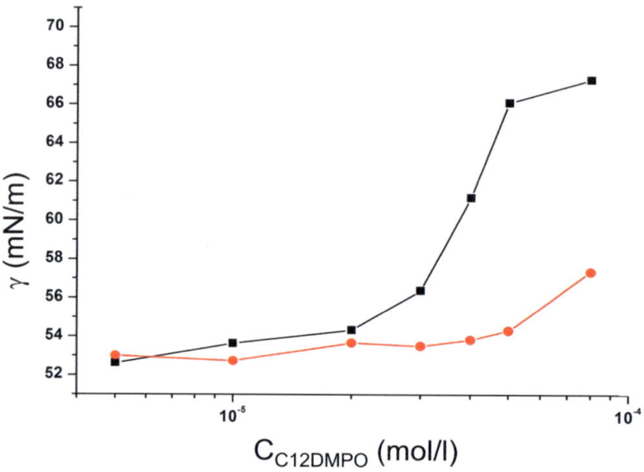


**Fig. (17).** Surface tensions as a function of the $C_{12}$DMPO surfactant solution after step V. in the sequential adsorption route (■) and after replacing the mixed protein-surfactant solution by pure buffer in the simultaneous adsorption route (●), for a $10^{-6}$ mol/l aqueous ß-casein solution; according to [40].

## Comparison of PAT for Drops and Bubbles as Direct Means of Estimating the Amount Adsorbed at the Interface

As shown in Fig. (**18**), when using different experimental methods, the obtained interfacial tension results are not always identical. However, the interfacial tension is a physical quantity and must not depend on the method used for its measurement. Hence, there must be explanations for such significant differences. In [41] it was explained why for strong surface active molecules, such as proteins, the measured interfacial tension values can significantly differ from each other. The reason is the large difference in the available reservoir of the adsorbing species. If the number of molecules adsorbed ($n_{ad}=\Gamma\times A$) is of a similar order of magnitude than the number of molecules in the solution ($n_{sol}=c\times V$) there will be a notable depletion of the solution bulk due to the adsorption. This loss of molecules in the solution can be estimated by $\Gamma\times(A/V)$ which has the dimension of a concentration. Using a Langmuir isotherm:

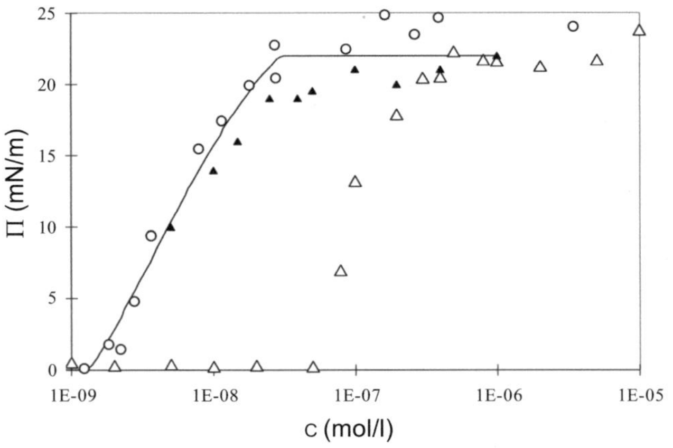


**Fig. (18).** Surface pressure Π as a function of ß-casein concentration; data from different methods: Wilhelmy plate method (▲), PAT with a bubble (○), PAT with a drop (Δ); according to [41].

$$\Gamma = \Gamma_{\infty} \frac{c}{a+c} \tag{15}$$

for small concentrations $c \ll a$ for which the surfactant or protein loss due to adsorption at the interface is most significant, the loss can be approximated by $c \times (\Gamma_{\infty}/a) \times (A/V)$. For a spherical drop $(A/V)$=3/r with r being the radius. Hence, for a small drop of few mm the loss is orders of magnitude larger than for an experiment with the Wilhelmy plate or with a bubble in a rather large cuvette.

Looking again at the data presented in Fig. (**18**) it becomes clear, that the reason for the shift of the surface tensions obtained by PAT with single drops toward higher concentrations is clearly caused by the discussed protein depletion from the solution due to adsorption at the surface of the small single drops. In contrast, for the PAT experiments with a bubble or using the Wilhelmy plate method the loss of protein molecules due to adsorption at the interface seems to be negligible.

For cases where this depletion effect is significant, parallel experiments with PAT using the bubble and the drop configuration, respectively, allow to determine the adsorbed amount, as it was demonstrated in [41] for ß-casein, bovine and human serum albumin.

## PAT for Studies of Interfacial Reactions

When 3-(Trimethoxysilyl)propyl methacrylate (TPM) comes into contact with water it undergoes a poly-condensation reaction at the interface due to a hydrolysis and poly-condensation of the alkoxy groups attached to silicon atom

[42]. This reaction can be controlled by the pH of the aqueous phase. The resulting surface active hydrolysation products can be used for the stabilization of emulsions for various applications, such as in tribology.

Using a PAT setup as shown in Fig. (**19**) the amount of the hydrolysed TPM can be studied. At first an air bubble is formed at the tip of a capillary immersed into the aqueous phase, and then TPM is carefully injected at the bottom of the cuvette, bringing it into contact with water and initiating the reaction.

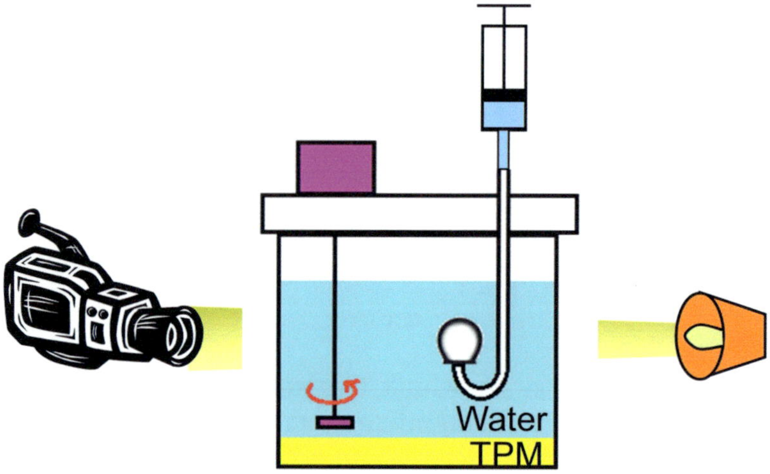


**Fig. (19).** The experimental set-up for measurements of the dynamic interfacial tension; classical PAT set-up with a U-shape capillary and a stirrer in the aqueous phase, the TPM phase is located at the bottom of the cell.

The reaction products of the hydrolysed TPM are produced at the water/TPM interface. It is surface active and water soluble and diffuses to the bubble surface. Here these surfactant molecules adsorb and change the measured interfacial tension. Hence, the air bubble immersed into the aqueous phase serves as a sensor for monitoring the reaction.

When using a stirrer, the reaction products do not need to diffuse toward the bubble but they are homogeneously distributed in the aqueous phase and reduce the reaction process by more than a factor of 10. In this way measurements of the interfacial tension allow recording the reaction kinetics at the water/TPM interface (Fig. **20**). As one can see, the highest rate of hydrolysis is observed at pH 3.

The properties of the formed polymeric reaction products can be further characterised *via* measurements of the interfacial visco-elasticity. Therefore, the bubble used as reaction sensor is forced to perform harmonic oscillations and the harmonic interfacial tension response can be analysed by Fourier transformation, which yields the dilational interfacial elasticity and viscosity.

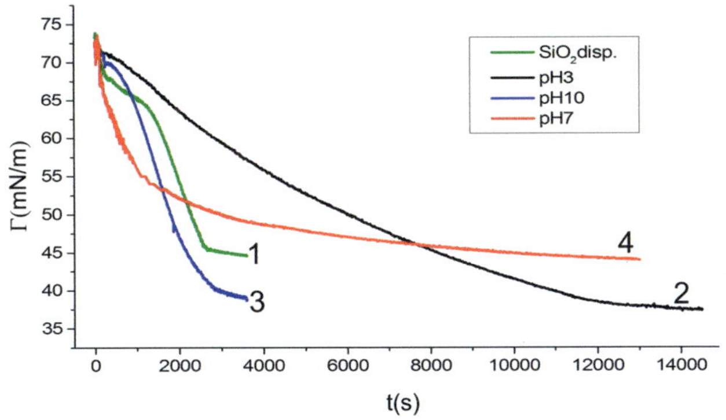


**Fig. (20).** Dynamic air/water interfacial tension in the presence of TPM on the bottom of the cuvette with continuously gentle stirring: green line–1.93% silica dioxide particles, black–buffer solution at pH 3, blue–buffer solution at pH 10, red–buffer solution at pH 7, all reactions performed at 25°C; according to [43].

The reaction at the TPM/water interface leads to a film formed at the bubble surface having highest values of the interfacial elasticity $E'$ at pH 3 and are lowest for pH 7 (see Fig. **21**). The dilational surface viscosity $E''$ provides information about the relaxation process of the adsorbed molecules. The data obtained at pH 7 show that only a small amount of less surface active compound is produced by the reaction. At pH 3 more polymeric surfactants are generated than at pH 10, but their molecular weight seems less as what is qualitatively obtained from the surface tension dynamics.

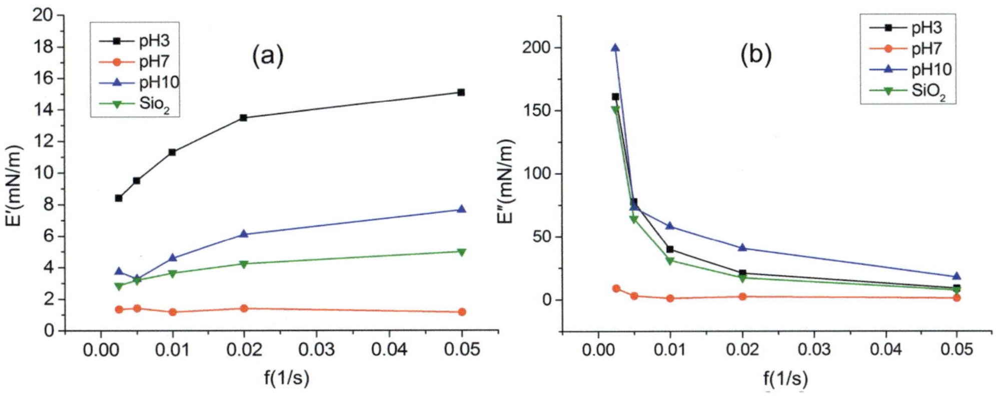


**Fig. (21).** Dilational interfacial elasticity $E'$ and viscosity $E''$ of the air/liquid interface 60 min after the injection of TPM on the bottom of cuvette and continuously gentle stirring; according to [43].

At pH 10 the poly-condensation process proceeds very fast (see Fig. **20**) and a more viscous film is formed. In presence of silica dioxide particles (pH 9) the elastic properties are less pronounced than in a buffer solution at pH 10. The very low viscosity and elasticity values at pH 7 pointing at a small concentration of surfactant are in agreement with the data published in [44]. Here it was shown that the hydrolysis rate coefficient is the lowest at pH 7, and increases significantly in both acidic and basic conditions, in agreement with our data obtained *via* our PAT measurements. A further data analysis and comparison with model substances can give a better quantitative estimation of the produced reaction products.

## PAT Applied for Matter Transfer Studies

Gas mass transfer across air-liquid interfaces is a key process in chemical engineering and physiological phenomena. Though there are many theories and mathematical methods based on a coefficient k to describe the "blocking" effect of the air-liquid interface on the mass transfer, however, there is nearly no focus on the effect of gas mass transfer on the interfacial characteristic properties.

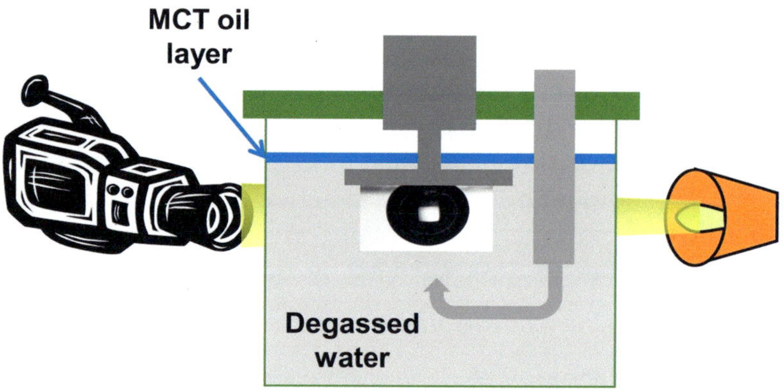


**Fig. (22).** Scheme of the measuring cell for gas mass transfer across the gas/water interface.

PAT provides a unique possibility of studying the interfacial characteristics during mass transfer. A gas bubble dissolution process can be performed with a recently developed measuring cell, as schematically shown in Fig. (**22**) [45]. The gas was injected by the dosing system of PAT, forming a bubble adhered to the solid surface on the top of the cell. This solid flat metal surface serves as a lid and has a slightly concave shape to keep the bubble in a centre position. On top of the aqueous phase a thin layer of oil (MCT–medium chain triglyceride) is deposited in order to avoid any solubility of air into the degassed water phase, as well as any loss of the gas dissolved from the bubble into the aqueous phase.

The bubble profile analysis system PAT recognizes the shape of the bubble and provides the corresponding bubble volume, bubble surface area and interfacial tension values as a function of time.

The results of a typical experiment are shown in Fig. (**23**) for the dissolution of the noble gas helium in pure water. Fig. (**23a**) shows the change of the bubble area and volume as a function of time and Fig. (**23b**) the corresponding dynamic interfacial tension.

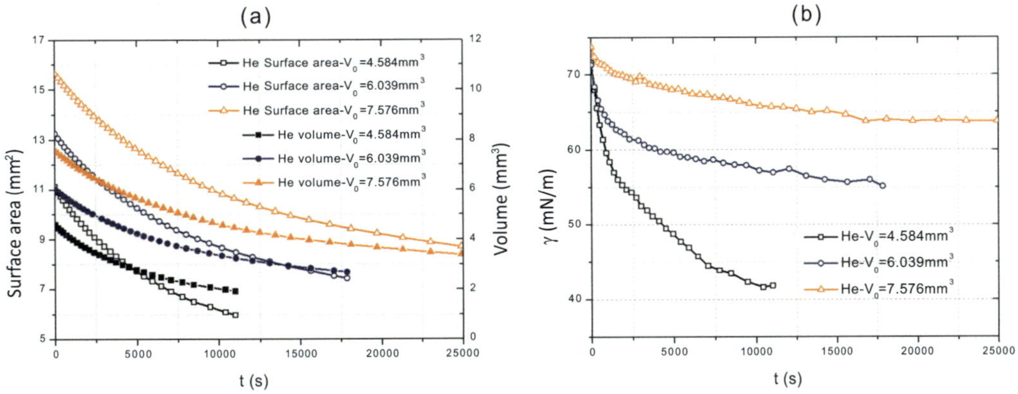


**Fig. (23).** Change of surface area, volume and surface tension with time during helium bubble dissolving in water at different initial bubble volumes as measured by PAT; according to [45].

Note, the aqueous phase can also contain various types of surface active molecules, such as surfactants or proteins, which then adsorb and can form a barrier for the gas molecules to transfer into the liquid.

## PAT as Tool for Investigating Spread Insoluble Monolayers

The drop profile analysis tensiometer can also be utilised for studies of insoluble monolayers spread on the surface of a single drop as it was first proposed by Kwok *et al* [46]. Using a very accurate microsyringe, the material solved in a respective solvent can be spread on a droplet formed by the respective subphase solution. Then, the drop size can be changed *via* the dosing system and by that the monolayer on the drop surface can be very homogeneously expanded and compressed so that a complete surface tension/area isotherm can be obtained. An example is given for the famous phospholipid DPPC in Fig. (**24**) [47]. The symbols in this graph correspond to the compression (■) and expansion (●) process, respectively. The solid lines have been obtained by measurements with a classical Langmuir trough. The most important features in the compression/expansion isotherm, *i.e.* the phase transitions from one monolayer state, are reproduced by PAT. The differences in the absolute values, however, can be caused by the curvature of the drop surface and the compression/expansion rates typically used in the two different experiments.

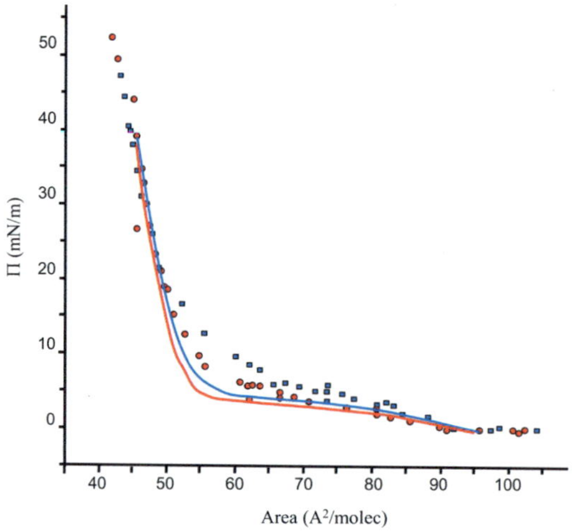


**Fig. (24).** Pressure/area isotherm of a DPPC monolayer using PAT (symbols) and a traditional Langmuir trough; ●-expansion, ■-compression; according to [47].

There are various advantages for using PAT as a mini-Langmuir trough. One important point is of course the extremely small amount of substance required for the experiments. This is particularly important for very expensive sample material. Another great advantage is the easy temperature control for the monolayer studies, essentially at temperatures which are almost impossible to be reached in a trough experiment. One more enormous advantage is the fact that PAT allows studying spread monolayer at the water/oil interface [48]. Such investigations are of course principally also possible as shown in [49], however, the enormous amount of very pure oil is a very large obstacle and the number of monolayer studies with the classical Langmuir trough technique are very rare.

## DROP PROFILE ANALYSIS FOR GROWING DROPS

Drop profile analysis tensiometry PAT (originally called axisymmetric drop shape analysis–ADSA [5]) is used in fundamental and applied research as an accurate method for measuring surface and interfacial tensions. It consists of the analysis of pendent drop or buoyant bubble profiles by the Gauss-Laplace equation of capillarity. Therefore, the application of PAT is limited to quasi-static profiles, *i.e.* when the profile is only affected by the physical properties of the interface and gravity/buoyancy (see also [5]). However, as long as the fitting of the Gauss-Laplace equation to an experimentally acquired profile is technically possible, PAT can be used to determine also the dynamic properties of interfacial layers.

For example, in oscillating drop experiments, the applied frequency range to measure interfacial visco-elastic properties is limited to certain values as reported in [50]. If one goes beyond this limit the measurements can lead to artefacts, such as a non-zero visco-elasticity of a pure water drop surface.

Nowadays, processes like drop formation and growth is an interesting field of research for many industrial applications (*ss.* food industry, pharmacy, biotechnology, petrochemical industry and extraction processes). Therefore, together with the importance of growing drop experiments for interfacial science, many investigations are performed to understand the complexities of such processes [51 - 55]. Thus, the application of PAT under dynamic conditions is very interesting.

In a recent study the limits of PAT for growing drop experiments was presented by Karbaschi *et al.* [56]. This investigation was based on the analysis of dynamic drop profiles by fitting the Gauss-Laplace equation to the experimentally obtained profiles of water drops growing in air and in hexane at different liquid flow rates. Figs. (**25** and **26**) represent the result of this study for water drops in air at a 0.45 mm circular capillary. As shown in Fig. (**25**), the standard deviation (Std) of profile fitting were sufficiently low for the studied flow rates. Hence, the drops had Laplacian shapes. This means the concept of fitting is valid even for high flow rates. Note, for large drops growing at high flow rates, a sharp increase in the Std values is observed which shows the limits for fitting the Gauss-Laplace equation to drop profiles. The deviations can be different for different systems and for smaller drops are expected at higher flow rates. In addition, small drops have spherical shapes and, therefore, are not suitable for the analysis *via* the Gauss-Laplace equation. In summary, the liquid flow rate and the total drop size are essential parameters which decide whether the Gauss-Laplace equation can be used to analyse dynamic drop profiles without any disturbing influence of hydrodynamics on the experimental results [56].

In Fig. (**26**), the apparent surface tension values obtained by fitting the classical Gauss-Laplace equation to the dynamic drop profiles are presented. For a pure water drops in air, the surface tension value must be 72 mN/m at room temperature. Therefore, any deviation of the obtained values from this physically expected value can be considered as a contribution of inertia to the deformation of the dynamic drop profiles during the growth of the water drops. Hence, the data for Re=11.9 and a drop volume V>6 $mm^3$ for the used capillary are acceptable, while for larger Reynolds numbers all drop sizes are too small for a reliable measurement of the surface tension.

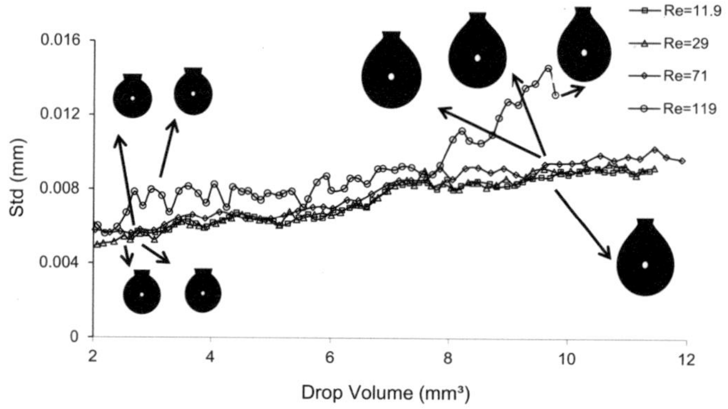


**Fig. (25).** Standard deviation of fitting the Gauss-Laplace equation to dynamic profiles of water drops in air growing at different Re numbers (varied only by changing flow rates); according to [56].

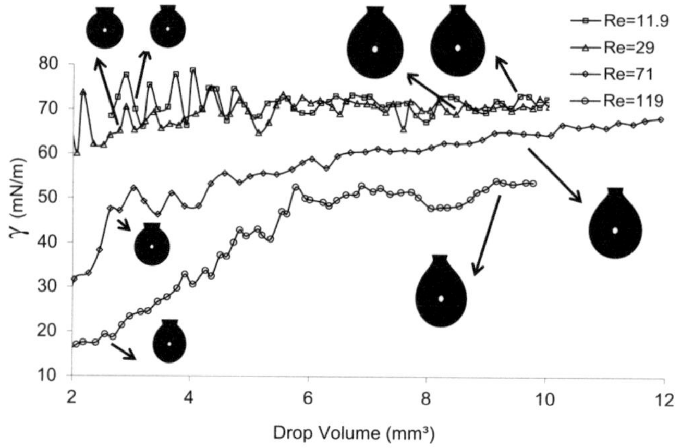


**Fig. (26).** Apparent surface tension of fitting the Gauss-Laplace equation to the dynamic profiles of water drops in air growing at different Re numbers (varied only by changing the liquid flow rates); according to [56].

Further attempts to numerically predict dynamic drop profiles were performed by Dieter-Kissling *et al.* [57] in order to more quantitatively investigate the inertia effects on the deformation of drop profiles during their growth. The use of an interface front-tracking approach for computational fluid dynamics simulations of growing drops gives access to the local properties of the flow field around the interface. The work was performed with the target to further develop the conventional Gauss-Laplace equation in such a way that it can be applied for measurements of surface properties with expanding or contracting drops as it was summarised in [58].

## SUMMARY AND OUTLOOK

Bubble and drop profile analysis tensiometry, introduced in 1984 as ADSA, is one of the most versatile tools for the characterisation of liquid interfaces. This technique is applicable to liquid/gas and liquid/liquid interfaces without significant modifications and provides the time dependence of surface and interfacial tension in a time range spanning over few seconds up to hours or even days. The obtained interfacial tensions are well suited for a quantitative characterisation of adsorption layers of surfactants, proteins, polymers, and of their mixtures at liquid interfaces.

Using modern instruments equipped with an accurate dosing system, PAT allows experiments under well controlled conditions, such as constant drop/bubble volume or surface area. Also pre-programed changes are feasible, for example drop volume changes in form of square pulses or harmonic oscillations in a frequency range between 1 mHz up to about 0.1 Hz. This type of experiments provides data that allow *via* Fourier transformation determination of the dilational visco-elastic properties of liquid interfacial layers.

For PAT there is some special equipment which makes it unique. One is the coaxial double capillary, providing experimental protocols hardly available by other experimental techniques. With this methodology, experiments on sequential and simultaneous adsorption routes can be performed. This allows analysing specific aspects of complex formation of proteins with low-molecular weight surfactants at the interface or in the solution bulk. Moreover, this arrangement is able to build up multilayers following a layer-by-layer technique. Due to the possibility of easily changing the volume of a drop serving as a template expansions and compressions can be produced and hence the 2D dilational rheology of the multilayers can be determined.

From a hydrodynamic point of view, the PAT methodology is limited to equilibrium drop/bubble profiles. Hence, when the size of a drop or bubble is changed too quickly their profiles are no longer Laplacian and the method would deliver wrong results. The limits of PAT for a specific system need to be determined experimentally. To validate the range of applicability or even expand it significantly toward larger deformation rates CFD simulations can be applied successfully. In [5] various examples of CFD simulations have been presented, for example the profile changes of a growing drop in dependence of the liquid flow rate were presented in order to show how reliable the calculated interfacial tensions are when obtained from dynamic drop profiles.

## CONFLICT OF INTEREST

The authors confirm that this chapter content has no conflict of interest.

## ACKNOWLEDGEMENTS

The work was financially supported by a project of the European Union (CoWet) and the COST actions CM1101 and MP1106.

## REFERENCES

[1] J.A. Plateau, *Statique expérimentale et théorique des liquides soumis aux seules forces moléculaires.* Gauthier-Villars, 1873.

[2] R. Shuttleworth, "The surface tension of solids", *Proc. Phys. Soc. A,* vol. 63, no. 5, p. 444, 1950. [http://dx.doi.org/10.1088/0370-1298/63/5/302]

[3] R. Miller, V.B. Fainerman, V. Pradines, V. Kovalchuk, N.M. Kovalchuk, E.V. Aksenenko, L. Liggieri, F. Ravera, G. Loglio, A. Sharipova, Y. Vysotsky, D. Vollhardt, N. Mucic, R. Wüstneck, J. Krägel, and A. Javadi, "Surfactant adsorption layers at liquid interfaces", In: *Surfactant Science and Technology.* CRC Press, 2014. [http://dx.doi.org/10.1201/b16802-7]

[4] A.F. Ward, and L. Tordai, "Time-dependence of boundary tensions of solutions I. The role of diffusion in time-effects", *J. Chem. Phys,* vol. 14, no. 7, pp. 453-461, 1946. [http://dx.doi.org/10.1063/1.1724167]

[5] M.T. Rahni, M. Karbaschi, and R. Miller, *Computational methods for complex liquid-fluid interfaces.* CRC Press, 2015. [http://dx.doi.org/10.1201/b19337]

[6] R. Miller, L. Liggieri, Ed., *Bubble and drop interfaces.* Brill: Leiden, Boston, 2011.

[7] Y. Rotenberg, L. Boruvka, and A.W. Neumann, "Determination of surface tension and contact angle from the shapes of axisymmetric fluid interfaces", *J. Colloid Interface Sci,* vol. 93, no. 1, pp. 169-183, 1983. [http://dx.doi.org/10.1016/0021-9797(83)90396-X]

[8] M. Ferrari, L. Liggieri, and R. Miller, *Drops and bubbles in contact with solid surfaces.* CRC Press, 2012.

[9] G. Quincke, "Ueber die Capillaritätsconstanten des Quecksilbers", *Ann. Phys,* vol. 181, no. 9, pp. 1-48, 1858. [http://dx.doi.org/10.1002/andp.18581810902]

[10] J.F. Padday, "Heights of sessile drops and meniscus properties", *Nature,* vol. 198, no. 4878, pp. 378-379, 1963. [http://dx.doi.org/10.1038/198378a0]

[11] P.S. de Laplace, *Mechanique celeste, vol. 1st Supp to*vol. 10. Impr. Imperiale: Paris, 1804.

[12] P.S. de Laplace, *Mechanique celeste, vol. 2nd Supp to*vol. 10. Impr. Imperiale: Paris, 1804.

[13] J.F. Padday, "Sessile Drop Profiles: Corrected Methods for Surface Tension and Spreading Coefficients", *Proc. R. Soc. Lond. Math. Phys. Eng. Sci,* vol. 330, 1972no. 1583, pp. 561-572 [http://dx.doi.org/10.1098/rspa.1972.0159]

[14] F. Bashforth, and J.C. Adams, *An attempt to test the theories of capillary action.* Cambridge University Press, 1883.

[15] A.M. Worthington, "On the error involved in Professor Quincke's method of calculating surface-tensions from the dimensions of flat drops and bubbles", *Philos. Mag. Ser. 5,* vol. 20, no. 122, pp. 51-66, 1885. [http://dx.doi.org/10.1080/14786448508627724]

[16] A.W. Porter, "The calculation of surface tension from experiment— Part I. Sessile drops", *Lond. Edinb. Dublin Philos. Mag. J. Sci,* vol. 15, no. 96, pp. 163-170, 1933. [http://dx.doi.org/10.1080/14786443309462172]

[17] S. Hartland, and R.W. Hartley, *Axisymmetric fluid-liquid interfaces: tables giving the shape of sessile and pendant drops and external menisci, with examples of their use.* Elsevier Scientific Pub. Co., 1976.

[18] C. Maze, and G. Burnet, "A non-linear regression method for calculating surface tension and contact angle from the shape of a sessile drop", *Surf. Sci,* vol. 13, no. 2, pp. 451-470, 1969. [http://dx.doi.org/10.1016/0039-6028(69)90204-0]

[19] C. Maze, and G. Burnet, "Modifications of a non-linear regression technique used to calculate surface tension from sessile drops", *Surf. Sci,* vol. 24, no. 1, pp. 335-342, 1971. [http://dx.doi.org/10.1016/0039-6028(71)90240-8]

[20] P. Cheng, D. Li, L. Boruvka, Y. Rotenberg, and A.W. Neumann, "Automation of axisymmetric drop shape analysis for measurements of interfacial tensions and contact angles", *Colloids Surf,* vol. 43, no. 2, pp. 151-167, 1990. [http://dx.doi.org/10.1016/0166-6622(90)80286-D]

[21] H.A. Wege, J.A. Holgado-Terriza, and M.A. Cabrerizo-Vílchez, "Development of a constant surface pressure penetration langmuir balance based on axisymmetric drop shape analysis", *J. Colloid Interface Sci,* vol. 249, no. 2, pp. 263-273, 2002. [http://dx.doi.org/10.1006/jcis.2002.8233] [PMID: 16290596]

[22] J.K. Ferri, C. Kotsmar, and R. Miller, "From surfactant adsorption kinetics to asymmetric nanomembrane mechanics: pendant drop experiments with subphase exchange", *Adv. Colloid Interface Sci,* vol. 161, no. 1-2, pp. 29-47, 2010. [http://dx.doi.org/10.1016/j.cis.2010.08.002] [PMID: 20810096]

[23] M. Hoorfar, and A.W. Neumann, "Axisymmetric drop shape analysis (ADSA)", In: *Applied Surface Thermodynamics.* CRC Press: New York, London, Boca Raton, 2011.

[24] S.A. Zholob, A.V. Makievski, R. Miller, and V.B. Fainerman, "Advances in calculation methods for the determination of surface tensions in drop profile analysis tensiometry", In: *Bubble and Drop Interfaces.* CRC Press: Leiden, 2011.

[25] K. Levenberg, "A method for the solution of certain non-linear problems in least squares", *Q. Appl. Math,* vol. 2, no. 2, pp. 164-168, 1944. [http://dx.doi.org/10.1090/qam/10666]

[26] D. Marquardt, "An algorithm for least-squares estimation of nonlinear parameters", *J. Soc. Ind. Appl. Math,* vol. 11, no. 2, pp. 431-441, 1963. [http://dx.doi.org/10.1137/0111030]

[27] G. Loglio, P. Pandolfini, A.V. Makievski, and R. Miller, "Calibration parameters of the pendant drop tensiometer: assessment of accuracy", *J. Colloid Interface Sci,* vol. 265, no. 1, pp. 161-165, 2003. [http://dx.doi.org/10.1016/S0021-9797(03)00138-3] [PMID: 12927178]

[28] G. Gochev, I. Retzlaff, E.V. Aksenenko, V.B. Fainerman, and R. Miller, "Adsorption isotherm and equation of state for β-Lactoglobulin layers at the air/water surface", *Colloids Surf. Physicochem. Eng. Asp,* vol. 422, pp. 33-38, 2013. [http://dx.doi.org/10.1016/j.colsurfa.2013.01.008]

[29] J.Y. Won, G.G. Gochev, V. Ulaganathan, J. Krägel, E.V. Aksenenko, V.B. Fainerman, and R. Miller, "Mixed Adsorption Mechanism for the Kinetics of BLG Interfacial Layer Formation at the Solution/Tetradecane Interface, Colloids Surfaces A", *Colloids Surf. Physicochem. Eng. Asp,* vol. 519, pp. 146-152, 2017.
[http://dx.doi.org/10.1016/j.colsurfa.2016.08.024]

[30] R. Miller, and L. Liggieri, *Interfacial rheology.* Brill: Leiden, 2009.
[http://dx.doi.org/10.1163/ej.9789004175860.i-684]

[31] J. Lucassen, and M. Van Den Tempel, "Dynamic measurements of dilational properties of a liquid interface", *Chem. Eng. Sci,* vol. 27, no. 6, pp. 1283-1291, 1972.
[http://dx.doi.org/10.1016/0009-2509(72)80104-0]

[32] J.Y. Won, G.G. Gochev, V. Ulaganathan, J. Krägel, E.V. Aksenenko, V.B. Fainerman, and R. Miller, "Dilational viscoelasticity of BLG Adsorption Layers at the Solution/Tetradecane Interface - Effect of pH and Ionic Strength", *Colloids Surf. Physicochem. Eng. Asp,* vol. 521, pp. 204-210, 2017.
[http://dx.doi.org/10.1016/j.colsurfa.2016.08.054]

[33] A. Javadi, N. Moradi, H. Möhwald, and R. Miller, "Adsorption of alkanes from the vapour phase on water drops measured by drop profile analysis tensiometry", *Soft Matter,* vol. 6, no. 19, pp. 4710-4714, 2010.
[http://dx.doi.org/10.1039/c0sm00367k]

[34] A. Javadi, N. Moradi, V.B. Fainerman, H. Möhwald, and R. Miller, "Alkane vapor and surfactants co-adsorption on aqueous solution interfaces", *Colloids Surf. Physicochem. Eng. Asp,* vol. 391, no. 1–3, pp. 19-24, 2011.
[http://dx.doi.org/10.1016/j.colsurfa.2011.08.002]

[35] V.B. Fainerman, E.V. Aksenenko, and R. Miller, "Influence of alkane and perfluorocarbon vapors on adsorbed surface layers and spread insoluble monolayers of surfactants, proteins and lipids", *Adv. Colloid Interface Sci,* 2017.
[http://dx.doi.org/10.1016/j.cis.2015.11.004] [PMID: 26656422]

[36] G. Decher, "Fuzzy nanoassemblies: toward layered polymeric multicomposites", *Science,* vol. 277, no. 5330, pp. 1232-1237, 1997.
[http://dx.doi.org/10.1126/science.277.5330.1232]

[37] J.K. Ferri, W-F. Dong, and R. Miller, "Ultrathin free-standing polyelectrolyte nanocomposites: a novel method for preparation and characterization of assembly dynamics", *J. Phys. Chem. B,* vol. 109, no. 31, pp. 14764-14768, 2005.
[http://dx.doi.org/10.1021/jp052672n] [PMID: 16852865]

[38] J.K. Ferri, P.A. Fernandes, J.T. McRuiz, and F. Gambinossi, "Elastic nanomembrane metrology at fluid–fluid interfaces using axisymmetric drop shape analysis with anisotropic surface tensions: deviations from Young–Laplace equation", *Soft Matter,* vol. 8, no. 40, pp. 10352-10359, 2012.
[http://dx.doi.org/10.1039/c2sm26604k]

[39] A. Dan, C. Kotsmar, J.K. Ferri, A. Javadi, M. Karbaschi, J. Krägel, R. Wüstneck, and R. Miller, "Mixed protein–surfactant adsorption layers formed in a sequential and simultaneous way at water–air and water–oil interfaces", *Soft Matter,* vol. 8, no. 22, pp. 6057-6065, 2012.
[http://dx.doi.org/10.1039/c2sm25463h]

[40] C. Kotsmár, D.O. Grigoriev, A.V. Makievski, J.K. Ferri, J. Krägel, R. Miller, and H. Möhwald, "Drop profile analysis tensiometry with drop bulk exchange to study the sequential and simultaneous adsorption of a mixed β-casein /C12DMPO system", *Colloid Polym. Sci,* vol. 286, no. 8–9, pp. 1071-1077, 2008.
[http://dx.doi.org/10.1007/s00396-008-1872-4]

[41] R. Miller, V.B. Fainerman, A.V. Makievski, M. Leser, M. Michel, and E.V. Aksenenko, "Determination of protein adsorption by comparative drop and bubble profile analysis tensiometry", *Colloids Surf. B Biointerfaces,* vol. 36, no. 3-4, pp. 123-126, 2004. [http://dx.doi.org/10.1016/j.colsurfb.2004.02.009] [PMID: 15276627]

[42] S. Savard, L-P. Blanchard, J. Léonard, and R.E. Prud'homme, "Hydrolysis and condensation of silanes in aqueous solutions", *Polym. Compos,* vol. 5, no. 4, pp. 242-249, 1984. [http://dx.doi.org/10.1002/pc.750050403]

[43] A. Tleuova, S. Aidarova, A. Sharipova, N. Bekturganova, and D. Grigoriev, "Using profile analysis tensiometry for monitoring auto-oscillations caused by the hydrolysis of 3-(trimethoxysilyl)propyl methacrylate when contacting water", *Colloids Surf. Physicochem. Eng. Asp,* vol. 505, pp. 18-22, 2016. [http://dx.doi.org/10.1016/j.colsurfa.2015.10.057]

[44] M. Pantoja, F. Velasco, D. Broekema, J. Abenojar, and J.C. del Real, "The influence of pH on the hydrolysis process of γ-Methacryloxypropyltrimethoxysilane, analyzed by FT-IR, and the silanization ofeElectrogalvanized steel", *J. Adhes. Sci. Technol,* vol. 24, no. 6, pp. 1131-1143, 2010. [http://dx.doi.org/10.1163/016942409X12586283821559]

[45] X.W. Hu, R. Miller, and L.J. Guo, "Experimental study on interfacial characteristics during bubble dissolution", *Colloids Surf. Physicochem. Eng. Asp,* vol. 505, pp. 179-185, 2016. [http://dx.doi.org/10.1016/j.colsurfa.2016.03.076]

[46] D.Y. Kwok, D. Vollhardt, R. Miller, D. Li, and A.W. Neumann, "Axisymmetric drop shape analysis as a film balance", *Colloids Surf. Physicochem. Eng. Asp,* vol. 88, no. 1, pp. 51-58, 1994. [http://dx.doi.org/10.1016/0927-7757(94)80085-5]

[47] J.B. Li, R. Miller, D. Vollhardt, G. Weidemann, and H. Möhwald, "Isotherms of phospholipid monolayers measured by a pendant drop technique", *Colloid Polym. Sci,* vol. 274, no. 10, pp. 995-999, 2016. [http://dx.doi.org/10.1007/BF00656630]

[48] J. Li, R. Miller, and H. Möhwald, "Characterisation of phospholipid layers at liquid interfaces 2. Comparison of isotherms of insoluble and soluble films of phospholipids at different fluid/water interfaces", *Colloids Surf. Physicochem. Eng. Asp,* vol. 114, pp. 123-130, 1996. [http://dx.doi.org/10.1016/0927-7757(96)03522-4]

[49] A. Bonfillon, and D. Langevin, "Viscoelasticity of monolayers at oil-water interfaces", *Langmuir,* vol. 9, no. 8, pp. 2172-2177, 1993. [http://dx.doi.org/10.1021/la00032a045]

[50] M.E. Leser, S. Acquistapace, A. Cagna, A.V. Makievski, and R. Miller, "Limits of oscillation frequencies in drop and bubble shape tensiometry", *Colloids Surf. Physicochem. Eng. Asp,* vol. 261, no. 1–3, pp. 25-28, 2005. [http://dx.doi.org/10.1016/j.colsurfa.2004.11.043]

[51] E. Ruckenstein, and D. Constanitnescu, "Mass or heat transfer with a change in interfacial area—I", *Int. J. Heat Mass Transfer,* vol. 12, no. 10, pp. 1249-1255, 1969. [http://dx.doi.org/10.1016/0017-9310(69)90169-0]

[52] Z. Che, T.N. Wong, N-T. Nguyen, Y.F. Yap, and J.C. Chai, "Numerical investigation of upstream pressure fluctuation during growth and breakup of pendant drops", *Chem. Eng. Sci,* vol. 66, no. 21, pp. 5293-5300, 2011. [http://dx.doi.org/10.1016/j.ces.2011.07.028]

[53] T. Kékesi, G. Amberg, and L. Prahl Wittberg, "Drop deformation and breakup", *Int. J. Multiph. Flow,* vol. 66, pp. 1-10, 2014. [http://dx.doi.org/10.1016/j.ijmultiphaseflow.2014.06.006]

[54] A. Javadi, D. Bastani, M. Taeibi-Rahni, M. Karbaschi, and R. Miller, "Interfacial mass transfer of growing drops in liquid-liquid systems", In: *Bubble and Drop Interfaces*vol. 2. Brill: Leiden, 2011, pp. 521-550.
[http://dx.doi.org/10.1163/ej.9789004174955.i-558.168]

[55] A. Javadi, D. Bastani, J. Krägel, and R. Miller, "Interfacial instability of growing drop: Experimental study and conceptual analysis", *Colloids Surf. Physicochem. Eng. Asp,* vol. 347, no. 1–3, pp. 167-174, 2009.
[http://dx.doi.org/10.1016/j.colsurfa.2009.04.001]

[56] M. Karbaschi, D. Bastani, A. Javadi, V.I. Kovalchuk, N.M. Kovalchuk, A.V. Makievski, E. Bonaccurso, and R. Miller, "Drop profile analysis tensiometry under highly dynamic conditions", *Colloids Surf. Physicochem. Eng. Asp,* vol. 413, pp. 292-297, 2012.
[http://dx.doi.org/10.1016/j.colsurfa.2012.04.027]

[57] K. Dieter-Kissling, M. Karbaschi, H. Marschall, A. Javadi, R. Miller, and D. Bothe, "On the applicability of Drop Profile Analysis Tensiometry at high flow rates using an interface tracking method", *Colloids Surf. Physicochem. Eng. Asp,* vol. 441, pp. 837-845, 2014.
[http://dx.doi.org/10.1016/j.colsurfa.2012.10.047]

[58] M. Lotfi, M. Karbaschi, A. Javadi, N. Mucic, J. Krägel, V.I. Kovalchuk, R.G. Rubio, V.B. Fainerman, and R. Miller, "Dynamics of liquid interfaces under various types of external perturbations", *Curr. Opin. Colloid Interface Sci,* vol. 19, no. 4, pp. 309-319, 2014.
[http://dx.doi.org/10.1016/j.cocis.2014.04.006]

CHAPTER 5

# Pendant Droplets: Overview of Dynamics and Applications

**Daulet Izbassarov** and **Metin Muradoglu***

*Department of Mechanical Engineering, Koc University, Rumelifeneri Yolu, Sariyer, Istanbul, Turkey*

**Abstract:** This chapter provides an overview of dynamics and applications of pendant drops. The pendant drop tensiometry is widely used in measurement of interfacial tension and thus it is described in details including the historical development and the state-of-the-art. The sensitivity of the method is discussed and the recent advancements are presented. The stability and breakup of the pendant drop are also discussed in the general context of jet instability. The multiphysics effects including the electric field, viscoelasticity and surfactants are briefly reviewed focusing on their influence on pendant drop instability.

**Keywords:** Drop breakup, Drop shape analysis, Electrical field, Non-Newtonian effects, Pendant droplet, Surface tension measurement, Surfactants.

## INTRODUCTION

Drop formation is a ubiquitous phenomenon in daily life, science, and technology. This is critically important in various applications such as ink-jet printing, DNA micro arraying, cell and organ printing, spray combustion, spraying of agricultural chemicals, and many others. The study of drop formation started from the late seventeenth century. In early studies, Mariotte [1] and Savart [2] observed that a liquid jet becomes unstable and eventually breaks into small droplets. Later, Plateau [3, 4] and Rayleigh [5, 6] did pioneering works on liquid jet instability and breakup, and consistently explained the reason for the jet instability. Early experimental investigations were mostly motivated by engineering applications thus focused on instability of fluid jets leaving a nozzle at high speed, slow dripping under gravity, and liquid bridges. Since the early times, the pendant droplets have been widely used to measure the interfacial tension and to study the transport properties of interfaces.

* **Corresponding author Metin Muradoglu:** Department of Mechanical Engineering, Koc University, Rumelifeneri Yolu, 34450 Sariyer, Istanbul, Turkey; Tel: +90(212)3381473; Fax: +90(212)3381548; E-mail: mmuradoglu@ku.edu.tr

**Mihail Lucian Pascu (Ed.)**

Pendant droplet is a droplet which hangs from a nozzle or some other surfaces as shown in Fig. (**1**). In an equilibrium state, the shape of a droplet is determined by the force balance of surface tension and gravity (Fig. **1d**). Surface tension tends to make drop spherical while the gravitational force tries to elongate it. In the absence of other external forces, the shape of a pendant drop depends on the Bond number (*Bo*.) and the contact angle ($\alpha$.). The Bond number measures the relative importance of the gravitational force compared to the interfacial force and is defined as $Bo = \Delta\rho g R_0^2/\gamma$, where $\Delta\rho$, $g$, $R_0$, and $\gamma$ are density difference across the interface, gravitational acceleration, characteristic size of the droplet (*e.g.*, drop radius at the apex), and surface tension coefficient, respectively. When *Bo* is small, *i.e.*, $Bo << 1$, the droplet takes a semi-spherical cap as shown in Fig. (**1a**). As *Bo* increases, the droplet elongates in the direction of gravitational acceleration (Fig. **1b**) and eventually breaks up when the gravitational force overcomes the surface tension (Fig. **1c**).

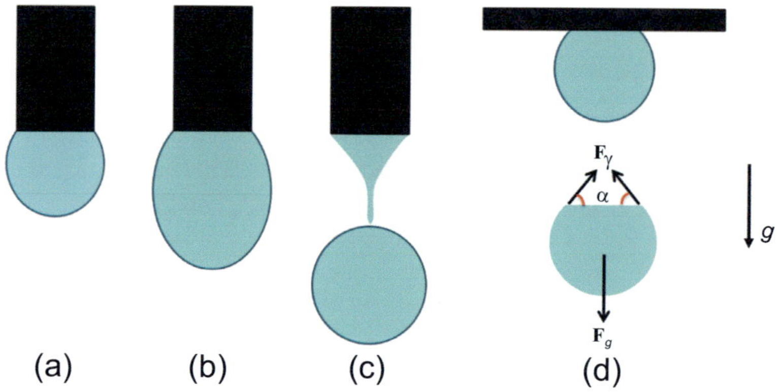


**Fig. (1).** Schematic illustration of a pendant droplet in various Bond number regimes. (**a**) A semi-spherical pendant droplet hanging from a nozzle at low *Bo*, *i.e.*, $Bo << 1$. (**b**) An elongated pendant droplet hanging from a nozzle at moderate Bond number. (**c**) Breakup of a pendant droplet ($Bo >> 1$). (**d**) Force balance for a pendant droplet hanging on a wall where $\alpha$, $F_\gamma$, and $F_g$ are the contact angle, the surface tension and the gravitational force, respectively.

There is a wide range of methodologies designed to measure surface and interfacial tension. Some common measurement methods are depicted in Fig. (**2**) [7 - 10]. Among these methods, the pendant drop tensiometry has been found to be the most versatile and robust method. Worthington was the first who proposed the measurement of interfacial tension using the shape of a pendant drop [11, 12]. Later Bashforth and Adams [13] produced tables for axisymmetric sessile drop profiles for different values of surface tension and radius of curvature. Then, Fordham [14] and Mills [15] developed similar tables for pendant drops. Maze and Burnet [16, 17] proposed a method using a nonlinear least-squares optimisation technique for the determination of surface tension of a sessile drop.

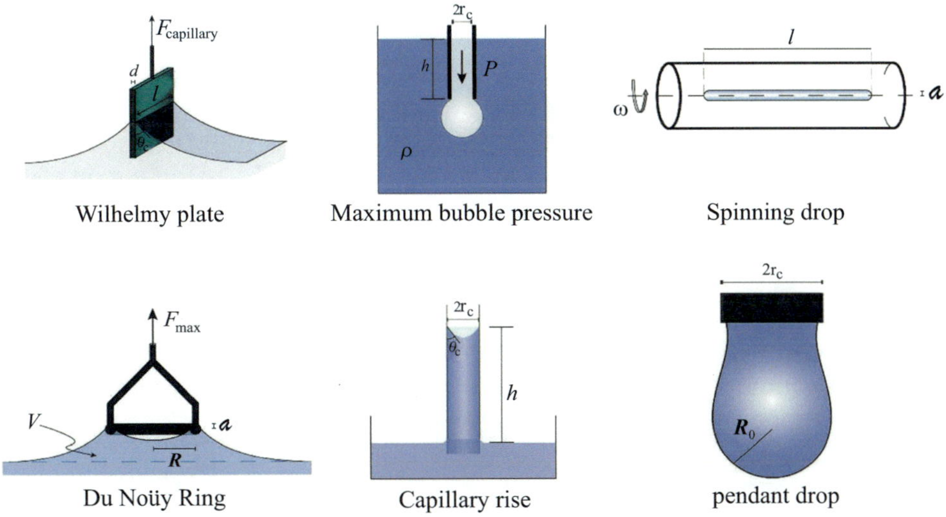


**Fig. (2).** Schematics of various experimental techniques used to measure interfacial tension (adapted from [10]).

Rotenberg *et al.* [18] developed a more robust technique called an axisymmetric drop shape analysis (ADSA), which fits the experimental shape with a theoretical drop profile. They used the Newton-Raphson method [19] for fitting the theoretical drop profile to the experimental data. Since then many adaptations and improvements have been carried out for the ADSA method [20]. Cheng and coworkers [21, 22] improved the ADSA method by utilizing an automatic image processing technique. del Rio and Neumann [23] found that using the Newton-Raphson method for optimisation is computationally expensive and furthermore the method may fail to converge if the initial guess is poor. Therefore, they combined the Newton-Raphson method with the Levenberg-Marquardt algorithm [19] to improve the likelihood of convergence. Hoorfar and Neumann [20] reviewed advances in computational methods of the ADSA algorithm including a close examination of potential sources of error and dynamic effects of experimental setup and fitting software.

In some cases, the edge detection of the experimental drop required by the ADSA is not possible mainly due to optical limitations. To overcome these difficulties, Cabezas *et al.* [24] proposed a new drop-shape methodology called theoretical image fitting analysis (TIFA) designed to avoid the edge detection in the ADSA. Unlike the ADSA method which involves a two-step procedure, TIFA achieves the interfacial tension measurement in one-step combining the image processing with the fitting procedure. In addition, TIFA matches the whole experimental image to the theoretical one. Thus, it yields a better accuracy than ADSA especially for the cases where acquisition of sharp image is not possible.

Hoorfar and Neumann [20] observed that the ADSA method suffers from a severe convergence problem at low Bond numbers. Particularly, the minimum Bond number for which the ADSA can provide a reliable measurement for the interfacial tension is generally found to be $Bo \sim 10^{-1}$ [25, 26]. To remedy this deficiency, Alvarez *et al.* [26] proposed a non-gradient algorithm based on the Nelder-Mead simplex method [27] for fitting theoretical drop profile to the experimental image. They found that the new algorithm can accurately measure surface tension for the Bond numbers as small as $Bo = 0.01$. Moreover, they observed that through the use of an averaging algorithm (averaging over multiple drop shapes), the method can yield reliable measurements even for $Bo \cong 0.002$. Note that, even though they significantly improved the measurement at low Bond numbers they could not overcome the zero Bond number problem. Recently, Neeson *et al.* [28] proposed to use a compound drop tensiometry for measurement of interfacial tension at low and even zero Bond numbers, which is discussed in more details in the next section.

Yeow *et al.* [29] proposed a new methodology for obtaining surface tension from a pendant droplet profile. In this method, a database is created using dimensionless volumes of pendant drop ($V_{pen}$) and spherical cap ($V_{cap}$), where $V_{cap}$ is the volume of the cap at the apex (see the shaded volume in Fig. **3**). Note that the cap has the same radius ($r_{scale}$) as the pendant droplet as sketched in Fig. (**3**). First, a database is created by calculating $V_{pen}$ and $V_{cap}$ from the Young-Laplace equation of pendant drop for different *Bo* numbers. Then, $V_{pen}$ and $V_{cap}$ are calculated from the measured profile and compared with the database to deduce Bond number and the surface tension.

Che *et al.* [30] performed numerical simulations to study upstream pressure fluctuations during growth and breakup of pendant drops. They observed a linear relationship between the surface tension and the pinch-off period. Moreover, they found that the pressure in the capillary fluctuates with the same period as the pinch-off. They proposed a novel method for surface tension measurement based on the period of pressure fluctuations. The method is concluded to be accurate, simple, and low cost.

In the following section, we first present the principles of the ADSA method and its sensitivity. After that, the new measurement technique based on the compound droplet is briefly described. Then the measurement of fast dynamic interfacial tension is also discussed.

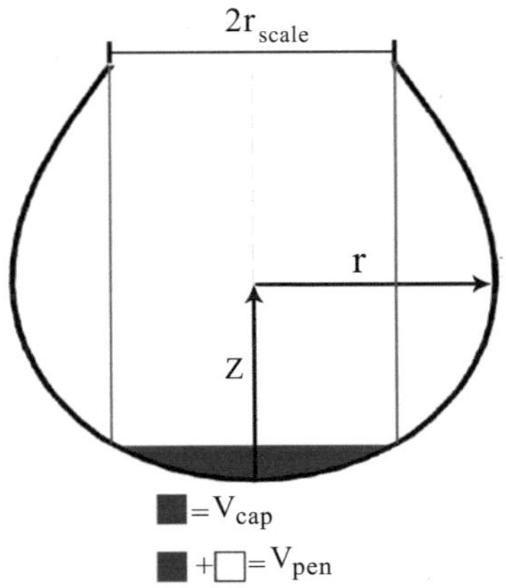


**Fig. (3).** Pendant drop tensiometry developed by Yeow *et al.* [29] (adapted from [29]). The volume of the whole droplet is $V_{pen}$ and the volume of spherical cap is $V_{cap}$ (shaded area). $V_{cap}$ is defined as the volume of the cap at the apex with the common radius of $r_{scale}$. See Yeow *et al.* [29] for details.

## PENDANT DROP TENSIOMETRY

Axisymmetric drop shape analysis (ADSA) is the most widely used method for measurement of surface tension coefficient due to its accuracy, flexibility and simplicity [31, 32]. The ADSA has been successfully employed in many areas ranging from biological systems to industrial applications [33, 34]. This approach was also the origin of the modern profile analysis tensiometers (PAT). It is a technique for measuring interfacial properties by fitting the theoretical curves to the experimentally determined shape of a pendant drop. The theoretical curves are defined by the classical Gauss-Laplace equation (equation 1) of capillarity. The ADSA is appropriate for a wide range of surface tensions and for any fluid-fluid system that can be represented by the Gauss-Laplace equation. The equation describes a mechanical force balance between two fluids across the interface. As shown in Fig. (**6**), it relates the pressure difference (Laplace pressure) to interfacial tension and curvature as:

$$\gamma\left(\frac{1}{R_1}+\frac{1}{R_2}\right)=\Delta P \quad (1)$$

where $\gamma$ is the interfacial tension coefficient, $R_1$ and $R_2$ are the principal radii of curvature, and $\Delta P$ is the pressure difference across the interface. When considering only gravity, $\Delta P$ can be expressed as:

$$\Delta P=\Delta P_0-\Delta\rho g z \quad (2)$$

where $\Delta P_0$ is the pressure difference at a reference plane and $z$ is the vertical coordinate of the drop measured from the reference plane. Thus, it is possible to determine the shape of a drop for a given $\gamma$, or inversely to calculate $\gamma$ from the shape [33].

The integration of the Gauss-Laplace equation (equation 1) is obvious for spherical shapes and cylindrical menisci but it becomes more complicated for general irregular shapes. Various numerical methods have been designed for the special case of an axisymmetric drop shape. It is possible to assume an axial symmetry for the most practical pendant and sessile drop systems. Thus, the method is described here only for the axisymmetric case.

## General Procedure

A general procedure for calculation of interfacial properties from the shape of pendant and sessile drops is summarized in Fig. (**4**). First, the coordinates of the experimental drop profile are extracted from the drop image (Fig. **5a**) using an image processing technique as shown in Fig. (**5d**). Then the surface tension is determined by fitting a series of theoretical curves to the experimental profile. The fitting process is schematically shown in Figs. (**5c** and **d**). This iterative procedure has proved to be successful especially for sufficiently large Bond numbers.

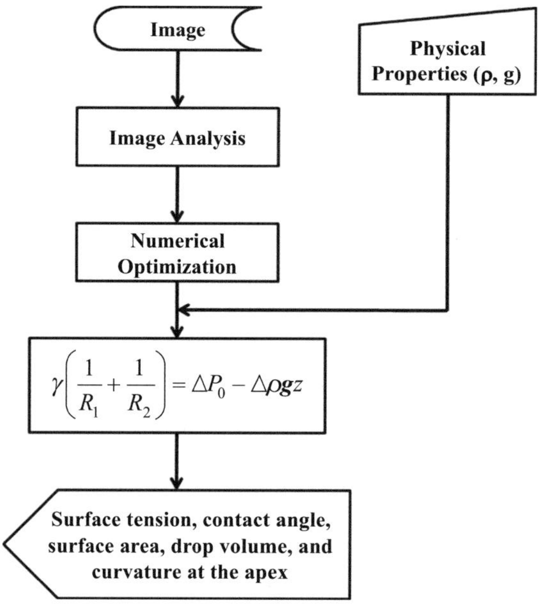


**Fig. (4).** A general procedure of the Axisymmetric Drop Shape Analysis (ADSA) (adapted from [20]).

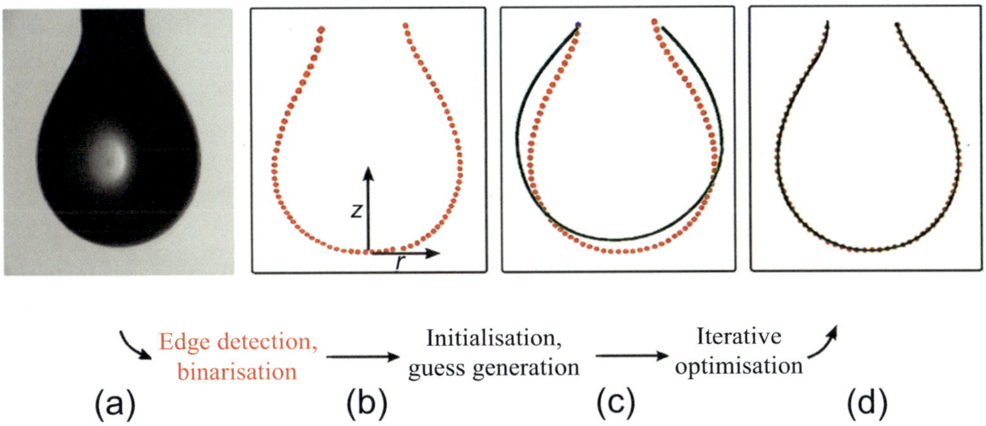


**Fig. (5).** A schematic illustration of the ADSA method, going from a raw experimental image to a fitted solution from which $\gamma$ can be obtained (adapted from [10]).

In the numerical method, the theoretical curves are generated by integrating the Gauss-Laplace equation (equation 1) for a given surface tension and curvature at the apex. Fig. (**6**) shows a typical coordinate system used in the ADSA method. Assuming that the droplet is symmetric about z-axis, two principle radii of curvature, $R_1$ and $R_2$, can be computed at any point ($x_i$, $z_i$) on the drop surface.

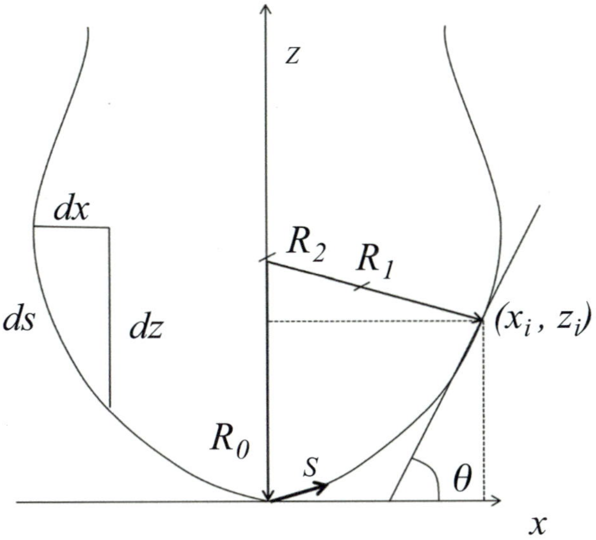


**Fig. (6).** Definition of coordinate system for a pendant drop in the ADSA method. Turning angle is $\theta$ at a point ($x_i$, $z_i$). The arc length, s, is evaluated along the drop. The two principal radii of curvature $R_1$ and $R_2$ are in the plane and perpendicular to the paper, respectively.

The first principal radius of curvature is given by:

$$\frac{1}{R_1} = \frac{d\theta}{ds} \tag{3}$$

where $s$ is the arc length and $\theta$ is the angle of inclination. The second radius of curvature can be calculated as:

$$\frac{1}{R_2} = \frac{sin\theta}{x} \tag{4}$$

where $x$ is the radial coordinate of the curve.

Due to axisymmetry, the curvature at the apex (*i.e.*, at $s = 0$) is constant in all directions so we have:

$$\left.\frac{1}{R_1}\right|_{s=0} = \left.\frac{1}{R_2}\right|_{s=0} = \frac{1}{R_0} = b \tag{5}$$

where $R_0$ and $b$ are the radius of curvature and the curvature at the origin, respectively.

Then using equation (1) the pressure difference at the origin is given by:

$$\Delta P_0 = 2b\gamma \tag{6}$$

Substituting equations (3), (4) and (6) into equation (1) yields:

$$\frac{d\theta}{ds} = 2b - cz - \frac{sin\theta}{x} \tag{7}$$

where

$$c = \frac{\Delta\rho g}{\gamma} \tag{8}$$

The geometric relations and the boundary conditions can be written as:

$$\frac{dx}{ds} = cos\theta \quad \textbf{(9)}$$

$$\frac{dz}{ds} = sin\theta \quad \textbf{(10)}$$

$$x(0) = z(0) = \theta(0) = 0 \quad \textbf{(11)}$$

Thus, the complete axisymmetric fluid-fluid interface can be obtained by simultaneously integrating this set of first order ordinary differential equations [equations (7), (9) and (10)] for given values of *b* and *c*. Once the theoretical curves are generated, an optimisation procedure is used to fit these curves to the experimental profile (see Fig. **5**).

**Sensitivity of ADSA Method**

The ADSA is accurate for a wide range of experimental conditions. However, it has been found that ADSA fails to predict the interfacial properties accurately in two regimes. In the first regime where $Bo >> 1$, the interfacial tension force is too low to preserve a stable hanging drop. Since the ADSA method is based on an equilibrium force balance as given by the Gauss-Laplace equation, the method fails to fit experimental drop profile with the theoretical curves. In the opposite regime of vanishing Bond number, *i.e.*, $Bo \rightarrow 0$, surface tension dominates and drop shape becomes nearly spherical. In this regime, only minor change occurs in drop shape even for a large increase in interfacial tension. Thus, the sensitivity of the method deteriorates dramatically for $Bo << 1$ due to uncertainties in experimental image as well as contamination of numerical solutions by truncation and round-off errors. Hoorfar and Neumann [20] proposed a shape parameter to determine the working range for the ADSA. The shape parameter is defined as the difference between the projected area of the drop and an inscribed circle with radius $R_0$, the radius of curvature at the apex, normalized by the projected area of the drop, *i.e.*,

$$P_s = \frac{A_{projected\_area} - A_{inscribed\_circle}}{A_{projected\_area}} = 1 - \frac{A_{inscribed\_circle}}{A_{projected\_area}} \quad \textbf{(12)}$$

where $A_{projected\_area}$ and $A_{projected\_circle}$ are the projected area of the drop and the area of inscribed circle, respectively, as shown in Fig. (7). The value of the shape parameter is restricted to a range from zero (nearly spherical drop) to one (well-deformed drop). The range of admissibility and critical shape parameter of the ADSA depend on the required accuracy. Hoorfar and Neumann [20] reported that for the ADSA to achieve desired accuracy, the shape parameter should be above a critical value. For instance, in their particular case of a cyclohexane drop in air, for a desired accuracy of 0.1 $mJ/m^2$ (a relative error of 0.004), the critical shape parameter was found as $P_{s\,(critical)} = 0.29$, *i.e.*, a value of shape parameter above this critical value guarantees the measurement error to be less than the desired accuracy. They concluded that, the critical shape parameter does not depend on the fluid but depends on the drop shape.

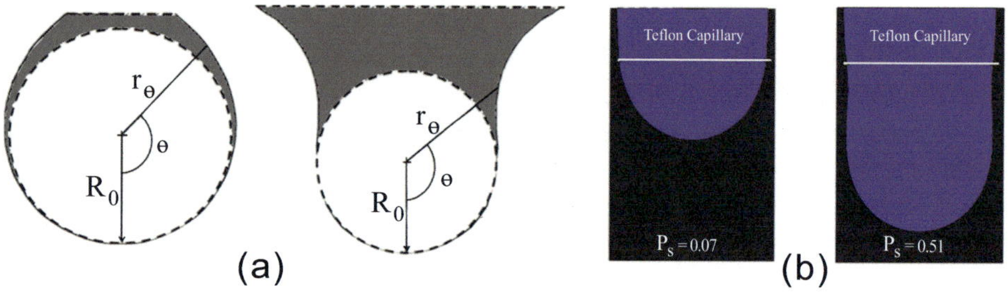


**Fig. (7).** Hoorfar and Neumann [20] proposed the shape parameter to characterize accuracy of ADSA (adapted from [20]). (**a**) The shape parameter $P_s$ corresponds to the shaded area between the drop profile and an inscribed circle with a radius of $R_0$. (**b**) The shape parameter $P_s$ computed for two drops of cyclohexane formed at the end of a Teflon capillary [20].

Ferrera *et al.* [25] analysed the sensitivity of pendant drops and liquid bridges for measurement of interfacial tension. They recommended to use liquid bridges instead of the pendant drops to enhance applicability of axisymmetric shape technique in adverse conditions, *i.e.*, liquids with similar densities and/or under microgravity. They reported that similar accuracy is found for both pendant drop and liquid bridge cases when $Bo \gtrsim 10^{-1}$, however liquid bridge case yields a better accuracy when $Bo \sim 10^{-2}$. Saad *et al.* [35] investigated in detail the design and accuracy of pendant drop methods. To quantify the accuracy of the results they proposed to use modified shape parameter. The new shape parameter is defined as the difference in the shapes between experimental drop and the circle that best fits to all the points on drop profile. It was found that the shape parameter depends on dimensionless volume and the Bond number while its critical value depends only on the Bond number. Recently, Berry *et al.* [10] discussed pendant drop tensiometry highlighting its features and limitations. They introduced a new non-dimensional number, the Worthington number ($Wo$), to characterize the measurement precision. The Worthington number (Fig. **8**) is defined as $Wo = \Delta\rho g V_d / \pi\gamma D_n$, where $\Delta\rho = \rho_d - \rho$ is the density difference and $\rho_d$ and

$\rho$ are density of drop and continuous phase, respectively, $V_d$ is the drop volume, and $D_n$ is the needle diameter. *Wo* can vary from 0 to 1, where a larger value indicates a better precision. They proposed to use *Wo* as a post-measurement 'sanity check' to give an indication of the measurement precision.

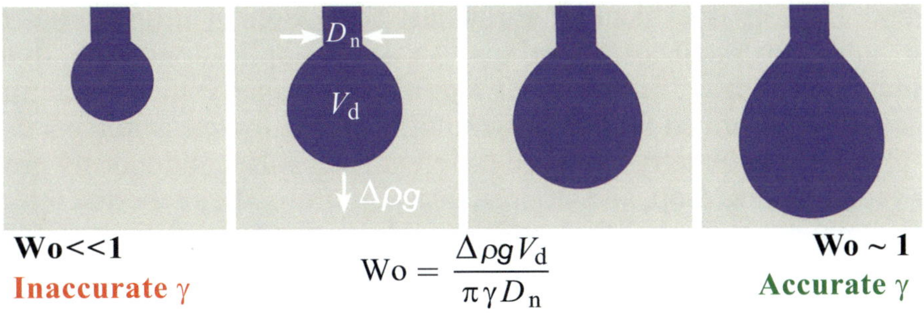


**Fig. (8).** Berry *et al.* [10] introduced a new parameter called the Worthington number (*Wo*) to characterize accuracy of pendant drop tensiometry (adapted from [10]). Larger values of Wo indicate better accuracy. Here $\Delta\rho = \rho_d - \rho$ is the density difference and $\rho_d$ and $\rho$ are density of drop and continuous phase, respectively, $V_d$ is the drop volume, and $D_n$ is the needle diameter.

## Compound Droplet Tensiometry

In the conventional pendant drop tensiometry, the vanishing Bond number constitutes the physical limitation. To overcome this limitation, Neeson *et al.* [28] proposed a new methodology to precisely measure interfacial tension at small and even zero Bond number cases. They developed a compound pendant drop technique in which a solid spherical particle of known size is added to the pendant drop as shown in Fig. (**9**). The interfacial tension is deduced by fitting the experimental profile to the theoretical curves obtained from the Young-Laplace equation similar to the ADSA method.

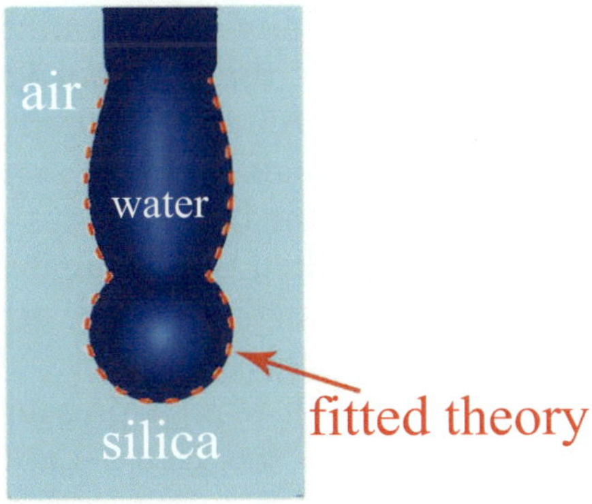


**Fig. (9).** Measurement of interfacial tension at low or even zero *Bo* using a compound pendant drop tensiometry (adapted from [28]).

## Fast Dynamic Interfacial Tension

In modern highly dynamic technologies, fast formation of large interfaces is widely used to produce emulsions and foams. In particular, adsorption and desorption kinetics of surfactants are measured by the dynamic interfacial tensiometry. Thus, it is critical to study fast formation of drop/bubble or its behaviour under external perturbations. In such applications, the dynamic surface/interfacial tension is the most significant property to be determined. Javadi *et al.* [36] performed fast dynamic interfacial tension measurement using a capillary pressure tensiometry *via* three different protocols: continuously growing drop, pre-aged growing drop, and stopped growing drop. They reported that these protocols complement each other and provide a good data set for dynamic interfacial tension at short adsorption times. Javadi *et al.* [37] experimentally and numerically investigated fluid flow and mass transfer in the coaxial double capillary set-up. They determined the most effective conditions for the coaxial double capillary in measuring adsorption kinetics and transport properties of bulk soluble surfactants. Javadi *et al.* [38] also examined the functionalities of a new experimental tool called the oscillating drop and bubble analyser (ODBA) for accurate measurement of fast dynamic interfacial tension. A schematic of the ODBA with its main components is shown in Fig. (**10**). The ODBA utilises the capillary pressure technique and is designed to measure the dynamic interfacial tension at short times as well as dilational visco-elasticity of interfacial layer at high frequencies. Their method was found to be capable of dealing with iso-dense systems. They reported accurate measurements of dynamic interfacial tension at frequencies ranging between 1 and 100 Hz, and elasticity values for liquid-liquid interface at high frequencies up to 80 Hz. The details of the experimental setup, protocols and data analysis technique can be found in [38].

Karbaschi *et al.* [39] examined applicability of profile analysis tensiometry (PAT) for dynamic surface tension using growing drop technique. In this approach, obtained drop profiles are analysed using the classical Gauss-Laplace equation. They found that the applicability of the PAT for dynamic measurement is limited by Reynolds number (*Re*) (growing rate) and the drop size. It was concluded that the error in PAT measurements is less for bigger drops and limited by $Re<21$. Recently, Dieter-Kissling *et al.* [40] studied the applicability of the PAT at high flow rates using an interface tracking method. They found that using Gauss-Laplace equation for fitting the drop profile is not valid at high flow rates ($Re>25$). In fact, they found that at high flow rates pressure inside the drop is not uniform as required in the derivation of the Gauss-Laplace equation, thus the shape does not represent the profile corresponding to theoretical surface tension coefficient. Recently, Karbaschi *et al.* [41] reviewed important experimental and computational findings related to the fast dynamic interfacial tension measurement.

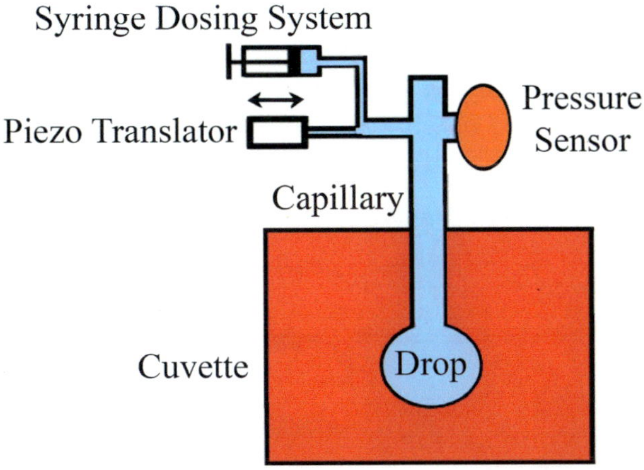


**Fig. (10).** Schematic of the oscillating drop bubble analyser ODBA (adapted from [38]).

## INSTABILITY OF PENDANT DROPS

Pendant drops become unstable and break up when the Bond number exceeds some critical value. This phenomenon has been extensively studied as part of the general topic of capillary instability and drop formation.

Formation of a drop is ubiquitous in daily life, science and technology. It is particularly important to produce monodispersed microdroplets for ink-jet printing, bioengineering, and various medical applications [42, 43]. The principal difficulty in these applications is to guarantee the production of monodispersed droplets and to prevent secondary satellite drop formation or breakup of the main drop. To understand these phenomena dynamics of drop formation has long been a topic of interest [44, 45].

The scientific investigation of drop formation started in the late seventeenth century. The drop formation problem was first mentioned by Mariotte [1]. He suggested that drops disintegrate from a stream of water due to gravity or other external forces. However, it is straightforward to show that drop formation cannot be a result of an action of a uniform force. It is obvious that surface tension is responsible for drop breakup process. Later Savart [2] studied disintegration of fluid jets. He observed that breakup process occurs as a result of growing waves on a jet. His results showed that the breakup must be a characteristic feature of the fluid flow. In addition, it was concluded that the instability of the jet was caused by amplification of small perturbations. Plateau [3] was the first who discovered the importance of surface tension on the drop formation and also provided a comprehensive description of the process. Plateau [4] and Rayleigh [5, 6] are the first who thoroughly studied dynamics of jet breakup problem. Plateau [4] showed that the surface area of jet is decreased due to long-wave perturbations facilitated by surface tension. Rayleigh [5, 6] examined cylindrical column of fluid perturbed with small sinusoidal perturbations. He discovered that as perturbations evolve the jet becomes unstable leading to a topological change. A typical liquid jet breakup process is shown in Fig. (**11**) where development of a capillary-driven instability of a fuel jet falling under the influence of gravity is illustrated [46].

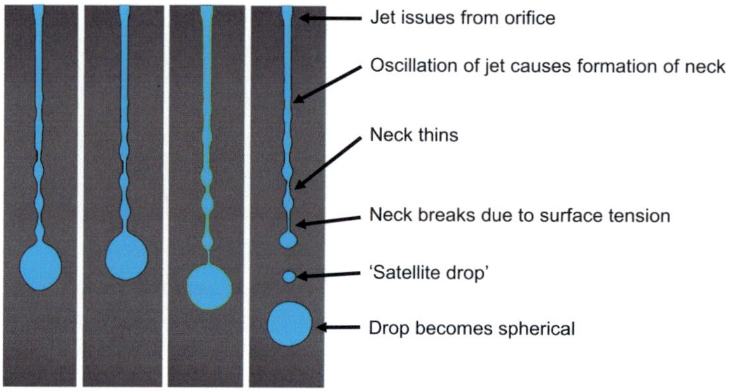


**Fig. (11).** The capillary-driven instability of a fuel jet falling under the influence of gravity (adapted from [46]).

Rayleigh [47] and later Edgerton *et al.* [48] have conducted early experimental studies on dynamics of free surfaces to examine the process of a drop breakup. Peregrine *et al.* [49] qualitatively described the bifurcation process of water drop in air using a high-speed photography. They provided photographic sequences of formation of main and satellite drops. Shi *et al.* [50] studied experimentally and computationally the effects of viscosity on drop formation. They observed that cases with higher viscosity elongate the filament during breakup. Zhang and Basaran [51] later performed extensive experiments to understand the dynamics of formation of a pendant droplet at low flow rates. They studied effects of geometrical and physical parameters on dynamics of drop formation. Basaran and co-workers [52 - 54] also studied the breakup process both computationally and experimentally in the region very close to a pinch-off. They observed that prior to a breakup the fluid interface can overturn forming a singularity even in the slightly viscous flows.

The studies mentioned above mainly considered formation of a single drop. However, formation of several drops from a nozzle in a sequence is of crucial importance in engineering applications and natural processes, and thus it has been also investigated. Basaran [43] reviewed various techniques that produce monodisperse drops. Eggers [42] and Clanet and Lasheras [55] found that different regimes can be attained depending on the flow rate (Fig. **12**). At low flow rates, periodic dripping regime occurs where main and satellite drops are formed periodically. As the flow rate increases, the dripping faucet regime occurs after the flow rate exceeds a critical value. In this regime, drops of different size are formed quasi-periodically or chaotically. When the flow rate is increased even more, drop breakup occurs in the jetting regime. In this regime jet emerges from the nozzle and drop is formed at its tip [56].

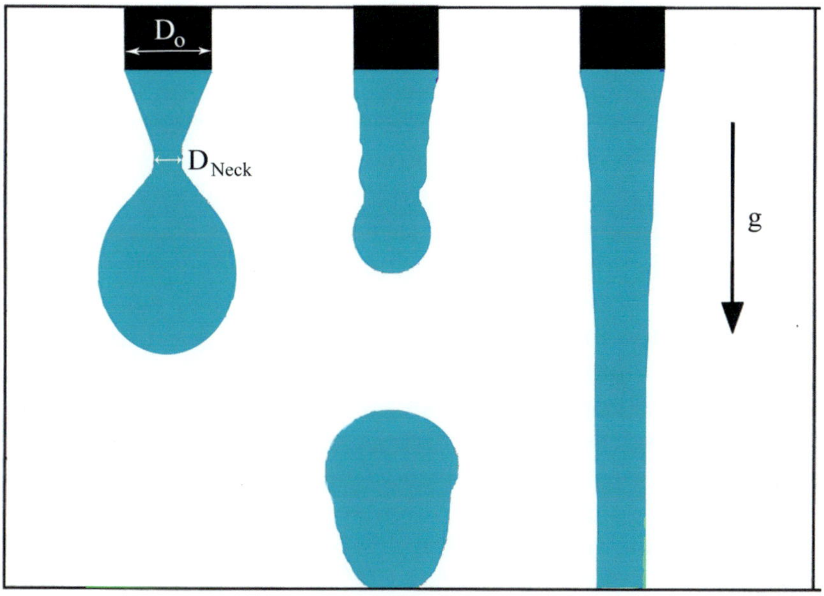


**Fig. (12).** Drop formation at the tip of a capillary tube. Different regimes are observed: (left) dripping, (middle) dripping faucet, and (right) jetting (adapted from [55]).

To study drop formation high speed and high-resolution video capturing is widely used [50 - 52, 57, 58], while to capture the dynamics of sequential drop formation, time interval between successive drop formations is usually measured using a laser-based drop-counter apparatus [59 - 61]. To analyse time interval data, time return maps have been used [62, 63]. In the early experimental work of Shaw [62], thousands of drops were formed in succession. Since Shaw's pioneering work, many experimental studies have been done to understand dripping faucet phenomenon. Experimental studies have been done to investigate the effects of the flow rate [61] and faucet radius [64] on dynamical response of dripping faucet. Ambravaneswaran *et al.* [65] have demonstrated that nonlinearity of the dripping dynamics can be decreased by using higher viscous fluids and decreasing radius of a thin walled faucet. D'Innocenzo *et al.* [66] experimentally studied effects of geometrical parameters on the dynamics of drop formation using nozzles with flat and obliquely shaped tips. Subramani *et al.* [67] computationally and experimentally investigated the dripping faucet regime. They provided significant insight into understanding of the effect of the range of the governing parameters on the dynamics of dripping. Shemesh *et al.* [68] used a capacitive device and demonstrated that the dripping faucet dynamics is determined by the degree of synchronization of drop oscillations and detachment. Recently, Taur *et al.* [69] explored phase diagram showing transition between the dripping modes for various angles of inclination of nozzle. They also studied correlation of drop volume over a wide range of parameters.

Comprehensive computational works have been also done on the drop formation. In particular, extensive simulations have been performed to investigate the nonlinear dynamics of the breakup in the vicinity of the singular point *i.e.*, the point where the breakup occurs. The reason for singularity in the governing equations at the bifurcation point is the fact that curvature increases unboundedly as bifurcation point is approached. Peregrine [70] introduced the first dynamical equation based on the lubrication theory and applied it to study the non-linear effects on drop formation in ink-jet printers. The linear stability theory predicts that jet disintegrates into uniform distributed droplets, while it actually breaks up in a main and a satellite drop. Lee [71] studied these phenomena for the inviscid case and was able to observe the formation of the satellite drop in addition to the main one. Another important concept used to study the nonlinear behaviour is the self-similarity [72]. The solution at the singular point has a time dependent length scale. In the context of the self-similarity, this time dependent length scale can be mapped on one value. Keller and Miksis [73] are the first who suggested self-similarity for the inviscid, surface tension-driven flow. Later, Kadanoff and his collaborators [74, 75] investigated breakup of a bubble in a Hele-Shaw cell using combined concepts of lubrication and self-similarity.

Eggers [76] and Eggers and Dupont [77] extended the idea of self-similarity to three-dimensional case. Eggers [76] established the scaling solutions for the pinching of an axisymmetric fluid neck. He used an expansion of the Navier-Stokes equations in the radial coordinate to investigate the self-similar behaviour of the necking region. To study the evolution of complete pendant drop, Eggers and Dupont [77] numerically solved the equations derived by Eggers [76]. Their method takes into account the viscous terms, which leads to improvement of the accuracy and stability of the numerical algorithm. The numerical results were found to be in qualitative agreement with the experimental observations. Schulkes [57] numerically studied the evolution and bifurcation of a pendant drop at the end of a nozzle assuming an axisymmetric, inviscid and irrotational flow. They found that the evolution is dependent on the growth rate of a drop and radius of the nozzle. The evolution of the pendant drop close to the breaking point revealed that after the main drop breaks away a secondary drop bifurcation occurs. It has been found that the time between two bifurcations is short for slow growth rates and gets longer as the growth rates increase. A detailed discussion about the instability and breakup of liquid jets can be found in the review paper by Eggers [42].

Full Navier-Stokes simulations have been also performed to examine the stability and breakup of pendant drops. Zhang [58] numerically studied the effects of physical and geometrical parameters on dynamics of growth and breakup of viscous pendant drops into air. The transient Navier-Stokes equations were solved for the axisymmetric free-surface problem using the volume of fluid-continuum surface force (VOF-CSF) method. The numerical method accounts for the effects of inertial, capillary, viscous and gravitational forces. Gueyffier *et al.* [78] developed both axisymmetric and three-dimensional numerical methods to study pinching of a pendant drop. The numerical simulations were compared with the experimental data and found to be in good agreement. Wilkes and Basaran [79] examined drop ejection from an oscillating rod using an axisymmetric Galerkin/finite element algorithm. They found that using appropriate forcing amplitude and/or frequency may restrict formation of long necks, which usually causes formation of satellite drops after the breakup. Ambravaneswaran *et al.* [80] compared one-dimensional and two-dimensional analysis for the drop formation from a capillary tube. They revealed that when surface tension force dominates over the other forces, the agreement between the models is within 2%. On the other hand, in the cases when capillary forces are much less than either the viscous or the inertial forces, the accuracy of the 1D model degrades. Ambravaneswaran *et al.* [65] numerically and experimentally studied dripping-jetting transitions for a dripping faucet. They created phase diagrams for transition from dripping to jetting regimes.

Xu and Basaran [81] studied drop-on-demand ink-jet printing. They performed computational analysis to study drop formation from a capillary tube by imposing a transient flow in the nozzle. They developed an operability diagram which consists of three attainable regimes based on the Weber number ($We$) that measures the relative importance of inertia to surface tension force. In the first regime of low Weber numbers, drop does not breakup and remains as a pendant drop attached to the nozzle. In the second regime, $We$ is large enough so drop breaks up but may move toward the nozzle after its formation. In the third regime, $We$ is further increased so that drop breakup occurs and the formed drop moves away from the nozzle. Pozrikidis [82] computationally studied the solution space and stability of sessile and pendant liquid drops. He found that various shapes such as compact, lightbulb, hourglass, and pearly shapes exist for a given contact angle and Bond number. Sample drop contours with compact, lightbulb, and hourglass shapes are shown in Fig. (**13**). Pozrikidis [82] also performed extensive boundary-integral simulations to explore evolution of unstable drop detachment. He considered both horizontal and inclined plane cases and found that the most stable configuration is pendant and sessile drop cases while the least stable case is the inclined configuration.

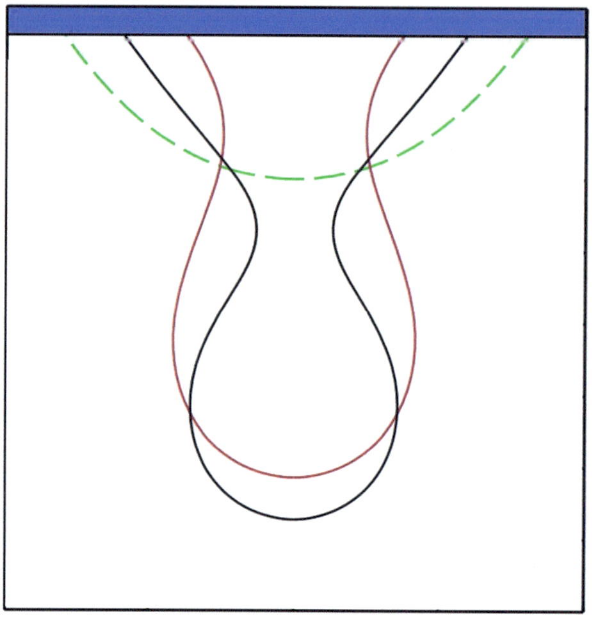

**Fig. (13).** Drop contours with compact, lightbulb, and hourglass shapes (adapted from [82]). All the drops have the same volume and contact angle.

## Effects of Electrical Field

Electrical field is widely used to manipulate and control the formation of drops in many applications including spray coating [83, 84], spraying of agricultural chemicals [85], inkjet printing [86] and separation processes [87]. Effects of electric field on liquid droplets are also an important topic in the other fields such as meteorology [88] and nuclear physics [89].

The equilibrium and stability of drops have been thoroughly investigated in an electrical field both for free and supported (pendant and sessile) cases. The early studies on the electrical field and charges on liquid drops were conducted by Benjamin Franklin [90] followed by Lord Rayleigh [91, 92]. Rayleigh's pioneering theoretical work on the stability of a charged drop marked the beginning of an era for investigation of stability of electrified interfaces and birth of electrohydrodynamics (EHD). Later Zeleny [93, 94] investigated a drop formation in an electrical field. He discovered that drop shapes become unstable above some critical field strength, resulting in a conical tip with a tiny streaming jet. Taylor [95] demonstrated that the conical tip is an equilibrium solution known as the Taylor cone. The Taylor cone is formed at the nozzle tip under the actions of surface tension and electrostatic stresses on the surface of the equipotential cone. A conical surface is formed during charge accumulation at the cone's apex. This phenomenon of charge accumulation leads to surface destabilisation and eventually results in drop ejection. Initial theoretical studies were restricted to the spherical, cylindrical, and planar cases or the cases that involve only small amount of deviation from these shapes.

Almost one hundred years after Rayleigh's pioneering work, a number of numerical methods were elaborated to study the equilibrium shapes and stability of electrified drops. The early work by Miksis [96] theoretically determined the equilibrium shapes of a free dielectric drop in a dielectric medium under an applied electric field. Later, Joffre *et al.* [97] computationally determined the equilibrium shapes of pendant conducting drops in a dielectric fluid under an electric field. Both Miksis [96] and Joffre *et al.* [97] have also performed experiments to verify their results. Wohlhuter and Basaran [98] studied axisymmetric shapes and stability of pendant and sessile dielectric drops in an electric field. Basaran *et al.* [99] also investigated nonlinear oscillations of inviscid conducting drops. They showed that, at sufficiently large field strength, the drop assumes a conical shape with a streaming jet issuing from its conical tip.

More recently, there have been a number of studies performed to understand the dynamics of electrified drop formation. It was observed that electrohydrodynamic (EHD) regimes of drop formation from a nozzle in the presence of electrical field depend on both strength of electrical field ($E$) and flow rate ($Q$) in the nozzle, as shown schematically in Fig. (**14**). At low values of $E$ and $Q$, the drop forms in a dripping regime mainly due to gravity [100]. When $E$ is increased while keeping $Q$ low, drop formation mode can be transferred from dripping to EHD jetting mode (Fig. **14**). It has been observed that during the EHD jetting mode the flow can be steady and axisymmetric, unsteady and axisymmetric, or unsteady and nonaxisymmetric [101]. For moderate values of $E$, cone-jet mode is achieved producing Taylor cone at the tip [102]. In between, pulsating liquid jet mode occurs where a Taylor cone continuously forms and relaxes. At higher $E$ values, more complex EHD jetting regimes occur resulting in an atomization mode at very high $E$ values [100]. When $E$ is kept low but $Q$ is increased, jetting mode is achieved. The EHD modes shown in Fig. (**14**) are widely used in different applications [43] such as spraying [103], fiber spinning [104], and ink-jet printing [105]. The extensive reviews of the EHD modes have been given by Cloupeau and Prunet-Foch [101] and by Jaworek and Krupa [106].

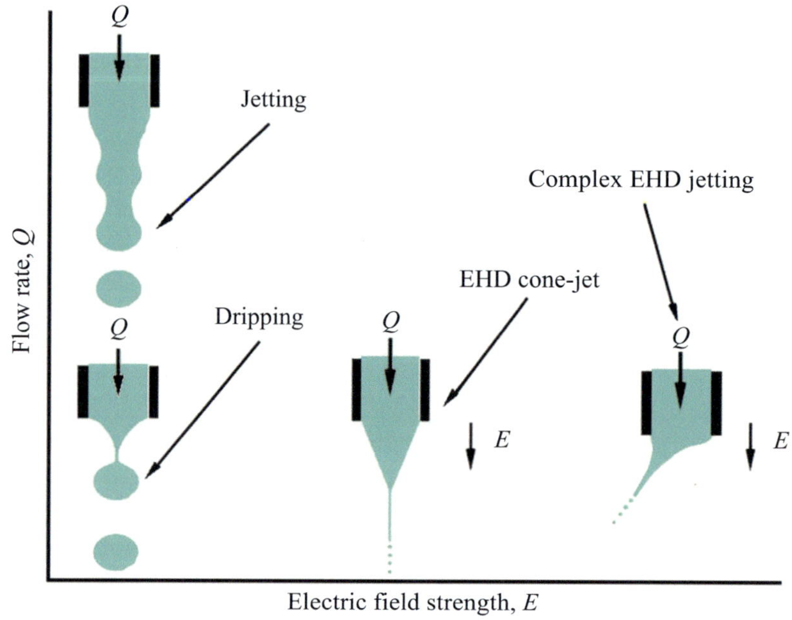


**Fig. (14).** Phase diagram for electrified drop formation (adapted from [100]).

Zhang and Basaran [107] have conducted experimental studies on the effects of electrical field on dynamics of liquid threads and generation of secondary drops. They showed that formation of monodispersed drops can be achieved when appropriate values of electrical field strength is used. Park *et al.* [108] experimentally studied drop emission from Taylor cone in an EHD jet printer. It was suggested that it is important to use sharp nozzle and low values of $E$ and distance between nozzle and substrate to achieve high-resolution printing. Lee *et al.* [109] introduced a systematic approach based on dimensional analysis to determine jetting modes of EHD printing. Different jetting modes which can be generated in the classical EHD jetting system by changing the electrical potential are shown in Fig. (**15**). They identified six important non-dimensional numbers for quantification of different jetting regimes. It was found that a pulsed electric field can be used to control the frequency of the jet and size of produced droplets [110 - 114]. Mishra *et al.* [110], Stachewicz *et al.* [111], Kim *et al.* [112], Stachewicz *et al.* [113], and Choi *et al.* [114] showed that the micro- and nanometer-sized droplets can be produced at frequencies reaching 10-50 kHz. It was also found that advanced methods such as sinusoidal voltage pulse shape [115] and pulse with more than two voltage values [116] can be used to control the jetting. Moreover, Kim *et al.* [117] and Lee *et al.* [118] observed that polarization of the electrical field, different electrode structures such as plane, hole, and pin types can also be used to control the production of droplet. Another important aspect for high-resolution production of droplets is related to properties of the nozzle [119]. It has been found that nozzle size and wetting properties of the nozzle surface are among the important parameters. Recently Hijano *et al.* [120] investigated for conditions of periodic microdripping regime of highly conductive and low-viscosity liquids. Sample snapshots of the microdripping regime are shown in Fig. (**16**). They were able to achieve emission of monodisperse droplets with diameter of one-tenth of the capillary at frequency of several kHz.

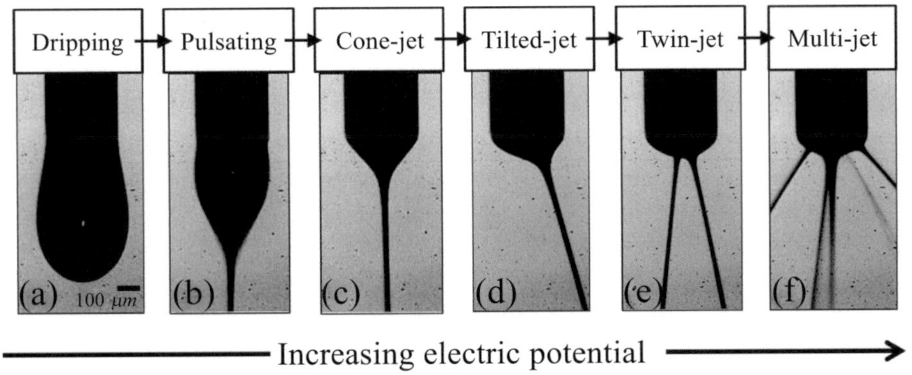


**Fig. (15).** EHD jetting modes with respect to electrical potential. Reprinted with permission from [109], copyright 2013 by the American Chemical Society.

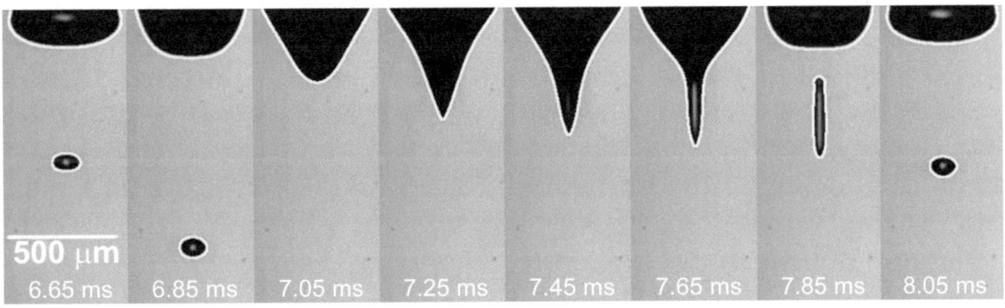


**Fig. (16).** Snapshots of the microdripping regime. Reprinted with permission from [120]. Copyright 2015 by the American Physical Society.

Notz and Basaran [121] investigated the effect of an electrical field on formation of pendant drop. Using Galerkin finite element method (FEM), they characterized the equilibrium shapes and the evolution of drop formation. Then Collins *et al.* [100] performed numerical investigations by solving the axisymmetric Navier-Stokes equation and studied the dynamics of perfectly conducting jet stressed by radial electric field. They reported breakup times and the sizes of main and secondary drops for a wide range of parameters. Later, Collins *et al.* [122] numerically and experimentally investigated drop formation from a film for different electrical conductivities. They developed a scaling law for prediction of size of a produced drop. Then Collins *et al.* [123] studied EHD streaming from a drop. They showed that depending on conductivity of the fluid there exist three regimes for the size and charge of a droplet. Also, they captured the details about formation of the Taylor cone with subsequent disintegration of small electrified drops, which is very difficult to do in experiments. Lee *et al.* [124] studied pinch-off behaviour of droplets from an EHD inkjet. The competing forces are the surface tension, viscous, electrostatic and gravitational forces. They investigated the effects of electrical conductivity, charge level, and surface tension on pinch-off. They showed that pinch-off topology depends on both the charge level and the electrical conductivity of the drop liquid. Kim *et al.* [125] studied the effect of wetting phenomena of the nozzle on EHD tip-streaming. They observed that monodispersed drops are ejected from the hydrophobic nozzle while unstable ejection occurred for hydrophilic case. Onses *et al.* [126] have recently reviewed the electrohydrodynamic jet printing.

## Non-Newtonian Effects on Drop Formation

Non-Newtonian fluids are ubiquitous in a wide range of engineering applications such as inkjet printing, microfluidics and polymer processing [127 - 131]. Their dynamics are significantly different from those of Newtonian fluids. In the Newtonian case, pinch-off is well described by the localised similarity solutions leading to a breakup in a finite time [132]. On the other hand, an inclusion of even a tiny quantity of polymers leads to formation of a nearly perfect cylindrical thread and suppression of a breakup. In the region close to the pinch point, elastic stresses are built up due to stretching of polymers hindering instability and breakup. As polymeric thread becomes sufficiently thin, it loses its uniformity due to instabilities leading to the formation of drops of different size on a polymeric filament. Most commonly occurring morphologies are shown in Fig. (**17**) [133]. Different types of instabilities have been reported by Sattler *et al.* [134].

Dynamics of thinning and break-up of viscoelastic filaments is governed mainly by two non-dimensional parameters: Ohnesorge (*Oh*) and Deborah (*De*) numbers. The Ohnesorge number measures the ratio of viscous to inertial forces and is defined as $Oh = \mu_0/\sqrt{\rho R \gamma}$, where $\mu_0$, $\rho$, $R$, and $\gamma$ are zero-shear-rate viscosity, density, characteristic length scale, and surface tension, respectively. The Deborah number ($De = \lambda/\tau_f$) represents the ratio of the relaxation time $\lambda$ of the polymer molecules to the characteristic process time $\tau_f$.

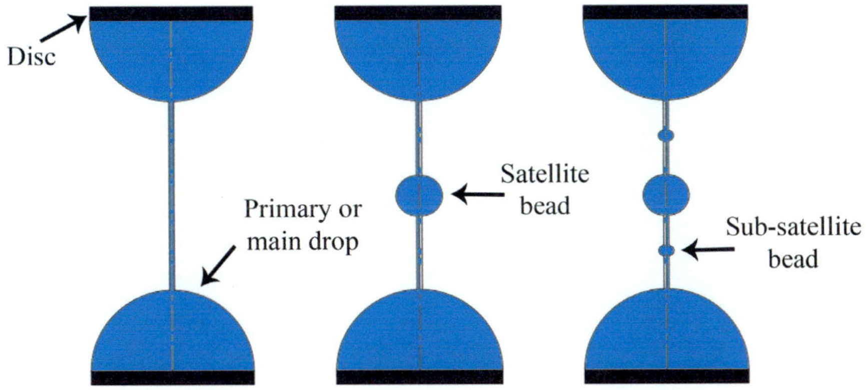


**Fig. (17).** Sketches of viscoelastic filaments exhibiting different beads-on-a-string (BOAS) morphologies reported by Bhat *et al* [133] (adapted from [133]).

Polymeric filaments have been observed in many applications [135 - 138]. Examples include liquid jets [135] (Fig. **18a**), drops falling from a capillary [136] (Fig. **18c**) and filament breakup devices [137] (Fig. **18b**). Note that there is a strong analogy between breakup of interface in cases of drop formation from a capillary tube and liquid bridge breakup as depicted in Fig. (**19**). Effects of shear-thinning liquids [139], yield-stress fluids [140] and sand [141] on filament breakup have been studied by various groups as reviewed by Eggers and Villermaux [132].

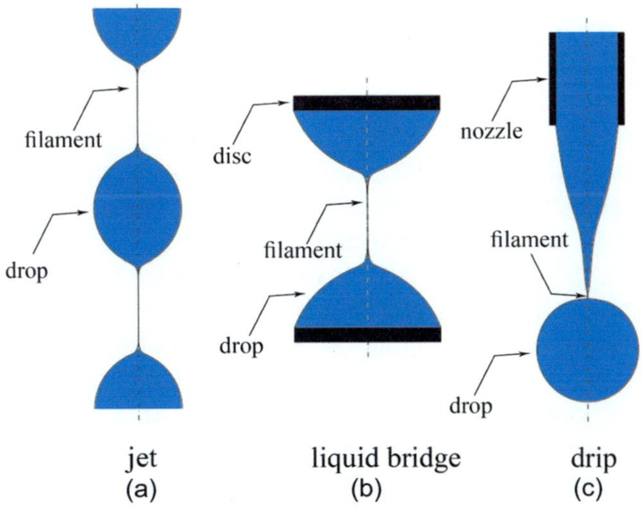


**Fig. (18).** Shapes of an interface for three common situations: jet breakup (left), liquid bridge between two disks (middle), and a drop dripping from a nozzle (adapted from [138]). The axis of symmetry is presented by dashed-dotted lines in the figure.

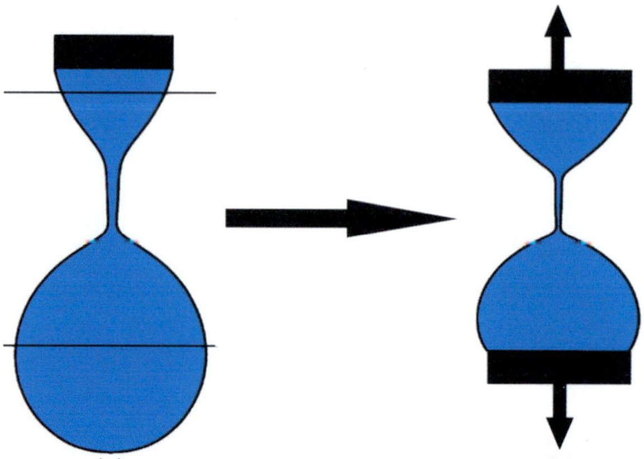


**Fig. (19).** The similarities between the dynamics of (**a**) liquid drops forming out of a capillary and (**b**) stretching liquid bridges make liquid bridges an ideal model system to study (adapted from [139]).

McKinley and Sridhar [142], McKinley [143] and Eggers and Villermaux [132] have extensively reviewed experimental and numerical works on filament thinning. Bhat *et al.* [133] numerically studied formation of beads-on-a-string (BOAS) structures during break-up of viscoelastic filaments. They concluded that a combined effect of viscoelasticity and inertia is required for formation of small beads. Also, it was found that viscoelasticity enhances growth of the bead and delays pinch-off. They created a phase diagram consisting of relevant parameters where different regimes of BOAS structure are determined. Bhat *et al.* [138] have studied self-similarity in the drop-filament corner region formed during the pinch-off of viscoelastic fluid threads. They showed that the corner region evolves self-similarly. They obtained new similarity solution which is valid in the filament, corner and drop region. Although their 1D model is able to determine self-similar dynamics it fails to predict steep interface profiles observed in experiments. Note that the obtained similarity solution is particularly useful in capillary-breakup rheometry to estimate fluid's extensional viscosity [132]. Morrison and Harlen [144] studied inkjet printing of viscoelastic fluids using finite-element analysis and presented its operating diagrams defining regimes with the BOAS structures along with the other regimes. Note that the BOAS structures are undesirable in the inkjet printing as it can lead to inaccuracy in printing due to formation of satellite drops. Recently, Eggers [145] studied instability of a polymeric thread. He reviewed previous proposed mechanisms for the instability and came up with a novel mechanism for the formation of beads. He proposed that the origin of the instability comes from the polymeric stress-density coupling that drives non-uniform polymer concentration along the thread.

Amarouchene *et al.* [136] have studied droplet fission of a polymeric fluid. They observed that addition of small amounts of polymer can inhibit finite-time singularity observed in the Newtonian case. The self-similar breakup is hindered and instead a long-lived cylindrical filament is created that thins exponentially in time as shown in Fig. (**20**). This characteristic feature is related to the dynamics of the formation of viscoelastic drops. Cooper-White *et al.* [146] investigated drop formation dynamics of elastic fluids. They have found that break-off length and time increase with increasing elasticity of the solutions and become proportional to the relaxation time of the solutions. Smolka and Belmonte [147] studied drop pinch-off and filament dynamics of wormlike micellar fluids. They have found that break-up dynamics of wormlike micellar drops is very different from Rayleigh-type (Newtonian fluids) or BOAS type instabilities. In fact, they observed that the falling drop can slow down or even stall before pinch-off. In addition, after break-up no satellite drop formation has been observed. McKinley and coworkers [148, 149] have studied transient extensional rheology and dynamics of elastocapillary thinning of viscoelastic filament, focusing on the later stages of evolution. They observed formation of BOAS structures in iterative manner. Tirtaatmadja *et al.* [150] investigated effects of molecular weight and concentration on drop formation and breakup of low viscosity elastic fluids. They observed that elasticity hinders formation of BOAS structure. Sattler *et al.* [134] have investigated the final stages of capillary break-up of polymer filament. They have described formation of different size drops along the filament. Moreover, at high polymer concentration no breakup of filament observed and instead nano solid polymeric fibers remain. Recently, Malkin *et al.* [151] reviewed extension of polymer filaments as well as instabilities occurring at the final stage of stretching.

**Fig. (20).** Pinch-off of Newtonian (water) (left) and polymeric (polyethylene-oxide solution) (right) fluids. Note the formation of a long-lived filament in the polymeric case. Reprinted with permission from [136], copyright 2001 by the American Physical Society.

## Effects of Surfactants

Surfactants change the surface tension and critically affect the dynamics of multiphase flow systems [152 - 157]. They may exist as impurities or are deliberately added to the drop or continuous phase to manipulate the flow. For instance, surfactants are regularly used as ink additives to control breakup of drops and jets in many applications such as inkjet printing [129], crop spraying [158], and DNA or protein microarraying [159].

Shin and Abbott [160] investigated effects of UV light on dynamic surface tension of mixed surfactant systems. They observed a decrease in dynamic surface tension due to illumination, which is used to manipulate drop formation from a capillary.

Liao *et al.* [161] investigated effects of soluble surfactants on the deformation and breakup of stretching liquid bridges. They found that dynamic shapes of liquid bridges strongly depend on the surfactant. Moreover, it was found that satellite formation can be avoided if the liquid bridge is stretched after adsorption equilibration. Liao *et al.* [162] computationally and experimentally examined deformation and breakup of a stretching liquid bridge covered with an insoluble surfactant monolayer. They observed that presence of surfactant results in lowering surface tension and reduction of the surfactant from the thinning region. Weidner [163] studied the effects of surfactant convection and diffusion on the evolution of an axisymmetric pendant droplet growing from an initially uniform liquid film. They found that, for sufficiently thin (low Bond number) and thick (high Bond number) films, droplet evolves in a similar manner as there is no surfactant present on the free surface. For intermediate Bond numbers, the convection of surfactant can significantly slow the growth of a droplet thus must be included in the analysis for accurate modelling of evolution of a typical pendant droplet. Javadi *et al.* [164] experimentally studied the inflow pattern of surfactants into a droplet and compared the experimental results with their own simulations. When compared with the experiments, simulations showed visual agreement and good compatibility with the surface tension evolution except for a considerable delay in the onset of Marangoni instability induced by inflow of the surfactant. They argued that the main reason for discrepancy on the onset of the instability is due to the mobility of the drop surface which can prevent formation of a quasi-static adsorbed layer.

## CONFLICT OF INTEREST

The authors confirm that this chapter content has no conflict of interest.

## ACKNOWLEDGEMENTS

This work was supported by the Scientific and Technical Research Council of Turkey, Grant No. 115M688. We also thank M. Toroghi for helpful discussions during the preparation of the chapter.

## REFERENCES

[1] E. Mariotte, *Traite du mouvement des eaux et des autres corps Fluides*. Michallet: Paris, 1686.

[2] F. Savart, "Mémoire sur la constitution des veines liquids lances par des orifices circulaires en mince paro", *Ann. Chim,* vol. 53, pp. 337-374, 1833.

[3] J. Plateau, "Mémoire sur les phénomènes que présente une masse liquide libre et soustraite de l'action de la pesanteur", *Mém L'Acad Roy Belg,* vol. 16, pp. 1-35, 1843.

[4] J. Plateau, *Sur les figures d'équilibre d'une masse liquide sans pésanteur*vol. 23. Mém. de l'Acad. Roy: Belgique, 1849.

[5] L. Rayleigh, "On the instability of jets", *Proc. Lond. Math. Soc,* vol. 10, pp. 4-13, 1879.

[6] L. Rayleigh, "On the capillary phenomena of jets", *Proc. R. Soc. Lond,* vol. 29, no. 196-199, pp. 71-97, 1879.
[http://dx.doi.org/10.1098/rspl.1879.0015]

[7] J.F. Padday, *Surface tension. Part II. The measurement of surface tension.* Surface and Colloid Science: New York, 1968.

[8] A.W. Neumann, and R.J. Good, "Techniques of measuring contact angles", In: *Surface and Colloid Science, R. J. Good and R. R. Stromberg.* Springer, 1979, pp. 31-91.

[9] J. Drelich, C.H. Fang, and C.L. White, "Interfacial tension measurement in fluid-fluid systems", In: *Encyclopedia of Surface and Colloid Science.* Marcel Dekker Inc: New York City, 2002, pp. 3152-3166.

[10] J.D. Berry, M.J. Neeson, R.R. Dagastine, D.Y. Chan, and R.F. Tabor, "Measurement of surface and interfacial tension using pendant drop tensiometry", *J. Colloid Interface Sci,* vol. 454, pp. 226-237, 2015.
[http://dx.doi.org/10.1016/j.jcis.2015.05.012]

[11] A.M. Worthington, "On Pendent Drops", *Proc. R. Soc. Lond,* vol. 32, no. 212-215, pp. 362-377, 1881.
[http://dx.doi.org/10.1098/rspl.1881.0032]

[12] A.M. Worthington, "Note on a point in the theory of pendent drops", *Philos. Mag. Ser. 5,* vol. 19, no. 116, pp. 46-48, 1885.
[http://dx.doi.org/10.1080/14786448508627638]

[13] F. Bashforth, and J.C. Adams, *An attempt to test the theories of capillary action by comparing the theoretical and measured forms of drops of fluid. With an explanation of the method of integration employed in constucting the tables which give the theoretical forms of such drops.* Cambridge University Press, 1883.

[14] S. Fordham, "On the Calculation of Surface Tension from Measurements of Pendant Drops", *Proc. R. Soc. Math. Phys. Eng. Sci,* vol. 149, no. 1036, pp. 1-16, 1948.
[http://dx.doi.org/10.1098/rspa.1948.0063]

[15] O.S. Mills, "Tables for use in the measurement of interfacial tensions between liquids with small density differences", *Br. J. Appl. Phys,* vol. 4, no. 8, p. 247, 1953.
[http://dx.doi.org/10.1088/0508-3443/4/8/305]

[16] C. Maze, and G. Burnet, "A non-linear regression method for calculating surface tension and contact angle from the shape of a sessile drop", *Surf. Sci,* vol. 13, no. 2, pp. 451-470, 1969.

[http://dx.doi.org/10.1016/0039-6028(69)90204-0]

[17] C. Maze, and G. Burnet, "Modifications of a non-linear regression technique used to calculate surface tension from sessile drops", *Surf. Sci,* vol. 24, no. 1, pp. 335-342, 1971. [http://dx.doi.org/10.1016/0039-6028(71)90240-8]

[18] Y. Rotenberg, L. Boruvka, and A.W. Neumann, "Determination of surface tension and contact angle from the shapes of axisymmetric fluid interfaces", *J. Colloid Interface Sci,* vol. 93, no. 1, pp. 169-183, 1983. [http://dx.doi.org/10.1016/0021-9797(83)90396-X]

[19] W.H. Press, Ed., *Numerical recipes: the art of scientific computing.*,1st ed. Springer-Verlag: New York, 1986.

[20] M. Hoorfar, and A.W. Neumann, "Recent progress in Axisymmetric Drop Shape Analysis (ADSA)", *Adv. Colloid Interface Sci,* vol. 121, no. 1-3, pp. 25-49, 2006. [http://dx.doi.org/10.1016/j.cis.2006.06.001]

[21] P. Cheng, *"Automation of Axisymmetric Drop Shape analysis using digital image processing"*, PhD Thesis, 1990.

[22] P. Cheng, D. Li, L. Boruvka, Y. Rotenberg, and A.W. Neumann, "Automation of axisymmetric drop shape analysis for measurements of interfacial tensions and contact angles", *Colloids Surf,* vol. 43, no. 2, pp. 151-167, 1990. [http://dx.doi.org/10.1016/0166-6622(90)80286-D]

[23] O.I. del Río, and A.W. Neumann, "Axisymmetric drop shape analysis: computational methods for the measurement of interfacial properties from the shape and dimensions of pendant and sessile drops", *J. Colloid Interface Sci,* vol. 196, no. 2, pp. 136-147, 1997. [http://dx.doi.org/10.1006/jcis.1997.5214]

[24] M.G. Cabezas, A. Bateni, J.M. Montanero, and A.W. Neumann, "A new drop-shape methodology for surface tension measurement", *Appl. Surf. Sci,* vol. 238, no. 1-4, pp. 480-484, 2004. [http://dx.doi.org/10.1016/j.apsusc.2004.05.250]

[25] C. Ferrera, J.M. Montanero, and M.G. Cabezas, "An analysis of the sensitivity of pendant drops and liquid bridges to measure the interfacial tension", *Meas. Sci. Technol,* vol. 18, no. 12, pp. 3713-3723, 2007. [http://dx.doi.org/10.1088/0957-0233/18/12/006]

[26] N.J. Alvarez, L.M. Walker, and S.L. Anna, "A non-gradient based algorithm for the determination of surface tension from a pendant drop: Application to low Bond number drop shapes", *J. Colloid Interface Sci,* vol. 333, no. 2, pp. 557-562, 2009. [http://dx.doi.org/10.1016/j.jcis.2009.01.074]

[27] J.A. Nelder, and R. Mead, "A simplex method for function minimization", *Comput. J,* vol. 7, no. 4, pp. 308-313, 1965. [http://dx.doi.org/10.1093/comjnl/7.4.308]

[28] M.J. Neeson, D.Y. Chan, and R.F. Tabor, "Compound pendant drop tensiometry for interfacial tension measurement at zero bond number", *Langmuir,* vol. 30, no. 51, pp. 15388-15391, 2014. [http://dx.doi.org/10.1021/la504406m]

[29] Y.L. Yeow, C.J. Pepperell, F.M. Sabturani, and Y-K. Leong, "New development in processing pendant droplet tensiometry data", *Langmuir,* vol. 24, no. 19, pp. 10942-10949, 2008. [http://dx.doi.org/10.1021/la801598w]

[30] Z. Che, T.N. Wong, N-T. Nguyen, Y.F. Yap, and J.C. Chai, "Numerical investigation of upstream pressure fluctuation during growth and breakup of pendant drops", *Chem. Eng. Sci,* vol. 66, no. 21, pp. 5293-5300, 2011. [http://dx.doi.org/10.1016/j.ces.2011.07.028]

[31] P. Cheng, and A.W. Neumann, "Computational evaluation of axisymmetric drop shape analysis

profile (ADSA-P)", *Colloids Surf,* vol. 62, no. 4, pp. 297-305, 1992. [http://dx.doi.org/10.1016/0166-6622(92)80055-7]

[32] J.W. Jennings, and N.R. Pallas, "An efficient method for the determination of interfacial tensions from drop profiles", *Langmuir,* vol. 4, no. 4, pp. 959-967, 1988. [http://dx.doi.org/10.1021/la00082a030]

[33] O.I. del Rio, P. Cheng, and A.W. Neumann, "Axisymmetric Drop Shape Analysis (ADSA)", In: *Applied Surface Thermodynamics,* S. Lahooti, Ed., vol. 20101338. Second Edition Marcel Dekker Inc: New York, 1996, pp. 441-507.

[34] P. Chen, D.Y. Kwok, R.M. Prokop, O.I. del Rio, S.S. Susnar, and A.W. Neumann, "Axisymmetric Drop Shape Analysis (ADSA) and its Applications", In: *Studies in Interface Science*vol. 6. Elsevier, 1998. [http://dx.doi.org/10.1016/S1383-7303(98)80019-7]

[35] S.M. Saad, Z. Policova, and A.W. Neumann, "Design and accuracy of pendant drop methods for surface tension measurement", *Colloid Surface A,* vol. 384, no. 1-3, pp. 442-452, 2011. [http://dx.doi.org/10.1016/j.colsurfa.2011.05.002]

[36] A. Javadi, J. Krägel, P. Pandolfini, G. Loglio, V.I. Kovalchuk, E.V. Aksenenko, F. Ravera, L. Liggieri, and R. Miller, "Short time dynamic interfacial tension as studied by the growing drop capillary pressure technique", *Colloid Surface A,* vol. 365, no. 1-3, pp. 62-69, 2010. [http://dx.doi.org/10.1016/j.colsurfa.2010.01.002]

[37] A. Javadi, J.K. Ferri, T.D. Karapantsios, and R. Miller, "Interface and bulk exchange: Single drops experiments and CFD simulations", *Colloid Surface A,* vol. 365, no. 1-3, pp. 145-153, 2010. [http://dx.doi.org/10.1016/j.colsurfa.2010.04.035]

[38] A. Javadi, J. Krägel, A.V. Makievski, V.I. Kovalchuk, N.M. Kovalchuk, N. Mucic, G. Loglio, P. Pandolfini, M. Karbaschi, and R. Miller, "Fast dynamic interfacial tension measurements and dilational rheology of interfacial layers by using the capillary pressure technique", *Colloid Surface A,* vol. 407, pp. 159-168, 2012. [http://dx.doi.org/10.1016/j.colsurfa.2012.05.026]

[39] M. Karbaschi, D. Bastani, A. Javadi, V.I. Kovalchuk, N.M. Kovalchuk, A.V. Makievski, E. Bonaccurso, and R. Miller, "Drop profile analysis tensiometry under highly dynamic conditions", *Colloid Surface A,* vol. 413, pp. 292-297, 2012. [http://dx.doi.org/10.1016/j.colsurfa.2012.04.027]

[40] K. Dieter-Kissling, M. Karbaschi, H. Marschall, A. Javadi, R. Miller, and D. Bothe, "On the applicability of Drop Profile Analysis Tensiometry at high flow rates using an interface tracking method", *Colloid Surface A,* vol. 441, pp. 837-845, 2014. [http://dx.doi.org/10.1016/j.colsurfa.2012.10.047]

[41] M. Karbaschi, M. Taeibi Rahni, A. Javadi, C.L. Cronan, K.H. Schano, S. Faraji, J.Y. Won, J.K. Ferri, J. Krägel, and R. Miller, "Dynamics of drops-Formation, growth, oscillation, detachment, and coalescence", *Adv. Colloid Interface Sci,* vol. 222, pp. 413-424, 2015. [http://dx.doi.org/10.1016/j.cis.2014.10.009]

[42] J. Eggers, "Nonlinear dynamics and breakup of free-surface flows", *Rev. Mod. Phys,* vol. 69, no. 3, pp. 865-930, 1997. [http://dx.doi.org/10.1103/RevModPhys.69.865]

[43] O.A. Basaran, "Small-scale free surface flows with breakup: Drop formation and emerging applications", *AIChE J,* vol. 48, no. 9, pp. 1842-1848, 2002. [http://dx.doi.org/10.1002/aic.690480902]

[44] A. Barrero, and I.G. Loscertales, "Micro- and nanoparticles via capillary flows", *Annu. Rev. Fluid Mech,* vol. 39, no. 1, pp. 89-106, 2007. [http://dx.doi.org/10.1146/annurev.fluid.39.050905.110245]

[45] E. Villermaux, "Fragmentation", *Annu. Rev. Fluid Mech,* vol. 39, no. 1, pp. 419-446, 2007. [http://dx.doi.org/10.1146/annurev.fluid.39.050905.110214]

[46] A. Atthasit, *Aircraft propulsion fuel injection,* 2014.

[47] L. Rayleigh, "Some Applications of Photography", *Nature,* vol. 44, no. 1133, pp. 249-254, 1891. [http://dx.doi.org/10.1038/044249e0]

[48] H.E. Edgerton, E.A. Hauser, and W.B. Tucker, "Studies in drop formation as revealed by the high-speed motion camera", *J. Phys. Chem,* vol. 41, no. 7, pp. 1017-1028, 1937. [http://dx.doi.org/10.1021/j150385a012]

[49] D.H. Peregrine, G. Shoker, and A. Symon, "The bifurcation of liquid bridges", *J. Fluid Mech,* vol. 212, pp. 25-39, 1990. [http://dx.doi.org/10.1017/S0022112090001835]

[50] X.D. Shi, M.P. Brenner, and S.R. Nagel, "A cascade of structure in a drop falling from a faucet", *Science,* vol. 265, no. 5169, pp. 219-222, 1994. [http://dx.doi.org/10.1126/science.265.5169.219]

[51] X. Zhang, and O.A. Basaran, "An experimental study of dynamics of drop formation", *Phys. Fluids,* vol. 7, no. 6, p. 1184, 1995. [http://dx.doi.org/10.1063/1.868577]

[52] E.D. Wilkes, S.D. Phillips, and O.A. Basaran, "Computational and experimental analysis of dynamics of drop formation", *Phys. Fluids,* vol. 11, no. 12, p. 3577, 1999. [http://dx.doi.org/10.1063/1.870224]

[53] P.K. Notz, A.U. Chen, and O.A. Basaran, "Satellite drops: Unexpected dynamics and change of scaling during pinch-off", *Phys. Fluids,* vol. 13, no. 3, p. 549, 2001. [http://dx.doi.org/10.1063/1.1343906]

[54] A.U. Chen, P.K. Notz, and O.A. Basaran, "Computational and experimental analysis of pinch-off and scaling", *Phys. Rev. Lett,* vol. 88, no. 17, 2002. [http://dx.doi.org/10.1103/PhysRevLett.88.174501]

[55] C. Clanet, and J.C. Lasheras, "Transition from dripping to jetting", *J. Fluid Mech,* vol. 383, pp. 307-326, 1999. [http://dx.doi.org/10.1017/S0022112098004066]

[56] S.P. Lin, and R.D. Reitz, "Drop and spray formation from a liquid jet", *Annu. Rev. Fluid Mech,* vol. 30, no. 1, pp. 85-105, 1998. [http://dx.doi.org/10.1146/annurev.fluid.30.1.85]

[57] R.M. Schulkes, "The evolution and bifurcation of a pendant drop", *J. Fluid Mech,* vol. 278, pp. 83-100, 1994. [http://dx.doi.org/10.1017/S0022112094003629]

[58] X. Zhang, "Dynamics of growth and breakup of viscous pendant drops into air", *J. Colloid Interface Sci,* vol. 212, no. 1, pp. 107-122, 1999. [http://dx.doi.org/10.1006/jcis.1998.6047]

[59] A. D'Innocenzo, F. Paladini, and L. Renna, "Experimental study of dripping dynamics", *Phys. Rev. E Stat. Nonlin. Soft Matter Phys,* vol. 65, no. 5, 2002.

[60] R.D. Pinto, W.M. Gonçalves, J.C. Sartorelli, and M.J. de Oliveira, "Hopf bifurcation in a leaky faucet experiment", *Phys. Rev. E Stat. Phys. Plasmas Fluids Relat. Interdiscip. Topics,* vol. 52, no. 6, pp. 6896-6899, 1995. [http://dx.doi.org/10.1103/PhysRevE.52.6896]

[61] K. Kiyono, T. Katsuyama, T. Masunaga, and N. Fuchikami, "Picture of the low-dimensional structure in chaotic dripping faucets", *Phys. Lett. A,* vol. 320, no. 1, pp. 47-52, 2003. [http://dx.doi.org/10.1016/j.physleta.2003.10.079]

[62] R. Shaw, *The dripping faucet as a model chaotic system.* Aerial Press: Santa Cruz, CA, 1984.

[63] P. Coullet, L. Mahadevan, and C. Riera, "Return map for the chaotic dripping faucet", *Prog. Theor. Phys,* vol. 139, suppl. Suppl., pp. 507-516, 2000. [http://dx.doi.org/10.1143/PTPS.139.507]

[64] J.C. Sartorelli, W.M. Gonçalves, and R.D. Pinto, "Crisis and intermittence in a leaky-faucet experiment", *Phys. Rev. E Stat. Phys. Plasmas Fluids Relat. Interdiscip. Topics,* vol. 49, no. 5, pp. 3963-3975, 1994. [http://dx.doi.org/10.1103/PhysRevE.49.3963]

[65] B. Ambravaneswaran, H.J. Subramani, S.D. Phillips, and O.A. Basaran, "Dripping-jetting transitions in a dripping faucet", *Phys. Rev. Lett,* vol. 93, no. 3, 2004. [http://dx.doi.org/10.1103/PhysRevLett.93.034501]

[66] A. D'Innocenzo, F. Paladini, and L. Renna, "Asymmetrical dripping", *Phys. Rev. E Stat. Nonlin. Soft Matter Phys,* vol. 69, no. 4, 2004.

[67] H.J. Subramani, H.K. Yeoh, R. Suryo, Q. Xu, B. Ambravaneswaran, and O.A. Basaran, "Simplicity and complexity in a dripping faucet", *Phys. Fluids,* vol. 18, no. 3, p. 32106, 2006. [http://dx.doi.org/10.1063/1.2185111]

[68] A. Shemesh, S. Akselrod, Z. Reich, D. Shahar, and R. Kapon, "Dripping faucet dynamics is determined by synchronization of drop oscillations and detachment", *Phys. Rev. E Stat. Nonlin. Soft Matter Phys,* vol. 86, no. 2, p. 026209, 2012. [http://dx.doi.org/10.1103/PhysRevE.86.026209]

[69] A. Taur, P. Doshi, and H.K. Yeoh, "Dripping dynamics of Newtonian liquids from a tilted nozzle", *Eur. J. Mech. BFluids,* vol. 51, pp. 8-15, 2015. [http://dx.doi.org/10.1016/j.euromechflu.2014.12.004]

[70] D.H. Peregrine, "Equations for water waves and the approximation behind them", In: *Waves on beaches and resulting sediment transport; proceedings of an advanced seminar conducted by the Mathematics Research Center, the University of Wisconsin, and the Coastal Engineering Research Center, U.S. Army, at Madison October 11-13,* R.E. Meyer, Ed., Academic Press: New York, 1972. [http://dx.doi.org/10.1016/B978-0-12-493250-0.50007-2]

[71] H.C. Lee, "Drop Formation in a Liquid Jet", *IBM J. Res. Develop,* vol. 18, no. 4, pp. 364-369, 1974. [http://dx.doi.org/10.1147/rd.184.0364]

[72] G.I. Barenblatt, *Scaling, self-similarity, and intermediate asymptotics.* Cambridge University Press: Cambridge, New York, 1996. [http://dx.doi.org/10.1017/CBO9781107050242]

[73] J.B. Keller, and M.J. Miksis, "Surface Tension Driven Flows", *SIAM J. Appl. Math,* vol. 43, no. 2, pp. 268-277, 1983. [http://dx.doi.org/10.1137/0143018]

[74] P. Constantin, T.F. Dupont, R.E. Goldstein, L.P. Kadanoff, M.J. Shelley, and S-M. Zhou, "Droplet breakup in a model of the Hele-Shaw cell", *Phys. Rev. E Stat. Phys. Plasmas Fluids Relat. Interdiscip. Topics,* vol. 47, no. 6, pp. 4169-4181, 1993. [http://dx.doi.org/10.1103/PhysRevE.47.4169]

[75] A.L. Bertozzi, M.P. Brenner, T.F. Dupont, and L.P. Kadanoff, "Singularities and Similarities in Interface Flows", In: *Trends and Perspectives in Applied Mathematics, L. Sirovich.* Springer: New York, 1994, pp. 155-208. [http://dx.doi.org/10.1007/978-1-4612-0859-4_6]

[76] J. Eggers, "Universal pinching of 3D axisymmetric free-surface flow", *Phys. Rev. Lett,* vol. 71, no. 21, pp. 3458-3460, 1993. [http://dx.doi.org/10.1103/PhysRevLett.71.3458]

[77] J. Eggers, and T.F. Dupont, "Drop formation in a one-dimensional approximation of the Navier-Stokes equation", *J. Fluid Mech,* vol. 262, p. 205, 1994. [http://dx.doi.org/10.1017/S0022112094000480]

[78] D. Gueyffier, J. Li, A. Nadim, R. Scardovelli, and S. Zaleski, "Volume-of-Fluid Interface Tracking with Smoothed Surface Stress Methods for Three-Dimensional Flows", *J. Comput. Phys,* vol. 152, no. 2, pp. 423-456, 1999. [http://dx.doi.org/10.1006/jcph.1998.6168]

[79] E.D. Wilkes, and O.A. Basaran, "Drop ejection from an oscillating rod", *J. Colloid Interface Sci,* vol. 242, no. 1, pp. 180-201, 2001. [http://dx.doi.org/10.1006/jcis.2001.7729]

[80] B. Ambravaneswaran, E.D. Wilkes, and O.A. Basaran, "Drop formation from a capillary tube: Comparison of one-dimensional and two-dimensional analyses and occurrence of satellite drops", *Phys. Fluids,* vol. 14, no. 8, p. 2606, 2002. [http://dx.doi.org/10.1063/1.1485077]

[81] Q. Xu, and O.A. Basaran, "Computational analysis of drop-on-demand drop formation", *Phys. Fluids,* vol. 19, no. 10, p. 102111, 2007. [http://dx.doi.org/10.1063/1.2800784]

[82] C. Pozrikidis, "Stability of sessile and pendant liquid drops", *J. Eng. Math,* vol. 72, no. 1, pp. 1-20, 2012. [http://dx.doi.org/10.1007/s10665-011-9459-3]

[83] R.L. Hines, "Electrostatic atomization and spray painting", *J. Appl. Phys,* vol. 37, no. 7, p. 2730, 1966. [http://dx.doi.org/10.1063/1.1782112]

[84] H. Yoon, J.H. Woo, B. Joshi, Y.M. Ra, S.S. Yoon, H.Y. Kim, S.J. Ahn, J.H. Yun, J. Gwak, K.H. Yoon, and S.C. James, "CuInSe2 (CIS) thin film solar cells by electrostatic spray deposition", *J. Electrochem. Soc,* vol. 159, no. 4, p. H444, 2012. [http://dx.doi.org/10.1149/2.jes113086]

[85] I.I. Inculet, "Industrial applications of static electricity", *J. Electrost,* vol. 16, no. 2-3, pp. 287-298, 1985. [http://dx.doi.org/10.1016/0304-3886(85)90052-X]

[86] U. Rothgordt, and G. Renelt, "Document printing", *Acta Electron. (Paris),* vol. 21, pp. 71-82, 1978.

[87] R.M. Wham, and C.H. Byers, "Mass transport from single droplets in imposed electric fields", *Sep. Sci. Technol,* vol. 22, no. 2-3, pp. 447-466, 1987. [http://dx.doi.org/10.1080/01496398708068963]

[88] J.D. Sartor, "Electricity and rain", *Phys. Today,* vol. 22, no. 8, p. 45, 1969. [http://dx.doi.org/10.1063/1.3035737]

[89] S. Cohen, F. Plasil, and W.J. Swiatecki, "Equilibrium configurations of rotating charged or gravitating liquid masses with surface tension. II", *Ann. Phys,* vol. 82, no. 2, pp. 557-596, 1974. [http://dx.doi.org/10.1016/0003-4916(74)90126-2]

[90] B. Franklin, *Experiments and observations on electricity, made at Philadelphia in America*. St. John's Gate: London, 1751. [http://dx.doi.org/10.5479/sil.211644.39088000092304]

[91] L. Rayleigh, "The influence of electricity on colliding water drops", *Proc. R. Soc. Lond,* vol. 28, no. 190-195, pp. 405-409, 1878. [http://dx.doi.org/10.1098/rspl.1878.0146]

[92] "On the equilibrium of liquid conducting masses charged with electricity", *Philos. Mag. Ser. 5,* vol. 14, no. 87, pp. 184-186, 1882. [http://dx.doi.org/10.1080/14786448208628425]

[93] J. Zeleny, "On the conditions of instability of liquid drops, with applications to the electrical discharge from liquid point", *Proc. Camb. Philos. Soc,* vol. 18, pp. 71-83, 1915.

[94] J. Zeleny, "Instability of Electrified Liquid Surfaces", *Phys. Rev,* vol. 10, no. 1, pp. 1-6, 1917. [http://dx.doi.org/10.1103/PhysRev.10.1]

[95] G. Taylor, "Disintegration of Water Drops in an Electric Field", *Proc. R. Soc. Math. Phys. Eng. Sci,* vol. 280, no. 1382, pp. 383-397, 1964. [http://dx.doi.org/10.1098/rspa.1964.0151]

[96] M.J. Miksis, "Shape of a drop in an electric field", *Phys. Fluids,* vol. 24, no. 11, p. 1967, 1981. [http://dx.doi.org/10.1063/1.863293]

[97] G. Joffre, B. Prunet-Foch, S. Berthomme, and M. Cloupeau, "Deformation of liquid menisci under the action of an electric field", *J. Electrost,* vol. 13, no. 2, pp. 151-165, 1982. [http://dx.doi.org/10.1016/0304-3886(82)90005-5]

[98] F.K. Wohlhuter, and O.A. Basaran, "Shapes and stability of pendant and sessile dielectric drops in an electric field", *J. Fluid Mech,* vol. 235, p. 481, 1992. [http://dx.doi.org/10.1017/S0022112092001198]

[99] O.A. Basaran, T.W. Patzek, R.E. Benner, and L.E. Scriven, "Nonlinear oscillations and breakup of conducting, inviscid drops in an externally applied electric field", *Ind. Eng. Chem. Res,* vol. 34, no. 10, pp. 3454-3465, 1995. [http://dx.doi.org/10.1021/ie00037a034]

[100] R. T. Collins, M. T. Harris, and O. A. Basaran, "Breakup of electrified jets", *J. Fluid Mech,* vol. 588, 2007. [http://dx.doi.org/10.1017/S0022112007007409]

[101] M. Cloupeau, and B. Prunet-Foch, "Electrohydrodynamic spraying functioning modes: a critical review", *J. Aerosol Sci,* vol. 25, no. 6, pp. 1021-1036, 1994. [http://dx.doi.org/10.1016/0021-8502(94)90199-6]

[102] M. Cloupeau, and B. Prunet-Foch, "Electrostatic spraying of liquids: Main functioning modes", *J. Electrost,* vol. 25, no. 2, pp. 165-184, 1990. [http://dx.doi.org/10.1016/0304-3886(90)90025-Q]

[103] B. Vonnegut, and R.L. Neubauer, "Production of monodisperse liquid particles by electrical atomization", *J. Colloid Sci,* vol. 7, no. 6, pp. 616-622, 1952. [http://dx.doi.org/10.1016/0095-8522(52)90043-3]

[104] J. Doshi, and D.H. Reneker, "Electrospinning process and applications of electrospun fibers", *J. Electrost,* vol. 35, no. 2-3, pp. 151-160, 1995. [http://dx.doi.org/10.1016/0304-3886(95)00041-8]

[105] R.G. Sweet, "High frequency recording with electrostatically deflected ink jets", *Rev. Sci. Instrum,* vol. 36, no. 2, p. 131, 1965. [http://dx.doi.org/10.1063/1.1719502]

[106] A. Jaworek, and A. Krupa, "Classification of the modes of EHD spraying", *J. Aerosol Sci,* vol. 30, no. 7, pp. 873-893, 1999. [http://dx.doi.org/10.1016/S0021-8502(98)00787-3]

[107] X. Zhang, and O.A. Basaran, "Dynamics of drop formation from a capillary in the presence of an electric field", *J. Fluid Mech,* vol. 326, p. 239, 1996. [http://dx.doi.org/10.1017/S0022112096008300]

[108] J-U. Park, M. Hardy, S.J. Kang, K. Barton, K. Adair, D. kishore Mukhopadhyay, C.Y. Lee, M.S. Strano, A.G. Alleyne, J.G. Georgiadis, P.M. Ferreira, and J.A. Rogers, "High-resolution electrohydrodynamic jet printing", *Nat. Mater,* vol. 6, no. 10, pp. 782-789, 2007. [http://dx.doi.org/10.1038/nmat1974]

[109] A. Lee, H. Jin, H-W. Dang, K-H. Choi, and K.H. Ahn, "Optimization of experimental parameters to determine the jetting regimes in electrohydrodynamic printing", *Langmuir,* vol. 29, no. 44, pp. 13630-13639, 2013.
[http://dx.doi.org/10.1021/la403111m]

[110] S. Mishra, K.L. Barton, A.G. Alleyne, P.M. Ferreira, and J.A. Rogers, "High-speed and drop-on-demand printing with a pulsed electrohydrodynamic jet", *J. Micromech. Microeng,* vol. 20, no. 9, p. 95026, 2010.
[http://dx.doi.org/10.1088/0960-1317/20/9/095026]

[111] U. Stachewicz, J.F. Dijksman, C.U. Yurteri, and J.C. Marijnissen, "Experiments on single event electrospraying", *Appl. Phys. Lett,* vol. 91, no. 25, p. 254109, 2007.
[http://dx.doi.org/10.1063/1.2826279]

[112] J. Kim, H. Oh, and S.S. Kim, "Electrohydrodynamic drop-on-demand patterning in pulsed cone-jet mode at various frequencies", *J. Aerosol Sci,* vol. 39, no. 9, pp. 819-825, 2008.
[http://dx.doi.org/10.1016/j.jaerosci.2008.05.001]

[113] U. Stachewicz, C.U. Yurteri, J.C. Marijnissen, and J.F. Dijksman, "Stability regime of pulse frequency for single event electrospraying", *Appl. Phys. Lett,* vol. 95, no. 22, p. 224105, 2009.
[http://dx.doi.org/10.1063/1.3269594]

[114] J. Choi, Y-J. Kim, S. Lee, S.U. Son, H.S. Ko, V.D. Nguyen, and D. Byun, "Drop-on-demand printing of conductive ink by electrostatic field induced inkjet head", *Appl. Phys. Lett,* vol. 93, no. 19, p. 193508, 2008.
[http://dx.doi.org/10.1063/1.3020719]

[115] H. Kim, J. Yang, and J. Chung, "Resonant oscillation phenomena of a sessile droplet on electrohydrodynamic jetting nozzle", *Jpn. J. Appl. Phys,* vol. 53, no. 583, p. 05HC03, 2014.
[http://dx.doi.org/10.7567/JJAP.53.05HC03]

[116] K. Rahman, A. Khan, N.M. Nam, K.H. Choi, and D-S. Kim, "Study of drop-on-demand printing through multi-step pulse voltage", *Int. J. Precis. Eng. Manuf,* vol. 12, no. 4, pp. 663-669, 2011.
[http://dx.doi.org/10.1007/s12541-011-0086-8]

[117] H. Kim, J. Song, J. Chung, and D. Hong, "Onset condition of pulsating cone-jet mode of electrohydrodynamic jetting for plane, hole, and pin type electrodes", *J. Appl. Phys,* vol. 108, no. 10, p. 102804, 2010.
[http://dx.doi.org/10.1063/1.3511685]

[118] S. Lee, J. Song, H. Kim, and J. Chung, "Time resolved imaging of electrohydrodynamic jetting on demand induced by square pulse voltage", *J. Aerosol Sci,* vol. 52, pp. 89-97, 2012.
[http://dx.doi.org/10.1016/j.jaerosci.2012.05.006]

[119] W.J. Scheideler, and C-H. Chen, "The minimum flow rate scaling of Taylor cone-jets issued from a nozzle", *Appl. Phys. Lett,* vol. 104, no. 2, p. 24103, 2014.
[http://dx.doi.org/10.1063/1.4862263]

[120] A.J. Hijano, I.G. Loscertales, S.E. Ibáñez, and F.J. Higuera, "Periodic emission of droplets from an oscillating electrified meniscus of a low-viscosity, highly conductive liquid", *Phys. Rev. E Stat. Nonlin. Soft Matter Phys,* vol. 91, no. 1, 2015.
[http://dx.doi.org/10.1103/PhysRevE.91.013011]

[121] P.K. Notz, and O.A. Basaran, "Dynamics of Drop Formation in an Electric Field", *J. Colloid Interface Sci,* vol. 213, no. 1, pp. 218-237, 1999.
[http://dx.doi.org/10.1006/jcis.1999.6136]

[122] R.T. Collins, J.J. Jones, M.T. Harris, and O.A. Basaran, "Electrohydrodynamic tip streaming and emission of charged drops from liquid cones", *Nat. Phys,* vol. 4, no. 2, pp. 149-154, 2008.
[http://dx.doi.org/10.1038/nphys807]

[123] R.T. Collins, K. Sambath, M.T. Harris, and O.A. Basaran, "Universal scaling laws for the

disintegration of electrified drops", *Proc. Natl. Acad. Sci. USA,* vol. 110, no. 13, pp. 4905-4910, 2013. [http://dx.doi.org/10.1073/pnas.1213708110]

[124] M.W. Lee, N.Y. Kim, and S.S. Yoon, "On pinchoff behavior of electrified droplets", *J. Aerosol Sci,* vol. 57, pp. 114-124, 2013. [http://dx.doi.org/10.1016/j.jaerosci.2012.09.003]

[125] Y-J. Kim, J. Choi, S.U. Son, S. Lee, X.H. Nguyen, V.D. Nguyen, D. Byun, and H.S. Ko, "Comparative study on ejection phenomena of droplets from electro-hydrodynamic jet by hydrophobic and hydrophilic coatings of nozzles", *Jpn. J. Appl. Phys,* vol. 49, no. 6, p. 60217, 2010. [http://dx.doi.org/10.1143/JJAP.49.060217]

[126] M.S. Onses, E. Sutanto, P.M. Ferreira, A.G. Alleyne, and J.A. Rogers, "Mechanisms, capabilities, and applications of high-resolution electrohydrodynamic jet printing", *Small,* vol. 11, no. 34, pp. 4237-4266, 2015. [http://dx.doi.org/10.1002/smll.201500593]

[127] C.D. Han, *Multiphase flow in polymer processing.* Academic Press: New York, 1981.

[128] M.S. Oliveira, M.A. Alves, and F.T. Pinho, "Microfluidic Flows of Viscoelastic Fluids", In: *Transport and Mixing in Laminar Flows,* R. Grigoriev, Ed., Wiley-VCH Verlag GmbH & Co. KGaA: Weinheim, Germany, 2011, pp. 131-174. [http://dx.doi.org/10.1002/9783527639748.ch6]

[129] O.A. Basaran, H. Gao, and P.P. Bhat, "Nonstandard Inkjets", *Annu. Rev. Fluid Mech,* vol. 45, no. 1, pp. 85-113, 2013. [http://dx.doi.org/10.1146/annurev-fluid-120710-101148]

[130] D. Izbassarov, and M. Muradoglu, "A front-tracking method for computational modeling of viscoelastic two-phase flow systems", *J. Non-Newt. Fluid Mech,* vol. 223, pp. 122-140, 2015. [http://dx.doi.org/10.1016/j.jnnfm.2015.05.012]

[131] D. Izbassarov, and M. Muradoglu, "A computational study of two-phase viscoelastic systems in a capillary tube with a sudden contraction/expansion", *Phys. Fluids,* vol. 28, no. 1, p. 12110, 2016. [http://dx.doi.org/10.1063/1.4939940]

[132] J. Eggers, and E. Villermaux, "Physics of liquid jets", *Rep. Prog. Phys,* vol. 71, no. 3, p. 36601, 2008. [http://dx.doi.org/10.1088/0034-4885/71/3/036601]

[133] P.P. Bhat, S. Appathurai, M.T. Harris, M. Pasquali, G.H. McKinley, and O.A. Basaran, "Formation of beads-on-a-string structures during break-up of viscoelastic filaments", *Nat. Phys,* vol. 6, no. 8, pp. 625-631, 2010. [http://dx.doi.org/10.1038/nphys1682]

[134] R. Sattler, S. Gier, J. Eggers, and C. Wagner, "The final stages of capillary break-up of polymer solutions", *Phys. Fluids,* vol. 24, no. 2, p. 23101, 2012. [http://dx.doi.org/10.1063/1.3684750]

[135] M. Goldin, J. Yerushalmi, R. Pfeffer, and R. Shinnar, "Breakup of a laminar capillary jet of a viscoelastic fluid", *J. Fluid Mech,* vol. 38, no. 4, p. 689, 1969. [http://dx.doi.org/10.1017/S0022112069002540]

[136] Y. Amarouchene, D. Bonn, J. Meunier, and H. Kellay, "Inhibition of the finite-time singularity during droplet fission of a polymeric fluid", *Phys. Rev. Lett,* vol. 86, no. 16, pp. 3558-3561, 2001. [http://dx.doi.org/10.1103/PhysRevLett.86.3558]

[137] S.L. Anna, and G.H. McKinley, "Elasto-capillary thinning and breakup of model elastic liquids", *J. Rheol. (N.Y.N.Y.),* vol. 45, no. 1, p. 115, 2001. [http://dx.doi.org/10.1122/1.1332389]

[138] P.P. Bhat, S. Appathurai, M.T. Harris, and O.A. Basaran, "On self-similarity in the drop-filament corner region formed during pinch-off of viscoelastic fluid threads", *Phys. Fluids,* vol. 24, no. 8, p. 83101, 2012.

[http://dx.doi.org/10.1063/1.4745179]

[139] O.E. Yildirim, and O.A. Basaran, "Deformation and breakup of stretching bridges of Newtonian and shear-thinning liquids: comparison of one- and two-dimensional models", *Chem. Eng. Sci,* vol. 56, no. 1, pp. 211-233, 2001. [http://dx.doi.org/10.1016/S0009-2509(00)00408-5]

[140] P. Coussot, and F. Gaulard, "Gravity flow instability of viscoplastic materials: The ketchup drip", *Phys. Rev. E Stat. Nonlin. Soft Matter Phys,* vol. 72, no. 3, 2005. [http://dx.doi.org/10.1103/PhysRevE.72.031409]

[141] M.E. Möbius, "Clustering instability in a freely falling granular jet", *Phys. Rev. E Stat. Nonlin. Soft Matter Phys,* vol. 74, no. 5, 2006. [http://dx.doi.org/10.1103/PhysRevE.74.051304]

[142] G.H. McKinley, and T. Sridhar, *Filament-stretching rheometry of complex fluids.* vol. 34. 2002, pp. 375-415.

[143] G.H. Mckinley, "Visco-elasto-capillary thinning and break-up of complex fluids", *Br. Soc. Rheol,* pp. 1-49, 2005.

[144] N.F. Morrison, and O.G. Harlen, "Viscoelasticity in inkjet printing", *Rheol. Acta,* vol. 49, no. 6, pp. 619-632, 2010. [http://dx.doi.org/10.1007/s00397-009-0419-z]

[145] J. Eggers, "Instability of a polymeric thread", *Phys. Fluids,* vol. 26, no. 3, p. 33106, 2014. [http://dx.doi.org/10.1063/1.4869721]

[146] J.J. Cooper-White, J.E. Fagan, V. Tirtaatmadja, D.R. Lester, and D.V. Boger, "Drop formation dynamics of constant low-viscosity, elastic fluids", *J. Non-Newt. Fluid Mech,* vol. 106, no. 1, pp. 29-59, 2002. [http://dx.doi.org/10.1016/S0377-0257(02)00084-8]

[147] L.B. Smolka, and A. Belmonte, "Drop pinch-off and filament dynamics of wormlike micellar fluids", *J. Non-Newt. Fluid Mech,* vol. 115, no. 1, pp. 1-25, 2003. [http://dx.doi.org/10.1016/S0377-0257(03)00116-2]

[148] M.S. Oliveira, and G.H. McKinley, "Iterated stretching and multiple beads-on-a-string phenomena in dilute solutions of highly extensible flexible polymers", *Phys. Fluids,* vol. 17, no. 7, p. 71704, 2005. [http://dx.doi.org/10.1063/1.1949197]

[149] M.S. Oliveira, R. Yeh, and G.H. McKinley, "Iterated stretching, extensional rheology and formation of beads-on-a-string structures in polymer solutions", *J. Non-Newt. Fluid Mech,* vol. 137, no. 1-3, pp. 137-148, 2006. [http://dx.doi.org/10.1016/j.jnnfm.2006.01.014]

[150] V. Tirtaatmadja, G.H. McKinley, and J.J. Cooper-White, "Drop formation and breakup of low viscosity elastic fluids: Effects of molecular weight and concentration", *Phys. Fluids,* vol. 18, no. 4, p. 43101, 2006. [http://dx.doi.org/10.1063/1.2190469]

[151] A.Y. Malkin, A. Arinstein, and V.G. Kulichikhin, "Polymer extension flows and instabilities", *Prog. Polym. Sci,* vol. 39, no. 5, pp. 959-978, 2014. [http://dx.doi.org/10.1016/j.progpolymsci.2014.02.006]

[152] H.A. Stone, "Dynamics of drop deformation and breakup in viscous fluids", *Annu. Rev. Fluid Mech,* vol. 26, no. 1, pp. 65-102, 1994. [http://dx.doi.org/10.1146/annurev.fl.26.010194.000433]

[153] M. Muradoglu, and G. Tryggvason, "A front-tracking method for computation of interfacial flows with soluble surfactants", *J. Comput. Phys,* vol. 227, no. 4, pp. 2238-2262, 2008. [http://dx.doi.org/10.1016/j.jcp.2007.10.003]

[154] S. Tasoglu, U. Demirci, and M. Muradoglu, "The effect of soluble surfactant on the transient motion of a buoyancy-driven bubble", *Phys. Fluids,* vol. 20, no. 4, p. 40805, 2008. [http://dx.doi.org/10.1063/1.2912441]

[155] U. Olgac, and M. Muradoglu, "Effects of surfactant on liquid film thickness in the Bretherton problem", *Int. J. Multiph. Flow,* vol. 48, pp. 58-70, 2013. [http://dx.doi.org/10.1016/j.ijmultiphaseflow.2012.08.007]

[156] U. Olgac, and M. Muradoglu, "Computational modeling of unsteady surfactant-laden liquid plug propagation in neonatal airways", *Phys. Fluids,* vol. 25, no. 7, p. 71901, 2013. [http://dx.doi.org/10.1063/1.4812589]

[157] M. Muradoglu, and G. Tryggvason, "Simulations of soluble surfactants in 3D multiphase flow", *J. Comput. Phys,* vol. 274, pp. 737-757, 2014. [http://dx.doi.org/10.1016/j.jcp.2014.06.024]

[158] W. Wirth, S. Storp, and W. Jacobsen, "Mechanisms controlling leaf retention of agricultural spray solutions", *Pestic. Sci,* vol. 33, no. 4, pp. 411-420, 1991. [http://dx.doi.org/10.1002/ps.2780330403]

[159] T. Goldmann, and J.S. Gonzalez, "DNA-printing: utilization of a standard inkjet printer for the transfer of nucleic acids to solid supports", *J. Biochem. Biophys. Methods,* vol. 42, no. 3, pp. 105-110, 2000. [http://dx.doi.org/10.1016/S0165-022X(99)00049-4]

[160] J.Y. Shin, and N.L. Abbott, "Using Light to Control Dynamic Surface Tensions of Aqueous Solutions of Water Soluble Surfactants", *Langmuir,* vol. 15, no. 13, pp. 4404-4410, 1999. [http://dx.doi.org/10.1021/la981477f]

[161] Y-C. Liao, H.J. Subramani, E.I. Franses, and O.A. Basaran, "Effects of Soluble Surfactants on the Deformation and Breakup of Stretching Liquid Bridges", *Langmuir,* vol. 20, no. 23, pp. 9926-9930, 2004. [http://dx.doi.org/10.1021/la0487949]

[162] Y-C. Liao, E.I. Franses, and O.A. Basaran, "Deformation and breakup of a stretching liquid bridge covered with an insoluble surfactant monolayer", *Phys. Fluids,* vol. 18, no. 2, p. 22101, 2006. [http://dx.doi.org/10.1063/1.2166657]

[163] D.E. Weidner, "The effect of surfactant convection and diffusion on the evolution of an axisymmetric pendant droplet", *Phys. Fluids,* vol. 24, no. 6, p. 62104, 2012. [http://dx.doi.org/10.1063/1.4729449]

[164] A. Javadi, M. Karbaschi, D. Bastani, J.K. Ferri, V.I. Kovalchuk, N.M. Kovalchuk, K. Javadi, and R. Miller, "Marangoni instabilities for convective mobile interfaces during drop exchange: Experimental study and CFD simulation", *Colloid Surface A,* vol. 441, pp. 846-854, 2014. [http://dx.doi.org/10.1016/j.colsurfa.2012.10.032]

# CHAPTER 6

# Multiple Drug Resistance: An Up-Date

**Ruxandra Pirvulescu[1,*], Mihaela Oana Romanitan[2] and Alina Popa-Cherecheanu[1]**

[1] *University of Medicine and Pharmacy "Carol Davila", Emergency University Hospital, Ophthalmology Clinic, Bucharest, Romania*

[2] *Södersjukhuset AB, Internmedicin kliniken, Sektion för Neurologi, Stockholm, Sweden*

**Abstract:** Multiple Drug Resistance (MDR) is acquired by bacteria, viruses, fungi, parasites and malignant tumours at interaction with antibiotics/medicines of old or new generation. Here is presented an up-date of MDR reported on bacteria, viruses, fungi, parasites and tumors, in this last field with emphasis on ophthalmology and neurology/neurosurgery. Multidrug resistant organisms exhibit *in vitro* resistance to one or more antimicrobial agents. One cause is the increasing use and misuse of antibiotics on humans and animals. Whereas particular bacteria are naturally resistant to some antibiotics, MDR occurs in other cases by accumulation of resistant plasmids and/or of genes, each gene determining resistance to a specific agent. The action of efflux pumps able to pump out more than one drug type is also a possible mechanism involved in MDR. In general, MDR is the most important "process" by which tumors acquire resistance to drugs during chemotherapy. Bacterial resistance to antibiotics used in ophthalmology has been reported since more than 10 years showing that several bacteria resistant to antibiotics were found in isolates from ocular infections. In neuroscience, development of new therapies to treat brain infections is more difficult. The most important cause of failure in developing new drugs for treating brain diseases is the existence and action of blood brain barrier (BBB). Brain tumors have usually poor prognosis and due to BBB, drug delivery to brain tumors is difficult. Some studies mention that BBB is involved in drug restriction to different brain neoplasias. The chapter concludes about the need to improve the arsenal conceived to overcome MDR acquired by different biological targets.

**Keywords:** Blood brain barrier, Cancer, Chemotherapy, Chemoresistance, DNA, Efflux pumps, Fungi, Gram-negative bacteria, Gram-positive bacteria, Hepatitis viruses, Herpes viruses, HIV, Malaria, Multiple drug resistance, Ophthalmology, Plasmodia species, *Pseudomonas aeruginosa*, *Saccharomyces cerevisiae*, *Staphylococcus aureus* (MRSA), Tuberculosis.

* **Corresponding author Ruxandra Pirvulescu:** University of Medicine and Pharmacy "Carol Davila", Emergency University Hospital, Ophthalmology Clinic, Bucharest, Romania; Tel: 0040 21 3180562; Fax: 0040 21 3180554; E-mail: ruxandrapascu78@gmail.com

**Mihail Lucian Pascu (Ed.)**

## DEFINITION

Multiple drug resistance (MDR), also known as Multiresistance or Multi-drug resistance is a condition developed by microorganisms such as viruses, bacteria, fungi or parasites, whereby they acquire resistance to distinct antimicrobials (antibiotics, antiviral, antifungal or antiparasitic drugs) [1]. There are different degrees of MDR, starting with extensively-drug resistant (XDR) and including pandrug-resistant (PDR) organisms [2]. The European Centre for Disease-Prevention and Control (ECDC) and the Center for Disease Control and Prevention (CDC) in the United States cooperated in finding a standardised international terminology for defining and describing acquired resistance profiles in microorganisms that are more likely to develop it (*Staphylococcus aureus, Enterobacteriaceae, Enterococcus spp., Pseudomonas aeruginosa* and *Acinetobacter spp.)* [2]. Epidemiologically, significant antimicrobial categories were built - up for each bacterium.

Consequently, the following definitions are used:

- MDR: acquired non-susceptibility to at least one agent in three or more antimicrobial categories.
- XDR: non-susceptibility to at least one agent in all but two or fewer anti-microbial categories (bacterial isolates remain susceptible to only one or two categories).
- PDR: non-susceptibility to all agents in all antimicrobial categories.

Besides the mentioned microorganisms, a special place is occupied by extended-spectrum β-lactamase (ESBLs) that induces multidrug resistance of Gram-negative bacteria and of tuberculosis. To ensure correct application of these definitions, is indicated that bacterial isolates are tested against all antimicrobial agents that belong to the mentioned antimicrobial categories (Fig. **1**). It is also recommended to avoid selective reporting and suppression of results [2].

## MULTIPLE DRUG RESISTANCE – GENERALITIES

Multiple drug resistance has become a serious public health issue, as more and more antimicrobial drugs became inefficient in treating the infections caused by infecting agents. When discussing about bacteria, there are microorganisms from both Gram negative and Gram positive category which are incriminated. Since this problem grows with each year, so does the need for accurate and complete definitions of MDR that describe and classify resistant bacteria, so that epidemiological surveillance data can be collected and compared across all healthcare settings and countries. Multi drug resistant organisms (MDROs) are labeled as such because of their *in vitro* resistance to one or more antimicrobial agents. Infections with MDROs can lead to inadequate treatments and can be associated with poorer results on patients. On the other hand, it is mention the restrained number of antimicrobial agents that are available or that are currently developed [2].

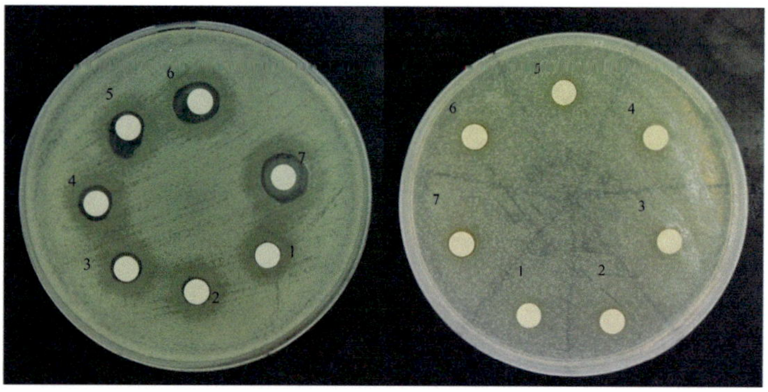


**Fig. (1).** Resistance test: bacteria are streaked on the culture media within Petri dishes. (Left) 30 min irradiated 20 mg/ml Chlorpromazine with 266 nm laser beam - bacteria susceptible to the antibiotic in each disc (dark, clear rings signify that bacteria have not grown); (Right) 240 min irradiated 20 mg/ml Chlorpromazine with 266 nm laser beam - bacteria are fully susceptible to solution; volume of solution applied on discs 1: 5 μl, 2: 10 μl, 3: 15 μl, 4: 20 μl, 5: 25 μl, 6: 30 μl, 7: 35 μl.

## MDR Acquired by Bacteria

Drug resistance is and has been an acquired pattern whereby microorganisms adapted, evolved and survived for thousands of years. During the Twentieth century, life expectancy has dramatically increased along with the discovery of antibiotics; antimicrobial therapy acquired critical importance in fighting infectious diseases produced by bacteria. However, even from the very beginning, resistance occurred and bacteria found more modalities to resist antibiotics and other antimicrobial drugs. One cause of the development of resistance to drugs is the increasing use and misuse of already produced antibiotics in human and veterinary medicine, as well as in agriculture and related fields.

In 1998, in the United States, 80 million prescriptions of antibiotics for human use were filled [3]. This means 12,500 tons of antibiotics per year. Animal and agricultural use of antibiotics is added to human use. On the other hand, agriculture accounts for more than 60% of antibiotics consumption in U.S. and this adds 18,000 tons per year to the antibiotic charge sent towards the environment [3]. Data recently available show an alarming picture: nowadays, 70% of bacteria that cause infections in hospitals are resistant to at least one drug that belongs to the most commonly used antibiotics for treatment. Some microorganisms are resistant to all approved antibiotics and can only be treated with experimental drugs that may be very toxic [3]. All these facts are continually leading to high investments in research, in order to develop new effective drugs against MDR bacteria. Recent studies show that another possibly effective method to combat MDR could be the use of non-antibiotic drugs, already developed for other purposes, their exposure to UV laser beams and then, their use to treat infected tissues as well as the target bacteria [4].

Since the discovery of penicillin in 1928, many other antibiotics were reported/discovered, produced and commercialised. Unfortunately, nowadays, the public general opinion considers that any infectious disease is curable if submitted to antibiotic therapy. Antibiotics are manufactured at a scale of more or less 100,000 tons per annum all over the world and in any case, their use had and currently has an important influence on the life of bacteria on earth. Due to their excessive use, more strains of pathogens are today antibiotic resistant, and some became resistant to a multitude of antibiotics and other chemical agents used in therapies. With respect to this phenomenon, the multidrug (or otherwise called multiple drug) resistance was defined [5].

There are some bacterial strains that in practical terms became resistant to all commonly available agents. A very well known case is the methicillin-resistant *Staphylococcus aureus* (MRSA) that is resistant not only to methicillin (which was initially developed to fight penicillinase-producing *S. aureus*) but also to macrolides, chloramphenicol, tetracycline, aminoglycosides, and lincosamides. Such strains are also resistant to disinfectants and MRSA is incriminated for a large part of hospital-acquired infections. Vancomycin, a relatively old antibiotic, was recovered for treatment of MRSA infections. However, transferable resistance to vancomycin is today quite common in *Enterococcus* and, in rare cases, even in MRSA [5].

Another very serious threat is represented by Gram-negative pathogens which are resistant to all available agents [6]. Fortunately, there is a lot of research concentrated on fighting this resistance, especially on MRSA. There are new developed agents that are active against vancomycin-resistant MRSA, such as linezolid and quinupristin/dalfopristin. However, "pan-resistant" Gram-negative strains were recently identified, particularly those belonging to *Pseudomonas aeruginosa* and *Acinetobacter baumanii*. This was observed after the decision of most major pharmaceutical companies to stop the development of new antibacterial agents. At the moment, agents that could be used against these strains are hardly available, since bacteria developed as defence "instruments" an outer membrane barrier of low permeability and an array of efficient multidrug efflux pumps which are combined with multitudes of specific resistance mechanisms [5].

Multidrug resistance in bacteria occurs by the accumulation of resistance (R) plasmids or transposons of genes, with each coding for resistance to a specific agent. The action of multidrug efflux pumps each of which can pump out more than one drug type [7] is another mechanism of action. Some bacteria are naturally resistant to certain antibiotics. Other bacteria may become resistant in two ways: (i) by genetic mutation and (ii) by DNA transfer, which involves acquiring resistance from another bacteria. By either of these two paths, antibiotics become inefficient. Seldom spontaneous changes of bacteria's genetic material may occur in approximately one in one million to one in ten million cells. Different genetic mutations are related to different types of acquired resistance. There are several known mechanisms that generate genetic mutations:

- Mutations enable bacteria to produce potent chemicals (enzymes) that inactivate antibiotics (enzymatic deactivation);
- Mutations modify glycoprotein cell wall; this decreases the cell wall permeability to antibiotics;
- Other mutations can alter target sites of antibiotic;
- Mechanisms that include the close-up of the entry ports which allow antibiotics into the cell and manufacture further pumping mechanisms that export antibiotic back outside cell (so that it never reaches its target), modifying the efflux pumps to remove antibiotics;
- The increase of the mutation rate as a stress response [1, 8].

By DNA transfer, bacteria are able to send copies of their DNA that code a mechanism of resistance to other bacteria even at distance, which then are also able to pass on the resistance genes by a similar mechanism; thus, generations of bacteria that resist to antibiotics may be produced (Fig. **2**).

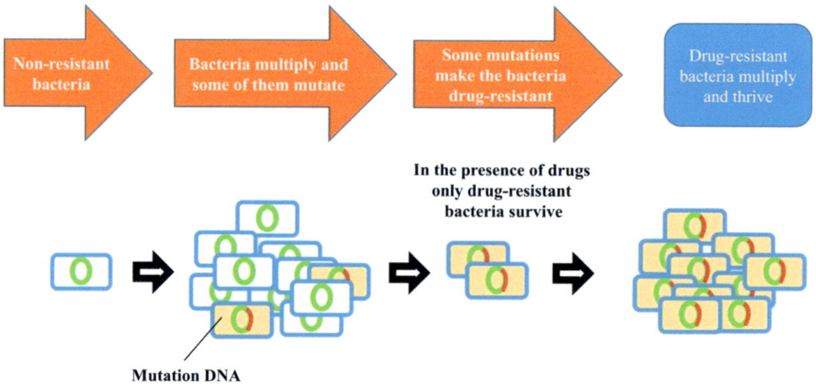


**Fig. (2).** How mutation takes place [6].

This process is known as horizontal gene transfer [1, 8]. By undergoing a simple mating process called "conjugation," bacteria can transfer genetic material, including genes encoding resistance to antibiotics (found on plasmids and transposons) from one member of bacteria community to another. Viruses' channels constitute another mechanism that may be activated to transfer resistance properties between bacteria. The resistance traits from one bacterium are packed into the head portion of the virus which then injects the resistance traits into any new bacteria it aims. Bacteria also have the ability to acquire naked, "free" DNA from their environment [8]. Any bacteria that acquire resistance genes, either by spontaneous mutation or by genetic exchange with other bacteria, have the ability to resist at least one antibiotic. Since bacteria can collect multiple resistance traits in time, they can become resistant to a lot of families of antibiotics (Fig. **3**) [8].

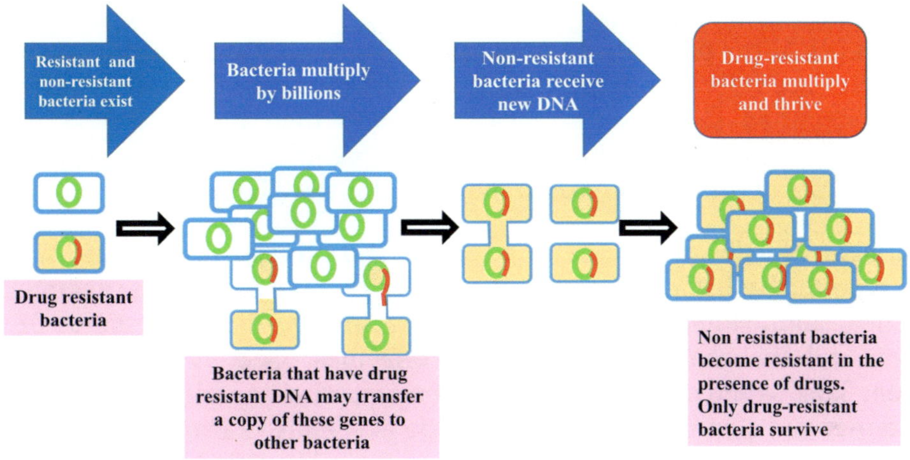


**Fig. (3).** DNA transfer and drug resistance [6].

Unless antibiotic resistance problems are detected as they emerge and actions are taken immediately to solve or limit them, human society could be faced with previously treatable diseases that have become again untreatable, as before the development of antibiotics.

## MDR Acquired by Viruses

Viral infections are treated with specific antiviral drugs. The purpose of these drugs is to prevent viruses from reproduction by inhibiting essential stages of their replication cycle in infected cells. The main known viruses that require antiviral treatment are HIV, hepatitis B, hepatitis C, herpes viruses (including varicella zoster virus), influenza, Epstein-Barr virus and cytomegalovirus. In each case, there are some strains that acquire resistance to the administered drugs [1].

One of the most problematic and complicated antiviral resistance is that of HIV. The use of antiretroviral therapy has led to a major improvement in the rates of morbidity and mortality of individuals infected with HIV. If only one antiviral drug is used, resistance emerges rapidly, whereas using three or more drugs together the problem may be fixed temporarily. However, new drugs are needed due to the fact that drug-resistant HIV strains are continuously emerging [1].

Viral fitness refers to the overall capacity of the virus to infect, replicate and produce mature infectious progeny in a given, well defined host environment [9]. When mutations leading to resistance appear within the reverse transcriptase (RT) and protease gene as a consequence of antiviral therapy, it is subsequently produced a reduction of viral fitness. Observations led to the hypothesis that transmission of resistant viruses occurs, in a proportion of approximately 80%, less frequently than expected if compared to the population of potential transmitters [9]. Virus populations with low replicative capacity seem however to be able to produce infection. It has also been shown that the drug resistant viruses may be transmitted significantly less frequently than the wild-type viruses. Furthermore, it is established that differences in treatment practices, if any, may influence the rate of drug resistance in the untreated population [9].

The treatment of HIV infections is complex and change fast in line with the progress reported by basic science research and with the experience accumulated from clinical applications. It is vital to understand the different stages of viral replication and the characteristics of the enzymes used by the virus for its replication. This would allow to identify the agents that can block the function of these enzymes and stop viral replication inside host cells. Various events that take place in HIV life cycle and in host factors have been identified as potential targets of applied antiretroviral therapy. The complexity of treatment dedicated to HIV infection increases continuously with the rising number of available compounds and with the accumulation of new information about their use [10].

Viral mutations occur daily and randomly, but many of them are harmless. As shown previously, most mutations actually create disadvantages for HIV, since they reduce the viral fitness and slow down virus ability to infect CD4 cells within the body. On the contrary, some mutations can give HIV an advantage to survive when specific medications are used, because these mutations can block drugs from working against HIV enzymes for which they are designed to target. This kind of mutations is of maximum concern in acquired drug resistance [11].

HIV counts on more enzymes to replicate inside a human cell; it also relies on proteins, such as gp41, to latch onto CD4 cells and infect them. Mutations may occur in any of these virus components causing drug resistance. Examples are:

- Reverse transcriptase: nucleoside reverse transcriptase inhibitors (NRTIs) and non-nucleoside reverse transcriptase inhibitors (NNRTIs) target this enzyme;
- Integrase: targeted by integrase inhibitors;
- Protease: targeted by protease inhibitors;
- gp 41- fusion inhibitors target it on HIV's outer wall.

HIV is a retrovirus that uses the enzyme reverse transcriptase to synthesise DNA from its RNA genome. There is not a mechanism developed to correct errors that may appear at the reproduction of its genome. Hence, HIV replicates its genome with the highest known mutation rate of any living organism. This is an ideal situation for natural selection to act on HIV population because genetic variations constitute the raw material for natural selection. The mutations that accumulate over many generations lead to great genetic variation within HIV populations. This produces an increased probability to develop evolutionary selective advantages with respect to other viruses. Natural selection then acts on HIV by selecting virions with higher fitness, as all others are eventually eliminated by treatments with drugs. The virions that escape the harmful effects of the drugs, form afterwards an entirely new, drug resistant population. They reproduce until the patient has a population of viruses as large as they originally had before the antiviral treatment reduced their numbers. This produces a cycle in which patients first experience success with treatment and their viral levels decreases, followed by the second stage when a decline in treatment effectiveness as the virus develops resistance and rebuilds its population [12].

In the case of people infected with HIV, the drug resistance can make drugs less effective or even completely ineffective against viruses, thus significantly reducing treatment options.

Drugs belonging to five different classes are currently available and several others are close to be released. Inhibitors of (i) HIV reverse transcriptase and of (ii) HIV protease are until now the most developed and most commonly used in clinical practice. Currently, newer drugs belonging to these categories are in development stages, having better efficacy against the virus and sustained activity against viral resistance. There are drugs with longer half-life that penetrate body compartments producing less systemic side effects to the host. They have higher antiviral potency, lower side effects and simpler tablet presentations. These may improve the long-term treatment adherence and assure longer patient survival [10].

## MDR Acquired by Fungi

Resistance of pathogenic fungi to antibiotics represents an issue of tremendous importance, when we refer to controlling the infections caused by these organisms.

Resistance to antibiotics acquired by pathogenic fungi requires special attention to control infections caused by them. Scientists were able to isolate mutant fungi in laboratory and in clinic, testing their resistance to a wide range of antibiotics besides those initially used for treatment. Unfortunately, the restrained number of antifungal drugs makes this phenotype an acute problem in the chemotherapeutic eradication of fungal infections.

Most data available on multidrug resistance in fungi originate from the studies on generally nonpathogenic yeast *Saccharomyces cerevisiae*, in which the multidrug-resistant phenotype is referred to as pleiotropic drug resistance or PDR. Genes that influence this phenotype are typically designated as PDR loci. The development and use of new genetic and molecular biology techniques opened new perspectives on the physiology of multidrug resistance based on experiments performed directly on pathogenic organisms such as *S. cerevisiae*, *Candida albicans*, *Candida glabrata*, or *Aspergillus fumigatus* [13].

Development of antifungal drugs has been slowed down by the striking similarity between the fundamental cell biology of a fungus and a typical mammalian cell. Antifungal drugs such as amphotericin B, one of the earliest developed, were efficacious in treating fungal infections but proved to induce high toxicity to the patient [14]. Drugs that are recently developed, like the azoles and echinocandins, are very efficient against pathogens but their use has led to the selection of highly resistant fungi to these antifungal agents [15]. Azole drugs are the primary antifungals used to treat fungal infections. Infections generated by azole resistant fungi dramatically lowered survival rates compared to those associated with azole susceptible organisms [16, 17]. Resistance to azoles is acquired by two different mechanisms: a direct one and an indirect mechanism. The direct mechanism refers to changes in the enzyme which is targeted by azole drugs, such as lanosterol 14α demethylase. This enzyme is encoded by *ERG11* gene in *Saccharomyces cerevisiae* with azole resistant alleles found to either alter expression levels of this gene product or change the sequence of this protein. Recent studies show that selection for azole resistant yeast elicits isolation of strains that overproduce ATP-binding cassette (ABC) transporter proteins. ABC transporter proteins act on efflux azole drugs found in the cell and prevent their accumulation at otherwise toxic levels [14].

Overproduction of ABC transporter is a well-known cause of the development of drug resistance in mammalian cells. Cancer patients submitted to chemotherapeutic treatment develop often tumours that, eventually, are proved to be resistant to multiple treatment. These tumours, selected *in vivo*, are refractory to growth inhibition produced by most chemotherapeutics, even for cases of drugs to which the patient was not previously exposed. It has been found that tumours resistant to multiple treatment may be associated with gene amplification events that produce an increase in the copy number of genes encoding ABC transporters with a concomitant elevation in protein level [14].

Multidrug resistant cells can be quite easily selected from fungal cells. These mutant strains do not have, usually, increased gene dosage but more probable show an increased expression of ABC transporter-encoding genes [18, 19]. High gene expression leads to an increase in the protein level of membrane transporters which is accompanied by a broad range drug tolerance. In the studied fungi, ABC transporters influence azole resistance making this a key issue in treatment of fungal infections.

There is a wide range of genes involved in producing multidrug resistance, but the understanding of how they regulate processes is incomplete, yet. There are cases where the identities of the transcription factors that control multidrug resistance genes are well-known, but a detailed picture of how the activity of these factors is regulated is not available. Thus, finding these factors and discovering their actions is an important future goal. There is a lot of progress in medical care regarding all kinds of drugs, which led to higher chronic reliance on antimicrobial chemotherapy, given the fact that fungal infections are more and more often encountered in medical practice.

*Candida* bloodstream infections are, at the moment, the 4th most common form of reported nosocomial infection. Aspergillosis is also rising and outcomes associated with azole tolerant forms of this pathogen are dramatically worsened with respect to azole susceptible fungi cases [14]. The continuous progress in understanding antifungal drug resistance at molecular level is an important step towards ensuring the sustainability of antifungal drug therapy and expecting a successful treatment of these microbial pathogens in the future.

## Multidrug Resistance Acquired by Tumours

Multidrug resistance is the main mechanism by which many tumours acquire resistance to drugs used in chemotherapy, and represents a major factor in the failure of many forms of chemotherapy. It affects patients who show blood cancers and/or solid tumours, like ovarian, breast, lung, and lower gastrointestinal tract cancers. Tumours include usually mixed populations of malignant cells, some of which are drug-sensitive while others are drug-resistant. Chemotherapy destroys drug-sensitive cells, but leaves behind a higher proportion of drug-resistant ones so that when a tumour begins to grow again, chemotherapy may fail because the remaining cells are now dominantly resistant [20].

Some studies revealed that resistance to therapy with drugs is correlated with the presence of at least two molecular "pumps" in tumour-cell membranes that actively evacuate chemotherapy drugs from within the cell (Fig. **4**). Thus, tumour cells avoid the drug toxic effects, as well as molecular processes within the nucleus or the cytoplasm. These two pumps that are responsible for chemoresistance in cancer are P-glycoprotein and the so-called multidrug resistance–associated protein (MRP). Because of their function and importance, they are the subjects of a significant number of anticancer researches [20].

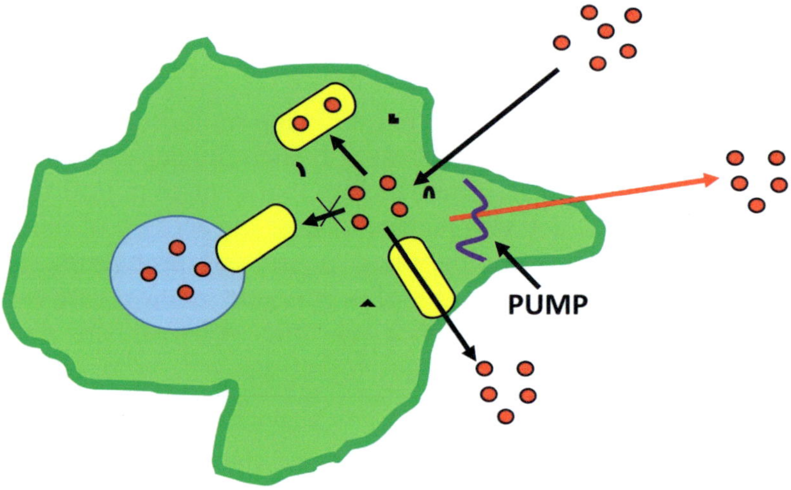


**Fig. (4).** MDR mechanisms. Drug molecules (small red circles) are entering the cell and, eventually, the cell nucleus. Vaults could be involved in MDR by: 1. Sequestering the drug; 2. Blocking the entering of drug in the nucleus; 3. Directing the drug out of cell; 4. Assisting drug efflux pumps (as P-gp) in drug efflux [26].

MDR has been also associated with the overexpression of transmembrane transporter molecules which act as pumps to remove toxic drugs from tumour cells. These include P-glycoprotein (P-gp), multidrug resistance protein (MRP), breast cancer resistance protein (BCRP), and Lung Resistance-related Protein (LRP) [21] isolated and having sequenced cDNA (complementary DNA) coding for the LRP gene and found to be human major vault protein. Since then, there have been reported many studies on the relationship between vaults and MDR. In studies on a variety of human cell lines and tumours, major vault protein (MVP) levels had the greatest individual value as a marker of *in vitro* and clinical resistance to chemotherapy. Some studies show that MVP could be a prognostic marker of response to chemotherapy [22], while others failed to show any such correlation [23]. Vaults seem also to be implicated in nucleocytoplasmic transport of cytotoxic drugs. Recent findings show that human vRNAs (viral RNAs) (hvg-1 and hvg-2) bind to mitoxantrone (a chemotherapeutic drug) [24]. Until now, vRNA is the only vault component that seems to bind directly to a chemotherapeutic compound. Despite the strong connection between vaults and MDR, there have been no studies demonstrating the direct involvement of vaults in imparting MDR to cells, except a single study reported in [25]. The study is using the SW620 colon carcinoma cells, and it shows that overexpression of MVP upon treatment of cells with sodium butyrate, conferred resistance to doxorubicin due to a rapid efflux of the drug from the nuclei of these cells. Using MVP-specific ribozymes to knockdown MVP expression in cells, the study demonstrated that doxorubicin was more rapidly removed from nuclei of cells that overexpressed MVP compared to those treated with ribozymes. Further, it showed that knocking down MVP levels resulted in the reversal of the drug-resistant phenotype of the sodium butyrate treated cells. However, reproduction of these results in other laboratories failed [20, 26]. On the other hand, unlike in [27, 28] is shown that disruption of the murine MVP gene is not conferring sensitivity to cytostatic drugs. Recently, in [29] was reported a demonstration that knocking down the expression of MVP using siRNAs (small interfering RNAs) did not change the ability of drug-resistant cells to efflux doxorubicin from the nucleus and it did not increase the drug-sensitivity of these cells. It has been also shown that upregulation of MVP in sensitive cells (by sodium butyrate) did not lead to a drug-resistant phenotype. In addition, other studies showed that more overexpression of MVP in cells does not confer drug resistance [30, 31]. All of these data therefore suggest that despite their upregulation in MDR cells and tumours, vaults do not play a direct role in establishing a MDR phenotype.

Development of MDR with connection to chemotherapy remains a major challenge in cancer treatment. Other known mechanisms by which tumour cells gain resistance are: decreased drug uptake, increased drug efflux, activation of detoxifying systems, activation of DNA repair mechanisms and evasion of drug-induced apoptosis. In the first part of this chapter, is briefly summarised the current knowledge regarding individual cellular mechanisms responsible for MDR, with a special emphasis on ATP-binding cassette transporters. Although extensive work has been done to characterise MDR mechanisms *in vitro*, the translation of this knowledge to the clinic has not been crowned with success. Therefore, identifying genes and mechanisms critical to the development of MDR *in vivo* and establishing a reliable method for analysing clinical samples might help predicting the development of resistance and lead to treatments designed to defeat it. The pleotropic response of cancer cells to chemotherapy is shown in an eloquent diagram (Fig. **4**) [20, 26].

## MDR Acquired by Parasites

The main example for MDR acquired against antiparasitic drugs is malaria which along with HIV and tuberculosis, is an important cause of morbidity and mortality, especially among children. This disease is caused by the protozoan parasite *Plasmodium*, and is transmitted by an Anopheline mosquito vector. The five *Plasmodia* species that affect humans are *Plasmodium falciparum*, *Plasmodium vivax*, *Plasmodium ovale*, *Plasmodium malariae*, and *Plasmodium knowlesi*, together causing approximately 225 million incident infections per year, leading to approximately one million deaths [32]. *Plasmodium vivax* has become a few decades ago resistant to chloroquine and sulfadoxine-pyrimethamine and emerged in 2012 as artemisinin-resistant *Plasmodium falciparum* in western Cambodia and western Thailand. *Toxoplasma gondii* may also become resistant to artemisinin, atovaquone and sulfadiazine, but is not usually multidrug resistant. Antihelminthic resistance is mainly reported in veterinary applications, for instance in connection with the practice of livestock drenching. It has been recently the focus of FDA regulation [1].

*P. falciparum* is the most prevalent malaria species worldwide, particularly in Africa, and it is responsible for triggering the most severe form of the disease responsible for over 90% of the deaths. *P. vivax* is the second most common species, located mainly in Asia and South America, and can cause a relapsing form of malaria. Resistance acquired by *Plasmodium* depends on various factors, such as: mutation rate of parasite, fitness costs associated with resistance mutations, overall parasite load, strength of drug selection, and treatment compliance [32].

The parasite's mutation rate has direct influence on the frequency of acquired resistance. Higher mutation rates enable faster emergence of resistance and may lead to accumulation of deleterious mutations. Generally, an increased mutation rate is advantageous for bacteria adaptation to a quickly changing environment. Reports are published on "accelerated resistance to multiple drugs" (ARMD) phenotype that is present in areas in Southeast Asia. The drug resistance is acquired at a higher rate than other geographically distinct strains in *in vitro* experiments [32].

Mutations associated with drug resistance often imply fitness costs and the selective advantage acquired by becoming drug-resistant is balanced by the biological cost arising from altered function of mutated protein. Such fitness costs can be mitigated by the acquisition of compensatory mutations during prolonged drug action [32].

It has also been shown that high parasitemias (*P. falciparum* infections rise to $10^{10}$–$10^{12}$ the parasites number within a host) can lead to faster elimination of deleterious mutations and enhance selection of compensatory mutations. On the other hand, the occurrence of drug-resistant parasites can also be accelerated by strong drug selection pressure, which decreases the prevalence of competing sensitive wild type parasites.

Inadequate drug exposure by improper dosing, poor pharmacokinetic properties, fake drugs, or infections acquired during drug elimination phase of a prior antimalarial treatment may be responsible for exposure of parasites to sub-optimal drug concentrations. In these cases, the probability for drug-resistant parasites to spring is high [32].

Transmission of parasites is another critical step for the emergence of drug-resistant parasites. The transmission intensity has an important role to insure that parasites are effectively transmitted during a mosquito blood meal. The drug-resistant parasites that emerge in a high-transmission area could most probably be present in a polyclonal infection. Drug resistance may enhance transmission if drug selection pressure lowers viability of sensitive gametocytes in a polyclonal infection, increasing the propensity for transmitting drug-resistant parasites [32].

There are more strategies that can be applied to ensure a maximum effect of currently used antimalarial compounds; one of them implies the use of multiple first-line treatments, that could lead to varying drug selection pressure on local parasite populations and could decrease the emergence and spread of parasites resistant to any particular antimalarial drugs. Another important factor is the use of combination therapies, which decreases the selection of drug-resistant parasites *via* the action of partner drugs. Another strategy could be the use of combinations of drugs that exert opposing drug resistance pressures, such is the case of chloroquine and lumefantrine. Adherence to the full treatment regimen is essential to delay the emergence of drug-resistant parasites and apply adequate treatment of individual infections.

A previously proposed strategy to extend the useful lifespan of currently available antimalarial list of procedures is the reintroduction of drugs after the prevalence of resistant parasites has waned. This assumes that drug-resistant parasites would carry a fitness cost, compared to sensitive parasites, and would consequently be outcompeted without the continued advantage during drug selection pressure [32]. This would also require that compensatory mutations had not already increased the fitness of resistant parasites.

All these strategies need sustained financial and political support as well as the understanding of the genetic mechanisms and pharmacological factors that influence treatment efficacy. In such cases are needed continuous efforts to monitor and flexibly address the emergence of drug-resistant parasites.

## Multiple Drug Resistance in Ophthalmology

When first appeared in the middle of the 20th century, antibiotics were considered "miracle" drugs. Though, in time, bacterial resistance emerged, and, unfortunately, nowadays resistance keeps increasing at an alarming rate. Bacteria such as methicillin-resistant *Staphylococcus aureus* (MRSA), *vancomycin-resistant S. aureus*, *Staphylococcus epidermidis*, *Enterococci*, and multidrug-resistant bacteria as well, represent a non-negligible challenge to the utilised treatments of infectious diseases.

There are many studies, such as Tracking Resistance in the U.S. Today (TRUST), Prospective Resistant Organism Tracking and Epidemiology for the Ketolide Telithromycin (PROTEKT), the international SENTRY Antimicrobial Surveillance Program, and Active Bacterial Core that document the increase in bacterial resistance. Some of them, such as PROTEKT, insisted to track the mechanisms of drug resistance [33]. One of the known mechanisms is the acquisition of -lactamases, which are enzymes that hydrolyse β-lactam antibiotics (*e.g.*, penicillin) and which have evolved rapidly and spread among bacteria. The adaptive changes in β-lactamases are, luckily, exclusive to one type of target; one antibiotic may be effective if another is destroyed by these enzymes [33].

As shown previously, bacteria are also able to alter the target site of a particular antibiotic. For example, mutations in bacterial ribosomes, the targets of aminoglycoside (*e.g.* streptomycin) and macrolide (*e.g.* erythromycin) antibiotics can decrease the ability of the antibiotic to bind the ribosomal subunit and may inhibit synthesis of proteins. In fluoroquinolone antibiotics, the major targets are DNA gyrase and topoisomerase IV which are enzymes affecting primarily DNA replication. If one or both enzymes got mutations, resistance occurs which is manifested by preventing the binding of a fluoroquinolone to the enzyme [33, 34].

Bacterial resistance increases in parallel to both systemic and topical antibiotics, antibiotics for ocular infections included, such as bacterial conjunctivitis.

In Table **1** are listed bacteria that are most commonly found in most frequent ocular infections [33].

**Table 1. List of bacteria found in most frequent ocular infections.**

| Keratitis | Conjunctivitis | Endophthalmitis |
|---|---|---|
| *Staphylococcus aureus* | *Staphylococcus aureus* | *Staphylococcus aureus* |
| Coagulase-Negative *Staphylococcus* | Coagulase-Negative *Staphylococcus* | *Staphylococcus epidermidis* |
| *Streptococcus pneumoniae* | *Staphylococcus epidermidis* | *Streptococcus pneumoniae* |

*(Table 1) contd.....*

| Keratitis | Conjunctivitis | Endophthalmitis |
|---|---|---|
| *Streptococci viridans* | *Streptococcus pneumoniae* | *Streptococci viridans* |
| *Corynebacterium* spp. | *Haemophilus influenzae* | *Bacillus cereus* |
| *Propionibacterium* spp. | *Moraxella* spp. | *Enterobacter* spp. |
| *Pseudomonas aeruginosa* | Gram-negative bacilli | *Pseudomonas aeruginosa* |
| *Serratia marcescens* | | *Propionibacterium acnes* |
| *Proteus mirabilis* | | *Haemophilus influenzae* |
| *Haemophilus influenzae* | | *Neisseria meningitidis* |
| *Moraxella* spp. | | |

Bacterial resistance to antibiotics used in ophthalmology is well documented worldwide for more than one decade. Studies made in 1990s in North and South America, India, Italy, and Japan found in isolates extracted from ocular infections, several bacteria resistant to antibiotics; *staphylococci* (resistant to oxacillin, gentamicin, tobramycin, and erythromycin), *S. pneumoniae* (penicillin resistant), coagulase-negative *staphylococci* (CoNS) (methicillin resistant), and *H influenzae* (-lactamase positive). Last years' studies show that, unfortunately, bacterial resistance in ophthalmic infections are more and more frequent; in [35] is reported an increase in MRSA isolates from pediatric patients presenting bacterial conjunctivitis from 4.4% in 1994 to 42.9% in 2003 (see also [33]). Studies have also shown the increase in resistance presented by *S. aureus* isolates to ciprofloxacin, erythromycin, and oxacillin. In [36] is likewise reported an increase of MRSA in ocular infections and it is shown that MRSA is often community acquired (CA-MRSA) [33]. Monitoring of pediatric cases of bacterial conjunctivitis has shown that 29% of *H influenzae* isolates were -lactamase positive, and 60% of *S. pneumoniae* isolates were penicillin resistant. A survey of ocular bacterial pathogens in 10 European countries found high rates of resistance of MRSA isolates to ciprofloxacin, ofloxacin, and gatifloxacin [32]; 31.3% of CoNS isolates were resistant to the older fluoroquinolones ciprofloxacin and ofloxacin, compared to 2.1% to the newer fluoroquinolone gatifloxacin. Multidrug resistance was also monitored by Ocular TRUST program that evaluates the annual change of *in vitro* antimicrobial susceptibilities in national samples of ocular isolates. The most recently published report on Ocular TRUST program shows that 75% to 85% of investigated MRSA isolates were resistant to ciprofloxacin, levofloxacin, and moxifloxacin [32]. Resistance to ciprofloxacin was found in 65.4% of MRSA isolates and in 45.9% of methicillin-resistant *S. epidermidis* isolates among recent clinical isolates from three bacterial conjunctivitis studies. The increase in multidrug resistance among ocular isolates requires now the development of new ophthalmic anti-infectives.

## Resistance to Fluoroquinolones

The mechanism of action of antibiotics against bacteria provides keys to fight bacterial resistance [37]. In Gram-positive bacteria, older fluoroquinolones such as ofloxacin and ciprofloxacin, while targeting both topoisomerase IV and DNA gyrase, preferentially target topoisomerase IV, while the newer moxifloxacin and gatifloxacin target both these enzymes. Therefore, in order to develop high-level resistance to newer fluoroquinolones, bacteria must generate resistant mutations in both enzymes. The newer fluoroquinolones also achieve high tissue concentrations relative to their minimum inhibitory concentration (MIC) values, which are generally two- to three-fold lower than the MIC of older fluoroquinolones [37]. Nevertheless, there are already reports of ocular pathogen resistance to the new fluoroquinolones, gatifloxacin and moxifloxacin.

In order to fight multiple resistance, it is necessary to understand the reasons for triggering bacterial resistance to topical antibiotics in the case of ocular infections, so that new strategies could be developed in order to minimise the occurrence of resistance. Topical fluoroquinolones, which are often first choice in treating ocular surface infections, need to be carefully administrated. Strategies should be considered to improve the activity of currently used fluoroquinolones, and new fluoroquinolones need to be considered, as well.

## Multiple Drug Resistance in Neuroscience

Human nervous system is incredibly complex; therefore, when one speaks about multidrug resistance, one should be aware that there can be exceptions for almost any general mechanism. Brain itself is the most delicate organ of human body. Some diseases like encephalitis, stroke, neurological disorders, multiple sclerosis and tumour lead to the damage of the brain function. The development of new therapies for these diseases is a difficult task and unfortunately, there is no effective treatment for most of the brain diseases. In most cases the major cause of failure in the development of new drugs for treatment of brain diseases is the presence of BBB (blood brain barrier). Brain tumours have usually poor prognosis, which varies according to the type and grade of the tumour. Due to the presence of BBB, drug delivery to brain tumours has been and still is a problematic issue. Some studies mentioned BBB as a controversial problem for brain tumour chemotherapy. They indicated that BBB is not the only factor responsible for impeding the success of brain tumour chemotherapy, but later, they revealed the involvement of BBB in drug restriction to different brain neoplasias [37].

BBB was discovered in 1880s; it protects neural tissues in the brain and works as a diffusion barrier that stops the influx of toxins and other compounds from blood to the brain. It took almost 70 years to successfully prove the existence of BBB by electron microscopic cytochemical studies. Later, in [38] is explained the initial understanding about the uniqueness of BBB tight junction and its physiology [37].

BBB is a system that includes two types of cellular junctions, the intercellular adherens junction and the paracellular tight junction. The functional integrity of BBB is maintained by adherens junction that is composed of vascular endothelium (VE), cadherin, actinin, and catenin. The major functionality of BBB is maintained by tight junctions, as they are primarily responsible for permeability through BBB. The BBB in adult is part of a complex cellular network. The main components of this system are brain endothelial cells, highly specialised basal membrane, pericytes embedded in the basal membrane and astrocytic end-feet [37].

In the brain of vertebrates one finds two types of cells that are known to bear TJs or TJs-like structures: vascular endothelial cells and oligodendrocytes [39]. TJs networks are formed by integral membrane proteins such as junctional associated membrane protein 1 [40], occludin [41] and claudin family [42]. Extensive research was dedicated to this special domain under the last 10 years and it has been shown that proteins such as occludin and claudins are expressed not only at the level of cerebral endhotelium, oligodendrocytes and astrocytes in human brains, but also in the neurons [43, 44] and they are overexpressed in neurological conditions that are affecting BBB such as Alzheimer disease or vascular dementia.

BBB may be crossed by endogenous compounds and drugs by different mechanisms such as passive diffusion, carrier-mediated transport (GLUT1 mediated transport), endocytosis, and active transport. Participation of various transport proteins is signalled in most of these transport systems. These different transport proteins of brain mediate the uptake and extrusion of various metabolites and compounds. The efflux and influx transporter systems of BBB are ATP-binding cassette (ABC) transporters and solute carrier (SLC) transporters [37].

The ABC transporters include the P-glycoprotein (P-gp), the breast cancer resistance protein, as well as the members of the multidrug resistance related proteins (MRPs).

P-gp plays a major role in the clinical resistance to chemotherapy, since it is encoded by multidrug resistant (MDR) genes. Two MDR isoforms have been identified in human tissues: MDR-1 and MDR-2. MDR1 encoded P-gp is a major efflux transporter of BBB, the expression of which evolved most probably to protect the brain from exposure to potentially neurotoxic xenobiotics. Thus, one considers that P-gp has a key role in maintaining accurate homeostatic environment necessary for proper neuronal function. The MDR1 gene product is 1280 amino acids long and has two homologous halves, each consisting of six transmembrane domains and ATP-binding site. On the first extracellular loop, two to four glycosylation sites are present. In the brain, P-gp is localized to the luminal and abluminal sides of BBB endothelium and to the apical plasma membrane of choroid plexus epithelial cells. Substrates of P-gp are usually nonpolar, weakly amphipathic compounds which significantly vary in molecular size. The different types of endogenous substrates of P-gp include cytokines, lipids, steroid hormones, and peptides. P-gp has a vast endogenous and exogenous substrate profile that renders difficulty in drug delivery across the BBB [37].

The MRP family has 9 homologues, generally designated as MRP1–9, and these isoforms have overlapping substrate profiles. Out of these, expression of MRP1–6 has been observed in human brain [45], whereas multiple MRPs like MRP1, 4, and 5 have been detected in the human BBB [46 - 48]. Existence of other MRPs, namely, MRP2, 3, and 6 and along with these MRP1, 4, and 5, has also been noticed in other vertebrates like rat, cow, pig, and fish. But presence of them in human BBB is still questionable. Structural similarity can be observed in MRP1, 2, 3, and 6, as each of them possesses 3 transmembrane domains (TMD) designated as TMD0, TMD1, and TMD2, respectively. TMD1 and TMD2 contain 6 alpha helices, whereas TMD0 contains only 5 alpha helices [49, 50]. It is believed that TMDs are assembled in the plasma membrane pore through which the transport of substrates occurs [49]. On the contrary, MRP4 and MRP5 have structural similarity with P-gp that lack TMD0 [49], but in all the MRP homologues, the conserved cytoplasmic linker (L0) portion is essential for transport function. MRP1, 4, and 5 are restricted to the luminal membrane of human brain capillary endothelial cells [50]. The localization of MRPs suggests that they play a crucial role in drug efflux transport through BBB.

The ability of drug to cross BBB depends on factors like molecular size (it should be less than 500 Da), charge (drugs should have low hydrogen bonding capabilities), and lipophilicity (drugs should have high lipophilicity) [49]. Thus, chemical modification of brain tumour drugs refers to the process of making an existing drug smaller in size, more perfectly charged, and more lipid soluble [50]. Existing brain tumour drugs may also be modified to make analogue of the ligand to the particular receptor present in the BBB or the ligand or a peptide can be linked to a drug against the cellular receptors of BBB. The drug melphalan has been modified by using this approach where melphalan nitrogen mustard (mechlorethamine) was linked to phenylalanine [51]. Another approach of drug modification is the use of lipid carriers for efficient transport through BBB. One example of such modification is incorporation of small drugs in fatty acids like N-docosahexaenoic acid (DHA) [52, 53]. Drugs are also modified so that they acquire increased capillary permeability, but after crossing BBB they undergo an enzymatic reaction and return to their active state. This approach is also known as prodrug therapy [54, 55].

The new era of brain cancer therapy is characterised by novel target specific drugs with efficient delivery strategies. Unfortunately, despite all progress made, the prognosis and median survival of the brain tumour patients are not satisfactory till date. One of the reasons may be the molecular heterogeneity of the brain tumours, the presence of CSCs, and the lack of effective drug delivery because of the presence of BBB. Rapid progress is needed in the sector of brain tumour characterisation and BBB research. Up to now, most effective drugs for brain tumour therapies are temozolomide, procarbazine, carmustine (BCNU), lomustine (CCNU), and vincristine. There are more and more studies involved in better modification of these drugs or identification of new chemical entities with enhanced efficacy and low side effects. Alternatively, identification of drugs which can modulate BBB components or transporter systems could be an effective future strategy. Another potential future approach is combinatorial therapy, where through BBB destruction/modification, tumour cells/CSCs could be targeted easily. Lately, there are modern techniques like nanotherapy that apparently facilitate this kind of approach. Therefore, future research is needed to focus on the development of more specific targeting strategies to cure brain cancer, overcoming the above-mentioned difficulties that arise in connection to the presence of BBB [56, 57].

## CONCLUSIONS

Multiple drug resistance is an issue of highest actuality given the more and more aggressive nature of microorganisms attacks on biological targets and in particular on human targets. A literature survey on the methods and means available to-date show that in fighting MDR acquired by bacteria, parasites, viruses, fungi and tumours one should consider that there are interactions between these resistant entities which enhance their individual adaptability and resistance (particularly if defined on statistical bases) to treating medicines and even procedures. In humans, another problem evidenced by MDR treatments is the systemic toxic effects of the utilised medicines on human organisms, so that the possible increase of dosages for treatment may destroy the target microorganisms and/or tumours but may have also negative impacts on the overall host treated organism. To avoid these possible consequences, it is recommended to find new transport vectors of medicines to targets so that the strictly necessary quantity of drug is delivered to the target. More, it is also recommended to use drugs that are very effective so that smallest quantities of substances recommended for treatments are yielded to targets such as infected volumes, MDR tumours or bacteria cultures. Therefore, having available very effective drugs in a rapid and flexible way in order to overpass MDR becomes a major research subject. One possible approach along this line is the use of medicines (that may be antibiotics or non-antibiotics) for which MDR was observed or which may be effective at very high doses that are in the toxic range, modified by exposure to laser radiation. This process leads to generation of photoproducts that either single or by synergistic action may be efficient against MDR microorganisms or tumours [58 - 61]. The new photoproducts, in mixtures or individually, may be kept as solutions in water or other biocompatible fluid and delivered to targets as microvolumetric or even nanovolumetric droplets produced in controlled ways [62].

In any case, fighting MDR remains a priority task in order to cope with infections produced by microorganisms resistant to treatment or with contamination of hospital spaces or other environmental media, either as a consequence of working procedures or by accident.

## CONFLICT OF INTEREST

The authors confirm that this chapter content has no conflict of interest.

## ACKNOWLEDGEMENTS

This work has been financed by the project PN-II-PT-PCCA-2011-3.1-1350.

## REFERENCES

[1] "Multiple drug resistance", *Wikipedia, the free encyclopedia,* 2016. Available at: https://en.wikipedia.org/w/index.php?title=Multiple_drug_resistance&oldid=737106264

[2] A.P. Magiorakos, A. Srinivasan, R.B. Carey, Y. Carmeli, M.E. Falagas, C.G. Giske, S. Harbarth, J.F. Hindler, G. Kahlmeter, B. Olsson-Liljequist, D.L. Paterson, L.B. Rice, J. Stelling, M.J. Struelens, A. Vatopoulos, J.T. Weber, and D.L. Monnet, "Multidrug-resistant, extensively drug-resistant and pandrug-resistant bacteria: an international expert proposal for interim standard definitions for acquired resistance", *Clin. Microbiol. Infect,* vol. 18, no. 3, pp. 268-281, 2012. [http://dx.doi.org/10.1111/j.1469-0691.2011.03570.x] [PMID: 21793988]

[3] T. Kenneth, *Bacterial Resistance to Antibiotics*. Available at: http://textbookofbacteriology.net/resantimicrobial.html

[4] R.A. Pascu, M. Trifu, M. Dumitrescu, A. Mahamoud, A. Staicu, B. Carstocea, and M.L. Pascu, "In vivo studies of the effects of alkyl substituted benzo[b]pyridinium ompounds exposed to optical radiation", *Rom. Rep. Phys,* vol. 60, no. 3, pp. 899-908, 2008.

[5] H. Nikaido, "Multidrug resistance in bacteria", *Annu. Rev. Biochem,* vol. 78, no. 1, pp. 119-146, 2009. [http://dx.doi.org/10.1146/annurev.biochem.78.082907.145923] [PMID: 19231985]

[6] C.P. Davis, "Antibiotic Resistance (Drug Resistance, Antimicrobial Resistance) Symptoms, Treatment, Causes - Causes of antimicrobial drug resistance", *MedicineNet*. Available at: http://www.medicinenet.com/antibiotic_resistance/article.htm

[7] D.M. Livermore, "The need for new antibiotics", *Clin. Microbiol. Infect,* vol. 10, suppl. Suppl. 4, pp. 1-9, 2004. [http://dx.doi.org/10.1111/j.1465-0691.2004.1004.x] [PMID: 15522034]

[8] *General Background: About Antibiotic Resistance.* Available at: http://emerald.tufts.edu/med/apua/about_issue/about_antibioticres.shtml

[9] A.M. Geretti, Ed., *Antiretroviral Resistance in Clinical Practice.* Mediscript: London, 2006.

[10] H. Taha, A. Das, and S. Das, "Clinical effectiveness of dolutegravir in the treatment of HIV/AIDS", *Infect. Drug Resist,* vol. 8, pp. 339-352, 2015.
[PMID: 26491363]

[11] "HIV Drug Resistance Explained", *POZ.* Available at: https://www.poz.com/basics/hiv-basics/hi--drug-resistance

[12] "HIV drug resistance", In: *Wikipedia, the free encyclopedia*2016.

[13] K. Gulshan, and W.S. Moye-Rowley, "Multidrug resistance in fungi", *Eukaryot. Cell,* vol. 6, no. 11, pp. 1933-1942, 2007.
[http://dx.doi.org/10.1128/EC.00254-07] [PMID: 17873085]

[14] S. Paul, and W.S. Moye-Rowley, "Multidrug resistance in fungi: regulation of transporter-encoding gene expression", *Front. Physiol,* vol. 5, p. 143, 2014.
[http://dx.doi.org/10.3389/fphys.2014.00143] [PMID: 24795641]

[15] G.D. Brown, D.W. Denning, and S.M. Levitz, "Tackling human fungal infections", *Science,* vol. 336, no. 6082, pp. 647-647, 2012.
[http://dx.doi.org/10.1126/science.1222236] [PMID: 22582229]

[16] M.A. Pfaller, "Antifungal drug resistance: mechanisms, epidemiology, and consequences for treatment", *Am. J. Med,* vol. 125, no. 1, suppl. Suppl., pp. S3-S13, 2012.
[http://dx.doi.org/10.1016/j.amjmed.2011.11.001] [PMID: 22196207]

[17] M.A. Pfaller, M. Castanheira, S.R. Lockhart, A.M. Ahlquist, S.A. Messer, and R.N. Jones, "Frequency of decreased susceptibility and resistance to echinocandins among fluconazole-resistant bloodstream isolates of Candida glabrata", *J. Clin. Microbiol,* vol. 50, no. 4, pp. 1199-1203, 2012.
[http://dx.doi.org/10.1128/JCM.06112-11] [PMID: 22278842]

[18] D. Sanglard, A. Coste, and S. Ferrari, "Antifungal drug resistance mechanisms in fungal pathogens from the perspective of transcriptional gene regulation", *FEMS Yeast Res,* vol. 9, no. 7, pp. 1029-1050, 2009.
[http://dx.doi.org/10.1111/j.1567-1364.2009.00578.x] [PMID: 19799636]

[19] J. Morschhäuser, "Regulation of multidrug resistance in pathogenic fungi", *Fungal Genet. Biol,* vol. 47, no. 2, pp. 94-106, 2010.
[http://dx.doi.org/10.1016/j.fgb.2009.08.002] [PMID: 19665571]

[20] "Cancer multidrug resistance", *Nat. Biotechnol,* vol. 18, suppl. IT18-20, 2000.
[PMID: 11221708]

[21] R.J. Scheper, G.L. Scheper, M.J. Flens, P. Van der Valk, C.J. Meijer, H.M. Pinedo, M.L. Sloval, L.H. Rome, R.H. Shoemarker, and M.A. Izquierdo, "Role of LRP-major vault protein in multidrug resistance", In: *Multidrug Resistance in Cancer Cells: Cellular, Biochemical, Molecular and Biological Aspects*John Wiley & Sons: New York, 1996.

[22] M.A. Izquierdo, A.G. van der Zee, J.B. Vermorken, P. van der Valk, J.A. Beliën, G. Giaccone, G.L. Scheffer, M.J. Flens, H.M. Pinedo, and P. Kenemans, "Drug resistance-associated marker Lrp for prediction of response to chemotherapy and prognoses in advanced ovarian carcinoma", *J. Natl. Cancer Inst,* vol. 87, no. 16, pp. 1230-1237, 1995.
[http://dx.doi.org/10.1093/jnci/87.16.1230] [PMID: 7563169]

[23] G.L. Scheffer, A.B. Schroeijers, M.A. Izquierdo, E.A. Wiemer, and R.J. Scheper, "Lung resistance-related protein/major vault protein and vaults in multidrug-resistant cancer", *Curr. Opin. Oncol,* vol. 12, no. 6, pp. 550-556, 2000.
[http://dx.doi.org/10.1097/00001622-200011000-00007] [PMID: 11085454]

[24] S.C. Gopinath, A. Matsugami, M. Katahira, and P.K. Kumar, "Human vault-associated non-coding RNAs bind to mitoxantrone, a chemotherapeutic compound", *Nucleic Acids Res,* vol. 33, no. 15, pp. 4874-4881, 2005.
[http://dx.doi.org/10.1093/nar/gki809] [PMID: 16150923]

[25] M. Kitazono, T. Sumizawa, Y. Takebayashi, Z.S. Chen, T. Furukawa, S. Nagayama, A. Tani, S. Takao, T. Aikou, and S. Akiyama, "Multidrug resistance and the lung resistance-related protein in human colon carcinoma SW-620 cells", *J. Natl. Cancer Inst,* vol. 91, no. 19, pp. 1647-1653, 1999.
[http://dx.doi.org/10.1093/jnci/91.19.1647] [PMID: 10511592]

[26] *Vaults - Pages - Scientists: Multidrug Resistance*. Available at: http://www.vaults.arc.ucla.edu/pages/sci-resistance

[27] M. Kitazono, H. Okumura, R. Ikeda, T. Sumizawa, T. Furukawa, S. Nagayama, K. Seto, T. Aikou, and S. Akiyama, "Reversal of LRP-associated drug resistance in colon carcinoma SW-620 cells", *Int. J. Cancer,* vol. 91, no. 1, pp. 126-131, 2001.
[http://dx.doi.org/10.1002/1097-0215(20010101)91:1<126::AID-IJC1018>3.0.CO;2-8] [PMID: 11149411]

[28] M.H. Mossink, A. van Zon, E. Fränzel-Luiten, M. Schoester, V.A. Kickhoefer, G.L. Scheffer, R.J. Scheper, P. Sonneveld, and E.A. Wiemer, "Disruption of the murine major vault protein (MVP/LRP) gene does not induce hypersensitivity to cytostatics", *Cancer Res,* vol. 62, no. 24, pp. 7298-7304, 2002.
[PMID: 12499273]

[29] K.E. Huffman, and D.R. Corey, "Major vault protein does not play a role in chemoresistance or drug localization in a non-small cell lung cancer cell line", *Biochemistry,* vol. 44, no. 7, pp. 2253-2261, 2005.
[http://dx.doi.org/10.1021/bi047948g] [PMID: 15709737]

[30] R.J. Scheper, H.J. Broxterman, G.L. Scheffer, P. Kaaijk, W.S. Dalton, T.H. van Heijningen, C.K. van Kalken, M.L. Slovak, E.G. de Vries, P. van der Valk, C.J. Meijer, and H.M. Pinedo, "Overexpression of a M(r) 110,000 vesicular protein in non-P-glycoprotein-mediated multidrug resistance", *Cancer Res,* vol. 53, no. 7, pp. 1475-1479, 1993.
[PMID: 7680954]

[31] A. van Zon, M.H. Mossink, M. Schoester, R.J. Scheper, P. Sonneveld, and E.A. Wiemer, "Efflux kinetics and intracellular distribution of daunorubicin are not affected by major vault protein/lung resistance-related protein (vault) expression", *Cancer Res,* vol. 64, no. 14, pp. 4887-4892, 2004.
[http://dx.doi.org/10.1158/0008-5472.CAN-03-3891] [PMID: 15256459]

[32] I. Petersen, R. Eastman, and M. Lanzer, "Drug-resistant malaria: molecular mechanisms and implications for public health", *FEBS Lett,* vol. 585, no. 11, pp. 1551-1562, 2011.
[http://dx.doi.org/10.1016/j.febslet.2011.04.042] [PMID: 21530510]

[33] M. McDonald, "Increasing bacterial resistance to ophthalmic antibiotics", *The newsmagazine of the American Society of Cataract & Refractive Surgery,* 2010. Available at: http://eyeworld.org/article-increasing-bacterial-resistance-to-ophthalmic-antibiotics

[34] N. Parry, "Multidrug Resistance Common Among Eye Pathogens, Study Finds", *Medscape,* 2015. Available at: http://www.medscape.com/viewarticle/853260

[35] K. Cavuoto, D. Zutshi, C.L. Karp, D. Miller, and W. Feuer, "Update on bacterial conjunctivitis in South Florida", *Ophthalmology,* vol. 115, no. 1, pp. 51-56, 2008.
[http://dx.doi.org/10.1016/j.ophtha.2007.03.076] [PMID: 17572497]

[36] D. Miller, M.G. Diaz, E.M. Perez, J. Newton, and E.C. Alfonso, "Prevalence of Community Acquired Methicillin Resistant Staphylococcus aureus among Ocular MRSA Isolates", *Am. J. Infect. Control,* vol. 34, no. 5, pp. E23-E24, 2006.
[http://dx.doi.org/10.1016/j.ajic.2006.05.180]

[37] A. Bhowmik, R. Khan, and M.K. Ghosh, "Blood brain barrier: a challenge for effectual therapy of brain tumors", *BioMed Res. Int,* vol. 2015, p. 320941, 2015.
[http://dx.doi.org/10.1155/2015/320941] [PMID: 25866775]

[38] P.A. Stewart, and M.J. Wiley, "Developing nervous tissue induces formation of blood-brain barrier characteristics in invading endothelial cells: a study using quailchick transplantation chimeras", *Dev. Biol,* vol. 84, no. 1, pp. 183-192, 1981.
[http://dx.doi.org/10.1016/0012-1606(81)90382-1] [PMID: 7250491]

[39] L.L. Rubin, "The blood-brain barrier in and out of cell culture", *Curr. Opin. Neurobiol,* vol. 1, no. 3, pp. 360-363, 1991.
[http://dx.doi.org/10.1016/0959-4388(91)90053-A] [PMID: 1821677]

[40] I. Martìn-Padura, S. Lostaglio, M. Schneemann, L. Williams, M. Romano, P. Fruscella, C. Panzeri, A. Stoppacciaro, L. Ruco, A. Villa, D. Simmons, and E. Dejana, "Junctional adhesion molecule, a novel member of the immunoglobulin superfamily that distributes at intercellular junctions and modulates monocyte transmigration", *J. Cell Biol,* vol. 142, no. 1, pp. 117-127, 1998.
[http://dx.doi.org/10.1083/jcb.142.1.117] [PMID: 9660867]

[41] M. Furuse, T. Hirase, M. Itoh, A. Nagafuchi, S. Yonemura, S. Tsukita, and S. Tsukita, "Occludin: a novel integral membrane protein localizing at tight junctions", *J. Cell Biol,* vol. 123, no. 6 Pt 2, pp. 1777-1788, 1993.
[http://dx.doi.org/10.1083/jcb.123.6.1777] [PMID: 8276896]

[42] N. Sonoda, M. Furuse, H. Sasaki, S. Yonemura, J. Katahira, Y. Horiguchi, and S. Tsukita, "Clostridium perfringens enterotoxin fragment removes specific claudins from tight junction strands: Evidence for direct involvement of claudins in tight junction barrier", *J. Cell Biol,* vol. 147, no. 1, pp. 195-204, 1999.
[http://dx.doi.org/10.1083/jcb.147.1.195] [PMID: 10508866]

[43] M.O. Romanitan, B.O. Popescu, B. Winblad, O.A. Bajenaru, and N. Bogdanovic, "Occludin is overexpressed in Alzheimers disease and vascular dementia", *J. Cell. Mol. Med,* vol. 11, no. 3, pp. 569-579, 2007.
[http://dx.doi.org/10.1111/j.1582-4934.2007.00047.x] [PMID: 17635647]

[44] M.O. Romanitan, B.O. Popescu, S. Spulber, O. Băjenaru, L.M. Popescu, B. Winblad, and N. Bogdanovic, "Altered expression of claudin family proteins in Alzheimers disease and vascular dementia brains", *J. Cell. Mol. Med,* vol. 14, no. 5, pp. 1088-1100, 2010.
[PMID: 20041969]

[45] A.T. Nies, G. Jedlitschky, J. König, C. Herold-Mende, H.H. Steiner, H-P. Schmitt, and D. Keppler, "Expression and immunolocalization of the multidrug resistance proteins, MRP1-MRP6 (ABCC1-ABCC6), in human brain", *Neuroscience,* vol. 129, no. 2, pp. 349-360, 2004.
[http://dx.doi.org/10.1016/j.neuroscience.2004.07.051] [PMID: 15501592]

[46] H. Bronger, J. König, K. Kopplow, H-H. Steiner, R. Ahmadi, C. Herold-Mende, D. Keppler, and A.T. Nies, "ABCC drug efflux pumps and organic anion uptake transporters in human gliomas and the blood-tumor barrier", *Cancer Res,* vol. 65, no. 24, pp. 11419-11428, 2005.
[http://dx.doi.org/10.1158/0008-5472.CAN-05-1271] [PMID: 16357150]

[47] D.S. Miller, S.N. Nobmann, H. Gutmann, M. Toeroek, J. Drewe, and G. Fricker, "Xenobiotic transport across isolated brain microvessels studied by confocal microscopy", *Mol. Pharmacol,* vol. 58, no. 6, pp. 1357-1367, 2000.
[PMID: 11093774]

[48] W. Zhang, J. Mojsilovic-Petrovic, M.F. Andrade, H. Zhang, M. Ball, and D.B. Stanimirovic, "The expression and functional characterization of ABCG2 in brain endothelial cells and vessels", *FASEB J,* vol. 17, no. 14, pp. 2085-2087, 2003.
[PMID: 12958161]

[49] C.A. Lipinski, F. Lombardo, B.W. Dominy, and P.J. Feeney, "Experimental and computational approaches to estimate solubility and permeability in drug discovery and development settings", *Adv. Drug Deliv. Rev,* vol. 46, no. 1-3, pp. 3-26, 2001. [http://dx.doi.org/10.1016/S0169-409X(00)00129-0] [PMID: 11259830]

[50] W.M. Pardridge, "Transport of small molecules through the blood-brain barrier: biology and methodology", *Adv. Drug Deliv. Rev,* vol. 15, no. 1, pp. 5-36, 1995. [http://dx.doi.org/10.1016/0169-409X(95)00003-P]

[51] D.R. Groothuis, B.E. Lippitz, I. Fekete, K.E. Schlageter, P. Molnar, O.M. Colvin, C.R. Roe, D.D. Bigner, and H.S. Friedman, "The effect of an amino acid-lowering diet on the rate of melphalan entry into brain and xenotransplanted glioma", *Cancer Res,* vol. 52, no. 20, pp. 5590-5596, 1992. [PMID: 1394182]

[52] V.E. Shashoua, and G.W. Hesse, "N-docosahexaenoyl, 3 hydroxytyramine: a dopaminergic compound that penetrates the blood-brain barrier and suppresses appetite", *Life Sci,* vol. 58, no. 16, pp. 1347-1357, 1996. [http://dx.doi.org/10.1016/0024-3205(96)00101-4] [PMID: 8614292]

[53] M.O. Bradley, N.L. Webb, F.H. Anthony, P. Devanesan, P.A. Witman, S. Hemamalini, M.C. Chander, S.D. Baker, L. He, S.B. Horwitz, and C.S. Swindell, "Tumor targeting by covalent conjugation of a natural fatty acid to paclitaxel", *Clin. Cancer Res,* vol. 7, no. 10, pp. 3229-3238, 2001. [PMID: 11595719]

[54] M.E. Sherman, Y.S. Erozan, R.B. Mann, A.A. Kumar, J.C. McArthur, W. Royal, S. Uematsu, and H.J. Nauta, "Stereotactic brain biopsy in the diagnosis of malignant lymphoma", *Am. J. Clin. Pathol,* vol. 95, no. 6, pp. 878-883, 1991. [http://dx.doi.org/10.1093/ajcp/95.6.878] [PMID: 2042597]

[55] T. Yoshikawa, T. Sugawara, K. Hirano, and V.J. Stella, "A novel chemical delivery system for brain targeting", *Adv. Drug Deliv. Rev,* vol. 36, no. 2-3, pp. 255-275, 1999. [http://dx.doi.org/10.1016/S0169-409X(98)00091-X] [PMID: 10837719]

[56] P. Borst, R. Evers, M. Kool, and J. Wijnholds, "A family of drug transporters: the multidrug resistance-associated proteins", *J. Natl. Cancer Inst,* vol. 92, no. 16, pp. 1295-1302, 2000. [http://dx.doi.org/10.1093/jnci/92.16.1295] [PMID: 10944550]

[57] S. Dallas, D.S. Miller, and R. Bendayan, "Multidrug resistance-associated proteins: expression and function in the central nervous system", *Pharmacol. Rev,* vol. 58, no. 2, pp. 140-161, 2006. [http://dx.doi.org/10.1124/pr.58.2.3] [PMID: 16714484]

[58] T. Alexandru, A. Staicu, A. Pascu, E. Radu, A. Stoicu, V. Nastasa, A. Dinache, M. Boni, L. Amaral, and M.L. Pascu, "Characterization of mixtures of compounds produced in chlorpromazine aqueous solutions by ultraviolet laser irradiation: their applications in antimicrobial assays", *J. Biomed. Opt,* vol. 20, no. 5, p. 051002, 2015. [http://dx.doi.org/10.1117/1.JBO.20.5.051002] [PMID: 25365041]

[59] M.L. Pascu, B. Danko, A. Martins, N. Jedlinszki, T. Alexandru, V. Nastasa, M. Boni, A. Militaru, I.R. Andrei, A. Staicu, A. Hunyadi, S. Fanning, and L. Amaral, "Exposure of chlorpromazine to 266 nm laser beam generates new species with antibacterial properties: contributions to development of a new process for drug discovery", *PLoS One,* vol. 8, no. 2, p. e55767, 2013. [http://dx.doi.org/10.1371/journal.pone.0055767] [PMID: 23405212]

[60] M.C. Morán, T. Tozar, A. Simon, A. Dinache, A. Smarandache, I.R. Andrei, M. Boni, M.L. Pascu, F. Cirisano, M. Ferrari, and M. Ferrari, "Toxicity study in blood and tumor cells of laser produced medicines for application in fabrics", *Colloids Surf. B Biointerfaces,* vol. 137, pp. 91-103, 2016. [http://dx.doi.org/10.1016/j.colsurfb.2015.06.041] [PMID: 26187648]

[61] I.R. Andrei, T. Tozar, A. Dinache, M. Boni, V. Nastasa, and M.L. Pascu, "Chlorpromazine transformation by exposure to ultraviolet laser beams in droplet and bulk", *Eur. J. Pharm. Sci,* vol. 81, pp. 27-35, 2016.
[http://dx.doi.org/10.1016/j.ejps.2015.09.017] [PMID: 26432595]

[62] M.L. Pascu, G.V. Popescu, C.M. Ticos, and I.R. Andrei, "Unresonant interaction of laser beams with microdroplets", *J. Eur. Opt. Soc. Rapid Publ.,* vol. 7, p. 12001, 2012.
[http://dx.doi.org/10.2971/jeos.2012.12001]

## CHAPTER 7

# Laser Beam Properties

**Mihail Lucian Pascu**[1,2,*]

[1] *National Institute for Laser, Plasma and Radiation Physics, Magurele, Ilfov, Romania*

[2] *Faculty of Physics, University of Bucharest, Magurele, Ilfov, Romania*

**Abstract:** Since the interaction of a laser beam or otherwise called laser radiation with droplets in pendant position is a key process in the data reported in the book, the current chapter is dedicated to the description of the main properties of a laser beam of the kind used in presented experiments. Basically, laser emits electromagnetic radiation with particular properties described briefly in this chapter, such as: coherence (spatial and temporal), directivity coupled with mode structure, narrow (lower than 0.1 nm) spectral width, time structure of the emission (pulsed or continuous wave-cw), polarisation state (which is in most cases linear or closed to linear), Gaussian distribution of intensity in beam cross section, high beam energy, power and brightness.

**Keywords:** Astigmatic beam, Asymmetric beam, Axial mode, Beam FWHM, Beam power, Beam TFW, Beam waist, Brightness, Coherence length, Coherence time, Directivity, Fluence, Gaussian laser beam, Laser active medium, Mode structure, Monochromaticity, Optical resonator, Polarised beam, Polarisation state.

## INTRODUCTION

The laser beam interaction with pendant droplets or bubbles depends on two main factors: (i) laser beam/radiation characteristics and (ii) composition and structure of the drop's/bubble's content such as: atoms, atomic and molecular ions, simple and complex molecules, molecular combinations, aggregates and supramolecular structures and in some cases, even more complex constituents.

In this chapter, the characteristics of a laser beam are described since they play a distinct role in producing the effects on pendant droplets in terms of shape and volume modifications, vibrations and material expulsion, as well as structural changes of substances/medicines contained in solutions presented in droplets form.

[*] **Corresponding author Mihail Lucian Pascu:** Laser Department, National Institute for Laser, Plasma and Radiation Physics, Magurele, Ilfov, Romania; Tel/Fax: 0040 214575739; E-mail: mihai.pascu@inflpr.ro

The laser is an emitter of electromagnetic radiation having specific properties, among which the wavelength in optical spectral range (*i.e.* in visible and near infrared). The laser extended theoretical bases were introduced by Einstein [1] and Kastler [2] who had contributions in early research stages and then by Basov and Prokhorov [3, 4], Gordon, Zeiger and Townes [5], Bloembergen [6], Schawlow and Townes [7] and, again, Schawlow [8]. Maiman [9] reported the operation of the first laser that was a ruby laser. The distribution of spectral ranges within which laser radiation is emitted by different types of lasers is shown in reports such as [10]. Nowadays, laser emission covers an electromagnetic range larger than, strictly speaking, optical domain, starting from $\gamma$ rays and going up to far infrared and even radio waves.

Beam characteristics at the output from a laser depend on emitting active medium, optical resonator (cavity) in which the active medium is embedded, some intra-cavity elements (such as pin-holes, optical filters, beam splitters, polarisers, prisms, diffraction gratings, optical Fabry-Perot etalons or interferometers, opto-mechanical systems *etc.*) and even energy delivery mode used to pump the active medium (which may work in pulsed or continuous wave-cw-regime) [10, 11]. Laser active media may be pure gases or mixtures of pure gases, liquids (containing either a single component or mixtures of solutions), solid state materials, semiconductors and plasma [12].

On the other hand, laser radiation used to interact with a target, such as a pendant droplet, has characteristics that derive from the laser beam emitted by laser itself, but they are finally controlled with optical and/or electro-optical components and systems used to process the beam and to tailor its properties to best respond experimental needs. In general, the processed beam properties at the place of interaction contain, in a significant degree, information included in it at emission from the laser. It is not possible to fully modify the properties of the beams outputting from laser and to avoid in this way some inconveniences regarding beam properties with respect to particular needs for applications and one has to always find the right compromise in terms of emitted beam characteristics *versus* beam properties required in a particular application.

## PROPERTIES OF THE LASER BEAM/RADIATION

The laser beam properties considered in the following are: monochromaticity and related bandwidth, coherence state, directivity and mode structure, energy, power and brightness, polarisation state and time structure.

## Monochromaticity and Related Bandwidth

The monochromaticity describes the spectral distribution of laser radiation correlated with its intensity that has a peak at a wavelength where the most part of radiation is emitted. In other words, the largest amount of laser beam energy is found at a restrained number of wavelengths around that of the maximum intensity of the beam. This intensity drops down to zero very fast near the peak's wavelength (but it does not behave like a δ-function) and this property shows the laser beam high spectral purity. As a consequence, laser beam spectral width is much narrower than in the case of incoherent optical emitters. An example of spectral distribution of laser radiation is shown in Fig. (**1**). In single axial mode amplification within the optical cavity, only the dominant line centered on $\lambda_0$ is emitted. Since stimulated emission is enhanced by successive reflections of radiation on the limiting mirrors of the optical cavity, the single longitudinal mode may coexist with other axial modes. The intensities of these modes decrease when the distance on wavelength scale increases (towards larger as well as lower values) with respect to the wavelength of the main peak of the laser line. The group of such adjacent lines generates a Lorentzian envelope which describes more correctly laser beam intensity distribution function of wavelength.

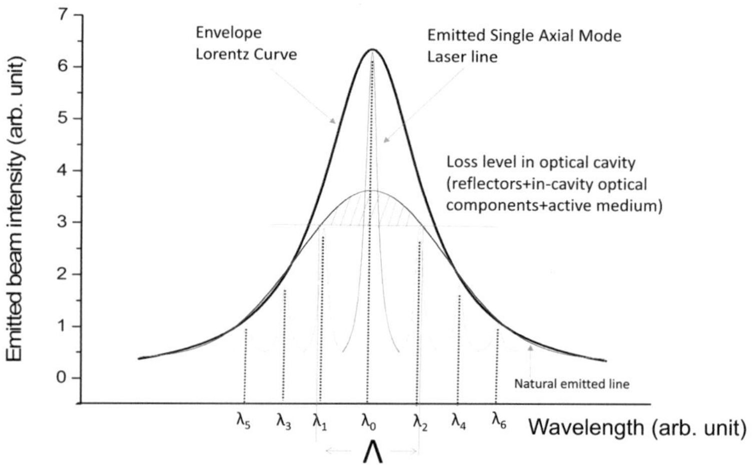


**Fig. (1).** Spectral distribution of the laser monochromatic beam; $\lambda_0$-laser beam wavelength; Λ-spectral bandwidth within which the gain is higher than losses.

At the same time, spectral distribution that describes the monochromaticity is determined by the shape of emitted line outputting from the active medium and by resonator axial modes. It is connected also to the relation between the gain and losses on the resonator and one may define a value Λ that is the spectral bandwidth within which the gain in the active medium is higher than losses. If the wavelengths of axial modes are $\lambda_1$, $\lambda_2$... $\lambda_M$ (frequencies $\nu_1$, $\nu_2$... $\nu_M$, respectively), as shown in Fig. (**1**), one may define the spacing in frequency units, $\nu_F$, between two adjacent modes; the number M of possible laser oscillation modes is:

$$M = \frac{\Lambda}{\nu_F} \quad \textbf{(1)}$$

where $\nu_F = c/(2d)$, "d" is the spacing between resonator mirrors and "c" is the speed of light. The maximum number of axial modes which are amplified by the optical cavity may be M.

Laser beam monochromaticity properties are evidenced by its spatial distribution *i.e.* looking at its experimentally produced divergence. If transverse electromagnetic modes are amplified by the optical cavity, for a mode that has a particular spatial distribution one may produce several longitudinal (axial) modes having the same spatial distribution but slightly different wavelengths. These modes appear separated on frequency scale having different $\nu_F$ values.

For applications, the laser line width is described by its FWHM (full width at half maximum) $\Delta\omega_{Laser}$ shown in equation (2):

$$\Delta\omega_{Laser} \sim \frac{8h\nu}{P_{out}} (\Delta\omega)^2 \quad \textbf{(2)}$$

where ν is the laser radiation frequency, Δω is the emitter's energy level width and $P_{out}$ is laser beam output power [7]. Typically, for a laser beam emitted in the visible spectral range a linewidth of $10^{-1}$ nm-$10^{-3}$ nm is obtained, but it may go down to $10^{-4}$ nm or even orders of magnitude lower.

## Coherence State

Coherence is an important and specific property of laser radiation. It is related to the mechanisms of stimulated emission that lead to laser radiation build-up, if one considers these mechanisms in their dynamics. The laser beam electric field $\vec{E}$ is described by:

$$\vec{E}(\vec{r}, t) = \vec{E}_0(\vec{r}, t) \exp[i(\omega t - \vec{k}\,\vec{r})] \quad \textbf{(3)}$$

where $\vec{r}$ is the vector that defines the position of the point where laser beam arrives, "t" is the time moment when observation is made, $\vec{E}_0$ is the electric field peak value in the considered point, ω the electric field angular frequency, $\vec{k}$ the beam wave vector and $i=\sqrt{-1}$ . According to (3), in a given point, at distance $\vec{r}$ from laser emitter, electric field may oscillate between $-E_0$ and $E_0$ with a constant angular frequency ω at any moment in time t. The reason for choosing the electrical field to describe coherence properties is that optical phenomena are related to electrical component of the electromagnetic field. Beam intensity is given by $|\vec{E}^2|$ and its polarisation state is described by the polarisation of the same field. Coherence property of a laser beam is related to its capability to produce interference fringes when interact [10, 12]:

### *Spatial Coherence*

Two rays originating from two different points of the same laser beam cross section interfere when superposed in the same point and produce interference fringes; the laser is called spatially coherent. The whole cross section of a laser beam is characterised by the same $\vec{E}(\vec{r},t)$. The beam is formed by a single plane wave and its spatial coherence is considered total. Spatial coherence is implied in laser beam divergence and in its focus properties as well.

### *Temporal Coherence*

Radiation emitted from a particular point of a laser source at a particular moment in time interferes with radiation emitted from the same point at a later time moment when both beams are incident in the same point. The time length within which interference is obtained is called "coherence time" and the laser that emits the radiation is defined as a source of temporally coherent radiation. Coherence time is the time interval Δt within which radiation is described by the same electric field $\vec{E}(\vec{r},t)$ given in equation (3). Based on coherence time one may define the coherence length of a laser beam which is that length in space within which interference phenomena may be obtained.

## Directivity and Mode Structure

Directivity is the property of laser radiation which allows it to be confined within a beam which propagates in such a way that its cross section dimensions are only slightly modified at short distance (*i.e.* of the order of some meters or tens of meters). Directivity is related to laser beam divergence property, this property describing its angular spreading and being measured in mrad. Directivity describes qualitatively the laser beam property to remain concentrated along and around a particular direction; complementary, divergence describes the same property quantitatively. Laser beam spatial distribution and divergence are function of optical cavity geometry, as well as of the shape of laser active medium. If, ideally, resonator mirrors are plane parallel and have unlimited dimensions, only axial modes are amplified and then the transverse spatial effects in laser active medium and resonator may be neglected.

In laser construction, spherical mirrors are used that have finite dimensions (the diameter of a mirror in most lasers is 20 to 50 mm long). This modifies laser beam divergence increasing it and makes possible the amplification of transverse modes which coexist with longitudinal (axial) modes.

Generally, an optical cavity that contains spherical mirrors produces more transverse electromagnetic modes $TEM_{lmq}$, where the pair $(l,m)$ defines a transverse mode which has a particular space distribution. In Fig. (**2**) are shown examples of transverse modes emitted by lasers in real working conditions where (0,0) transverse mode is specific to a Gaussian beam. In general, $l$ indicates the number of dark spaces contained in the beam cross section and appearing between spots on horizontal and $m$ indicates the same number on vertical dimension [13].

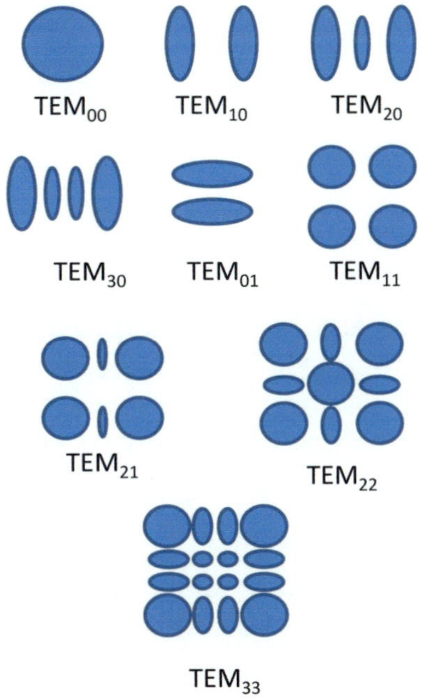


**Fig. (2).** Examples of laser beam intensity distributions in beam cross section, for different TEM.

For a pair (*l,m*), *q* shows the number of compatible axial modes that have the same spatial distribution, but different frequencies; separation on frequency scale between longitudinal modes is $\nu_F = c/(2d)$.

If laser beam propagation is evidenced in space, TEM structure is shaping the beam as depicted in Fig. (**3**).

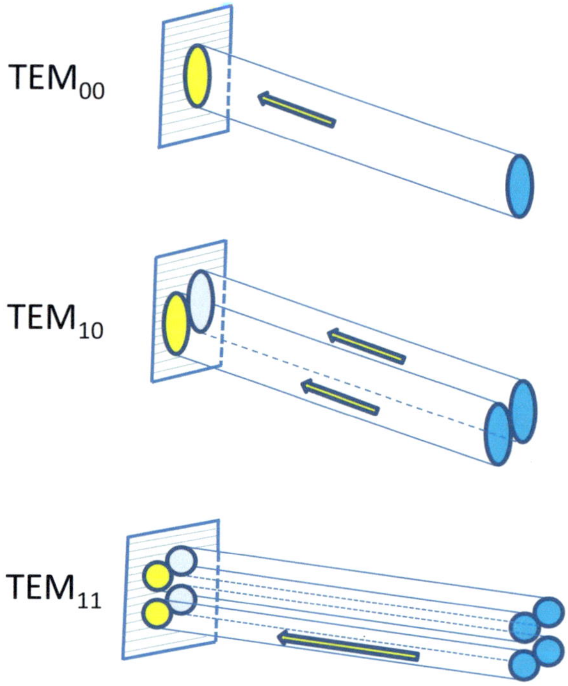


**Fig. (3).** Shapes of laser beams in free propagation for different transverse mode structures. The arrows represent the propagation direction and sense.

In working with laser beams and in processing them with optical components, their mode structure is of outmost importance since the divergence may be different from one mode to another; experimentally, two or more modes may coexist. When laser beam divergence is specified, usually this is mentioned for $TEM_{00}$ (Gaussian) mode extracted from laser as a whole.

A real Gaussian laser beam is somewhat close to an ideal Gaussian, exhibiting asymmetries along beam propagation direction. In single plane longitudinal section of laser beam, two most probable intensity distributions may be obtained as shown in Fig. (**4**). In Fig. (**4a**) the waist $\Phi_0$ for $TEM_{00}$ mode is the waist of real laser beam and in Fig. (**4b**) it is higher than $\Phi_0$. In experiments, real laser beams may be more complicated than shown in Fig. (**4**). If a three-dimensional picture of a propagating beam is considered, one may define a system of orthogonal axes:

Ox and Oy in a plane perpendicular to the beam propagation direction and Oz along it. The beam may exhibit three cases of asymmetries or combinations of them as shown in Fig. (**5**).

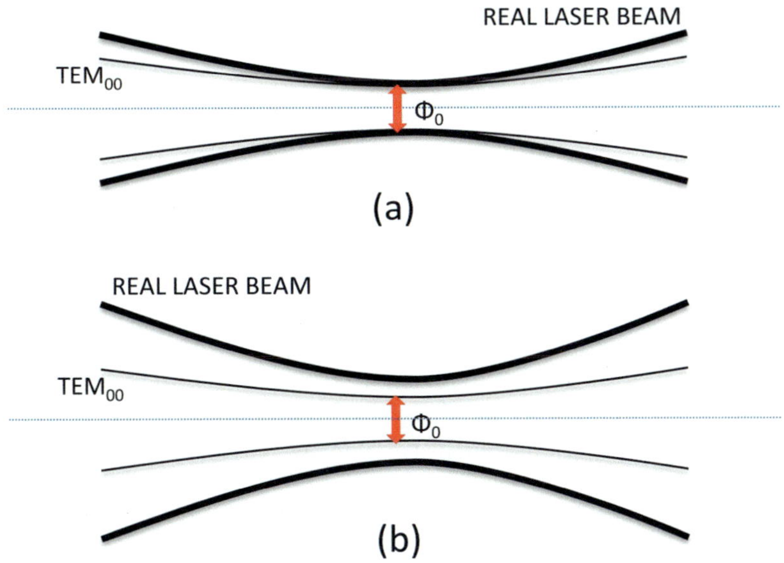


**Fig. (4).** Laser beam longitudinal cross section showing intensity distribution in real Gaussian mode.

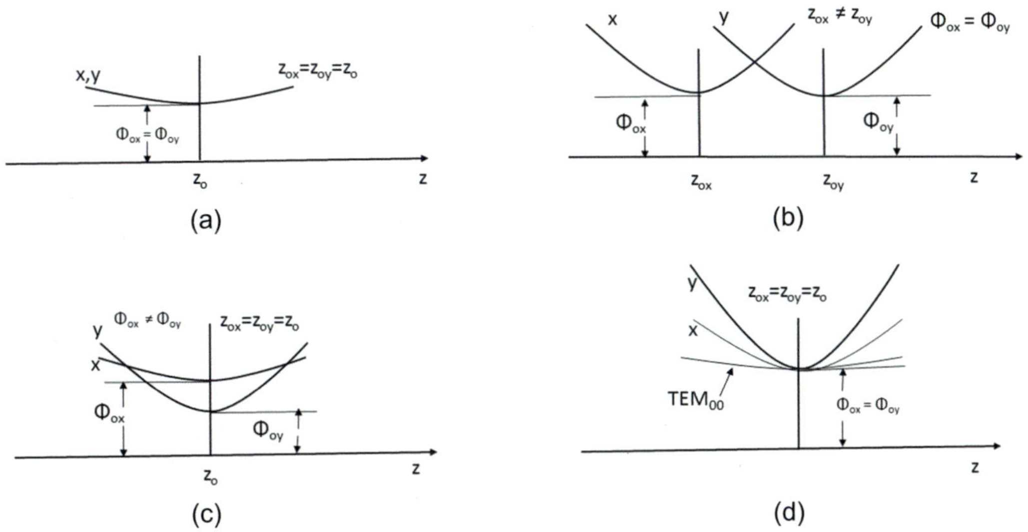


**Fig. (5).** Asymmetry cases of a laser beam: (**a**) no asymmetry; (**b**) astigmatism; (**c**) asymmetric waists; (**d**) asymmetric divergence.

Another case takes place when the laser beam has $\Phi_{0x} = \Phi_{0y}$, at $z_{0x} = z_{0y} = z_0$ but the divergence is asymmetric (it is different along Ox with respect to Oy, Fig. **5d**).

If the beam does not have asymmetry, waists dimensions on Ox and Oy are equal ($\Phi_{0x} = \Phi_{0y}$) and take place in the same plane, *i.e.* at the same $z_0$ ($z_{0x} = z_{0y}$, Fig. **5a**). If the beam is astigmatic the two waists dimensions are equal ($\Phi_{0x} = \Phi_{0y}$), but they are obtained in different planes (at different z values: $z_{0x} \neq z_{0y}$, Fig. **5b**). The waist may be asymmetric as in Fig. (**5c**), where $\Phi_{0x} \neq \Phi_{0y}$ and $z_{0x} = z_{0y} = z_0$.

Despite these structural characteristics, real laser beams have better directivities and significantly lower divergences if compared to incoherent light beams. The described mode structures should be considered if applications regarding interaction of focused laser beams with droplets are performed since the intensity distribution in the cross section of the laser beam is kept, even if at lower dimensions after focusing, within the focus waist at the interaction point/place with the target. More, sub-millimetric dimensions of droplets require an accurate focusing of the laser beam on the droplet. In any case, a very well fixed and reproducible position of focus with respect to droplet generated in pendant position is needed.

### Time Structure

Laser radiation emission may be continuous, *i.e.* the beam is emitted continuously or otherwise said in continuous wave regime (cw). It is considered that the laser emits continuously if emission is longer than or equal to 0.25 s [10].

If the laser has pulsed emission, it is accepted that the beam consists of a single light pulse, a train of pulses or repetitive pulses, the time length of one pulse being less than 0.25 s (according to IEC regulations). For one pulse, one may define the full time width (FTW) which is the total duration of the laser beam measured at the lowest laser line intensity points. The time width at half maximum (TWHM) may be defined as the time length of the pulse at half maximum intensity. Nowadays pulsed lasers emit laser beams having FTW that may vary, function of laser type, from some ms to fs.

If the laser has pulsed emission one may also define pulse repetition rate measured in pulses per second which may vary from single pulse to millions of pulses per second (in this case the pulse repetition rate is usually expressed in Hz, kHz, MHz or GHz).

### High Energy, Power and Brightness

Laser beam brightness is defined based on its intensity which is given by the number of photons emitted out of the resonator, reported to unit surface of beam cross section. At the same time, the brightness is due to laser beam directivity coupled with its power and with the monochromaticity of the radiation.

Laser beam brightness is derived from its high power density (W/m$^2$) or high energy density (J/m$^2$) and observed in a cross section of it. The energy density is also referred to, as fluence that is measured in J/cm$^2$. For a laser beam one may also define the energy density reported to the unit spectral linewidth, as the radiant exposure of a laser beam reported to the spectral linewidth of that laser beam (units: J m$^{-2}$ nm$^{-1}$). Function of the time characteristics of beam emission, for applications at interaction with droplets, laser beam energy may vary from some μJ to 10-20 J for cw lasers. For ns pulsed lasers the laser beam average energy is utilized within the same range of values.

Brightness is controlled by the transmission of the extraction mirror of optical cavity that allows to obtain a compromise between the gain inside laser active medium and losses, while the needed optical output from the laser is considered.

If the transmission through laser extraction mirror would be the only source of optical loss, the total output photon flux $N_0$ (the number of the photons leaving the optical cavity in one second) is given by:

$$N_0 = (R - Rt)V \quad \textbf{(4)}$$

where R is the rate at which the atoms of the active medium are pumped on the superior laser level, Rt is the pumping rate at threshold and V is the volume of active medium.

If there are other loss mechanisms besides those on the extraction laser mirror, the output photon flux $N_0$ may be written:

$$N_0 = L_e(R - Rt)V \quad \textbf{(5)}$$

where $L_e$ is the emission efficiency defined as the ratio of the loss due to light extraction to total losses in the optical cavity.

**Polarisation State**

Polarisation is not a specific property of laser radiation since any electromagnetic radiation exhibits it. But it is important to know it because it coexists with other specific beam properties. The polarisation state of laser radiation is given by the trajectory described by its electric field. A more detailed graphic presentation of possible polarisation states of laser beams is given in [10] and [13]. The polarisation of laser radiation is related to its property to have the electric and magnetic vectors vibrating in a single plane. Polarisation direction is the direction of the electric field vector which is always perpendicular to propagation direction. The beam may be unpolarised when polarisation direction varies randomly, linearly polarized when it remains constant and elliptically polarized when polarisation direction has two coherent, linearly polarized components whose directions make a right angle. If the axes of the ellipse are equal, the beam is circularly polarized [14].

Generally speaking, the properties of laser beams recommend them for applications in microfluidics and optofluidics, as well as in the development of new techniques in micro- and nano-spectroscopy.

## CONFLICT OF INTEREST

The authors confirm that this chapter content has no conflict of interest.

## ACKNOWLEDGEMENTS

This work has been financed by the National Authority for Research and Innovation in the frame of Nucleus programme-contract 4N/2016. The author would like to acknowledge the help extended by Dr. Tatiana Tozar in tracing the graphics of some of the figures embedded in this chapter.

## REFERENCES

[1] A. Einstein, "Zur Quantentheorie der Strahlung", *Phys. Z.,* vol. 18, pp. 121-128, 1917.

[2] A. Kastler, "Applications of polarimetry to infra-red and micro-wave spectroscopy", *Nature,* vol. 166, no. 4211, pp. 113-113, 1950.
[http://dx.doi.org/10.1038/166113a0] [PMID: 15439165]

[3] N.G. Basov, and A.M. Prokhorov, "The use of molecular bundles for radiospectroscopic study of the band spectra of molecules (in russian)", *Zh Experim Teor Fiz,* vol. 4, no. 10, pp. 431-438, 1954.

[4] N.G. Basov, and A.M. Prokhorov, "Possible methods of obtaining active molecules for a molecular oscillator", *J. Exp. Theor. Phys.,* vol. 1, no. 1, pp. 184-185, 1955.

[5] J.P. Gordon, H.J. Zeiger, and C.H. Townes, "The Maser—new type of microwave amplifier, frequency standard, and spectrometer", *Phys. Rev.,* vol. 99, no. 4, pp. 1264-1274, 1955.
[http://dx.doi.org/10.1103/PhysRev.99.1264]

[6] N. Bloembergen, "Proposal for a New Type Solid State Maser", *Phys. Rev.,* vol. 104, no. 2, pp. 324-327, 1956.
[http://dx.doi.org/10.1103/PhysRev.104.324]

[7] A.L. Schawlow, and C.H. Townes, "Infrared and Optical Masers", *Phys. Rev.,* vol. 112, no. 6, pp. 1940-1949, 1958.
[http://dx.doi.org/10.1103/PhysRev.112.1940]

[8] A.L. Schawlow, "Infrared and Optical Masers", In: *Quantum Electronics,* C.H. Townes, Ed., Published by Columbia University Press: New York, 1960.

[9] T.H. Maiman, "Stimulated Optical Radiation in Ruby", *Nature,* vol. 187, no. 4736, pp. 493-494, 1960. [http://dx.doi.org/10.1038/187493a0]

[10] L. Dănăilă, and M.L. Pascu, *Lasers in neurosurgery.* Editura Academiei Române: Bucureşti, 2001.

[11] M.L. Pascu, "Lasers in Medicine and dentistry", In: *Lasers in medicine and dentistry: basic science and up-to-date clinical application of low energy-level laser therapy,* Z. Simunovic, Ed., Vitagraf: Rijeka, 2000.

[12] M.L. Pascu,"Laser physics for biomedical application",In: *Laser Manual.* vol. 1. Officina Editoriale Oltrano: Firenze, 2014.

[13] A. Yariv, *Introduction to optical electronics.,* 2nd ed. Holt, Rinehart and Winston: New York, 1976.

[14] T.V. Higgins, *Wave nature of light shapes its many properties.* Laser Focus, 1994, pp. 59-65.

**CHAPTER 8**

# Unresonant Interaction of Laser Beams with Pendant Droplets

**Ionut Relu Andrei**[1], **Mihai Boni**[1,2] and **Mihail Lucian Pascu**[1,2,*]

[1] *National Institute for Laser, Plasma and Radiation Physics, Magurele, Ilfov, Romania*

[2] *Faculty of Physics, University of Bucharest, Magurele, Ilfov, Romania*

**Abstract:** The unresonant interaction between a single droplet and a single laser beam takes place when the beam is not absorbed by droplet's material; it leads to specific effects induced in the droplet. The experiments reported here are made on droplets consisting in water solutions; 532 nm laser beam is used to produce unresonant interaction since water absorbs very little in green. The results on unresonant interaction, evidence several effects: (i) damped deformations and vibrations, (ii) expulsion of micro-jets of water that propagate with supersonic speeds and production of micro- and nano-droplets that are springs from mother droplet with different lower speeds, (iii) production of horizontal pillars of liquid and of cylindric channels within the droplet leading to appearance inside it of air bubbles. The water droplet interaction with a laser beam focused in its equatorial plane on droplet's surface, produces a velocity gradient of water inside the droplet that may be experimentally evidenced. The velocity gradient increases with increasing beam energy and does not depend on droplet volume. The surface tension and liquid viscosity characteristics influence the effects produced by the laser beam on droplet and the propagation speeds of liquid material formations emitted out of them. Studies on unresonant processes have shown that if produced on a droplet that contains a laser dye, they may be accompanied by effects induced when the interaction is resonant, *i.e.* the laser radiation is absorbed by droplet material. Such a "combination" depends on laser beam energy, solvent, viscosity and liquid surface tension.

**Keywords:** Laser ablation, Laser dye, Microfluidics, Microdroplets, Nano-droplets, Non-resonant interaction, Optical radiation pressure, Pendant droplet, Resonant interaction, Rhodamine 6G, Resonant interaction.

## INTRODUCTION

Milli-, micro- and nano-droplets are used more and more for technological and biomedical applications and this triggers complex experiments regarding new methods to produce and more accurately characterise them. The optical and, in

* **Corresponding author Mihail Lucian Pascu:** Laser Department, National Institute for Laser, Plasma and Radiation Physics, Magurele, Ilfov, Romania; Tel/Fax: 0040 21 4575739; E-mail: mihai.pascu@inflpr.ro

particular, laser methods utilised to study or even generate droplets constitute the bases of instruments conceived to get new data about small and very small dimension droplets. These techniques are a significant part of optofluidics that is a relatively new and dynamic field developed mainly in the last 10-15 years [1, 2]. Recent literature reports reveal new results about laser beam interaction with micro-volumetric droplets which contain single liquids or mixtures of immiscible compounds. The droplets are generated in different surrounding environments such as liquids or gaseous media, in particular in air [3 - 6].

In most studies, droplets are quasi-spherical or hemi-spherical. They contain materials in liquid phase positioned on transparent surfaces that may be chosen for hydrophobic or superhydrophobic properties, or may have repulsion properties with respect to other liquids such as oil, in which case they are defined as oleophobic or superoleophobic. When sessile, a droplet interacts with a laser beam that is transmitted through its supporting piece [7]. Sometimes, droplets are produced, fixed and/or moved at the interior of microchannels organized as structures that are built for their manipulation by light. In other cases, the droplets are introduced in single- or two-phase flows for use in optical chips [8, 9].

Other data are reported on micro-droplets that are free falling in air, levitating, hanging on a capillary tip or embedded in other liquid with respect to which the droplet material is immiscible. Most part of these results is related to fluorescence emission of droplets or even to lasing when droplets are optically pumped by laser beams [10 - 16]. The droplet's components such as dyes, or more general, fluorophores and their solvents shape the behaviour of droplets at interaction with laser beams. Other factors that may influence droplet's evolution at interaction with a laser beam are: (i) the environment in which it is embedded, and (ii) its structure (the droplet may contain one compound only, or a core and one or more covering thin layers of immiscible materials [17]). The most used solvent is ethanol, but in some cases polymeric materials are also utilised [15]. Water is a quite common solvent used in droplets since its properties (optical, physico-chemical, fluidic) are relatively well known. Though, water behavior at interaction with laser beams is not entirely known and studies of its interaction with them when irradiated in bulk [8, 18, 19], or as droplets are needed.

As mentioned above, the laser beam-droplet interaction may be resonant or unresonant [4]. Here, resonant means that the laser beam is absorbed by molecular species included in the droplet; these are excited, normally, on the first excited singlet state by absorption of one or more photons of incoming beam. If absorbent concentration is properly chosen, photons are absorbed immediately after entering in the droplet and no more photons are available to propagate within it. A result of exciting molecules on singlet states could be droplet's temperature increase produced by (i) nonradiative transfers of molecules within the singlet states or (ii) intersystem (singlet states -triplet states) crossing with slight energy losses.

Unresonant means that the beam wavelength is such that it is not absorbed by droplet's molecular compounds or the beam-droplet interaction processes are not dominated by absorption phenomena [4]. This takes place in droplets made of a single material, such as water, the interaction producing: (i) effects originating in light pressure, (ii) application of electrostrictive forces on droplet's material as a result of laser action and (iii) thermal effects within droplet's mass [20, 21].

Hanging droplets make possible to individualise the effects of a single laser beam on a single droplet in different environments. They may present a free modification of their shapes and, at the same time and within some limits related to their material's viscosity, surface tension and weight, they may remain connected to the capillary. In this chapter, in the first place are shown results about the interaction of liquid aqueous microliter droplets with 532 nm pulsed laser beams. At this wavelength, the interaction is dominated by unresonant phenomena produced on the micro-droplet. The beam average energy is stepwise changed and the pulse full time width at half maximum (FTWHM) is 6 ns. Laser beam impact with a pendant micro-droplet produces droplet deformations as well as its mechanical vibrations. Droplets may loose some material after being collided by the laser beam if its energy is high enough and conditions in which such phenomena are produced are also described in this chapter. The effects are studied on single droplets and pulse by pulse. They are function of (i) laser beam energy, (ii) laser beam focus dimensions and position, (iii) exposure geometry, (iv) content of the droplet, (v) interaction of droplet's material with the capillary's surface on which it is pendant. The droplet shape and its changing due to collision with the laser beam are registered with a high speed recording camera used in tandem with a continuous, cold, white light source that illuminates a diffusing screen opposed to the camera. Following droplet interaction with laser beams are produced (i) nanodroplets (nl volumes) propagating at high speed, and (ii) micro-jets and micro-droplets that move at lower speeds. After interaction, the pendant droplet may have smaller volume/diameter than the parent droplet and micro-bubbles of gas (most probable air) are generated within its material/volume. Emission of nanodroplets that propagate freely around the parent is used in nanoscience, nanotechnology and nanomedicine where nanodroplets of materials are displaced within controlled gaseous media and fall on target surfaces that have characteristics which should be modified in controlled and reproducible ways.

Another kind of experiments reported here is the investigation of the behaviour of microdroplets made of solutions of laser dyes in solvents, at resonant interaction with laser pulses. The results show new ways to produce combined modifications on droplets when they interact with laser beams. They are produced by controlling the absorption level of laser radiation in the droplet when adjustment of absorbent concentration between negligible to high values is made. This provides information about the reciprocal influence between resonant and unresonant interaction of a single laser pulse with a single droplet, *i.e.* about the way in which unresonant interaction that produces mechanical effects on droplets may be used simultaneously with resonant interaction that leads, among others, to laser induced fluorescence emission by droplets.

## MATERIALS AND METHODS

The general scheme describing the experimental arrangement is shown in Fig. (**1**). Hanging droplets were produced in open air, one by one, using a computer controlled droplet generator dual syringe unit which is composed of a commercial pumping system, automatic and manual valves, capillary tubes, and calibrated capillary tips that produce droplets with controlled volume using stock solutions. A capillary tip can be a needle with one capillary inside or a needle with two inner capillaries either concentric or sided. The system resolution is 0.05% of the utilized syringe volume, *i.e.* 0.0025 µl (2.5 nl) for a 50 µl total volume of the syringe; the accuracy in generating a given volume is ±1% (when droplet volume is 10% of syringe volume) and precision is ±0.2% (indicating the variation of generated volume from droplet to droplet). The capillary material is hydrophobic so that the droplet surface is tangent to the capillary tip and droplet's material does not adhere to capillary inner and outer solid surface.

Representation of a droplet in connection with its generating assembly that shows it in contact with the needle and suggests the valve utilised to avoid capillarity reabsorption of liquid is shown in Fig. (**1b**). The outer/inner diameter of the capillary is 510/210 µm. The capillary can be accurately moved, so that the vertical position of the droplet as well as its place in horizontal plane can be changed. So, droplet's translation in horizontal plane along laser beam and perpendicular to it may be achieved with an increment of 250 µm (Fig. **1c**). For all recorded images the capillary outer diameter is used as reference length.

Droplets with different volumes, 3, 3.5 and 4 µl labeled $V_1$, $V_2$, and $V_3$ respectively, were generated. Solvents such as distilled water, ethyl alcohol, or dimethyl sulfoxide (DMSO), as well as mixtures of them with laser dyes rhodamine 6G (Rh6G) and 4-(Dicyanomethylene)-2-methyl-6-(4- dimethyl-aminostyryl)-4H-pyran (DCM) were used.

Laser beam was the second harmonic (SHG) of Nd:YAG laser fundamental radiation at 1.064 µm, emitted at 532 nm. Its energy made available on the droplet was controlled with a neutral filter that filters 532 nm and cuts-off the most important part of remaining fundamental beam.

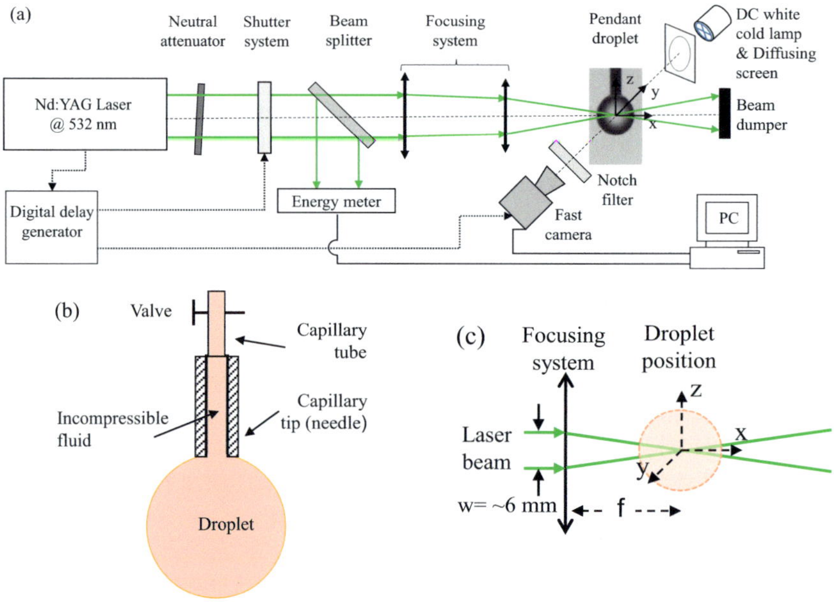


**Fig. (1).** (**a**) Experimental set-up; (**b**) Scheme of utilised pendant droplet system; (**c**) Scheme of irradiation geometry: w, beam waist; f, focus length. Laser, shutter and fast camera are synchronized by a digital delay generator according to the indicated signal path in (**a**); droplet position can be adjusted on all three axes.

The 532 nm laser beam had average energies per pulse: $E_1$=0.25 mJ, $E_2$=0.4 mJ, $E_3$=0.7 mJ or $E_4$=1 mJ. The beam was focused on droplet, using a lens system $L_1$-$L_2$ in which $L_2$ is a microscope objective (magnification 20X) that reflects 0.09% from incoming beam at 532 nm. Focus length of the system was 8 mm and the cross section of the focal spot had a minimum diameter of 90 µm. The beam configuration around focus assured that the diameter of 90 µm was kept unchanged over 250 µm along laser beam propagation direction. In order to avoid air currents around droplet and to minimise water vaporisation and heat exchanges with environmental atmosphere, the droplet was screened using an isolating box.

A Digital Shutter System giving way to laser beam towards droplet was synchronised with laser pulse generator and with fast camera by a Digital Delay Generator. The fast camera was adjusted to record a picture while the shutter was open a time interval of 100 µs. The beam energy was monitored pulse by pulse extracting 10% of it through a beam splitter and using an energy meter equipped with a suitable sensor. The droplet's behaviour under laser beam and after interaction with it is followed in real time with the Drop Shape Analysis System which utilises a High Speed Camera taking images at a speed of 10,000 frames per second (10 kfps). To do this a diffusing screen lightened by a cold white lamp is used, that is placed in opposite position with respect to the camera. The droplet hangs between the screen and the camera. The measurements regarding unresonant interaction were made on drops containing aqueous solutions; water absorbs poorly at 532 nm since the absorption coefficient is ~0.0447 $m^{-1}$ and one may consider that the beam absorption in droplet's volume is negligible [22, 23]. The focus of the laser beam was placed in different positions on droplet surface or inside its volume by moving the droplet with respect to focus point (Fig. **1c**). For droplets of distilled water only, the full range of available volumes and beam energies were utilized.

In the following sections, are shown figures in which the effects of the pulsed laser beam on a single droplet are presented together with the relative positions of a droplet with respect to the beam. In most cases, a Movie is associated to a corresponding figure, keeping figure numbering.

## PRESSURE EFFECT OF LASER BEAMS ON DROPLETS

In Fig. (**2**) are shown some cases of interactions of a laser beam with a single pendant droplet. The droplet irradiation geometry and the light pressure applied on it are drawn when the pulsed laser beam is sent (i) in droplet's equatorial plane, focused on its front face (Fig. **2a**), (ii) upper position - off-axis, focused on its front face (Fig. **2b**), (III) the South Pole (Fig. **2c**), (iv) in the equatorial plane, focused in its centre (Fig. **2d**), and (v) in the equatorial plane, focused on its back face (Fig. **2e**). For each figure, in the upper image is drawn the geometry of interaction; in the image at bottom the mechanical effects associated with it for a 3.5 μl droplet at energy $E_1$ = 0.25 mJ in Fig. (**2a**) and $E_4$ = 1 mJ in Figs. (**2b-e**) are shown. At 1 mJ beam energy, beam mechanical effects on droplet are quite violent producing explosive phenomena. When it falls on droplet, laser beam is refracted at its interface with air (see Chapter 3, Fig. **1**), once at input and secondly at output. In both positions, laser beam applies pressure on droplet.

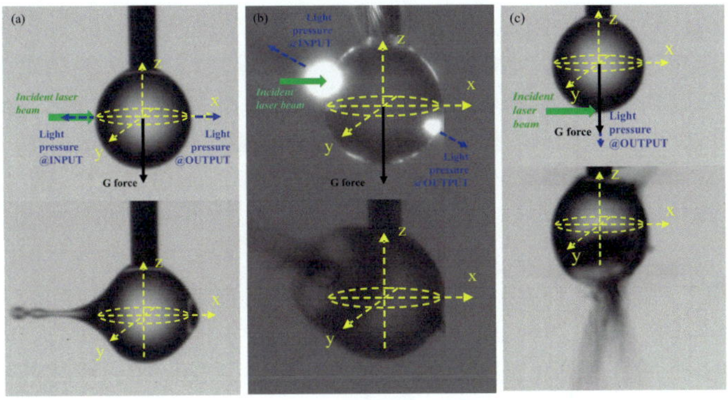


*Fig. 2 contd.....*

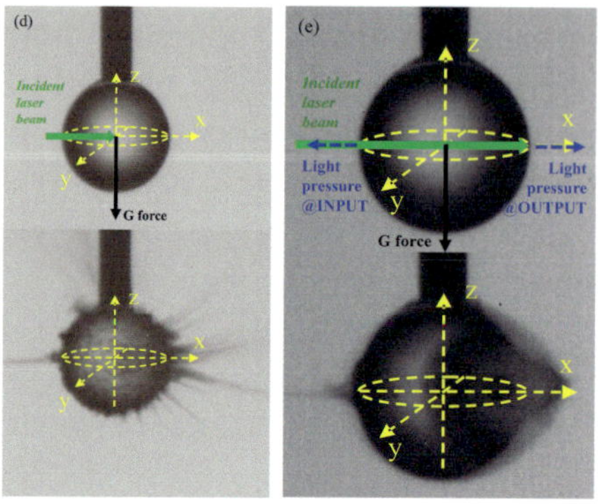


**Fig. (2).** Geometry of droplet irradiation and mechanical effects associated with it for laser beam sent in (**a**) equatorial plane and focused on its front face (Movie **2a**), (**b**) off-axis, upper position and focused on its front face (Movie **2b**), (**c**) off-axis, tangent at South Pole (Movie **2c**), (**d**) equatorial plane, focused in its centre (Movie **2d**), (**e**) equatorial plane, focused on its back face (Movie **2e**).

According to general theory, the pressure acts perpendicularly to droplet surface and light force is always oriented from water (that has higher refractive index) towards air (lower refractive index) [8, 24 - 26].

This holds regardless the beam incident angle on droplet at input/output on/from it, and of the position of laser beam focus on droplet surface. The laser light pressure applied on a pendant droplet that is approximately spherical, is always oriented towards the exterior of the droplet and along its radius of curvature traced at the contact point with the beam. This is why, at input, the force exerted by the beam on droplet is oriented approximately counter the laser beam propagation sense. At output, the pressure is oriented from droplet towards air, more or less in the same sense with laser beam propagation. The modification of droplet shape and some loss of its material are produced by light pressure only; it interacts with viscosity forces specific to droplet's material, with droplet surface tension and with other possible forces exerted on droplet and originating in electrostriction [18, 27] and thermal effects, like local overheating and vaporization of the liquid [3]. The focusing system produces a low divergence laser beam. The waist diameter is 90 μm and at approximately 1 mm before or after the focus it is about ten times larger. The used beams at 1.064 μm and 532 nm are said to be single mode, but the intensity in the waist is neither rigorously Gaussian, nor constant as results from Fig. (**3**). In Fig. (**3a**), is given the 2D image of the laser beam intensity in the waist. Fig. (**3b**) presents 3D intensity distribution corresponding to the 2D image, recorded in the same conditions. The images show that (i) laser beam intensity has not the peak on optical axis, but off-axis and (ii) the intensity distribution contains two unequal and quite intense and very close (spatially) peaks. Dimensions of focal spot are circumscribed in the yellow ellipse in Fig. (**3a**) and are given in the inserted table.

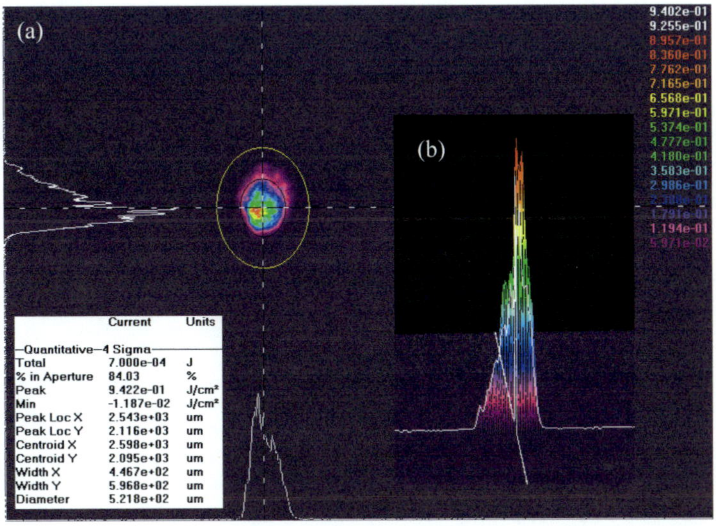


**Fig. (3).** Profile of laser beam intensity in focus, measured with SPIRICON, LW 230 profiler. Incident beam filtered: $10^{-7}$ of its intensity sent to detector; (**a**) 2D and (**b**) 3D distribution corresponding to (**a**).

The laser beam longitudinal section is constant for more than 250 μm along propagation direction to avoid breakdown at the contact with droplet's surface, since the set of conditions specified in [14], are not met. These qualities of the processed laser beam influence the unresonant effects produced on droplet while setting the position of the beam focus with respect to droplet surface or its centre. Further, is presented a synthesis of the results obtained on unresonant interaction. They are displayed for each experimental case as a series of selected frames (frame sequence). Each image or frame contains superimposed events recorded with the fast camera along a 100 μs time interval. In Fig. (**4**) are presented two frame sequences containing images recorded in the same focusing conditions of laser radiation, but under different recording conditions. In Fig. (**4a**) if compared to Fig. (**4b**) it was not used the Notch filter which attenuates the laser radiation scattered on droplet surface. Frame 2 of each sequence in Fig. (**4**) shows the superposition of images recorded during the action of laser pulse (scattered light) and mechanical effects that occur at later stages during laser irradiation. In these cases, about 100 μs after interaction are recorded. The halo around droplet in Frame 2, sequence (a) is produced by the reflection of 532 nm green laser beam on droplet's external thin layer [18, 28, 29]. Mechanical effects are distinguished in Fig. (**4b**) where scattered light is attenuated by Notch filter that cuts-off more than 99% out of the radiation in the band of 533±17 nm. Analysing the droplet contour in Fig. (**4b**), Frame 2 that is uniformly illuminated, one may see that during laser pulse action, the droplet does not show significant mechanical deformations.

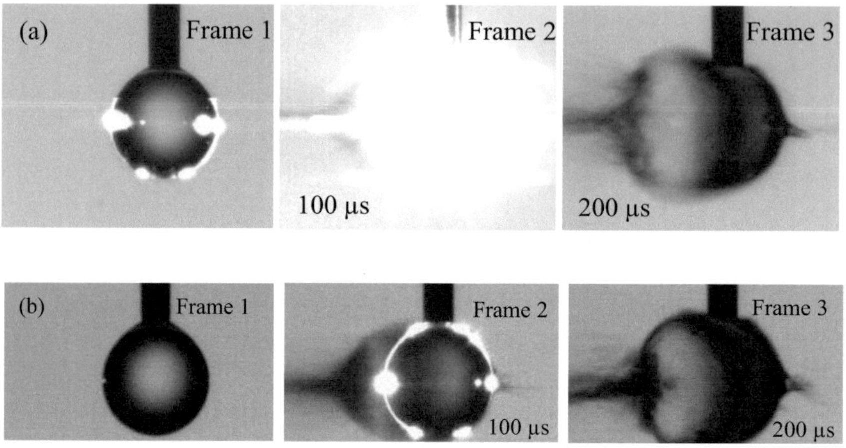


**Fig. (4).** Effects produced on a single droplet by a laser beam sent in its equatorial plane and focused on front surface. Images recorded (**a**) in absence (Movie **4a**) and (**b**) in presence of a Notch filer (Movie **4b**). Droplet volume $V_2$=3.5 μl; beam energy $E_4$=1 mJ, frame rate 10 kfps (frame capture time 100 μs). Time specified within each frame represents the interval that passed from interaction moment, considering that the laser pulse falls on the droplet at the beginning of Frame 2 in both cases.

Also, to reduce or totally eliminate the scattered radiation recorded into a frame capture, the time delay of triggering the fast camera can be adjusted with respect to laser pulse so that the moment of interaction is positioned between two successive frames (Fig. **5**). The time interval between two successive frames in the same sequence resulting from global shutter working limitations is 1.25 μs, independent with respect to the frame rate selected for fast camera recording. In each frame are recorded events produced along 100 μs. From Fig. (**5**) it results that for a difference in time delay of 1.25 μs, the beam-droplet interaction is recorded in different frames, namely Frames 2 and 3, corresponding to sequences 5 and 1, respectively. For time delay difference of about 1 μs (recorded between sequences 2 and 4), the interaction moment is not recorded being placed between displayed frames. In Fig. (**5**) a sequence is related to a single droplet, so that for each sequence a new single droplet interacts with a new laser beam single pulse.

The values of the time delays specified in Fig. (**5**) were reported to a reference value $t_o$ of the time delay that depends on selected synchronization chain. In Fig. (**5**) the recording conditions were as in Fig. (**4b**). In sequence 1, Frame 3 mechanical effects during about 100 μs are recorded and in sequence 5, Frame 2 only effects during 0.25 μs are superimposed. The microdroplet contains a Newtonian liquid and this is demonstrated by the experiments in which water droplets continue to have the same fluidic evolution during and after interaction with the pulsed laser beams [24].

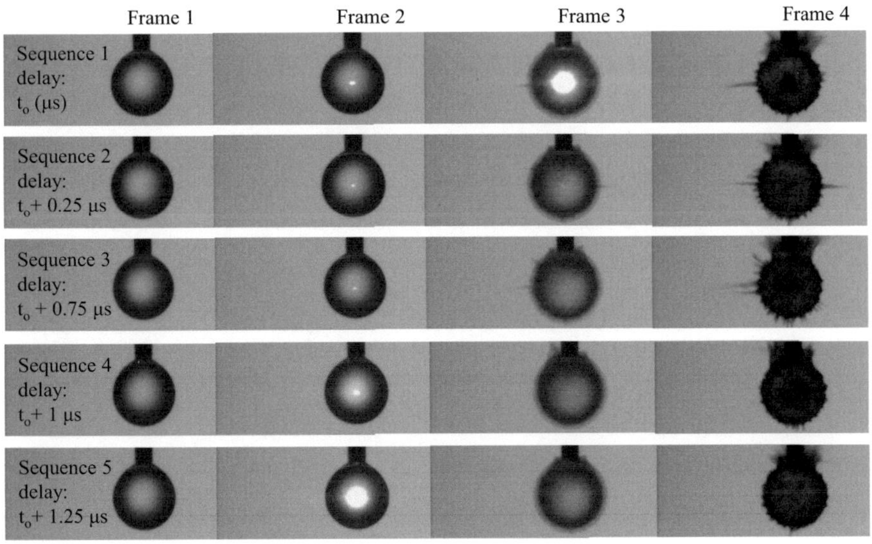


**Fig. (5).** Effects of time delay variation shown in frames recorded for a laser beam sent in droplet equatorial plane with focus placed in its geometrical centre. Sequences are recorded with a difference in time delay of 0.25, 0.5, 0.75, 1 and 1.25 µs between them (Movie **5 seq. 1, seq. 2, seq. 3, seq. 4, seq. 5,** respectively). Droplet volume $V_2$=3.5 µl; laser beam energy E=1 mJ, recording repetition rate 10 kfps; $t_o$ is a fixed value of time delay introduced by the measuring chain.

Since the droplet content is a Newtonian incompressible fluid, the system can be theoretically and computationally investigated by applying the analytical model of droplet in partial contact whith a solid support developed by Strani and Sabetta [30]. This model was improved for nonlinear oscillations of pendant droplets by Basaran and DePaoli [31] and for forced oscillations by Wilkes and Basaran [32].

In such conditions, one challenge is to understand mechanical effects (droplet vibrations, liquid expulsion as nanodroplets and microjets, and droplets explosion and detachment) produced by laser beam on droplet and its shape evolution.

## Droplet Vibrations

In Fig. (**6** and **7**) is shown a recorded sequence of movements of a droplet after interaction with a laser pulse at different time moments after impact. The laser beam has 0.4 mJ energy, or 1.83 GW/cm$^2$ irradiance and propagates in the equatorial plane of the droplet the focus being placed in its geometrical centre. Significant inhomogeneities are produced around focus, which after 200 µs generate shock waves in droplet that propagate at ~0.3 m/s [4]. These waves are, most probably, reflected on the capillary and at droplet's surface, which fact produces the modification of droplet's shape. The waves are then damped down in relatively short time intervals.

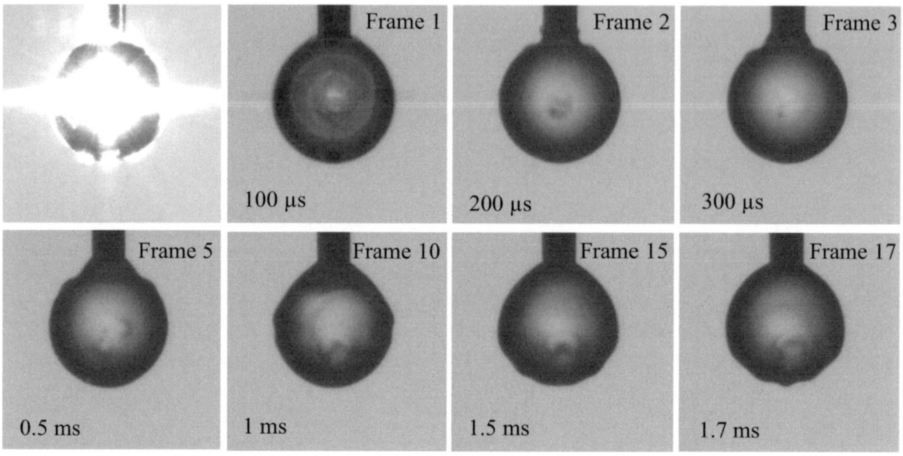


**Fig. (6).** Effects produced on a single pendant droplet by one laser pulse sent in its equatorial plane and focused in droplet's centre. (Movie **6** and **7**); $V_2$=3.5 µl, $E_2$=0.4 mJ, frame repetition rate 10 kfps (frame capture time 100 µs). Specified time represents the time interval that passed from the interaction moment.

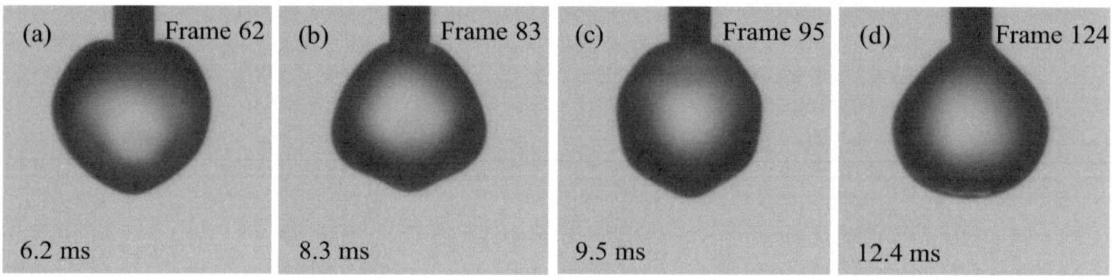


**Fig. (7).** Vibration modes identified in droplet evolution associated to resonance frequencies reported in Strani and Sabetta [30] of order (**a**) II, (**b**) III, (**c**) IV, and (**d**) I (Movies **6** and **7**). The same frame sequence is taken as in Fig. (**6**).

The power density corresponding to laser beam energy (irradiance 1.83 GW/cm$^2$) is an order of magnitude smaller than the values reported in literature for breakdown threshold produced in water by 532 nm beam which is 10 GW/cm$^2$ [14, 21, 33, 34]. Thus, the inhomogeneities produced in the focus region can be induced by electrostrictive effects combined with the possible local thermal effect. If produced, local overheating of the liquid to vapor phase followed by sudden expansion by increasing the vapor pressure in the formed microcavity (precursory to breakdown effect) may contribute to the "emission" of micro- and/or nano-droplets by the parent droplet [21]. This heating process strongly depends on laser beam energy. Literature reports show that the lowest absorption of water takes place at 420 nm where the absorption coefficient is 0.0062±0.0006 m$^{-1}$ at 25°C and at 532.5 nm it is 0.0447 m$^{-1}$ with a standard deviation of 0.0017 m$^{-1}$ [22, 35]. Consequently, the heating process is only local in droplet, in focus, and does not have a significant influence on droplet's thermodynamic temperature as a whole.

The shock waves propagate to surface causing droplet deformation and the generated surface waves are reflected from capillary tip on which droplet is hanging. As it results from Fig. (**6**), surface waves are built at around 100 μs from the impact with laser beam and they dominate the interaction scene from 300 μs onward. They propagate on droplet to its South Pole where merge after 1.7 ms engaging in oscillation the entire volume of liquid droplet.

Axi-symmetrical deformation of droplet due to the movement of fluid along droplet vertical axis (elongation or flattening) produces axi-symmetrical oscillations of it (mechanical vibrations). In the first stages of droplet deformation, the oscillations are chaotic and there is a large number of produced resonant frequencies associated with oscillation directions of droplet [30]. At about 6.2 ms after interaction (Fig. **7** contains frames that follow those given in Fig. **6**), vibration modes can be seen as those determined experimentally by Bisch *et al.* [36] and by computer simulations reported by Strani and Sabetta [30]. In these studies, it was considered as initial perturbation (initial conditions) the axi-symmetrical deformation (elongation) of the droplet along its vertical axis.

In Bisch *et al.* [36], the investigated physical system consisted of a droplet in partial contact with a solid surface, respectively a sessile droplet on the top of a cylinder with ratio $r_0/R = 0.75$, where $r_0$ is the outer radius of the cylinder, and $R$ the radius of undistorted droplet. Theoretically, have been investigated other cases such as $r_0/R = 0.17$ and $r_0/R = 10^{-4}$ [30].

In the experiments, starting from different initial conditions of droplet perturbation were observed vibration modes corresponding to the first four resonant frequencies identified in Strani and Sabetta [30]. Thus, for a ratio $(r_0/R) = 0.27$, where $r_0 = 0.25$ mm is the radius of capillary tip and $R = 0.94$ mm, the radius of 3.5 μl droplet, modes of vibration highlighted are associated in order of their sequence to the II$^{nd}$ order at 6.2 ms, to III$^{rd}$ order at 8.3 ms, to IV$^{th}$ order at 9.5 ms, and to I$^{st}$ order at 12.4 ms, respectively (Figs. **7a-d**).

It was also observed that immediately after interaction, the resonance frequency increases with the increase of the amplitude of droplet oscillations and, at the same time, the order of associated vibration modes increases. Then, in time, the oscillations are damped under the action of gravitational force.The vibration mode of the I$^{st}$ order may be observed as shown in Fig. (**7d**).

## Droplet Explosion and Detachment

After impact, the shock waves reflected by capillary tip generate surface waves at droplet's surface leading to medium or strong deformation function of laser beam energy. Such inhomogeneities are shown in Fig. (**8**) where higher pumping energy was used. The scattered light and the shock waves apply pressure on water droplet, so that water jets are further generated, as well as water cylinders. Droplet explosion may be also produced. All these effects are usually followed by partial or total detachment of droplet from the capillary tip.

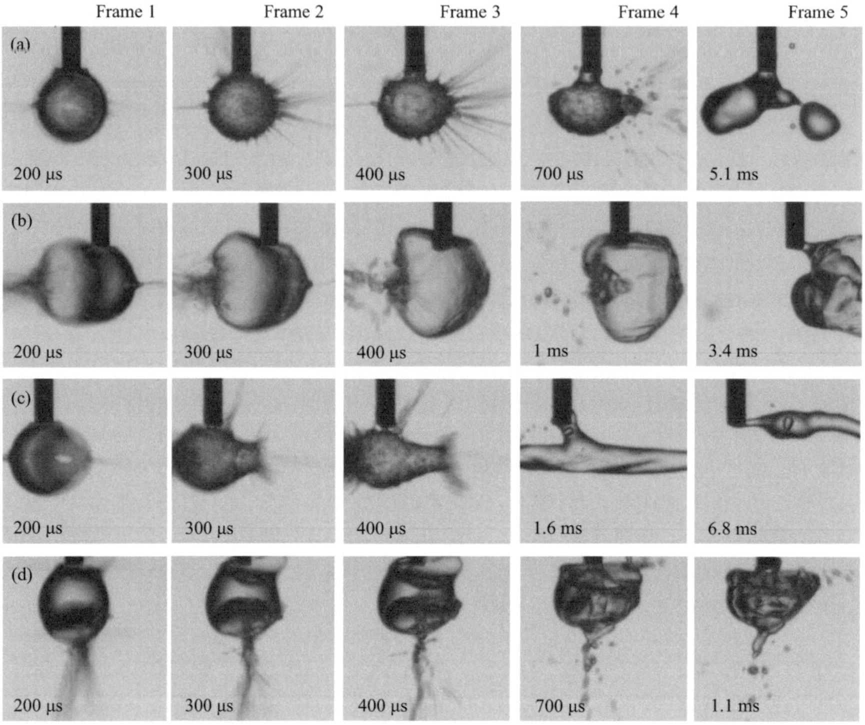


**Fig. (8).** Droplet deformation after laser beam-droplet interaction function of focus point position. Laser beam sent in droplet's equatorial plane and focused in (**a**) its centre (Movie **8a**), (**b**) front face (Movie **8b**), (**c**) back face (Movie **8c**) and (**d**) its South Pole (Movie **8d**) according to Fig. (**2**). Drop volume $V_2$=3.5 μl, beam energy $E_4$=1 mJ, frame repetition rate 10 kfps (frame capture time 100 μs). Specified time represents the time interval passed from interaction moment.

Measurements made in parallel with a thermocamera evidence that the droplet's overall temperature (defined as a thermodynamic property of droplet's material) is not increased during or after interaction with a single laser pulse when its energy ranges from 0.25 to 1 mJ [27]. Actually, the droplet temperature is constant within the measuring errors with the thermocamera. But, around focus point, the local temperature can increase suddenly, overheating droplet material. The liquid begins to boil if temperatures exceed the temperature of saturated vapor and an internal vapor cavity is generated [21]. At the used power densities breakdown, effects at droplet surface or within its volume are avoided [3]. So, the inhomogeneities produced in the interior of the droplet in focus area may be dominantly produced by the electrostrictive effect induced by the pulsed laser beam on water that is a dielectric medium [37, 38] or/and by the ablation due to local boiling of droplet's liquid [3].

It has been demonstrated that after interaction with a laser beam that has the energy larger than a given threshold, instabilities are produced in bulk liquid water that look and behave like acoustic instabilities produced in plasma [38]. This is due to the fact that water being a dielectric fluid flows under interaction with laser beam from the space where it has lower energy density towards the volume where this density is higher. The process is converse if compared to water natural behaviour which normally flows from high to low pressures. These inhomogeneities produce in the droplet when thermodynamic temperature is constant, stimulated Brillouin scattering (SBS) and stimulated Rayleigh scattering. Spontaneous Brillouin and Rayleigh scattering are not significant in this case since the laser beam power is higher than the threshold above which stimulated scattering phenomena are produced. According to [39], the threshold beam power to produce SBS, is:

$$P_{th} = (G_{th}\lambda)/2g \quad \textbf{(1)}$$

where $G_{th}$ = 25 is the gain parameter at threshold, $\lambda$ the beam wavelength (532 nm) and g the SBS gain factor, having here the value $g$ = 0.048 m/GW. The resulting $P_{th}$ is 13.85 kW that is lower than the laser beam power (of the order of hundreds of kW) used to obtain the effects shown in Fig. (**8**).

On the other hand, in the case of laser beam focused in droplet's geometrical centre, light scattered is angularly distributed around droplet (see Fig. **6**). Normally, SBS is produced in bulk samples in the beam backward sense if the sample is long enough [38]. Here, in a droplet irradiated with 1 mJ beam energy (Fig. **8a**) this effect has lower intensity since its diameter is small (~2 mm), with the consequence that the most light is scattered around it. This result may confirm the electrostriction mechanism leading to light scattering by the instabilities produced in droplet in the focus point. As a matter of fact, the mechanism is already suggested for compressible fluids and is extended to droplets of water that interact with higher energy laser beams slightly absorbed by them [38].

In Fig. (**8a**), electrostrictive inhomogeneities induced in one single droplet by a pulsed laser beam apply a force on water that has a peak at the contact point between light and droplet surface. Micro- and nano-droplets as well as microjets of water material are expelled along droplet radiuses in the respective points. The jet emitted at impact point with droplet propagates faster than the other jets, as may be observed in Frame 1, Fig. (**8a**) which records events along 100 μs. This microjet is due to pressure generated by inhomogeneities induced in droplet and favoured by light pressure exerted on it. Slower motions within droplet are also produced that lead to detachment of liquid "portions" from it or, in some cases, of the droplet itself from the capillary. This effect is concomitant with emission in 4π around droplet of micro- and/or nano-droplets which are not forming jets and may coalesce (Fig. **8a**, Frame 4 at 700 μs). In Fig. (**8a**) the time necessary for droplet or parts of it to leave the capillary is around 5.1 ms.

If the laser beam is sent in droplet's equatorial plane, focused on its front surface at 1 mJ pulse energy, one generates explosive effects on droplet that destroy it as results from Fig. (**8b**). The process is produced in a few milliseconds by shaping air bubbles in the water droplet and by detaching it from the capillary on which it hangs. In Fig. (**8c**) interaction effects are shown for a laser beam sent in droplet's equatorial plane and focused on its back surface. Here, besides a cone structure formation, a picture similar to electrostrictive effect is obtained of the kind shown in Fig. (**8a**) where the beam is focused in droplet's centre. Important expulsions of water jets as well as droplets are produced along other directions than backward or forward and the associated collars. In Fig. (**8c**), first frame shows droplet's shape at the beginning of the conic structure construction. A strong deformation of the second droplet's hemisphere is observed, followed by cone and associated collar production and movement. Droplet detachment time from capillary is approximately 6.8 ms. Another observation about this kind of interaction is that the capillary once touched by shock waves produced in the droplet, contributes to expulsion of liquid under the form of jets or micrometric droplets, as results from Fig. (**8c**). In Fig. (**8d**) is shown the result of laser beam interaction with droplet; the pulsed beam propagates horizontally being focused on South Pole of droplet. In this case, it is critical to control the focus position on surface considering that the 90 μm spot diameter may be entirely included in droplet under its surface, or only a fraction of it. If focus volume is included in droplet, in its surface proximity, refraction phenomena [40] are produced. The total reflection of laser beam on droplet's surface may take place, and in this case the light wave propagates under droplet surface within a thin layer [4, 18]. The interaction of the shock wave generated in focus point with the capillary produces violent emission of material in the neighborhood of the capillary tip. Once the laser beam penetrated inside the droplet, the inhomogeneities induced by electrostrictive effects and laser ablation near surface, produce emission of nanodroplets which propagate with relatively high speed in a direction along the vertical axis of the droplet, respectively orthogonally on the laser beam axis (Fig. **8d**). After 300 μs (Fig. **8d**, Frame 2) a pillar is generated following the same direction as the expulsed nanodroplets. The pillar generates finally micrometric droplets. The capillary plays a key role in enhancing jets emission around it and in droplet detachment from it [4].

Focused beams induce relatively important electrostrictive inhomogeneities that produce intense light scattering in droplet and associated emission of nanodroplets, microjets, pillars and droplet detachment. The time interval of droplet detachment from capillary varies function of focus point position and pumping beam energy.

## Emission of Nanodroplets and Microjets

A process evidenced in unresonant interaction experiments for all irradiation geometries is generation of small droplets, microjets and pillars along beam propagation direction both in backward and forward sense. The intensity of these effects, depend on laser beam energy, focus point position and droplet volume [4].

In Fig. (**9**) are shown the results of laser beam interaction with a 3 µl droplet for three focusing positions, each at two energies that are lower than those used in Fig. (**8**). Sequence 1 in each figure shows the effects for $E_1$=0.25 mJ, and sequence 2 for $E_2$=0.40 mJ. Time delay of frames recording is set so that the first 100 µs frame shows the effects produced in the first 10 µs after interaction, when only an intense light scattering is present. In Frame 1 of all sequences some inhomogeneities are produced, but no important mechanical effects are evidenced. The next frames show mechanical deformations that have shapes and intensities which depend on beam focus position, respectively beam energy.

Strong mechanical effects are absent when low energies are used and laser beam is focused in droplet's geometrical centre (Fig. **9a**). Fig. (**9a**), sequence 2, shows phenomena in the same category as those presented in Alexander *et al.* [41] and Utsunomiya *et al.* [42]. Following Fig. (**9a**), sequence 2, Frame 2, one may conclude that emission of nanodroplets which form jets is produced more rapidly at input of the beam in the droplet than at output. Jet speeds were computed by measuring the length of nanodroplets trajectories in the image and performing time-of-flight measurements. The jet speed at input is 13.6 m/s and 5.1 m/s at output.

In all cases of impact of laser beam with droplet in its equatorial plane on front surface (Fig. **9b**), at contact point between the beam and the droplet, a jet of water nanodroplets is emitted backward with respect to beam propagation sense; the same is produced forward, at output from the droplet. After interaction, nanodroplets and microjets are evidenced firstly in backward direction having a very high propagation speed (Fig. **9b**, Frames 1). At a later stage and backwards, lower speed pillars (water cylinders) are emitted which have a central cone structure surrounded by a liquid crown (Fig. **9b**, Frames 3). The form of these jets strongly depends on laser beam energy, focusing conditions and droplets dimensions. Their propagation speeds vary from 4.4 m/s to 20 m/s.

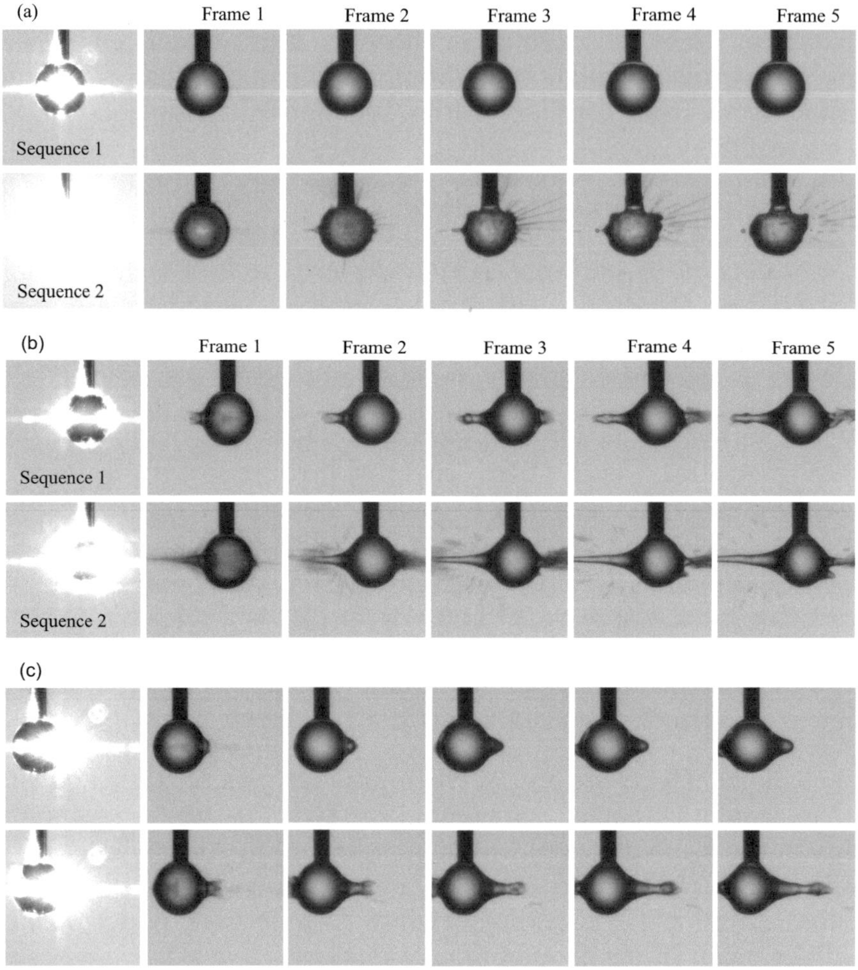


**Fig. (9).** Nanodroplets, microjets or pillars generation function of laser beam focus point position and energy. Laser beam sent in droplet equatorial plane, focused in (**a**) its centre, (**b**) front surface and (**c**) back face. Irradiation energy $E_1$=0.25 mJ in sequence 1, $E_2$=0.40 mJ in sequence 2; sequences show consecutive frames from 10 μs after impact in start frame to100 μs in the other frames up to 510 μs in Frame 5. Droplet volume $V_1$=3 μl; frame repetition rate 10 kfps (frame capture time 100 μs).

In all images, starting with Frame 2 backward jets spread at lower speeds and have variable shapes function of beam-droplet impact conditions, beam energy and droplet's volume [4].

Other expulsions of liquid are produced at latter stages at output of laser beam from droplet. This takes place between 100 and 400 μs following the beam penetration of the droplet. Expulsion speeds vary in this case from about 4.4 to 10.2 m/s. These speeds are lower than corresponding backward jets speeds.

In all studied cases of laser beam interaction with droplet at its front face, the central jet is surrounded by a conical crown constituted of small diameter droplets. The crowns have a somewhat regulated form (Fig. **9b**, sequence 2, Frame 2), function of beam power, focus place and droplet volume. According to Thoroddsen *et al.* [7], such a cone-crown system shows-up function of the actual focus deepness position inside droplet which is in these cases of the order of 10 - 30 µm. This may be the case in the presented experiments although the conditions reported in Thoroddsen *et al.* [7] are quite different from those utilise here. The formation of a crown could originate in light pressure exerted by the rays distributed at the periphery of focus spot on water.

If the beam focus is positioned in the equatorial plane perpendicularly on the back surface (Fig. **9c**), one produces the same kind of effects as if the beam is focused on droplet's front face.

In Fig. (**9c**) the interaction consequences of a 0.4 mJ beam with a 3 µl droplet are shown. The effects at front surface have a lower intensity whereas at the back, the formation of a central jet surrounded by smaller collar is evidenced (Fig. **9c**, sequence 2, Frame 1). The jet and the collar propagate with the same speeds as in front face interaction and this is finalized with the production of micrometric size droplets and even the parent droplet detachment from capillary if high energy levels are used (Fig. **8c**).

The deformations produced by the pressure exerted by the laser beam on droplet, as shown in Fig. (**9c**), have the same forms as described in Chraibi *et al.* [18] for fluid-fluid interface deformation by exposure of one fluid to a cw laser beam. In the experiments reported here, similar effects were obtained while using a pulsed laser beam slightly absorbed by water at fluences between 4 and 17 $J/cm^2$ when the beam is focused in different points on the surface of a droplet that hangs in air. The jets emitted in forward direction have characteristics somewhat similar with those described for front face interaction.

On the other hand, in Fig. (**8c**), the droplet is detached from capillary, the detachment time being around 6.8 ms. It is shorter than in Fig. (**9c**) which starts after 10 ms, as shown in Pascu *et al.* [4], and this may be produced by the higher energy of the pumping laser beam utilized in Fig. (**8c**) combined with larger droplet volume. The observation related to faster production of detachment for larger droplets could be an argument in favor of the hypothesis that water behaves in these experiments as a Newtonian fluid with normal inertial properties. The effects described above may be understood supposing that water has unchanged viscosity during experiments [43].

The ratio between the pressure applied by laser beam on droplet, water viscosity within droplet and its surface tension should be considered. During the interaction laser beam-water droplet, a shear stress is produced on water and it tends to flow so that layers of water within droplet move with various speeds that are function of water viscosity. The shear stress F, may be calculated considering the model described in [44] that supposes that water is an homogeneous and isotropic liquid in which two plane plates are introduced placed at distance y between them and having surface area in water A larger than the area of beam impact spot on droplet. If plate/droplet boundary conditions are neglected, a shear stress F applied on one moving plate produces a water shear flow between plates with the velocity gradient u; if μ is water viscosity, the shear stress is given by:

$$F = \mu A u / y \tag{2}$$

For water, at 20°C, $\mu$ = 1.002 mPa·s, or $1.002 \cdot 10^{-3}$ Ns/m$^2$.

At the same time, *F* is the pressure force $F_p$ given by:

$$F_p = E/(c\Delta t) \tag{3}$$

where *E* is beam energy, *c* is light speed in water (225,407,863 m/s) and $\Delta t$ =6 ns (laser pulse full time width at half maximum) [45, 46].

Considering that upper, immobile plate is the capillary that is always tangent to droplet's surface, one may compute the velocity gradients u shown in Table **1**, for beam energies and droplet diameters used in the experiments; this is made for a laser beam focused on droplet's front face and propagating in its equatorial plane. The mobile surface with area A is described by laser beam in droplet, and A is given by the product between focus diameter, 90 μm, and droplet diameter.

**Table 1. Water velocity gradients produced in droplets by the laser beam emitted at 532 nm.**

| Energy $E \times 10^{-3}$ (J) | Force exerted by laser beam $F \times 10^{-3}$ (N) | Volume (μl) | Diameter × $10^{-3}$ (m) / $y \times 10^{-3}$ (m) | u (m/s) |
|---|---|---|---|---|
| $E_1 = 0.25$ | $F_1 = 0.220$ | 3 | $D_1 = 1.79$ <br> $y = 0.895$ | $u_1^{(1,2,3)} = 1{,}219.8$ |
| | | 3.5 | $D_2 = 1.88$ <br> $y = 0.94$ | |
| | | 4 | $D_3 = 1.97$ <br> $y = 0.985$ | |

*(Table 1) contd.....*

| Energy $E\times10^{-3}$ (J) | Force exerted by laser beam $F\times10^{-3}$ (N) | Volume (µl) | Diameter × $10^{-3}$ (m) $y\times10^{-3}$ (m) | u (m/s) |
|---|---|---|---|---|
| $E_2 = 0.40$ | $F_2 = 0.355$ | 3 | $D_1 = 1.79$ $y = 0.895$ | $u_2^{(1,2,3)} = 1{,}968.3$ |
| | | 3.5 | $D_2 = 1.88$ $y = 0.94$ | |
| | | 4 | $D_3 = 1.97$ $y = 0.985$ | |
| $E_3 = 0.70$ | $F_3 = 0.621$ | 3 | $D_1 = 1.79$ $y = 0.895$ | $u_3^{(1,2,3)} = 3{,}443.1$ |
| | | 3.5 | $D_2 = 1.88$ $y = 0.94$ | |
| | | 4 | $D_3 = 1.97$ $y = 0.985$ | |
| $E_4 = 1.0$ | $F_4 = 0.887$ | 3 | $D_1 = 1.79$ $y = 0.895$ | $u_4^{(1,2,3)} = 4{,}917.9$ |
| | | 3.5 | $D_2 = 1.88$ $y = 0.94$ | |
| | | 4 | $D_3 = 1.97$ $y = 0.985$ | |

The computed $u$ values show that the speed of water layers moved inside the droplet by interaction with the beam has supersonic gradients covering the approximate range 1,200 m/s to 4,900 m/s which depends on beam energy and droplet's volume. From calculated values (Table **1**), taking into account that the capillary's material is hydrophobic it results that, for a given droplet volume, the velocity gradient depends mainly on pressure force/beam energy. The computed values show a good agreement between the emission supersonic speeds of jets or nanodroplets in the first phase of laser beam-droplet interaction.

At the same time, the interaction speed of laser beam with a single water droplet is very high. Considering the light speed in water (refractive index n = 1.33) at room temperature and normal air pressure, c = 225,407,863 m/s and the time intervals in which light crosses droplets are: for 3 µl ($D_1$ = 1. 79 mm) $t_1$ = 7.94 ps, for 3.5 µl ($D_2$ = 1.88 mm) $t_2$ = 8.34 ps and for 4 µl ($D_3$ = 1.97 mm) $t_3$ = 8.74 ps. In 8 ps, for instance, the propagation length of the wave due to velocity gradient varies from 9.75 nm ($E_1$ = 0.25 mJ) to 39.3 nm ($E_4$ = 1 mJ) so that at the recording speeds used, one may consider that during propagation of laser beam through droplet, this one is not changed. All fluidic processes are produced at lower speeds and are measurable after the laser beam interaction with droplet's material comes to an end, *i.e.* the beam already crossed the droplet. More, considering that the laser pulse duration in droplet is 5 ns, resulting from its time width, after the first photons leave droplet, there are still incoming photons for a remaining time interval of around 4.992 ns. Due to velocity gradients shown in Table **1**, the time necessary for the wave produced by laser beam in droplet to propagate in it, is approximately 1.5 μs for $E_1 = 0.25$ mJ and 0.4 μs for $E_4 = 1$ mJ and there is no time for the laser beam to interact directly within droplet with the pressure waves produced by it.

Laser beam may induce electrostriction in water droplet [7, 29, 37, 47] which leads to formation of well-defined bulge and jet. If the focus is precisely positioned on droplet surface layers, after few hundreds of μs the front jet emits microdroplets in backward sense even if energy is small (Fig. **10**) and the droplet is not destroyed. During experiments, more bulges were noticed in some cases; an individual bulge is a cone that evolves in such a way that at the tip, a cylinder is formed with a cross section diameter of some microns. In most cases, function of laser energy and focus condition, liquid continues to flow inside (in the core) of formed cone with a higher speed than that of neighbour surface layers [48]. Laminar flow of liquid stretches the cone and continuously accumulates at the end of it forming a new bulge. Surface tension can no longer keep the bulge linked of cone since the elastic tendency of fluid surface is to acquire the least surface area possible [43]. The surface tension pulls the bulge, forming a droplet with a diameter of few tens of micrometers which is detached from cone; at the same time, the remaining long cone having the basis in droplet evolves towards reintegration in it (Fig. **10**). This process lasts between a few hundreds of μs to some ms.

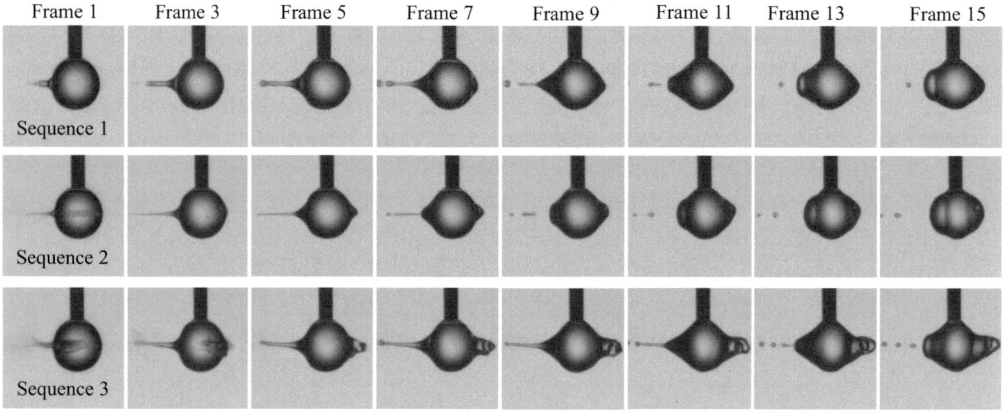


**Fig. (10).** Backward generation of nanodroplets and microjets (cones) function of deepness of focus point on droplet front face (Movie **10 seq. 1**, **seq. 2**, **seq. 3**, respectively). Sequences shown the odd consecutive frames from 100 μs to 1,500 μs obtained in the same experimental conditions. Pulse energy $E_1$=0.25 mJ; droplet volume $V_2$=3.5 μl; frame repetition rate 10 kfps (frame capture time 100 μs).

Fig. (**10**) shows the same experiment repeated three times (sequences 1 to 3). The only difference between experiments is that droplets are slightly different from each other in volume and shape since they are generated by a computer controlled syringe system; for the droplets generated in this experiment the difference is approximately 0.05 mm [4]. Variable effects on droplet may be produced by the relatively small differences of laser beam energy from pulse to pulse, due to laser jitter functioning. The intensity distribution of the laser beam in its cross section may be also associated with jitter fluctuations.

According to [37] the surface tension force applied at the tip of cone originating from the parent droplet is:

$$F = 2\pi r \sigma \sin\theta \quad \textbf{(4)}$$

where r is the radius of the cone section measured from the symmetry axis of the cylinder, σ is water surface tension and θ is the angle made by the normal at cone surface with its symmetry axis.

The force F may be considered as the threshold starting from which droplets may be detached from the cone. Knowing F, r may be calculated considering it as the radius of the produced droplet (Fig. **10**). In equation (4) the value $\sigma = (72.4 \pm 0.6) \times 10^{-3}$ N/m is extracted from [17] and sinθ is around 1. The forces that contribute to the cone shaping at surface are the pressure force exerted by laser beam on droplet and the shear stress force related to the water viscosity that has the computed values shown in Table **1** and may be opposed to light pressure force given the fact that water is isotropic. The light pressure force may be calculated as shown above considering that c is the speed of light in air, *i.e.* 299,102,690 m/s [45]. For the beam energies utilised in experiments, laser beam exerted forces are $0.166 \times 10^{-3}$ N for $E_1 = 0.25$ mJ, $0.266 \times 10^{-3}$ N for $E_2 = 0.40$ mJ, $0.466 \times 10^{-3}$ N for $E_3 = 0.70$ mJ, and $0.667 \times 10^{-3}$ N for $E_4 = 1.0$ mJ.

Using these values and those given in Table **1**, the corresponding r values are 0.118 mm for energy $E_1$, 0.196 mm for $E_2$, 0.34 mm for $E_3$, and 0.486 mm for $E_4$. The measured diameters of the microdroplets expulsed from parent droplets and shown in pictures presented here are found in this dimensions range. According to images presented in Fig. (**10**), the cones that are deforming droplets are developed in, at least, three stages. In the first, produced very fast, micro- and nano-droplets are expulsed from parent droplet at supersonic speeds or even a very thin liquid cylinder is produced.

In the second, produced slower, a larger cone with basis of hundreds of μm is shaped. Towards the end of this stage one or, usually, more droplets with larger diameters are expulsed from the tip of cone, at much lower speeds, as results from Fig. (**10**), Frames 11-15. The diameters and volumes of expelled droplets are different from case to case. Their propagation speeds are of the order of some tens of m/s. The third stage contains cone retraction in water droplet mass which is in fact the damping of the oscillations generated by the interaction with the laser beam. It may take milliseconds to reach the droplet static equilibrium state.

Static and dynamic dimensions of nanodroplets/microjets can be determined by comparing them with outer diameter of the capillary tip. This is known with precision (510 µm) and it can be used as a benchmark for determining the actual length corresponding to a pixel in the image. Thus, by graphical analysis on each frame, the length of a segment is determined by adding the pixel number corresponding to that segment (Fig. **11**). Propagation velocity of a microjet is calculated by determining the propagation length during a frame exposure from two consecutive frames, one used as reference as in Fig. (**11c**). Knowing the frames recording rate, 10 kfps, to each frame corresponds an exposure time of 100 µs. Events recorded during frame exposure overlap in the respective image. Thus, knowing the distance traveled by jet and the travel time one can determine the propagation velocity of the jet.

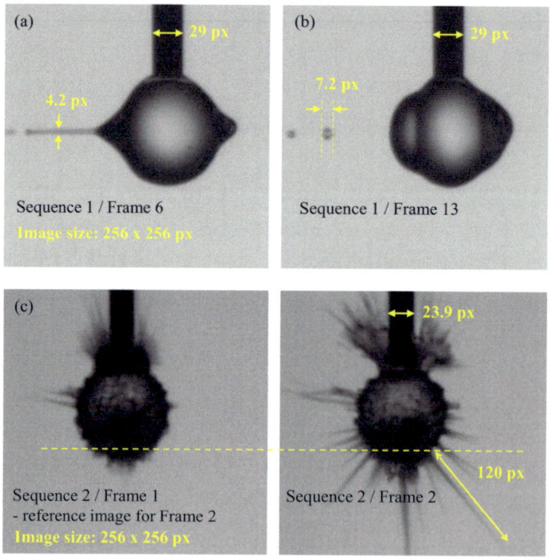


**Fig. (11).** Dimensional characteristics of liquid structures ejected from droplet by a laser pulse that propagates in droplet's equatorial plane being focused on (**a**)-(**b**) its front face, and (**c**) its centre. Pulse energy $E_1$ = 0.25 mJ in (**a**) and (**b**), and $E_4$ = 1 mJ in (**c**); droplet volume $V_2$ = 3.5 µl; capillary outer diameter 0.51 mm; frame repetition rate 10 kfps (frame capture time 100 µs); dimensions are specified in pixel units. Images (**a**) and (**b**) are extracted from the same sequence at different time intervals with respect the interaction moment.

In the experiments reported here were identified individual droplets with linear dimensions of the order of tens of microns, and volumes of 1 nl (Fig. **11b**). These have relatively low speeds, being generated at about 1 ms after interaction (Fig. **10**, Frame 11) compared to aerosols and microjets, which are ejected immediately after interaction (Fig. **10**, Frame 1).

In Figs. (**11a** and **b**) are shown two frames recorded at 0.6 ms, respectively 1.3 ms after the interaction of a laser pulse with droplet, and focused on front face of it. After interaction, a jet of liquid is generated in backward direction followed by its segmentation due to the action of surface tension force and formation of individual droplets with volumes of nl order, as is the case in Fig. (**10**).

Thus, using a graphic image analysis program, Image J free software [49], it was determined the dimension in pixel units of outer diameter of capillary tip, as well as those of nanodroplets and microjets generated after interaction (Fig. **11**). The generated liquid jet has the waist around 0.07 mm (4.2 pixels) as shown in Fig. (**11a**); droplets have diameters around 0.13 mm (7.2 pixels) and corresponding volumes of 1 nl (Fig. **11b**); propagation speeds of jets after impact are approximately 25 m/s (Fig. **11c**).

When the focus of the pulsed laser beam is fixed with accuracy on droplet's surface, the values of the beam intensity in the cross section of the focal plane may produce effects on droplet like those shown in Fig. (**12a**). The two liquid cylinders expelled from droplet in the forward sense may be due to the "single mode" laser beam that, nevertheless, does not have a precise Gaussian intensity distribution in the cross section and consequently, in the focus. In cross section, there are two intensity peaks that propagate through the droplet and produce two divergent jets related to their propagation directions. As a consequence of the energy densities of light beam within droplet, water is expulsed by the light force originating in the two peaks. This effect is also related to the laser beam waist of 90 μm. The effect may be correlated with results reported in Utsunomiya *et al.* [42] who measured the pulse laser ablation at a liquid surface in a different geometry than pendant droplet hanging in air and in other focusing conditions. In Utsunomiya *et al.* [42] it was demonstrated that the shape and behaviour of the expulsed material at water surface when interaction with a focused laser beam sent from water onto air is produced, depend on the intensity distribution within the beam cross section: top-hat, Gaussian and torus distributions were studied.

The deformations have conical forms that may propagate backward or forward if compared with the laser beam propagation sense; they are function of the intensity distribution in the beam focus and of the accuracy with which the focus is placed on droplet's surface. If there are more intensity peaks in focus waist, one may obtain more conical deformations as shown in Fig. (**12a**). The collision with the photons of the laser beam may have also explosive effects without droplet detachment, which are described as ablation processes [3, 21].

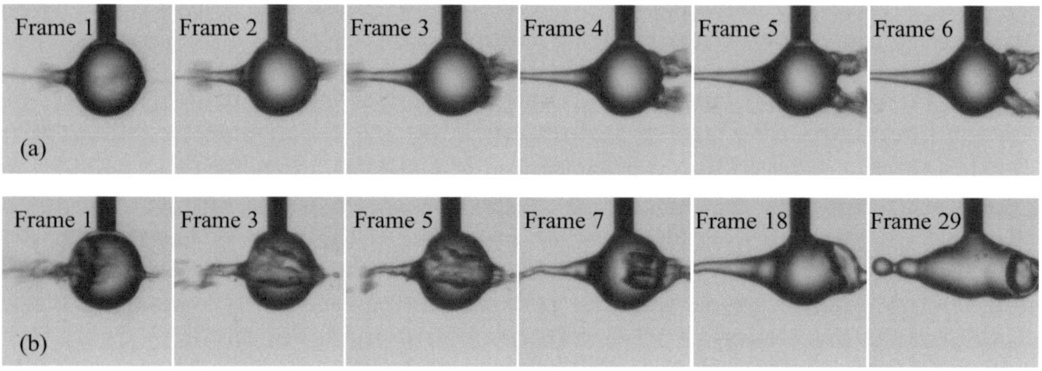


**Fig. (12).** Effects produced on droplet by laser beam that propagates in droplet's equatorial plane and is focused on its front face. Pulse energy (**a**) $E_2 = 0.4$ mJ (Movie **12a**), and (**b**) $E_3 = 0.7$ mJ (Movie **12b**); droplet volume $V_3 = 4$ µl; frame repetition rate 10 (frame capture time 100 µs).

First, even for nonviolent effects the appearance of backward central cones originating from the focus placed on front surface, accompanied by conic collars is supplied from surrounding water. Consequently, in droplet's volume is shaped an air/gas cylinder that is coaxial with the cone and propagates the more within droplet the longer the cone (Fig. **12b**).

After laser beam penetrates in the second half of droplet passing through its centre, a forward cone is produced that also extracts water from its neighborhood in droplet (Fig. **12b**). So, depending of beam energy, focus position and droplet volume, the cylinder may break through droplet for a short time or, at least, a large air bubble and possibly smaller diameters air bubbles are produced within the droplet.

From Fig. (**8**) it results that for a given pulse energy and droplet volume, the direction and shape of liquid ejected from droplet (aerosols, nanodroplets or microjets) can be controlled by changing the position of focus point of the laser beam inside droplet and on its surface. The microjet emission directions can be controlled by a fine positioning of beam focus position around droplet's centre (Fig. **13**). From these images it results that the position of focus with respect to droplet's geometrical centre is critical. This is true for all droplet volumes and for all laser beam powers.

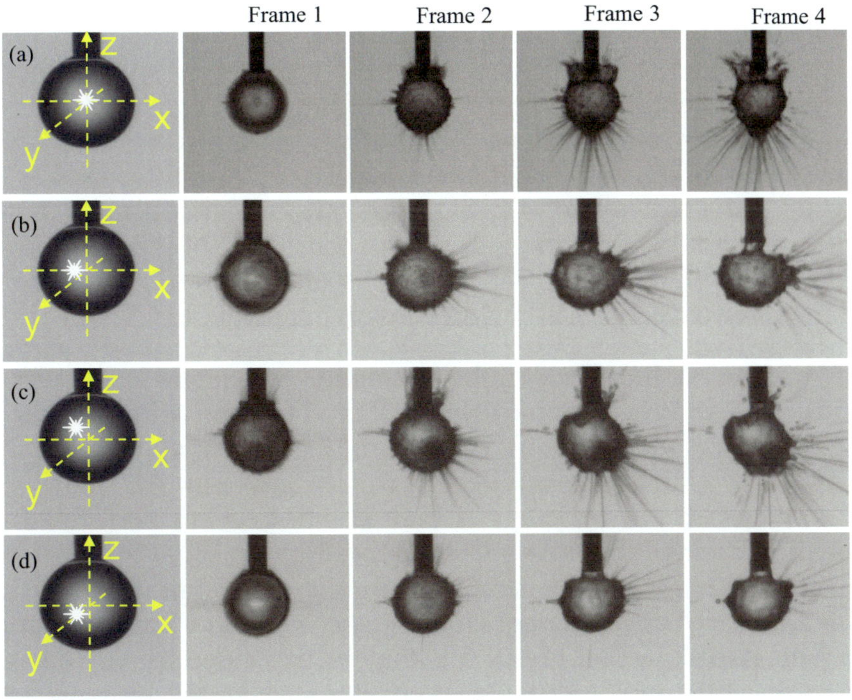


**Fig. (13).** Control of nanodroplets and microjets emission directions by a proper positioning of the focus around droplet's geometrical centre; (**a**) on-axis, in centre (Movie **13a**), (**b**) on-axis, in front of it (Movie **13b**); (**c**) off-axis, in front-of it and in upper position (Movie **13c**); (**d**) off-axis, before it and in lower position (Movie **13d**). Sequences show consecutive frames from Frame 1 to 4. First image in each sequence shows the scheme of local irradiation geometry. Pulse energy $E_4$ = 1 mJ in (**a**)-(**c**) cases and $E_3$ = 0.7 mJ in (**d**) case; droplet volume $V_2$ = 3.5 μl for (**a**)-(**b**), and $V_1$ = 3 μl for (**c**)-(**d**); frame repetition rate 10 kfps (frame capture time 100 μs).

## Dynamics of Droplet Deformation Containing Laser Dyes

In the experiments reported here it was analysed the dynamics of droplets deformation at resonant interaction with a laser pulse when a single droplet contains only solvent (distilled water or alcohol) or a solution that includes a solvent doped with a laser dye at $10^{-3}$ M concentration.

Measurements were performed on individual droplets containing solution of water doped with Rh6G laser dye, of ethyl alcohol doped with Rh6G, of ethyl alcohol doped with DCM, and of DMSO doped with DCM.

Rh6G is one dye mostly utilized in tunable lasers being quite completely characterized from optical, spectral and thermal points of view [27, 50]. Its quantum efficiency that governs absorption and emission of radiation is relatively high which makes it one of the most efficient fluorophores in water or alcohol solvents. Rh6G solutions in water strongly absorb in green, at 532 nm.

The water is used as solvent because it has low evaporation rate and the droplets do not significantly loose material by vaporization, during experiments [13]. Ethyl alcohol is one of the most widely used solvents for laser dyes because they exhibit a quantum yield higher in alcohol than in combination with water [50]. In order to emphasize the influence of viscosity and surface tension of solvent and dye on mechanical effects produced by interaction with laser pulse, were used as solvents distilled water, ethyl alcohol and DMSO. Ethyl alcohol has viscosity 0.98 mPa·s, close to that of water (1.0 mPa·s) and surface tension (22.39 mN/m) lower than that of water (72 mN/m). DMSO has a viscosity (1.99 mPa·s) greater than that of water and a surface tension 43.54 mN/m at room temperature.

In Fig. (**14**) are shown the effects of the interaction of a single laser pulse of 1 mJ energy with a 3.5 μl single water droplet (Fig. **14a**) and with another of 4 μl containing ethyl alcohol (Fig. **14b**). For volumes smaller than 4 μl, ethyl alcohol droplets tend to climb on capillary due to low surface tension. Thus, measurements carried out on ethyl alcohol solutions were performed on 4 μl droplets to allow maintaining it in a hanging position. In Fig. (**14b**) it is observed that for 1 mJ laser beam energy, droplet deformation is explosive with a higher intensity in the case of ethyl alcohol although the drop has a volume that is larger with 33% with respect to Fig. (**14a**); this may be explained by the lower surface tension than that of water, *i.e.* the weaker resistance to the pressure generated from the focus region.

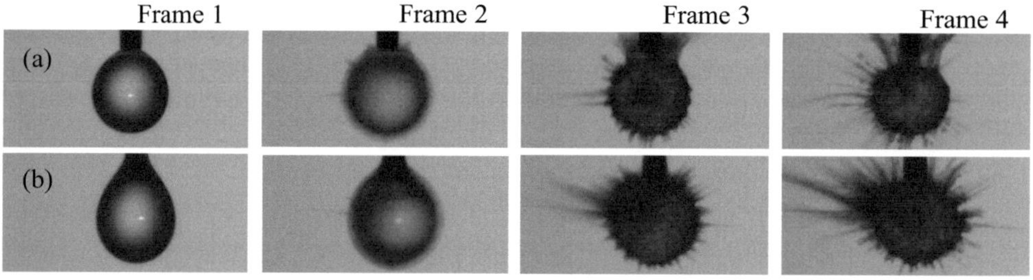


**Fig. (14).** Dynamic of droplet deformation for (**a**) water (Movie **14a**) and (**b**) ethanol (Movie **14b**) for a laser pulse that propagates in droplet's equatorial plane and is focused in its geometrical centre. Laser beam energy $E_4$ = 1 mJ, droplet volume $V_2$ = 3.5 μl for water and $V_3$ = 4 μl for ethyl alcohol; frame repetition rate 10 kfps (frame capture time 100 μs).

For droplet containing Rh6G water solution at $10^{-3}$ M, the amplitude of mechanical effects produced after interaction is attenuated (Fig. **15**). The reason is that the dye strongly absorbs the beam at 532 nm, so that most of the photons are used to excite dye molecules on excited singlet states.

The absorption takes place in the superficial layers of the droplet so that photons are not available anymore to further transfer their moments to droplet's content and to build significant light pressure effects on it. Also, effects are different on droplet containing aqueous solution (Fig. **15**) compared to those having ethyl alcohol or DMSO as solvent.

In Fig. (**15**) a Notch filter was used to filter scattered radiation at 532 nm, with the exception of Fig. (**15a**) used as control for comparison. In Fig. (**15**) the laser beam interaction at $E_4$ =1 mJ with the droplet is produced and the results are shown when the focus is placed in droplet's centre. The Frame 1 which captures only the first 10 μs after laser beam-droplet interaction shows an intense laser induced fluorescence following the absorption of 532 nm beam. The next frames show that only droplets containing water (Fig. **15a**) and alcohol (Fig. **15b**) doped with Rh6G suffer deformations (Fig. **15b**) or loss of material (Fig. **15a**, Frame 2, and Fig. **15b**).

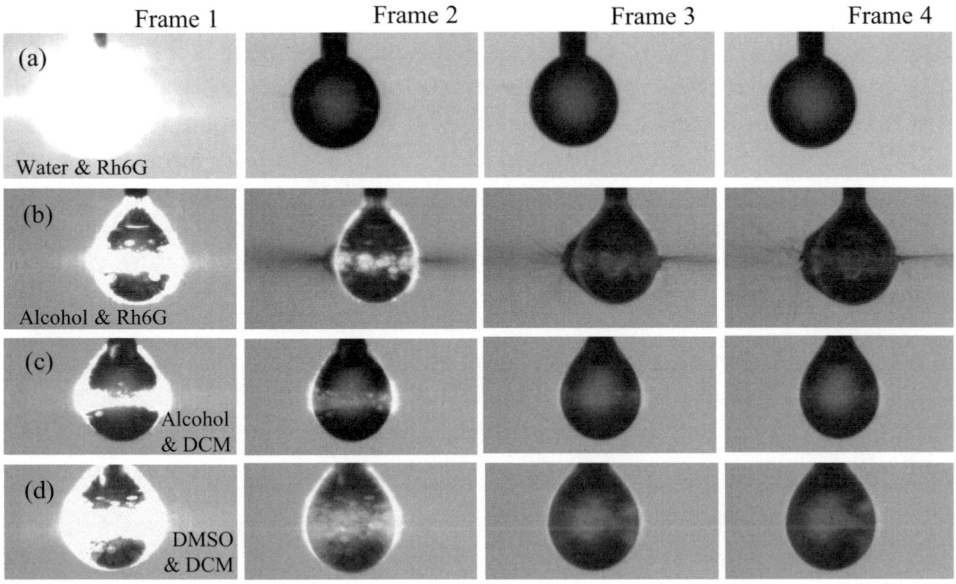


**Fig. (15).** Dynamics of droplet deformation containing $10^{-3}$ M Rh6G doped solution for a laser pulse that propagates in droplet equatorial plane, focused in its geometrical centre. Solutions contain (**a**) distilled water and Rh6G (Movie **15a**), (**b**) ethanol and Rh6G (Movie **15b**), (**c**) ethanol and DCM (Movie **15c**), (**d**) DMSO and DCM (Movie **15d**). Laser beam energy $E_4$ = 1 mJ; aqueous droplet volume $V_1$ = 3 μl; ethyl alcohol and DMSO containing droplet volumes $V_3$ = 4 μl; frame repetition rate 10 kfps (frame capture time 100 μs).

For solutions doped with DCM some inhomogeneities are produced, and only surface waves are observed (Fig. **15d**). These effects have not significant amplitudes and the next frames are not evidencing important droplet's deformations or vibrations. It seems that absorption characteristics of dye molecules play an important role since for the same solvent and irradiation conditions (Figs. **15b** and **c**) mechanical effects are totally different. Though, in Fig. (**15a**), Frame 2, a weak backward structured emission is noticed at input surface which is either a tinny solution jet or a succession of micro- and nano-droplets which propagate at lower speed. The strong absorption of green radiation by the Rhodamine molecules allows to propagate further in the droplet a relatively small number of photons able to apply pressure on droplet. One should also consider that the shape of laser beam in the neighborhood of the waist, makes available relatively constant photon densities 250-500 μm along the beam propagation direction.

On the other hand, from Fig. (**15**) one may conclude that surface tension plays an important role in observed mechanical effects, but properly choosing the laser dye with specific absorption properties one may control and reproduce these effects.

## CONCLUSIONS

Literature does not present too many data on laser beams interactions with droplets in suspended positions in open atmosphere. In studying a droplet in pendant position in air, there are experimental difficulties such as contamination of droplet's content with impurities and their association with surface tension and contact angles modifications, vaporisation of droplet's material and the resulting changes of their volumes, dimensions, shapes and components concentrations. In this chapter, results about the unresonant interaction of droplets with pulsed laser beams focused on them are reported.

Resonant interaction is also considered to outline the ratio between resonant and unresonant effects when droplets contain fluorophores such as laser dyes, but the conclusions are true for medicines solutions, as well. Basically, a droplet is an optical body.

At interaction with a laser beam a distilled water droplet has the refractive index $n_{water} = 1.33 > n_{air} = 1.003$ at room temperature and normal atmospheric pressure. The laser beam is refracted at droplet's surface that separates it from the surrounding air and once penetrating in the droplet it may be internally reflected up to total reflection, if the incidence angle allows it. Function of its wavelength, the laser beam is not absorbed in droplet at unresonant interaction or absorbed, at resonance.

The interaction beam-droplet is very fast, since the light needs about 8 ps to cross a distilled water droplet having approximately a 2 mm diameter. So, during propagation in droplet, the light finds its material in a stable, which may be considered at macroscopic scale "frozen", state from the fluidic point of view. All effects produced on droplet take place at later stages. The exception is the emission of laser induced fluorescence by fluorophores at resonant interaction, that is produced in a time interval comparable to laser pulse FTWHM.

The focal length of the set-up and the divergence of the laser beam at impact with droplet are critical during interaction process. It seems that if the focal length is longer and the divergence in interaction zone is lower, the area around focus directly affected by the beam is more confined and the beam power needed to obtain light pressure effects on droplet is smaller. This is critical as well because the errors in positioning the focus on droplet at its front surface, in the centre of it, or on its back surface may have a diminished influence on interaction results. The position in $4\pi$ of the beam impact point with respect to droplet's surface is also critical and it may promote several kinds of light pressure effects.

Comparing the images, it seems that the beam effects strongly depend on volume dimensions at the same pumping energy, when focused in droplet's centre. The smaller the volume, the more important the emission of micro- and nano-droplets by parent droplet. Although the waist of laser beam is not produced steeply, the focus position with respect to droplet surface is relatively critical and positioning errors even lower than 100 μm may prove to be important. In general, inertial effects are important in behaviour of water droplets at unresonant interaction with laser beams.

In the water droplet, a velocity gradient is produced by interaction with laser beam focused in its equatorial plane on its surface and due to water viscosity. The gradient increases with the increasing of the beam energy and does not depend directly on droplet volume. The computed velocity gradients are in agreement with the measured expulsion speeds of micro- and nano-droplets and the water jets emission immediately after interaction with laser beam. The surface tension and viscosity are influencing the effects produced on droplets and the speeds of emitted liquid. The supersonic speeds are quite relevant examples that are not obtainable peak values of such speeds resulting from unresonant irradiation of water droplet with laser beams. There are effects, particularly electrostrictive inhomogeneities and ablation produced by focusing the beam in droplet's centre, that may be produced only above a given threshold. The results reported for the case of unresonant interaction show several effects on distilled water droplet: damped deformations and vibrations, liquid jets emissions at supersonic speeds and emission of micro- and nano-droplets at higher and lower speeds, production of horizontal pillars of liquid and cylinder channels in the droplet that leads to formation of air bubbles inside it. The effects are somewhat similar with those reported at the application of electric or magnetic fields on dielectric fluids that are introduced in combinations/ mixtures of immiscible fluids.

These conclusions need to be completed by a larger recording frame rate of the droplet in order (i) to set the right mechanism leading to jetting from ns pulses (bulk bubble nucleation and/or electrostriction as a function of the fluence is another possible mechanism) and (ii) to access to early stage dynamics of induced fluid flow which builds liquid jet. The experiments have shown that introduction of an obstacle within droplet or at its surface, such as the capillary on which droplet hangs, may influence the kind, quality and dimensions of effects obtained at interaction with a pulsed laser beam. The results at resonance have demonstrated that the effects of a pulsed laser beam on a droplet may be mixed with effects obtained at unresonant interaction and this depends not only on droplet's materials but also on laser beam energy or power density, liquid surface tension and viscosity. One may deal with beam absorption without producing mechanical effects on droplet since after absorption there are not available photons to build light force. At higher beam energies, when there is an excess of photons in droplet, some effects specific to unresonant interaction may be produced. Regulating laser beam absorption in droplets (by varying absorbent concentration for example) one may control the number and characteristics of mechanical effects produced by light pressure on them.

## CONFLICT OF INTEREST

The authors confirm that this chapter content has no conflict of interest.

## ACKNOWLEDGEMENTS

This work has been financed by Romanian ANCS in the frame of Nucleus programme-contract 4N/2016 and the projects PN-II-ID-PCE-2011-3-0922. This work was also supported by COST networks MP1106 and MP1205. The authors acknowledge helpful discussions with Prof. J.P. Delville on data interpretation.

## REFERENCES

[1] D. Psaltis, S.R. Quake, and C. Yang, "Developing optofluidic technology through the fusion of microfluidics and optics", *Nature,* vol. 442, no. 7101, pp. 381-386, 2006.

[2] V. Horowitz, D. Awschalom, and S. Pennathur, "Optofluidics: field or technique?", *Lab Chip,* vol. 8, no. 11, p. 1856, 2008.

[3] A.L. Klein, W. Bouwhuis, C.W. Visser, H. Lhuissier, C. Sun, J.H. Snoeijer, E. Villermaux, D. Lohse, and H. Gelderblom, "Drop shaping by laser-pulse impact", *Phys. Rev. Appl.,* vol. 3, no. 4, 2015.

[4] M.L. Pascu, G.V. Popescu, C.M. Ticos, and I.R. Andrei, "Unresonant interaction of laser beams with microdroplets", *J. Eur. Opt. Soc. Rapid. Publ,* vol. 7, p. 12001, 2012. Rapid Publ

[5] M.L. Pascu, I.R. Andrei, M. Ferrari, A. Staicu, A. Smarandache, A. Mahamoud, V. Nastasa, and L. Liggieri, "Laser beams resonant interaction with micro-droplets which have a controlled content", *Colloid Surface A,* vol. 365, no. 1, pp. 83-88, 2010.

[6] V. Pratap, N. Moumen, and R.S. Subramanian, "Thermocapillary motion of a liquid drop on a horizontal solid surface", *Langmuir,* vol. 24, no. 9, pp. 5185-5193, 2008.

[7] S.T. Thoroddsen, K. Takehara, T.G. Etoh, and C-D. Ohl, "Spray and microjets produced by focusing a laser pulse into a hemispherical drop", *Phys. Fluids,* vol. 21, no. 11, p. 112101, 2009.

[8] J-P. Delville, M. Robert de Saint Vincent, R.D. Schroll, H. Chraïbi, B. Issenmann, R. Wunenburger, D. Lasseux, W.W. Zhang, and E. Brasselet, "Laser microfluidics: fluid actuation by light", *J. Opt. Pure Appl. Opt.,* vol. 11, no. 3, p. 34015, 2009.

[9] H-R. Jiang, and M. Sano, "Stretching single molecular DNA by temperature gradient", *Appl. Phys. Lett.,* vol. 91, no. 15, p. 154104, 2007.

[10] A.L. Klein, C.W. Visser, W. Bouwhuis, H. Lhuissier, C. Sun, J.H. Snoeijer, E. Villermaux, D. Lohse, and H. Gelderblom, "Laser impact on a drop", *Phys. Fluids,* vol. 27, no. 9, p. 91106, 2015.

[11] H. Azzouz, L. Alkhafadiji, S. Balslev, J. Johansson, N.A. Mortensen, S. Nilsson, and A. Kristensen, "Levitated droplet dye laser", *Opt. Express,* vol. 14, no. 10, p. 4374, 2006.

[12] M. Boni, V. Nastasa, A. Militaru, A. Smarandache, I.R. Andrei, A. Staicu, and M.L. Pascu, "Laser beams interaction with liquids in optofluidic experiments", *Rom Rep Phys,* vol. 64, pp. 1179-1194, 2012.

[13] M. Boni, V. Nastasa, I.R. Andrei, A. Staicu, and M.L. Pascu, "Enhanced fluorescence emitted by microdroplets containing organic dye emulsions", *Biomicrofluidics,* vol. 9, no. 1, p. 14126, 2015.

[14] Y. Utsunomiya, T. Kajiwara, T. Nishiyama, K. Nagayama, S. Kubota, and M. Nakahara, "Laser ablation of liquid surface in air induced by laser irradiation through liquid medium", *Appl. Phys., A Mater. Sci. Process.,* vol. 101, no. 1, pp. 137-141, 2010.

[15] R. Sahay, C.J. Teo, and S.T. Thoroddsen, "Laser-induced onset of electrospinning", *Phys. Rev. E Stat. Nonlin. Soft Matter Phys.,* vol. 81, no. 3, 2010.

[16] A. Braun, C. Kornmesser, and V. Beushausen, "Simultaneous spatial and spectral imaging of lasing droplets", *J. Opt. Soc. Am. A Opt. Image Sci. Vis.,* vol. 22, no. 9, p. 1772, 2005.

[17] V. Nastasa, V. Pradines, I.R. Andrei, M. Boni, M.L. Pascu, and R. Miller, "Studies about the generation and characterisation of microdroplets with a controlled content", *Optoelectron. Adv. Mater.-Rapid Commun.,* vol. 4, no. 11, pp. 1916-1919, 2010.

[18] H. Chraibi, D. Lasseux, R. Wunenburger, E. Arquis, and J-P. Delville, "Optohydrodynamics of soft fluid interfaces: Optical and viscous nonlinear effects", *Eur. Phys. J. E,* vol. 32, no. 1, pp. 43-52, 2010.

[19] A. De Giacomo, M. Dell'Aglio, O. De Pascale, and M. Capitelli, "Spectroscopic investigation of laser-water interaction beyond the breakdown threshold energy", *Spectrochim. Acta B At. Spectrosc.,* vol. 62, no. 2, pp. 87-93, 2007.

[20] D. Rivière, B. Selva, H. Chraibi, U. Delabre, and J-P. Delville, "Convection flows driven by laser heating of a liquid layer", *Phys. Rev. E Stat. Nonlin. Soft Matter Phys.,* vol. 93, no. 2, 2016.

[21] K. Volkov, and V. Emelyanov, "Interaction of laser pulse with liquid droplet", *28$^{th}$ International Symposium on Shock Waves,* 2012

[22] F.M. Sogandares, and E.S. Fry, "Absorption spectrum (340-640 nm) of pure water. I. Photothermal measurements", *Appl. Opt.,* vol. 36, no. 33, p. 8699, 1997.

[23] W.S. Pegau, D. Gray, and J.R. Zaneveld, "Absorption and attenuation of visible and near-infrared light in water: dependence on temperature and salinity", *Appl. Opt.,* vol. 36, no. 24, p. 6035, 1997.

[24] B. Kirby, *Micro- and nanoscale fluid mechanics: transport in microfluidic devices.* University Press: Cambridge, 2010.

[25] A. Ashkin, and J.M. Dziedzic, "Radiation pressure on a free liquid surface", *Phys. Rev. Lett.,* vol. 30, no. 4, pp. 139-142, 1973.

[26] R. Wunenburger, B. Issenmann, E. Brasselet, C. Loussert, V. Hourtane, and J-P. Delville, "Fluid flows driven by light scattering", *J. Fluid Mech.,* vol. 666, pp. 273-307, 2011.

[27] M.L. Pascu, A. Smarandache, M. Boni, J. Kristiansen, V. Nastasa, and I.R. Andrei, "Spectral properties of some molecular solutions", *Rom Rep Phys,* vol. 36, pp. 1267-1284, 2011.

[28] S.K. Tang, R. Derda, Q. Quan, M. Lončar, and G.M. Whitesides, "Continuously tunable microdroplet-laser in a microfluidic channel", *Opt. Express,* vol. 19, no. 3, p. 2204, 2011.

[29] H-M. Tzeng, K.F. Wall, M.B. Long, and R.K. Chang, "Laser emission from individual droplets at wavelengths corresponding to morphology-dependent resonances", *Opt. Lett.,* vol. 9, no. 11, p. 499, 1984.

[30] M. Strani, and F. Sabetta, "Free vibrations of a drop in partial contact with a solid support", *J. Fluid Mech.,* vol. 141, no. 1, p. 233, 1984.

[31] O.A. Basaran, and D.W. DePaoli, "Nonlinear oscillations of pendant drops", *Phys. Fluids,* vol. 6, no. 9, pp. 2923-2943, 1994.

[32] E.D. Wilkes, and O.A. Basaran, "Forced oscillations of pendant (sessile) drops", *Phys. Fluids,* vol. 9, no. 6, p. 1512, 1997.

[33] D.A.Cremers, and L.J. Radziemski, *Handbook of laser-induced breakdown spectroscopy.,* 2nd ed. John Wiley & Sons Ltd., Oxford, 2013.

[34] P.K. Kennedy, D.X. Hammer, and B.A. Rockwell, "Laser-induced breakdown in aqueous media", *Prog. Quantum Electron.,* vol. 21, no. 3, pp. 155-248, 1997.

[35] R.M. Pope, and E.S. Fry, "Absorption spectrum (380-700 nm) of pure water. II. Integrating cavity measurements", *Appl. Opt.,* vol. 36, no. 33, p. 8710, 1997.

[36] C. Bisch, A. Lasek, and H. Rodot, "Comportement hydrodynamique de volumes liquides spheriques semi-libres en apesanteur simulee", *J Mec Theor Appl,* vol. 1, no. 165, 1982.

[37] J-Z. Zhang, and R.K. Chang, "Shape distortion of a single water droplet by laser-induced electrostriction", *Opt. Lett.,* vol. 13, no. 10, p. 916, 1988.

[38] S.M. Chitanvis, "Acoustic instability induced in compressible, transparent fluids by electrostrictive effects", *Opt. Lett.,* vol. 15, no. 14, pp. 763-765, 1990.

[39] R.W. Boyd, *Nonlinear optics.,* 3rd ed. Academic Press, 2008.

[40] T.N. Buch, W.B. Pardo, J.A. Walkenstein, M. Monti, and E. Rosa, "Experimental issues in the observation of water drop dynamics", *Phys. Lett. A,* vol. 248, no. 5-6, pp. 353-358, 1998.

[41] D.R. Alexander, J.P. Barton, S.A. Schaub, and G.M. Holtmeier, "Nonlinear interaction of KrF laser radiation with small water droplets", *Appl. Opt.,* vol. 30, no. 12, p. 1455, 1991.

[42] Y. Utsunomiya, T. Kajiwara, T. Nishiyama, K. Nagayama, and S. Kubota, "Pulse laser ablation at water-air interface", *Appl. Phys., A Mater. Sci. Process.,* vol. 99, no. 3, pp. 641-649, 2010.

[43] C.C. Mei, *Lecture notes on fluid dynamics-lecture 25 (1.63J/2.21J).* Available at: https://www.coursehero.com/file/9481833/lecture25/

[44] V.L. Streeter, E.B. Wylie, and K.W. Bedford, *Fluid mechanics.,* 9th ed. McGraw-Hill Companies: Boston, 1998.

[45] R. Haglund, "The properties of light", In: *Springer Handbook of Lasers and Optics, F.T. Prof.* Springer: New York, 2007.

[46] O. Svelto, S. Longhi, G. Valle, S. Kück, G. Huber, M. Pollnau, H. Hillmer, S. Hansmann, R. Engelbrecht, H. Brand, J. Kaiser, A. Peterson, R. Malz, S. Steinberg, G. Marowsky, U. Brinkmann, D. Lo†, A. Borsutzky, H. Wächter, M. Sigrist, E. sSaldin, E. Schneidmiller, M. Yurkov, K. Midorikawa, J. Hein, R. Sauerbrey, and J. Helmcke, "Lasers and coherent light sources", In: *Springer handbook of lasers and optics,* F. Träger, Ed., Springer: New York, 2007.

[47] P.T. Rakich, P. Davids, and Z. Wang, "Tailoring optical forces in waveguides through radiation pressure and electrostrictive forces", *Opt. Express,* vol. 18, no. 14, p. 14439, 2010.

[48] R.W. Fox, A.T. McDonald, and P.J. Pritchard, *Introduction to fluid mechanics,* $6^{th}$ ed. Wiley: New York, 2003.

[49] *ImageJ Software*. Available at: https://imagej.nih.gov/ij/

[50] F.P. Schäfer, Ed., *Dye lasers.*vol. 1. Springer: Berlin, Heidelberg, 1973.

# CHAPTER 9

# Resonant Interaction of Laser Beams with Pendant Droplets

**Mihail Lucian Pascu**[1,2,*], **Mihai Boni**[1,2], **Tatiana Tozar**[1], **Adriana Smarandache**[1], **Alexandru Stoicu**[1] and **Ionut Relu Andrei**[1]

[1] *National Institute for Laser, Plasma and Radiation Physics, Magurele, Ilfov, Romania*

[2] *Faculty of Physics, University of Bucharest, Magurele, Ilfov, Romania*

**Abstract:** Resonant interaction of a laser beam with a single droplet takes place if the beam is absorbed in the droplet. Results are shown regarding the resonant interactions of pulsed laser beams with microliter droplets in pendant position which contain solutions of a laser dye-rhodamine 6G, phenothiazines (chlorpromazine, promazine), hydantoin derivative [(5Z)-5-(3-chlorobenzylidene)-2-thioxo-4-imidazolidinone)] and antibiotics (vancomycin), respectively, in water. A description of the electronic structure of rhodamine 6G and chlorpromazine using Gaussian09 and GaussView 5.0 software is made. Exposure of droplets to laser radiation leads to generation of new photoproducts that may have different properties if compared to parent compounds. Such modifications are evidenced by laser induced fluorescence and thin layer chromatography. In irradiated chlorpromazine water solution, out of hundreds of photoreaction products, 5 were identified: promazine (PZ), promazine sulfoxide (PZ-SO), 2-hydroxy promazine (PZ-OH), 2-hydroxy promazine sulfoxide (PZ-OH-SO), chlorpromazine sulfoxide (CPZ-SO). For rhodamine 6G solutions, the main results include the increasing intensity of LIF spectra in droplet with respect to bulk, even if the volume of the cuvette for bulk measurements is much higher than droplet's volume. This behaviour may be explained by the confinement of light inside droplet, where total internal reflection at surface occurs, the droplet being associated with a spherical micro-optical resonator.

**Keywords:** Chlorpromazine, Droplet, Fluorescence, Hydantoins, Laser dye, Lasing, Laser induced fluorescence, Promazine, Rhodamine 6G, Vancomycin, Whispering gallery modes, WGM.

## INTRODUCTION

In Chapter 8, the unresonant impact of droplets generated in pendant position, with laser beams of different energies is presented. Deformations and mechanical

* **Corresponding author Mihail Lucian Pascu:** Laser Department, National Institute for Laser, Plasma and Radiation Physics, Magurele, Ilfov, Romania; Tel/Fax: 0040 214575739; E-mail: mihai.pascu@inflpr.ro

vibrations are produced and losses of material under the form of nanodroplets and/or microdroplets may be produced. Nanodroplets may propagate at high (possibly hyper- or super-sonic) speeds and microdroplets propagate at lower, although high enough, speeds. Pendant droplets of smaller dimensions than the initial/parent one as well as micro/nano gas bubbles within the microdroplets volume may be also produced. These effects could constitute a basis of a new method to deliver medicines to targets.

Resonant interaction of a laser beam with a microdroplet (microliter volume droplet) in pendant position takes place when it is absorbed by droplet's components. Actually, radiation is absorbed by molecules contained in the microdroplet and this process has important consequences on their content. In order to produce resonant interaction, the laser beam must have a wavelength at which it is absorbed by one or more types of molecules contained in the droplet. To properly study the resonant interaction, solvents in droplets must not absorb the laser beam, so that only the active substance(s) molecules interact with pumping radiation. As a consequence of beam absorption, a molecule passes from fundamental singlet state on excited singlet states, either by direct absorption of one or more photons or by fast and/or slow internal rearrangements. Another consequence of absorption of laser radiation is the dissociation of absorbing molecules and the production of new substances deriving from "parent" molecules. Interaction of photoreaction products with solvent and with the same laser pumping beam may be also important. Once excited, molecules fall back on fundamental singlet state either by fluorescence emission or by nonradiative de-excitation; in this last case, the temperature of microdroplet may increase with some degrees. On the other hand, generation of photoreaction products in microdroplets following resonant interaction with laser beams is produced only as long as laser beam is on.

In this chapter, results are presented on the resonant interaction of a pulsed laser beam with liquid droplets in pendant/hanging position that contain solutions of different fluorophores (laser dye, phenothiazines, hydantoin derivatives, antibiotics) [1]. The studies are made at energy levels which do not produce at all, or produce insignificant unresonant interactions (mechanical vibrations, droplet shape deformations and liquid expulsion *etc.*) [2].

## MATERIALS

Rhodamine 6G (Rh6G), [9-(2-ethoxycarbonylphenyl)-6-(ethylamino)-2,7-dimethylxanthen-3-ylidene]-ethylazanium, $C_{28}H_{31}N_2O_3^+$, is a fluorescent dye from xanthene family. The physical properties of Rh6G such as absorption, subsequent fluorescence emission and fluorescence lifetime as well as quantum yield are reported in literature [3 - 6]. Due to its high quantum yield and high solubility in a large number of solvents (among which ethyl alcohol and water) it is used in a large area of applications and studies. Rh6G is intensely used as a laser dye (active medium in dye lasers) but it is also used as marker for a wide range of applications in biology and as photosensitiser. Its 3D molecular structure is presented in Fig. (**1**).

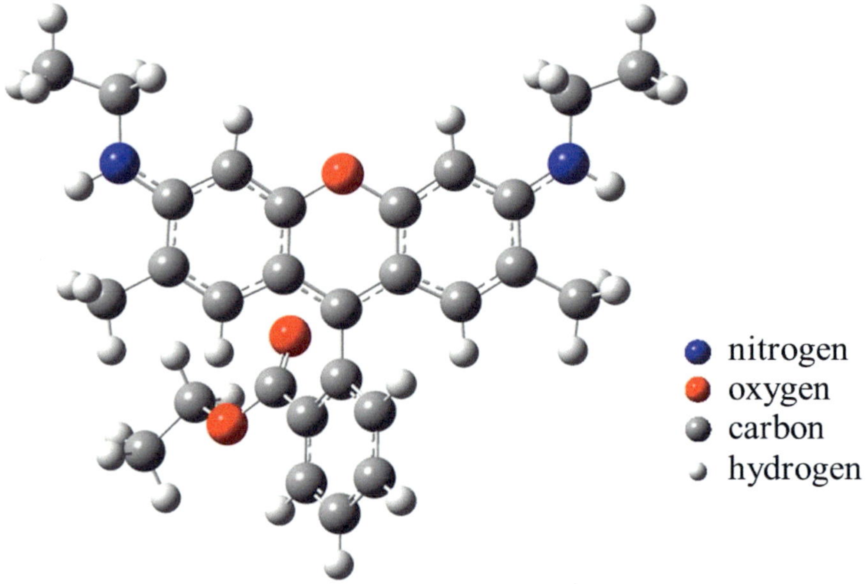


**Fig. (1).** 3D optimised chemical structure of Rh6G.

A phenothiazine is an organic compound having a heterocyclic system of two rings linked in the center of the ring by an -NH-group and a sulfur atom (-S-), and represents the parent molecule of many heterocyclic bioactive derivatives that have interesting biological and pharmaceutical properties. Chlorpromazine (CPZ) is a phenothiazine derivative, its primary use being in antipsychotic medication [7]. Beside this property, it was discovered that CPZ can be used with success in cancer treatment, where it induces apoptosis or inhibits proliferation in cultured cells like leukemia [8] or melanoma [9] cells. CPZ molecular structure is presented in Fig. (**2**).

Recent reports show that CPZ, dissolved in ultrapure water, exposed to ultraviolet (UV) laser beam undergoes a variety of structural modifications resulting in formation of photoproducts [10 - 12].

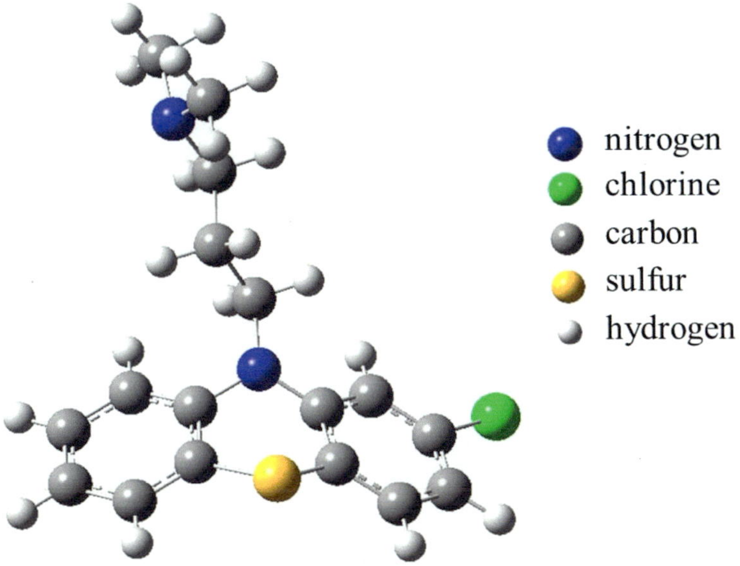


**Fig. (2).** 3D optimised chemical structure of CPZ.

# METHODS

## Experimental Set-up for Irradiation and Laser Induced Fluorescence (LIF) Measurements

The measurements were performed, essentially, using the same experimental set-up as in Chapter 8 (Fig. **3**). The set-up contains an Nd:YAG laser system which emits at 532, 355 or 266 nm, the second, third or fourth harmonics of the fundamental beam, respectively, and a target which is a single droplet that contains solution of fluorophores at different concentrations. The laser emits pulses at 10 pps repetition rate with 6 ns full time width at half-maximum. A difference with respect to previous chapter is that the fast camera was replaced with a spectrograph (Acton Research, Spectra Pro 750) and an intensified CCD camera (Princeton Instruments, IMAX 1024 iCCD) with 0.08 nm optical resolution which collects fluorescence signal through an optical fiber. The fiber can be translated in vertical and horizontal planes and rotated horizontally around droplet's circumference with increments of 10 μm and 1°, respectively. The capillary through which droplet is generated can be translated in horizontal plane along laser beam and in vertical plane with an increment of 250 μm. The laser beam was sent on droplet in a quasi-parallel beam geometry, $d>>w>D$, slightly divergent ($d<f$) or convergent ($d>f$) through a lens with focal length $f=150$ mm (Fig. **3b**), where d is the distance lens-droplet, w is the laser beam waist at incidence on lens, and D is the droplet diameter.

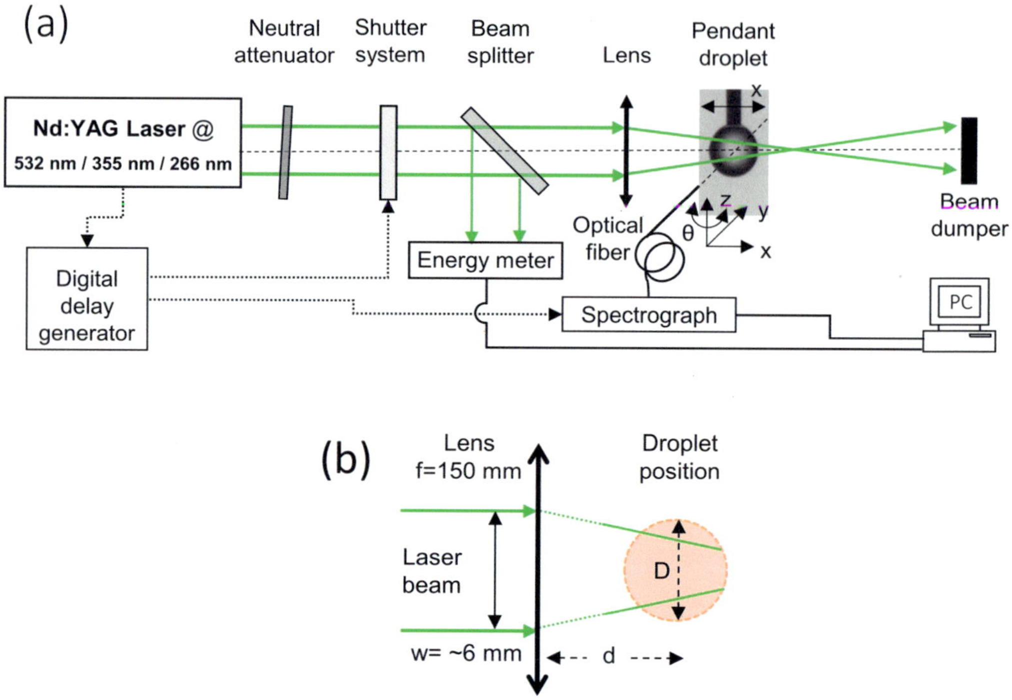


**Fig. (3).** Experimental set-up, (**a**) general set-up to characterize LIF emitted by fluorophore-doped liquid droplets, (**b**) sketch of irradiation geometry; D: droplet diameter, w: beam waist, d: lens-droplet distance.

Droplet position can be chosen with respect to the lens focal point in such a way that the laser beam spot covers, at the limit, droplet cross section or it is smaller than this. Beam waist has around 2.5 mm diameter in the interaction plane with the droplet, *i.e.* close to droplet's diameter that is around 2.67 mm at 10 μl volume (d = 220 mm), or around 2 mm diameter in the interaction plane with droplet of 7 μl volume and diameter around 2.4 mm (d = 100 mm).

## Gaussian Software

The molecular structures of Rh6G (Fig. **1**) and CPZ (Fig. **2**) were obtained using electronic structure calculation software Gaussian09 [13] and GaussView 5.0 [14]. The structures were subjected to geometry optimisation or energy minimization calculations, the term "energy minimisation" implying that the molecular structure which results from these calculations is the depiction of the chemical structure in its fundamental state, when internal energy is at its lowest possible level with respect to environment surrounding it. All of the resulting structures have also been subjected to stability calculations in order to check and ensure that the final result truly represents the fundamental state of each molecule. The computing scheme used for these calculations included factors such as temperature, which was set to 22°C/295.15 K (the temperature at which experiments were performed), pressure at 1 bar and water used as solvent. In order to replicate solvation effects of water on molecular structure IEFPCM (Integral Equation Formalism Polarizable Continuum Model) module for solvation was used. Calculation is performed in presence of a solvent by placing solute in a cavity within solvent reaction field. This method creates the solute cavity *via* a set of overlapping spheres [15]. The program computes energy in solution by making solvent reaction field self-consistent with solute electrostatic potential (the latter being generated from computed electron density with specified model chemistry) [16]. Also, the calculation scheme employed several steps that used different methods due to software inability to perform high level calculations (*i.e.* Density Functional Theory) on roughly drawn structures. As such, every structure was first computed using the Hartree-Fock method with a 3-21G basis set (functions combined in linear combinations to create molecular orbitals) but with no restraints concerning temperature and environment. After completion of calculation, Density Functional Theory was applied along with B3LYP functional and 6-31G basis set, this method being known to result in stable and reliable organic structures description of molecules. Besides the used basis sets, polarisation functions (d, p) were added to calculation in order to add more flexibility to the 6-31G basis set [17].

## Thin Layer Chromatography

Chromatography is a class of methods that are used for separation, isolation, identification and quantification of compounds from a mixture, thin layer chromatography (TLC) being a subdivision of it. Egon Stahl invented the term “thin layer chromatography” at the end of 1950’s and had an important contribution in standardising materials and protocols and in description of selective solvent systems that are used in the separation of many classes of compounds [18]. TLC is considered a simple, rapid and inexpensive method used for separation, identification and visual semiquantification of a wide variety of compounds. This method requires the application of a small quantity of sample, 1-10 μl drop, as a spot at the origin of a thin absorber. Mobile phase moves along the stationary phase due to capillary action. In TLC case, separation of each component has the same migration time but different migration distances.

TLC mobile phase consists of a single solvent or a mixture of solvents. The sorbent can be silica gel, cellulose, alumina, and polyamides. The polar and non-polar phases chemically linked can be impregnated on a suitable support plate (glass, polyester or aluminium foil). The variety of mobile and stationary phases provides considerable versatility in choosing a system that has a good resolution for almost all types of compounds.

TLC can be used as, both, analytical and preparative technique. The once separated compounds that are identified by TLC can be eluted for further classification using micro-analytical methods, such as nuclear magnetic resonance or mass spectrometry, gas chromatography and/or high pressure liquid chromatography [19].

The mobile phase (solvent) is the transport medium for the sample to be separated where substance migrates due to capillary forces along the stationary phase. The movement of the compound on TLC plate is the result of two opposing forces: the driving force of mobile phase and the resistive force of sorbent. The driving force tends to move substances in the direction given by the solvent flow and the resistive force prevents the movement of substances by extracting them from the mobile phase into sorbent. At the end of experiment, each zone has migrated a specific distance characterised by the compounds in the mixture. Substances that moved slower present a greater attraction to the layer, while those moving quickly spend less time in sorbent due to low affinity towards the layer and more solubility in mobile phase. The capacity to have a good separation of the mixture is a consequence of the selectivity, efficiency and capability of the chromatographic system [19].

The protocol used for performing a TLC experiment consists in the following steps:

1. The sample is applied at 1 cm from the origin of the base plate and 1 cm from the right side of the plate; the distance between spots is 1 cm;
2. Solvent present in the sample is allowed to evaporate from the plate;
3. Proper mixture of solvents for the particular sample is prepared, poured in developing tank, stirred slightly and left at rest 5-10 min to obtain a chamber saturated with vapors;
4. The TLC plate is placed in the chamber, taking care that the mobile phase does not exceed the spots of the sample to be tested;
5. Keep TLC plate in developing tank until solvent is at 1 cm below the upper side of the plate;
6. Plate is removed from developing tank and the solvent front is marked;
7. Mobile phase is removed from the plate by air flow or heat drying;
8. TLC plates are visualised using an UV illumination cabinet and photographed;
9. Obtained images are analysed using TLC dedicated software in order to acquire qualitative and quantitative information about compounds present in the mixture.

## MODIFICATION OF THE MOLECULAR CONTENT OF LIQUID DROPLETS BY EXPOSURE TO LASER RADIATION

The resonant effect between pendant droplet aqueous solutions and laser radiation can be analysed by LIF and TLC in order to identify the modifications induced in the molecular content of liquid droplet.

### Laser Induced Fluorescence

In the following, is shown how LIF measurements can be used to study substances placing the focus on LIF measurements performed on small quantities of solutions that are prepared as pendant droplets.

In a first instance, a comparison is made between LIF spectra recorded for bulk and for a pendant droplet; the samples contain a laser dye, Rh6G solution in distilled water, at $10^{-5}$ M. The same energy of 2.2 mJ has been used to irradiate both samples, while the emission spectra were different as presented in Fig. (**4**). One of the major differences consists in the increased intensity of fluorescence spectra (an order of magnitude higher) in the case of droplet with respect to bulk, even if the volume of the cuvette is much higher than droplet's volume.

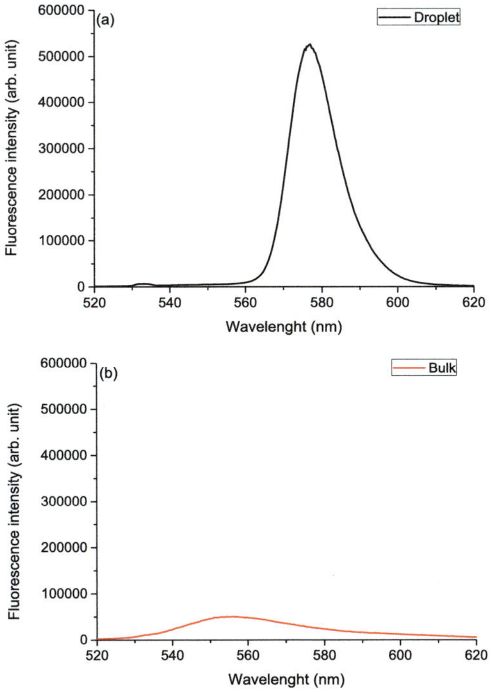


**Fig. (4).** LIF measured for Rh6G solution in distilled water on (**a**) pendant droplets and (**b**) spectrophotometric cuvette.

Another different characteristic is the shape of the spectra and of the fluorescence peak which is sharper and is obtained at longer wavelengths for droplets than for bulk samples. Most probably, this behaviour appears due to the confinement of light inside the droplet, where total internal reflections at droplet surface occur. Such phenomena are reported in literature for spherical micro-resonators [20 - 22].

In the last decade, similar studies were conducted on LIF obtained on sessile droplets, pendant droplets and even levitating droplets that contain laser dyes characterised by high quantum yield. The aim of this kind of experiments is to obtain a micro- and/or nano-spherical laser, with emission in 4π or to obtain LIF amplification making possible detection of small quantities of molecules. A detailed study of amplification of fluorescence obtained in pendant droplets containing Rh6G is presented in Chapter 17.

In other studies [23], the measurement of LIF emitted by pendant droplets that contain medicines is used to monitor the evolution of droplet's content during irradiation.

In general, resonant interaction between laser beams and hanging/pendant droplets that contain solutions triggers different effects such as excitation of an absorbing molecule from its fundamental singlet state $S_0$ on excited singlet states ($S_1$, the first excited singlet state), or production of new derivative substances by dissociations of the parent molecules. De-excitation of molecules back on fundamental state occurs either by emission of fluorescence (or, at longer times, phosphorescence) or by nonradiative reaction pathways. By exposure of parent substances to laser beams, their chemical structure can be fundamentally changed and new compounds may be produced. These transformations may be monitored by measuring, in real time, LIF emitted by that irradiated solution which contains a cocktail of photoproducts and changes profoundly in time its composition. The technique has a great advantage due to the possibility to produce and follow the modifications in real time. On this line, the following results show how structural changes of medicine molecules take place during irradiation by measuring their LIF in real time. For this purpose, were used solutions of a hydantoin derivative in dimethyl sulfoxide (DMSO), SZ-2, whose chemical, physical and biological properties are described in Chapter 11. Solutions of SZ-2 in DMSO were prepared at $10^{-3}$ M and LIF was measured in bulk (quartz spectrophotometric cuvette, volume 1.5 ml) and in pendant droplet (5.5 μl volume). The experimental set-up is shown in Fig. (**3**). For irradiation, 266 nm beam was used. The pumping beam average energy was 1 mJ and the fluorescence spectra were collected as an average over 10 pumping pulses.

In Fig. (**5**), LIF spectra are comparatively presented collected from pendant droplet and from bulk that contain SZ-2 solution, for different acquisition times. The spectra measured on bulk sample exhibit two peaks. The first is located at 435 nm and presents a bathochromic shift of 20 nm when exposure time is increased. The second arises at 485 nm and has a bathochromic shift of 15 nm in time. The evolution of the intensities of first and second fluorescence peaks are shown in Fig. (**6**).

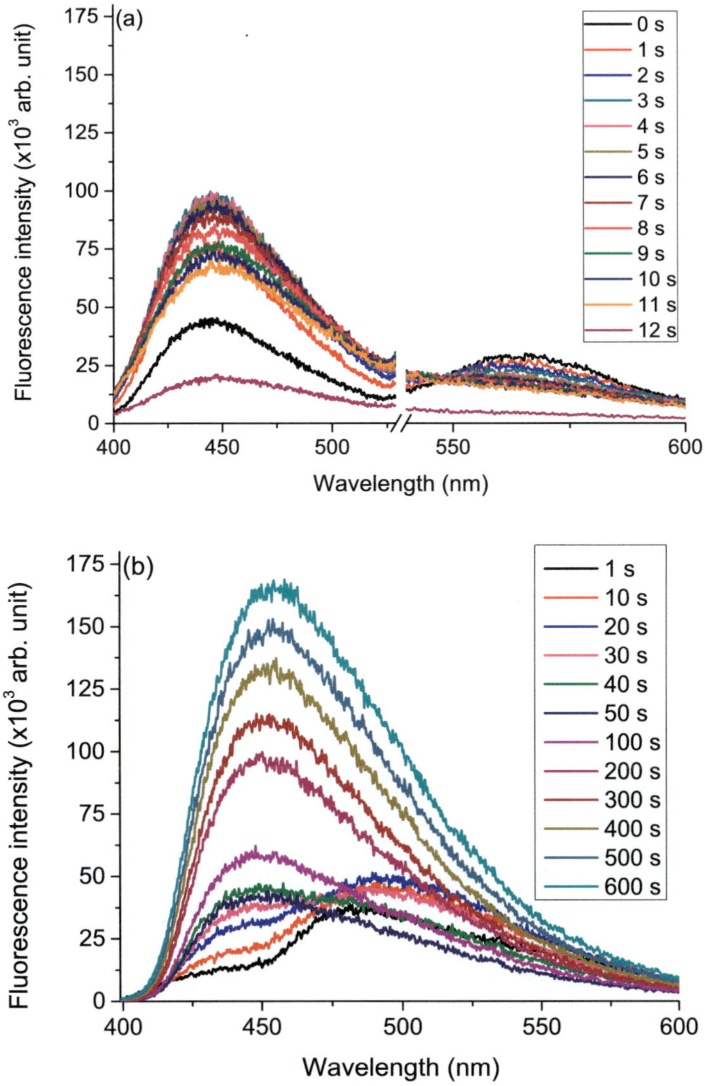


**Fig. (5).** LIF measured for SZ-2 solution in DMSO on (**a**) pendant droplet and in (**b**) spectrophotometric cuvette.

The intensity of the first peak increases throughout the measurement and reaches a plateau after 500 s, but the second peak presents a slow increase in the first 20 s and thereupon it has a decreasing behaviour and is vanishing after 50 s. These simultaneous changes of intensities of the two peaks could be correlated with the modifications of the parent molecules induced by irradiation with laser beams and the generation of new fluorescent photoproduct(s).

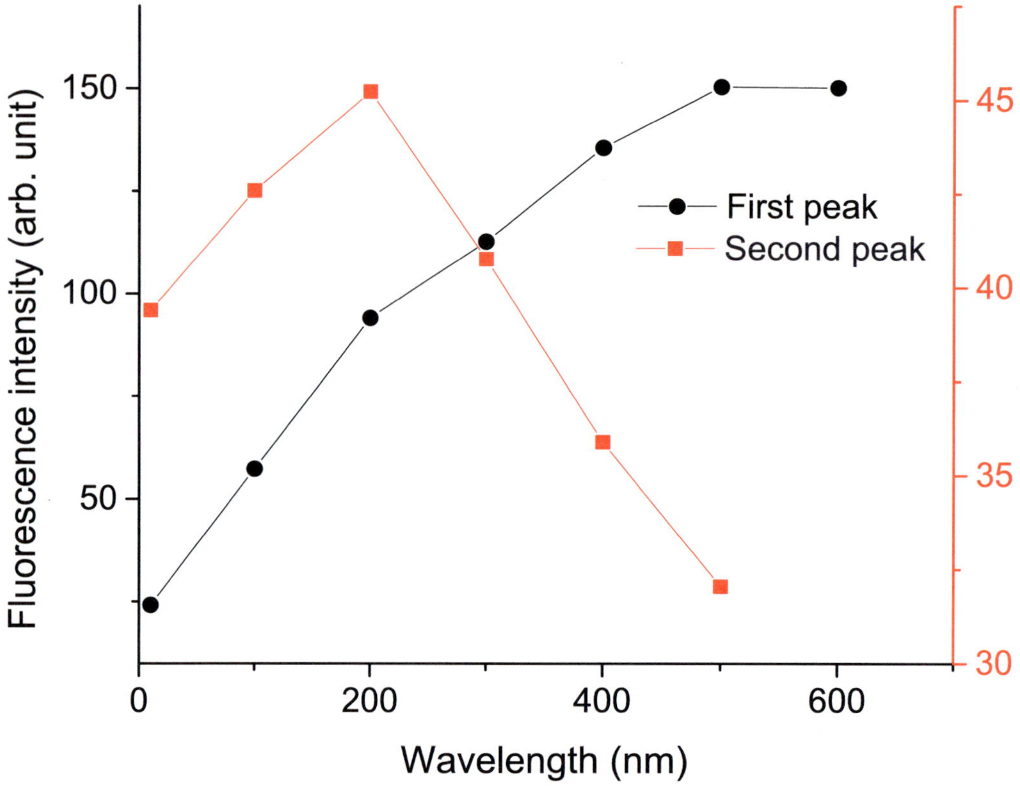


**Fig. (6).** The evolution of SZ-2 peak fluorescence intensities in time for bulk measurements.

The same characteristics are evidenced for measurements performed on droplets, observing that the peaks appear at about 448 nm and 565 nm. Following the time evolution of the intensities of the two peaks measured in droplet, it can be observed that they have the same behavior as for bulk, but these changes are produced faster.

On the other side, the misuse of antibiotics had the consequence that development of bacteria which are resistant to antibiotics took place. This is usually called multiple drug resistance (MDR). According to general, evidence based, experience there is an urgent need to find new ways to fight MDR acquired by resistant strains of Gram-positive and/or Gram-negative bacteria. An example of a possible solution in this respect is the use of medicines exposed to laser radiation that generate, under exposure, new photoproducts which may have individually or in cocktail mixtures, bactericide effects. Phenothiazines are an example of such medicines. It was shown that some of them have activity against a wide range of bacteria [24 - 26]. An important number of drugs employed currently in medicine have their origins in the chemical manipulation of phenothiazines [27, 28]. Chemical manipulation of a compound to create new derivatives is limited by the existing methods of organic chemistry, is time consuming and the percentage yield of an active product is quite small.

The need to use millimeter-, micro- and even nano-droplets as vectors to transport substances to targets in technological and biomedical applications triggered research related to new methods for their production and more accurate characterisation [29].

The use of droplets modified by resonant interaction with laser beams in biomedicine includes possible application of phenothiazines prepared as nano and/or microdroplets exposed to laser radiation and modified by it. The droplets may originate from either simple or structured pendant droplets. This would allow to develop new treatment procedures with particular use in curing eye infections, for instance.

Studies of laser beams interaction with microdroplets of solutions generated and kept in pendant position are performed in order to modify fast, in a first step, the most part of molecules in the droplet that absorb laser radiation and to produce, in a second step, by unresonant interaction with laser beams, modifications of the shapes of the droplets. Bubbles generated in the droplet may be also produced and emission of nanodroplets and microdroplets with smaller dimensions than the initial, parent droplet may take place.

Combining resonant and unresonant interaction of laser beams with droplets to produce micro- and nano-droplets that contain modified phenothiazine molecules and spreading them on animal eye tissues targets infected with Gram-positive and Gram-negative bacteria could constitute a method to combat MDR.

In the following, a detailed study is presented that is dedicated to monitoring the decomposition of CPZ, promazine (PZ) and vancomycin (VCM) in water solutions irradiated in droplet or bulk by showing modifications of emitted LIF spectra measured in real time. Spectra were recorded with a spectrometer (Ocean Optics, HR4000, optical resolution 0.64 nm) and a spectrograph (Princeton Instruments, Acton Spectra Pro750, optical resolution of 0.08 nm) [30, 31].

Results concerning the interaction of 266 nm and 355 nm pulsed laser radiation with solutions of 10 mg/ml CPZ in ultrapure water, and interaction of 266 nm pulsed laser radiation with 2 mg/ml VCM in distillated water exposed in microdroplet and bulk forms are shown. For CPZ, LIF measurements have shown that parent compound transformation starts in a time interval that is shorter for exposure at 355 nm with respect to 266 nm at the same laser beam fluencies. Also, the modifications are similar in bulk and droplet at the same exposure dose, with higher rates at 355 nm. In irradiated CPZ solution, 5 were identified out of hundreds of photoreaction products: promazine (PZ), promazine sulfoxide (PZ-SO), 2-hydroxypromazine (PZ-OH), 2-hydroxypromazine sulfoxide (PZ-OH-SO), chlorpromazine sulfoxide (CPZ-SO) [10, 12, 31, 32].

Results evidenced on CPZ samples (droplet and bulk) at 10 mg/ml were compared with those obtained for two commercially available pure substances which were identified as photoproducts generated in exposed CPZ: PZ and CPZ-SO.

In droplet experiments, the irradiation energy was 3 mJ in order to obtain fast resonant effects and at the same time to avoid mechanical effects specific to unresonant interaction [2]. The droplet was positioned at 220 mm from the lens so that the beam fully covered the droplet cross section.

Bulk experiments were performed using the same experimental set-up, but replacing the droplet from its position with a standard spectrophotometric cell placed at 425 mm with respect to the lens. In this case, laser beam waist was at the input on the cell around 11 mm. The beam energy was 7 mJ in agreement with previous experiments [10, 12, 33] performed on 10 and 20 mg/ml CPZ in ultrapure water.

The irradiation time intervals were chosen so that irradiation doses were the same for droplet and bulk. An irradiation dose was calculated considering the average laser power incident on exposed area (average power density) during irradiation time. Irradiation area was 0.056 $cm^2$ in droplet (circular area described by the droplet radius) and about 1 $cm^2$ in bulk. So, the average power density calculated for droplet was 524 $mW/cm^2$ whereas for bulk it was 70 $mW/cm^2$, *i.e* the irradiation dose was about 7.5 times lower in bulk than in droplet for the same exposure time. Consequently, the bulk samples were irradiated 7.5 times longer to obtain the same irradiation dose as in droplet [34].

LIF emission at laser beam-droplet interaction (Fig. **7a**) was compared with results obtained on bulk for the same irradiation time intervals (Fig. **7b**) and irradiation doses (Fig. **7c**).

LIF measurements performed on pendant microdroplets (Fig. **7a**) at the two irradiation wavelengths, 266 nm (Fig. **7aI**) and 355 nm (Fig. **7aII**) have shown that a LIF peak is obtained in about 20 s at 266 nm laser exposure and in less than 10 s at 355 nm (Fig. **7aIII**). Actually, measurements made at a shorter integration time than 10 s on droplets irradiated at 355 nm show that LIF peak appears in the first seconds of exposure. So, one can assume that the laser-induced molecular modifications are faster for irradiation at 355 nm than at 266 nm. A similar evolution of LIF intensity is observed in bulk when irradiated at the same time intervals (Fig. **7b**). LIF peak shows-up after 4 min and 30 s of exposure at 266 nm and 1 min and 10 s at 355 nm (Figs **7bI-III**). In bulk, irradiated at the same dose as droplet (Fig. **7c**), LIF intensity increases up to 4 min and 30 s of laser exposure at 266 nm and then it is followed by uniform decreasing (Fig. **7cI**); it shows a decreasing trend for laser exposure at 355 nm (Fig. **7cII**). The behavior of LIF intensity for droplet and bulk irradiated at the same dose for both wavelengths is almost similar (Figs. **7aIII** and **cIII**). This means that there are no important laser-induced molecular differences between the two exposures, but processes are faster (i) in droplet than in bulk and (ii) at exposure to 355 nm with respect to 266 nm laser beam.

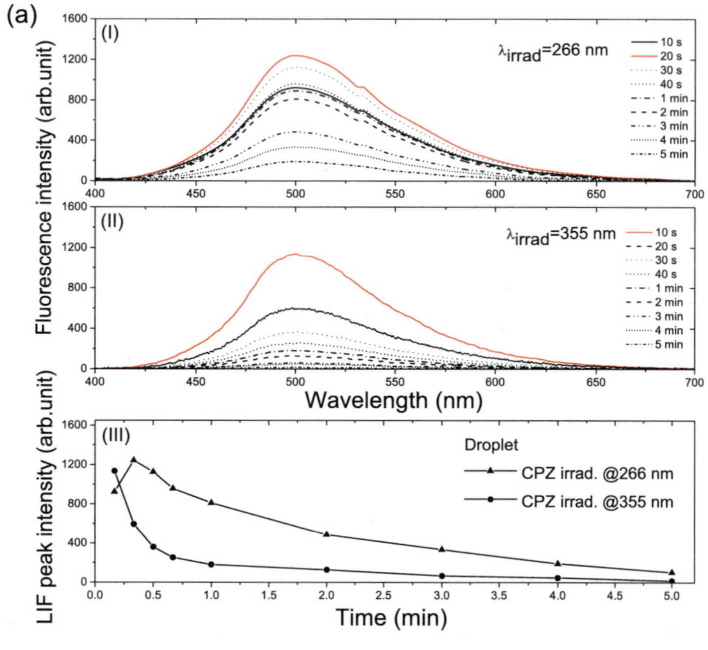


*Fig. 7 contd.....*

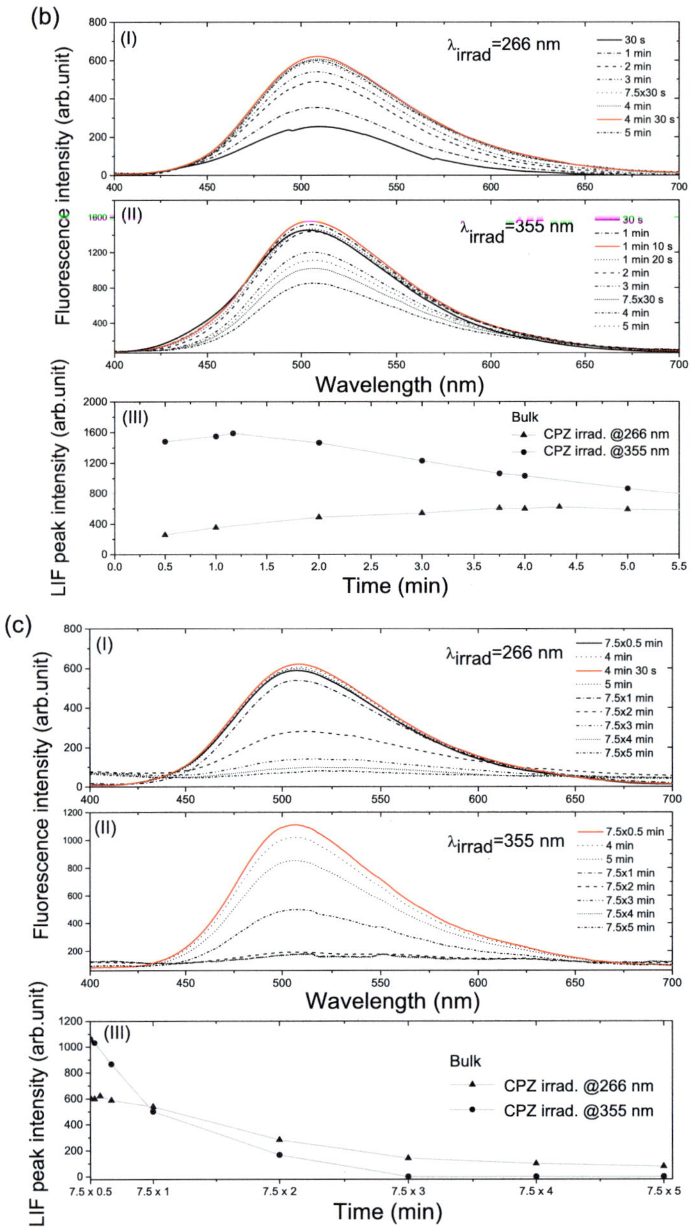

**Fig. (7).** LIF spectra of CPZ solutions in ultrapure water at 10 mg/ml. (**a**) 10 µl droplet at laser irradiation energy of E = 3 mJ, (**b**) bulk of 1 ml at E = 7 mJ and the same irradiation time intervals as in droplet, and (**c**) bulk of 1 ml at E = 7 mJ for the same irradiation dose as in droplet; exposure at (**I**) 266 nm and (**II**) 355 nm laser beams; (**III**) LIF intensity peaks correlation. LIF spectra in (**b**) and (**c**) were selected from the same irradiation session, *i.e.* the exposure was continuous from the beginning of the session to 7.5×5 min. A bold curve corresponds to the shortest irradiation time interval; red lines represent LIF peak intensities obtained along the entire irradiation time interval.

The behaviour of LIF peak at irradiation with 355 nm may be due to the fact that $S_1$ singlet state is close to $T_1$ triplet state of the CPZ molecule and in the first stage molecules decay radiatively on $S_0$ starting to accumulate slightly on $T_1$. In a second stage, after more than 10 s, molecules accumulate on $T_1$ and the LIF signal starts to decrease [35 - 37].

On the other hand, a direct correlation between the evolution of LIF peak at the two irradiation wavelengths in droplet and bulk and the evolution of CPZ photodegradation process was not made, but LIF behaviour of irradiated CPZ solution seems to be influenced by LIF activity of newly generated photoproducts. Since some photoproducts were identified [11, 32, 35, 38], *e.g.* PZ and CPZ-SO, reference grade substances of PZ (Sigma-Aldrich, DE, purity 98%) and CPZ-SO (British Pharmacopoeia Commission Laboratory, Teddington, UK) were used, as received, to evaluate their LIF activity. PZ and CPZ-SO were dissolved each in ultrapure water at 0.2 mg/ml. Then, 1 ml of each solution was exposed to laser radiations in the same conditions as CPZ solution in bulk.

All LIF spectra recorded for PZ and CPZ-SO irradiated at 266 nm or 355 nm show two spectral bands (Fig. **8**), one of higher intensity centered on 388 nm for CPZ-SO (Fig. **8a**) and 384 nm for PZ (Fig. **8b**) and the other of lower intensity centered on about 475 nm for both. LIF intensity of the second band slightly increases during irradiation in CPZ-SO and decreases in PZ case.

The LIF peak time evolutions of the first band in CPZ-SO and PZ at both irradiation wavelengths were compared with those obtained for the CPZ solution (Fig. **9**). For LIF spectra intensities, the measuring units are arbitrary and the differences in intensities for CPZ spectra and PZ and CPZ-SO spectra, respectively, are due to the fluorescence measuring systems which were used. This is not relevant since the trends in LIF evolution of samples are followed and not the quantitative ratio between LIF signals. As a data to be used for comparison, the fluorescence quantum yield of CPZ calculated in [10] is 15% at excitation with 266 nm.

In Fig. (**9**) is shown the time correlation of LIF intensity peaks for all spectra. LIF studies have shown that photodegradation process of CPZ depends on irradiation wavelength in droplet or bulk.

At 0.2 mg/ml, LIF studies on CPZ-SO and PZ (this concentration is 50 times lower than that of CPZ) were performed. LIF spectral ranges are slightly shifted to UV with respect to CPZ. If the photoproducts concentrations increase, as demonstrated recently in [10] and [39] (where 2 ml of PZ at 2 mg/ml, respectively 20 mg/ml in ultrapure water were irradiated at 6.5 mJ), the spectral range shifts towards 500 nm.

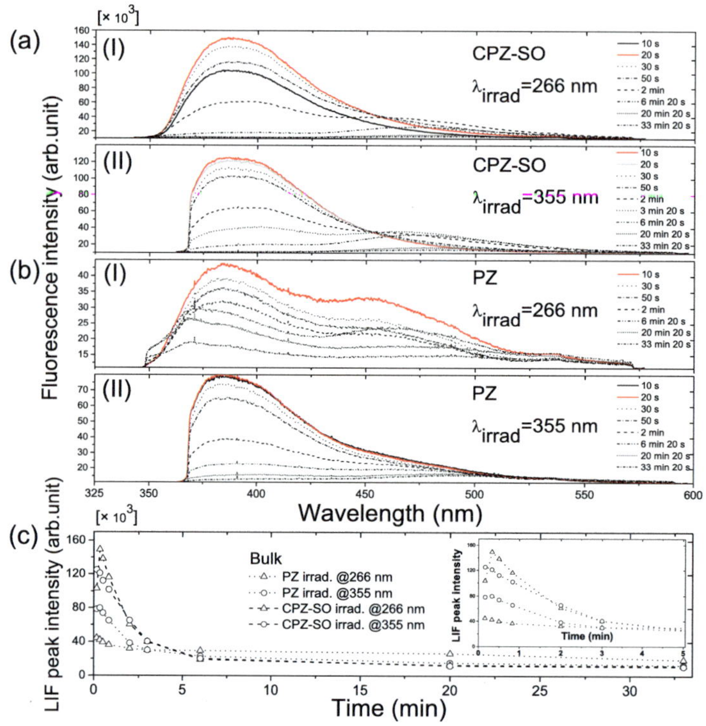

**Fig. (8).** LIF spectra of two pure photoproducts (that are also commercially available) obtained from CPZ after irradiation. (**a**) CPZ-SO and (b) PZ solutions in ultrapure water exposed at (**I**) 266 nm and (**II**) 355 nm laser beams in bulk; (**c**) LIF intensity peaks correlation for all spectra. Laser beam energy E=7 mJ and exposure time interval were the same as for CPZ. Bold lines correspond to the shortest exposure time; red lines denote the peak of LIF intensity obtained along the entire exposure. Inset graph in (**c**) shows intensity peaks correlation for the first 5 min of irradiation. CPZ-SO and PZ solutions concentrations was 0.2 mg/ml in ultrapure water.

In bulk experiments the decreasing trend of the LIF intensity of CPZ-SO is similar to that of CPZ in droplet (Fig. **9a**) if compared for the same exposure wavelength. This is because CPZ-SO is one of the first photoproducts obtained in irradiated CPZ samples and it has one of the highest generation rates together with PZ in the first minutes of irradiation [10]. It is shown that for the 20 mg/ml CPZ solution in ultrapure water irradiated at 266 nm with 6.5 mJ, CPZ-SO reaches the maximum relative concentration of 6% in the first 6 min of exposure. PZ as photoproduct has about the same concentration after 5 min, but reaches the maximum relative concentration of 40% after 60 min of irradiation. Due to the exposure of droplet at high irradiation energy that is delivered in a short time interval, CPZ-SO activity is dominant in comparison with that of PZ in an irradiated droplet.

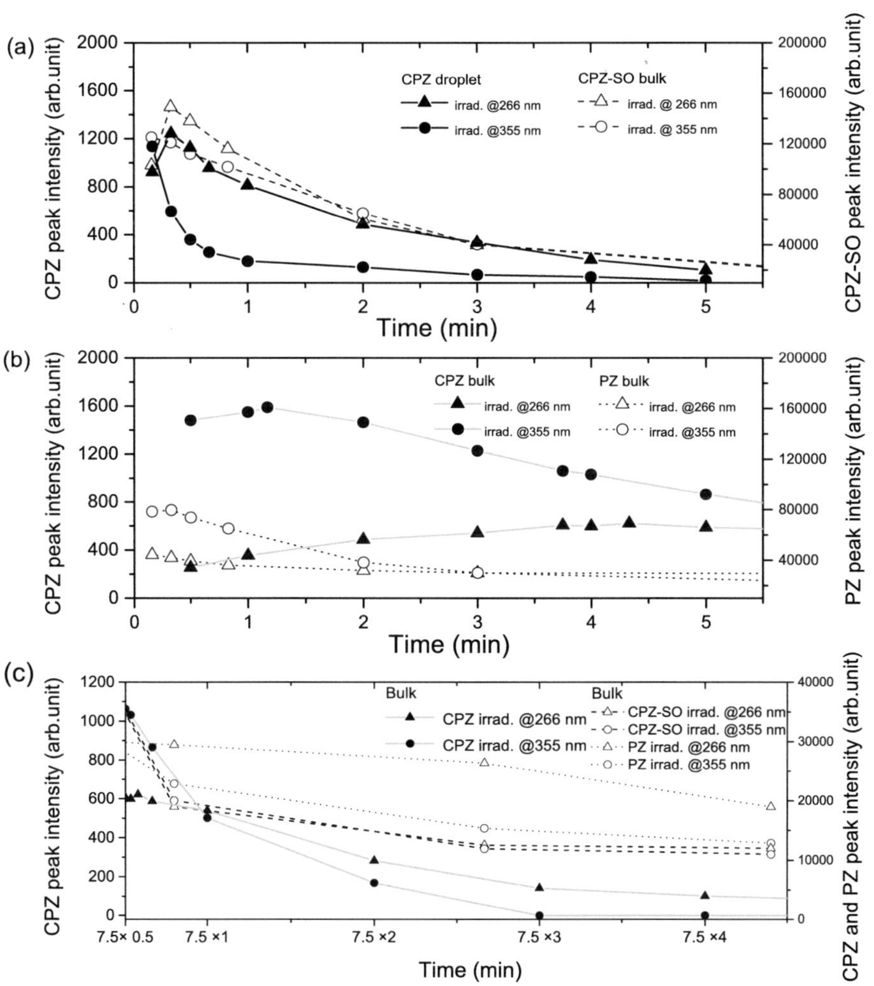


**Fig. (9).** LIF peaks evolution during irradiation. (**a**) CPZ solution in droplet and bulk at 10 mg/ml; (**b**) PZ and CPZ-SO in bulk at 0.2 mg/ml, irradiated at 266 nm and 355 nm; (**c**) time correlation of LIF peaks for CPZ, PZ and CPZ-SO in the first 5 min of irradiation. The beam energies were 3 mJ for droplet and 7 mJ for bulk; samples volumes were 10 µl for droplet and 1 ml for bulk.

When the bulk volume increases and the laser beam energy is kept the same as in droplet, CPZ LIF intensity has a similar trend as for PZ (Fig. **9b**) at the same exposure wavelength. This happens due to PZ concentration increase with exposure of CPZ [10] and therefore, CPZ fluorescence intensity appears to be influenced by PZ LIF intensity.

On the other hand, these correlations are better evidenced if LIF peak evolutions at the same irradiation doses in bulk (Fig. **9c**) are compared. In this case CPZ LIF intensities for both wavelengths have an almost similar trend as CPZ-SO due to

CPZ-SO activity which is dominant in comparison with PZ for the same irradiation conditions.

LIF intensities differences observed between CPZ and CPZ-SO on one side and PZ on the other, appear to be mainly due to the influences of the photoproducts generated in CPZ irradiated solutions. PZ LIF peaks have the same behaviour as described in [39].

LIF effects of the kind mentioned above for microdroplets that contain CPZ are obtained on other medicines such as VCM, as well. Here, LIF was measured for VCM dissolved in distilled water at 2 mg/ml and exposed to laser radiation as individual microdroplets in pendant position.

In Fig. (**10**) is shown the LIF spectrum of a 2 ml sample of 2 mg/ml VCM in distilled water exposed in bulk. The excitation beam is a superposition of 266 nm and 532 nm (residual beam) at overall 10 mJ average energy emitted by laser. One may observe a first fluorescence peak at about 358 nm and a second one at around 490 nm. This is in agreement with previous reports [40].

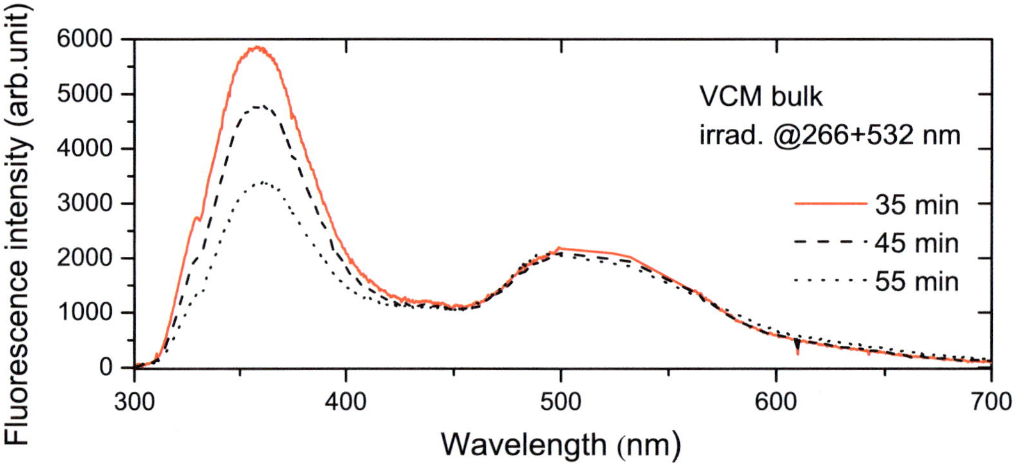


**Fig. (10).** LIF spectra of VCM solution in bulk; 2 ml solution exposed to a combination of two laser beams emitted at 266 nm and 532 nm, respectively. Dependence of LIF spectrum of irradiation time is evidenced between 35 and 55 min. Overall average energy of laser beams is 10 mJ. Solution concentration is 2 mg/ml VCM in distillated water.

If VCM solution is exposed to this combination of laser radiations longer time intervals, *i.e.* tens of minutes up to one hour, LIF spectrum measured in real time shows a modification of the molecular structure of the medicine evidenced by a relatively steep decrease of the peak at 358 nm and a less important decrease of the peak intensity at 490 nm.

A LIF spectrum which reproduces that obtained from bulk samples when excitation is made at 266 nm is emitted by pendant microdroplets such as shown in Fig. (**11**), where the solution volume is 10 μl.

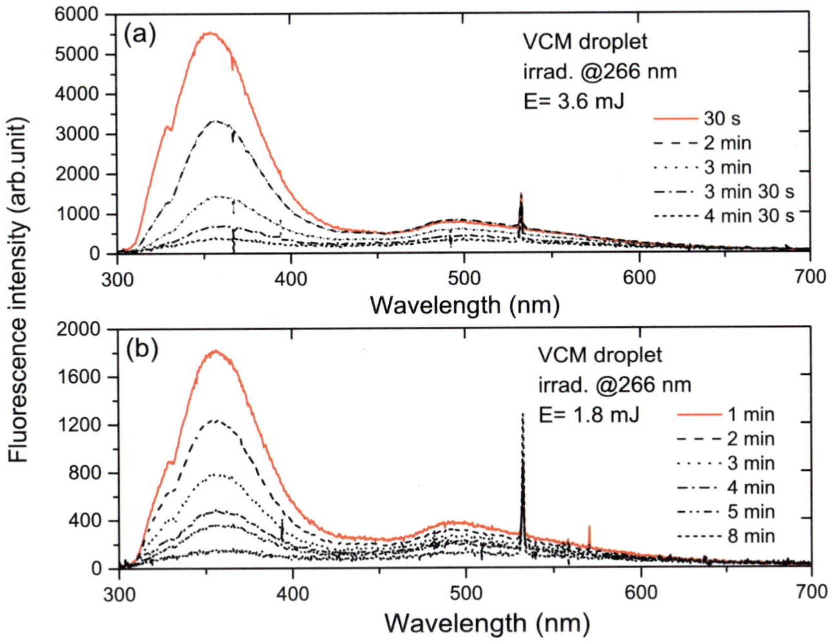


**Fig. (11).** LIF spectra of VCM solution in droplet. Droplet's volume: 10 μl; solution of 2 mg/ml VCM in distilled water. Irradiation made at 266 nm, (**a**) up to 4 min and 30 s with a pulsed laser beam of 3.6 mJ average energy, (**b**) up to 8 min at 1.8 mJ average beam energy. In (**a**) and (**b**) remnant 532 nm laser radiation is also detected.

As in CPZ case, it is observed that in microdroplets, VCM solution in water is much faster modified than in bulk, *i.e* in time intervals lasting from some seconds to up to 5 min. The rapidity with which modification of the molecular content of microdroplets is produced depends on the energy of the pumping laser beam. The modifications in bulk are produced for longer exposure time intervals than in microdroplets in which the upper limit of the allowed pumping laser beam energy depends on focusing conditions and exposure geometry to avoid breakdown (mechanical effects) at beam input in the droplet.

The exposure of medicine solution in droplet is of interest for biomedical applications since the molecular structure of it may be modified in some seconds by resonant interaction and then it may be spread on a biological target using a laser beam which interacts unresonantly with the microdroplet.

LIF shows that laser-induced molecular modifications of CPZ are faster at 355 nm irradiation than at 266 nm in droplet and bulk for the same exposure time intervals. LIF results obtained on CPZ (droplet and bulk) at 10 mg/ml were compared with those obtained for two commercially available pure substances which were identified as photoproducts generated in exposed CPZ, namely PZ and CPZ-SO [10]. Similarities between LIF peaks behaviour of CPZ in droplet with those of CPZ-SO and between CPZ in bulk with those of PZ at the same irradiation wavelength were observed. In general terms, these results contain complementary data with respect to TLC experiments.

**Thin Layer Chromatography Analysis of CPZ Solutions**

TLC method and associated analysis software are useful to investigate the molecular modification of a solution as result of interaction with UV laser beams when the solution is presented in pendant droplet form. CPZ dissolved in ultrapure water at 10 mg/ml was irradiated with 266 nm or 355 nm laser beams both in bulk and pendant droplet and the modification of the molecular content of droplets and bulk solution were compared for both wavelengths. The experimental set-up is as shown in Fig. (**3**).

TLC allows a direct comparison of the photoproducts generated by irradiation with a 266 nm or 355 nm laser beams emitted by a Nd:YAG laser for different exposure times [10, 12, 20]. The mobile phase was a mixture of acetone: methanol: 25% ammonia (150:50:5/V:V:V). The plate (DC-Alufolien Kieselgel 60F254, Merck, Darmstadt, DE) size was 14×10 cm. From each unirradiated and irradiated sample, 1 μl was applied on a plate which was afterwards placed inside the developing tank that contains mobile phase. After drying, each plate was exposed to 254 nm and 365 nm in an UV illumination cabinet [Chromo-Vue® Cabinet C-65 (UVP)] and photographed. Using a TLC dedicated software, namely CP ATLAS 2.0 freeware, each photograph of the plate was converted into a graph representing the intensity of the spot (each spot represents a photoproduct or class of photoproducts with the same polarity) as a function of position (expressed in pixels) on the plate.

The TLC plates corresponding to the irradiated solution in pendant droplet, visualised at 254 nm and 365 nm are shown in Fig. (**12**). In this case, the average energy per pulse was 3 mJ, and exposure times were 0.5, 1, 2, 3, 4 and 5 min. Each column in TLC plate corresponds to a specific irradiation time: the first column is the unirradiated sample, then from left to right are shown irradiated CPZ samples for different irradiation times. The red dotted lines indicate the unirradiated CPZ and the quantity of CPZ left in each sample after irradiation. All observed photoproducts are more polar than CPZ, their polarity increasing from top to bottom. The intensity of the spots decreases as irradiation time increases. When irradiating with 355 nm it can be observed a faster photodegradation of parent compound compared to exposure at 266 nm. At the same time, using for plate visualisation two wavelengths, 254 nm and 365 nm, it is possible to detect photoproducts, some of which absorb at 254 nm and other at 365 nm.

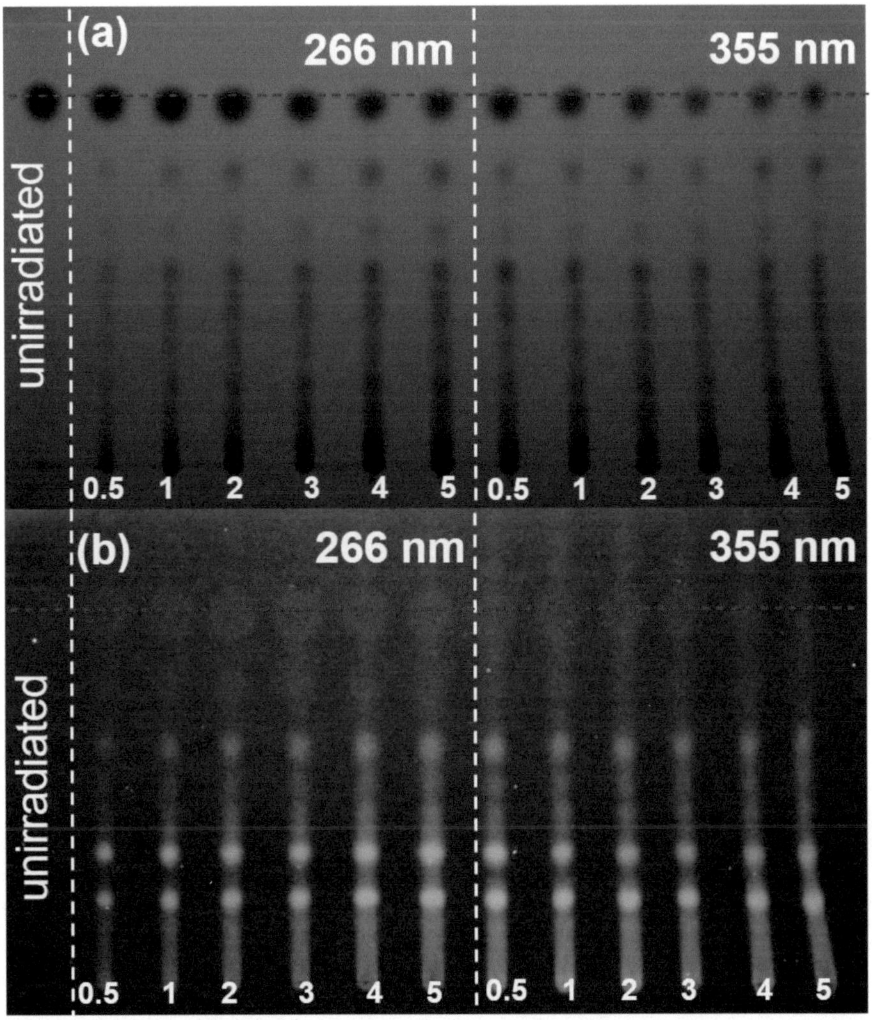


**Fig. (12).** Unirradiated 10 mg/ml, and irradiated up to 5 min 10 mg/ml CPZ water solution in pendant droplet, with 266 nm or 355 nm laser beams. Plates visualised at (**a**) 254 nm and (**b**) 365 nm.

In Fig. (**13**) is shown the TLC plate, corresponding to irradiation in bulk (1 ml), of 10 mg/ml CPZ solution with a 266 nm or 355 nm laser beam having a mean energy of 7 mJ, for exposure intervals of 0.5, 1, 2, 4 and 5 min.

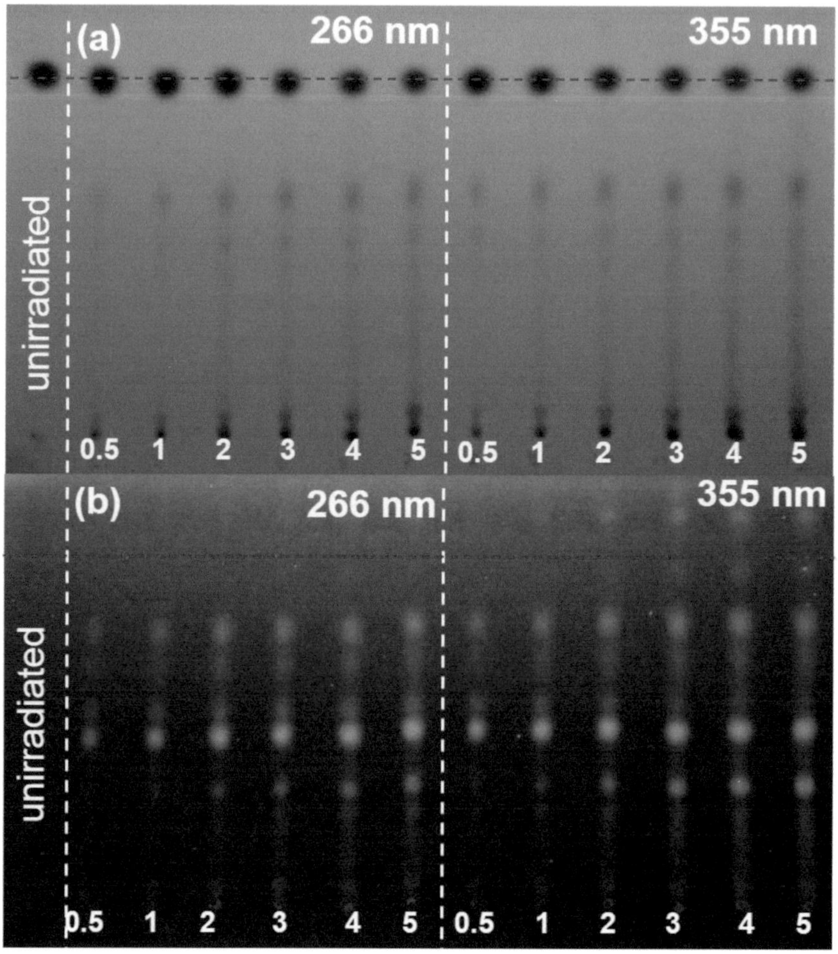


**Fig. (13).** CPZ, 10 mg/ml: unirradiated and irradiated up to 5 min, in bulk with 266 nm or 355 nm laser beams; visualised at (**a**) 254 nm and (**b**) 365 nm.

At irradiation in bulk, the volume used is significantly greater than in droplet, so that the photodegradation of the parental compound for the same exposure time as in the previous experiment is not as fast as in pendant droplet. The photoproducts observed on TLC plate in Fig. (**13**) are less than those observed in Fig. (**12**). Also, in this case one can distinguish compounds that absorb at 254 nm or 365 nm observing that at 365 nm more photoproducts are visible.

In Fig. (**14**) is presented the TLC plate image visualised at 254 nm and 365 nm for bulk solution irradiated with the same dose as the pendant droplet, that is, increasing the exposure time interval by 7.5 times.

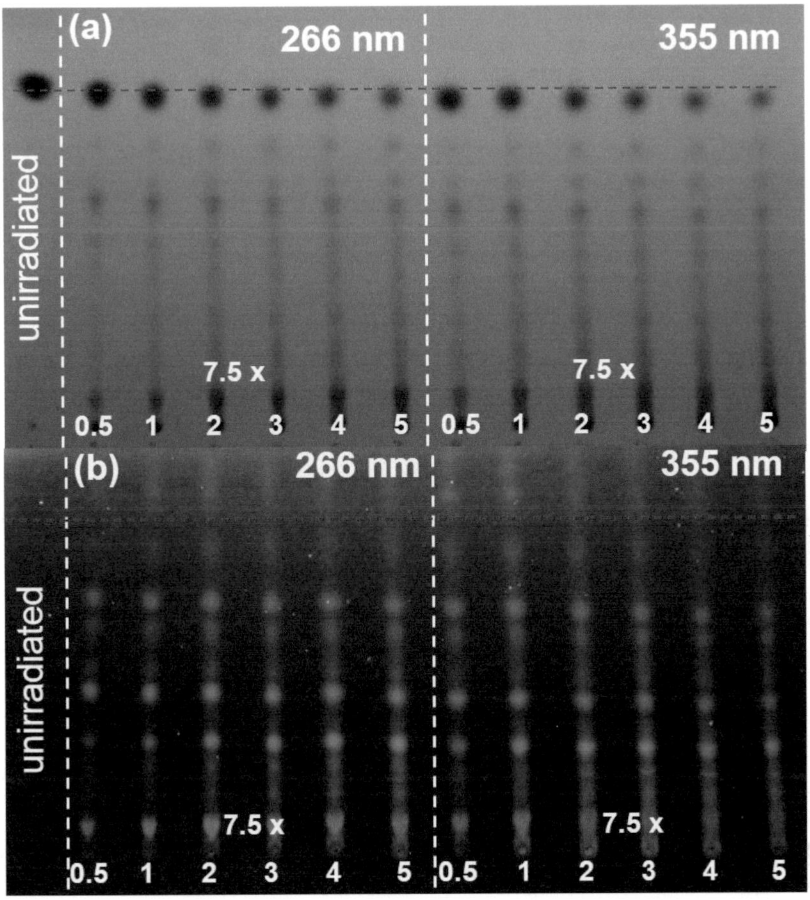


**Fig. (14).** Unirradiated and irradiated up to 7.5×5 min 10 mg/ml CPZ in bulk with the same dose as in pendant droplet with 266 nm or 355 nm. Plates visualised at (**a**) 254 nm and (**b**) 365 nm.

The mean energy per pulse was 7 mJ for exposure times to UV laser radiation of 7.5×0.5, 7.5×1, 7.5×2, 7.5×3, 7.5×4 and 7.5×5 min. More photoproducts can be observed compared to irradiation in bulk at the same energy. The irradiation effects depend on exposure time, laser beam wavelength, irradiation dose and sample's volume.

Figs. (**15-17**) show the distribution of the quantity of unirradiated CPZ and remaining CPZ in the samples at the end of exposure process with 266 nm and 355 nm in pendant droplet (Fig. **15**), in 1 ml bulk solution (Fig. **16**) and in bulk solution at the same dose of radiation as in droplet (Fig. **17**). These results were extracted by analysing the images from Figs. (**12a-14a**) with CP ATLAS 2.0 freeware software.

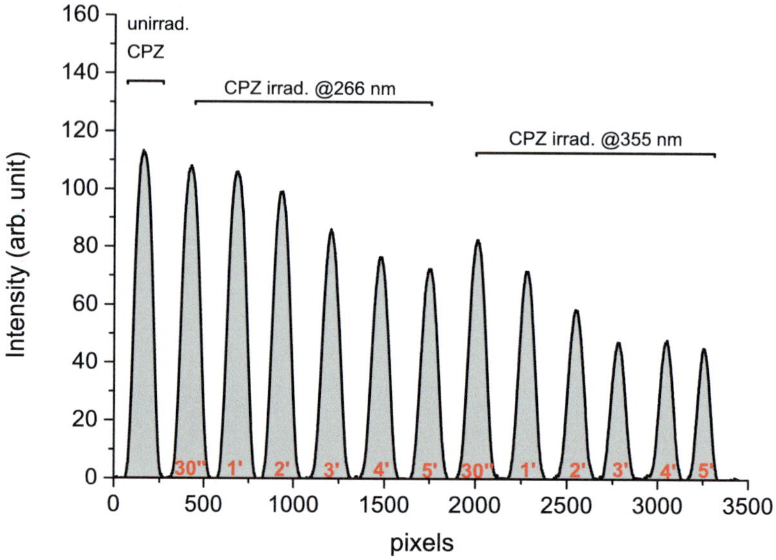


**Fig. (15).** Distribution of spots intensities of unirradiated and irradiated 10 mg/ml CPZ solution after exposure to UV laser radiation in pendant droplet.

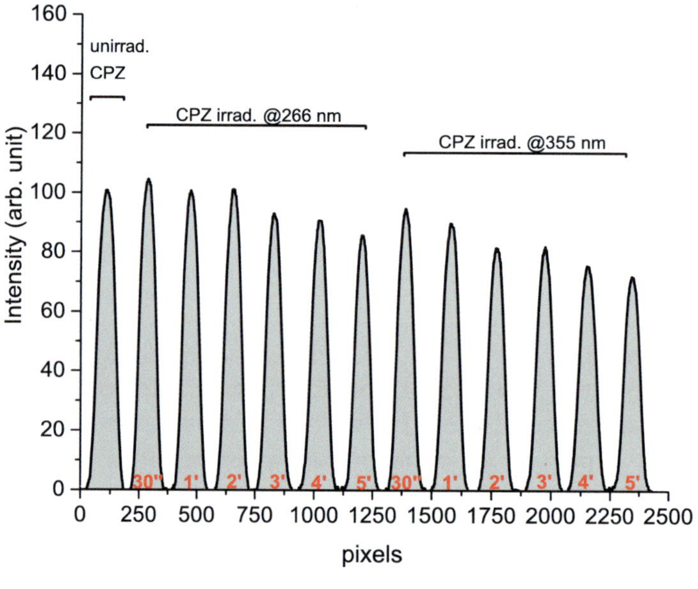


**Fig. (16).** Distribution of spots intensities of unirradiated and irradiated 10 mg/ml CPZ after exposure to UV laser radiation in bulk.

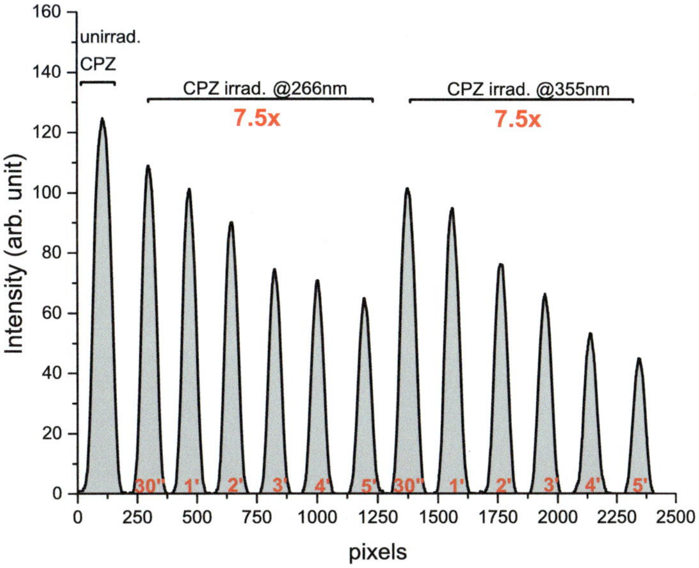


**Fig. (17).** Distribution of spots intensities of unirradiated and irradiated 10 mg/ml CPZ after exposure to UV laser radiation in bulk at the same dose as in droplet.

The results in Fig. (**15**) show a more rapid photodegradation of the parental compound if it is irradiated with 355 nm. The intensity of CPZ spot for a sample irradiated 5 min with 266 nm laser beam is the same with that of the solution irradiated 1 min with 355 nm. So, photodegradation of parental compound in the case of exposure to 355 nm is 5 times faster than for irradiation with 266 nm.

In Fig. (**16**) are shown the results on TLC plate for unirradiated and irradiated CPZ in bulk, up to 5 min with 266 nm and 355 nm. CPZ photodegradation takes place faster when using 355 nm, as well. The same magnitude of CPZ spot can be observed when sample is irradiated with 266 nm 4 min or with 355 nm 1 min.

The time evolution during irradiation, with the same dose as in pendant droplet at 266 nm and 355 nm, of a CPZ bulk solution presented in Fig. (**17**), shows that photodegradation of the parental compound is faster in droplet.

The spot intensity of remaining CPZ in the irradiated solution is the same for the sample irradiated 1 min with 266 nm and 30 s with 355 nm. Also, the same observation can be made for the sample irradiated 5 min with 266 nm and 3 min with 355 nm laser beam.

The intensity distributions of the photoproducts are presented in Figs. (**18-23**) for TLC plates shown in Figs. (**12-14**). The analysis was performed for the 254 nm visualisation of TLC plates containing CPZ solution irradiated 5 min in pendant droplet (Fig. **18**), in bulk (Fig. **19**) and in bulk at the same irradiation dose as the pendant droplet (Fig. **20**). For 365 nm visualisation of TLC plates in the same conditions, but at an exposure time of CPZ solution up to 30 s the graphs are shown in Figs. (**21-23**).

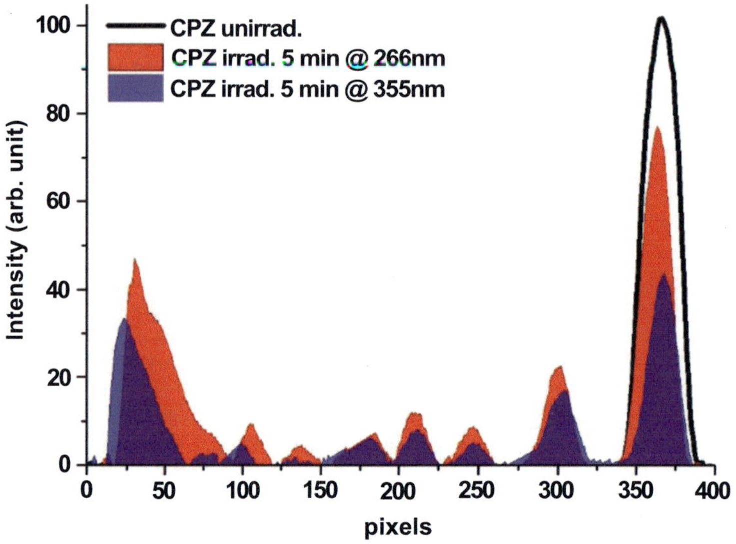


**Fig. (18).** Distribution of spots intensities of unirradiated and irradiated 10 mg/ml CPZ applied on TLC plates. Visualisation at 254 nm and analysis with ATLAS 2.0 freeware software; solutions irradiated 5 min in pendant droplet.

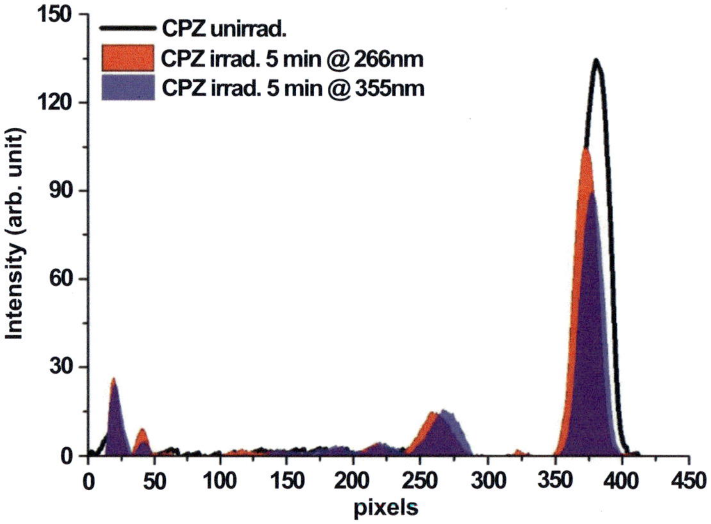


**Fig. (19).** Distribution of spot intensity of unirradiated and irradiated 10 mg/ml CPZ applied on TLC plates. Visualisation at 254 nm; analysis with ATLAS 2.0 freeware software; solutions irradiated 5 min in bulk.

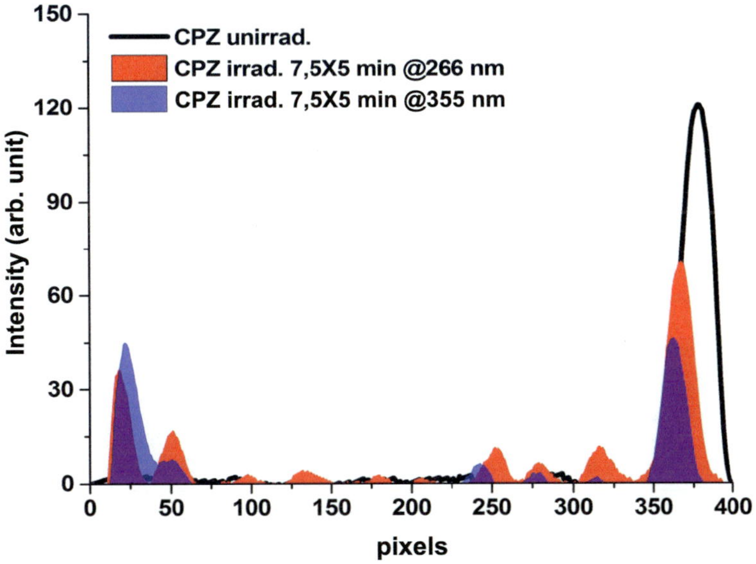


**Fig. (20).** Distribution of spots intensities of unirradiated and irradiated 10 mg/ml CPZ applied on TLC plates, visualised at 254 nm and analysed with ATLAS 2.0; solutions exposed 5 min in bulk at the same irradiation dose as for droplet.

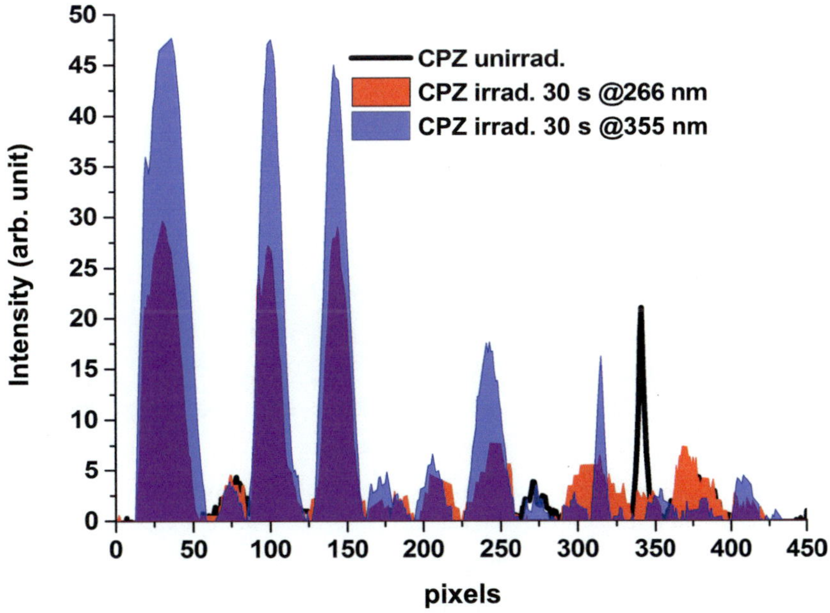


**Fig. (21).** Distribution of spots intensities of unirradiated and irradiated 10 mg/ml CPZ applied on TLC plates, visualised at 365 nm and analysed using ATLAS 2.0; the solutions were irradiated 30 s in pendant droplet.

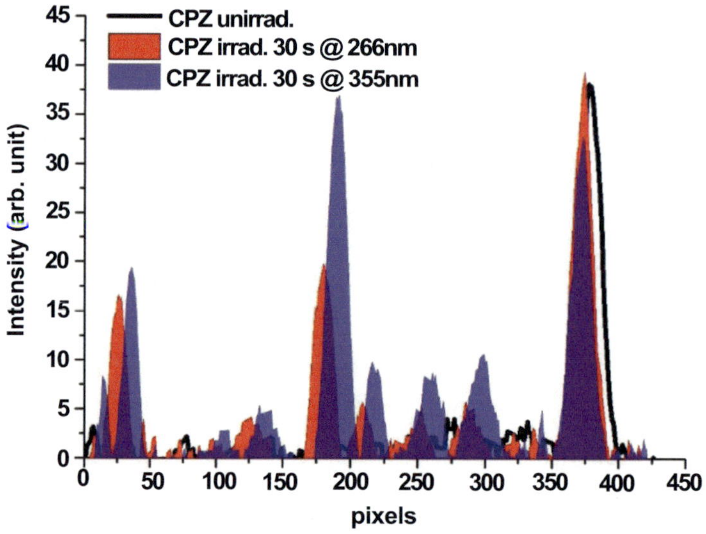


**Fig. (22).** Distribution of spots intensities of unirradiated and irradiated 10 mg/ml CPZ applied on TLC plates. Visualised at 365 nm and analysed using ATLAS 2.0; solutions irradiated 30 s in bulk.

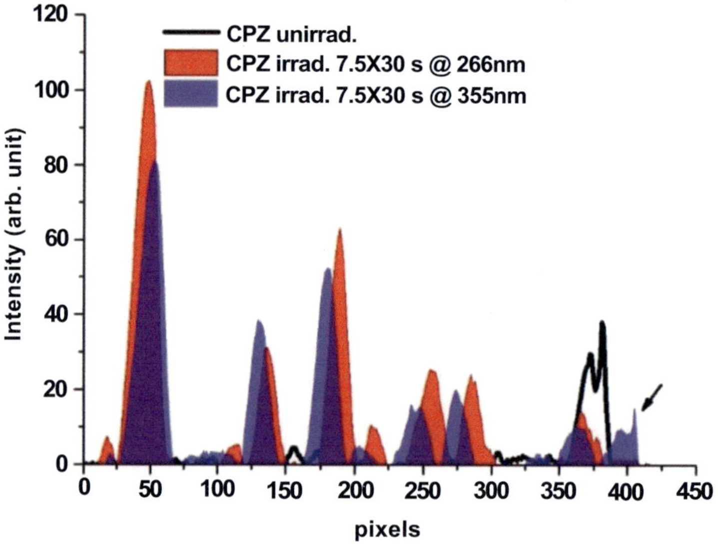


**Fig. (23).** Distribution of spots intensities of unirradiated and irradiated 10 mg/ml CPZ applied on TLC plates. Visualisation at 365 nm. Images analysed using ATLAS 2.0. Solutions exposed 30 s in bulk at the same irradiation dose as for droplet.

In general, it was observed that during irradiation with the two wavelengths the same photoproducts are present. This suggests that the molecular changes are the same for irradiation in pendant droplet and bulk.

The analysis of TLC plates visualised at 254 nm indicates that photoproducts have approximately the same concentration for exposure at 254 nm or 355 nm.

The parental compound, CPZ, centered at 375 pixels, has a lower concentration when exposed to laser beam at 355 nm than at 266 nm when exposed in droplet and in bulk at the same dose of radiation (Figs. **18** and **20**).

The analysis of TLC plates visualised at 365 nm shows significant differences in comparison with the plate visualised at 254 nm. Therefore, for CPZ solution irradiated 5 min in droplet and bulk (with the same dose) the main photoproducts present approximately the same intensity. However, irradiating 30 s, in pendant droplet, the main photoproduct has a higher intensity when the sample is irradiated with 355 nm (Fig. **21**).

The evolution of the spots intensities on TLC plates, that represents the photoproducts or classes of photoproducts having the same polarity, shows different dynamics of photodegradation process of the parental compound when comparing the irradiation in bulk and in pendant droplet, for both wavelengths used in exposure processes to UV laser beams.

When visualising the TLC plates at 365 nm, as they originate from CPZ solutions irradiated 30 s at 355 nm, it can be observed a photoproduct with lower polarity than CPZ (above CPZ in TLC palate), highlighted by the back arrow in Figs. (**21** and **23**), at about 405 pixels. This photoproduct is formed only when irradiation is made with the laser beam emitted at 355 nm, in pendant droplet or bulk solution at the same dose as in the droplet case.

## CONCLUSIONS

In this chapter is presented the resonant interaction between pendant droplets containing different solutions such as laser dyes or medicines and laser beams. Usually, the interaction takes place in the regime one single pendant droplet -one single laser pulse. The resonant interaction consists in absorption of the pumping beam by droplet's components, the amplification of laser induced fluorescence emitted by them or/and in modification of molecular structure of medicines and formation of new photoproducts.

For Rh6G solutions in water, the main result consists in (i) increased intensity of fluorescence spectra in droplet with respect to bulk, even if the volume of the cuvette is much higher than droplet's volume and in (ii) obtaining a sharper fluorescence peak at longer wavelengths. This behavior is explained by the confinement of light inside droplet, where total internal reflections at the surface of the droplet occur, the droplet acting as a spherical micro-resonator.

An extended part of this chapter presents how the exposure of parent substances to laser beams can fundamentally change their chemical structure and new compounds may be produced. These transformations are monitored in real time by performing LIF measurements during irradiation. In the case of SZ-2 solution, the results of LIF measured at different moments of irradiation, both for bulk and for droplet present two peaks whose intensities evolve differently/complementary in time: intensity of the first peak increases while the intensity of the second decreases until it is vanished. These changes are correlated with modifications of parent molecules induced by irradiation with laser beams and generation of new fluorescent photoproducts. Another interesting result is that in droplets, these changes are produced faster.

These studies contained also the monitoring results about decompositions of CPZ, PZ and VCM in water solution while irradiation of droplet or bulk is produced. For CPZ, LIF measurements reveal that the parent compound evolution during irradiation depends on the wavelengths of pumping laser beam, the modification being faster at 355 nm with respect to 266 nm at the same laser beam fluencies. In irradiated CPZ solution, out of hundreds of photoreaction products, 5 were identified, namely: PZ, PZ-SO, PZ-OH, PZ-OH-SO, and CPZ-SO.

For both CPZ and VCM solution in water, the modifications are much faster in droplets than in bulk and depend on pumping laser beam energy.

Modifications in bulk are produced for longer exposure time intervals than in microdroplets in which the upper limit of allowed pumping laser beam energy depends on focusing conditions and on exposure geometry to avoid breakdown (mechanical effects) at beam input in the droplet.

Exposure of medicine solution presented as droplet is of interest for biomedical applications since the molecular structure of the medicine may be modified in some seconds by resonant interaction and then it may be spread on the biological target using a laser beam which interacts unresonantly with the microdroplet.

## CONFLICT OF INTEREST

The authors confirm that this chapter content has no conflict of interest.

## ACKNOWLEDGEMENTS

This work has been financed by the National Authority for Research and Innovation in the frame of Nucleus programme-contract 4N/2016 and COST network MP1106.

# REFERENCES

[1] M.L. Pascu, I.R. Andrei, M. Ferrari, A. Staicu, A. Smarandache, A. Mahamoud, V. Nastasa, and L. Liggieri, "Laser beams resonant interaction with micro-droplets which have a controlled content", *Colloid. Surface. A.,* vol. 365, no. 1-3, pp. 83-88, 2010. [http://dx.doi.org/10.1016/j.colsurfa.2010.02.036]

[2] M.L. Pascu, G.V. Popescu, C.M. Ticos, and I.R. Andrei, "Unresonant interaction of laser beams with microdroplets", *J. Eur. Opt. Soc. Rapid Publ.,* vol. 7, p. 12001, 2012. [http://dx.doi.org/10.2971/jeos.2012.12001]

[3] K. Kolmakov, V.N. Belov, J. Bierwagen, C. Ringemann, V. Müller, C. Eggeling, and S.W. Hell, "Red-emitting rhodamine dyes for fluorescence microscopy and nanoscopy", *Chemistry,* vol. 16, no. 1, pp. 158-166, 2010. [http://dx.doi.org/10.1002/chem.200902309] [PMID: 19950338]

[4] V.A. Kuznetsov, N.I. Kunavin, and V.N. Shamraev, "Spectra and quantum yield of phosphorescence of rhodamine 6G solutions at 77°K", *J. Appl. Spectrosc.,* vol. 20, no. 5, pp. 604-607, 1974. [http://dx.doi.org/10.1007/BF00607454]

[5] D. Magde, G.E. Rojas, and P.G. Seybold, "Solvent Dependence of the Fluorescence Lifetimes of Xanthene Dyes", *Photochem. Photobiol.,* vol. 70, no. 5, pp. 737-744, 1999. [http://dx.doi.org/10.1111/j.1751-1097.1999.tb08277.x]

[6] K. H. Drexhage, "Fluorescence Efficiency of Laser Dyes", *J. Res. Natl. Bur. Stand. - Phys. Chemestry,* vol. 80 A, no. 3, 1976. [http://dx.doi.org/10.6028/jres.080A.044]

[7] W.W. Good, M. Sterling, and W.H. Holtzman, "Termination of chlorpromazine with schizophrenic patients", *Am. J. Psychiatry,* vol. 115, no. 5, pp. 443-448, 1958. [http://dx.doi.org/10.1176/ajp.115.5.443] [PMID: 13583246]

[8] Z. Zhelev, H. Ohba, R. Bakalova, V. Hadjimitova, M. Ishikawa, Y. Shinohara, and Y. Baba, "Phenothiazines suppress proliferation and induce apoptosis in cultured leukemic cells without any influence on the viability of normal lymphocytes. Phenothiazines and leukemia", *Cancer Chemother. Pharmacol.,* vol. 53, no. 3, pp. 267-275, 2004. [http://dx.doi.org/10.1007/s00280-003-0738-1] [PMID: 14663628]

[9] I. Gil-Ad, B. Shtaif, Y. Levkovitz, J. Nordenberg, M. Taler, I. Korov, and A. Weizman, "Phenothiazines induce apoptosis in a B16 mouse melanoma cell line and attenuate in vivo melanoma tumor growth", *Oncol. Rep.,* vol. 15, no. 1, pp. 107-112, 2006. [PMID: 16328041]

[10] T. Alexandru, A. Staicu, A. Pascu, E. Radu, A. Stoicu, V. Nastasa, A. Dinache, M. Boni, L. Amaral, and M.L. Pascu, "Characterization of mixtures of compounds produced in chlorpromazine aqueous solutions by ultraviolet laser irradiation: their applications in antimicrobial assays", *J. Biomed. Opt.,* vol. 20, no. 5, p. 051002, 2015. [http://dx.doi.org/10.1117/1.JBO.20.5.051002] [PMID: 25365041]

[11] A. Simon, T. Alexandru, M. Boni, V. Damian, A. Stoicu, V. Dutschk, and M.L. Pascu, "Interaction of solutions containing phenothiazines exposed to laser radiation with materials surfaces, in view of biomedical applications", *Int. J. Pharm.,* vol. 475, no. 1-2, pp. 270-281, 2014. [http://dx.doi.org/10.1016/j.ijpharm.2014.08.031] [PMID: 25148730]

[12] M.L. Pascu, B. Danko, A. Martins, N. Jedlinszki, T. Alexandru, V. Nastasa, M. Boni, A. Militaru, I.R. Andrei, A. Staicu, A. Hunyadi, S. Fanning, and L. Amaral, "Exposure of chlorpromazine to 266 nm laser beam generates new species with antibacterial properties: contributions to development of a new process for drug discovery", *PLoS One,* vol. 8, no. 2, p. e55767, 2013. [http://dx.doi.org/10.1371/journal.pone.0055767] [PMID: 23405212]

[13] M.J. Frisch, G.W. Trucks, H.B. Schlegel, G.E. Scuseria, M.A. Robb, J.R. Cheeseman, G. Scalmani, V. Barone, B. Mennucci, G.A. Petersson, H. Nakatsuji, M. Caricato, X. Li, H.P. Hratchian, A.F. Izmaylov, J. Bloino, G. Zheng, J.L. Sonnenberg, M. Hada, M. Ehara, K. Toyota, R. Fukuda, J. Hasegawa, M. Ishida, T. Nakajima, Y. Honda, O. Kitao, H. Nakai, T. Vreven, J.A. Montgomery Jr, J.E. Peralta, F. Ogliaro, M. Bearpark, J.J. Heyd, E. Brothers, K.N. Kudin, V.N. Staroverov, R. Kobayashi, J. Normand, K. Raghavachari, A. Rendell, J.C. Burant, S.S. Iyengar, J. Tomasi, M. Cossi, N. Rega, J.M. Millam, M. Klene, J.E. Knox, J.B. Cross, V. Bakken, C. Adamo, J. Jaramillo, R. Gomperts, R.E. Stratmann, O. Yazyev, A.J. Austin, R. Cammi, C. Pomelli, J.W. Ochterski, R.L. Martin, K. Morokuma, V.G. Zakrzewski, G.A. Voth, P. Salvador, J.J. Dannenberg, S. Dapprich, A.D. Daniels, Ö. Farkas, J.B. Foresman, J.V. Ortiz, J. Cioslowski, and D.J. Fox, *Gaussian 09, Revision D.01*Gaussian, Inc.: Wallingford, CT, 2009.

[14] Roy Dennington, *Todd Keith, and John Millam, GaussView 5.0.* Semichem Inc., Shawnee Mission, 2009.

[15] M. Cossi, V. Barone, R. Cammi, and J. Tomasi, "Ab initio study of solvated molecules: a new implementation of the polarizable continuum model", *Chem. Phys. Lett.,* vol. 255, no. 4-6, pp. 327-335, 1996.
[http://dx.doi.org/10.1016/0009-2614(96)00349-1]

[16] R. Improta, V. Barone, G. Scalmani, and M.J. Frisch, "A state-specific polarizable continuum model time dependent density functional theory method for excited state calculations in solution", *J. Chem. Phys.,* vol. 125, no. 5, p. 054103, 2006.
[http://dx.doi.org/10.1063/1.2222364] [PMID: 16942199]

[17] J-G. Lee, C. Sagui, and C. Roland, "Quantum Chemistry Simulations of Glycopeptide Antibiotics", In: *New Algorithms for Macromolecular Simulation,* B. Leimkuhler, C. Chipot, R. Elber, A. Laaksonen, A. Mark, T. Schlick, C. Schütte, R. Skeel, Eds., Springer: Berlin, Heidelberg, 2006.
[http://dx.doi.org/10.1007/3-540-31618-3_18]

[18] E. Stahl, "History of the Development of Thin-Layer Chromatography", In: *Thin-Layer Chromatography,* E. Stahl, Ed., Springer Berlin Heidelberg, 1965.
[http://dx.doi.org/10.1007/978-3-662-01031-0_1]

[19] B. Fried, and J. Sherma, *Thin-Layer Chromatography, Revised And Expanded.* CRC Press, 1999.
[http://dx.doi.org/10.1201/9780203910214]

[20] V. Sandoghdar, F. Treussart, J. Hare, V. Lefèvre-Seguin, J. Raimond, and S. Haroche, "Very low threshold whispering-gallery-mode microsphere laser", *Phys. Rev. A,* vol. 54, no. 3, pp. R1777-R1780, 1996.
[http://dx.doi.org/10.1103/PhysRevA.54.R1777] [PMID: 9913762]

[21] M. Pelton, and Y. Yamamoto, "Ultralow threshold laser using a single quantum dot and a microsphere cavity", *Phys. Rev. A,* vol. 59, no. 3, pp. 2418-2421, 1999.
[http://dx.doi.org/10.1103/PhysRevA.59.2418]

[22] A. Chiasera, Y. Dumeige, P. Féron, M. Ferrari, Y. Jestin, G. Nunzi Conti, S. Pelli, S. Soria, and G.C. Righini, "Spherical whispering-gallery-mode microresonators", *Laser Photonics Rev.,* vol. 4, no. 3, pp. 457-482, 2010.
[http://dx.doi.org/10.1002/lpor.200910016]

[23] A. Smarandache, V. Nastasa, M. Boni, A. Staicu, J. Handzlik, K. Kiec-Kononowicz, L. Amaral, and M.L. Pascu, "Laser beam resonant interaction of new hydantoin derivatives droplets for possible biomedical applications", *Colloid. Surface. A,* vol. 505, pp. 37-46, 2016.
[http://dx.doi.org/10.1016/j.colsurfa.2015.12.002]

[24] J-M. Pagès, and L. Amaral, "Mechanisms of drug efflux and strategies to combat them: challenging the efflux pump of Gram-negative bacteria", *Biochim. Biophys. Acta,* vol. 1794, no. 5, pp. 826-833, 2009.
[http://dx.doi.org/10.1016/j.bbapap.2008.12.011] [PMID: 19150515]

[25] J.E. Kristiansen, and L. Amaral, "The potential management of resistant infections with non-antibiotics", *J. Antimicrob. Chemother.*, vol. 40, no. 3, pp. 319-327, 1997. [http://dx.doi.org/10.1093/jac/40.3.319] [PMID: 9338482]

[26] J.E. Kristiansen, O. Hendricks, T. Delvin, T.S. Butterworth, L. Aagaard, J.B. Christensen, V.C. Flores, and H. Keyzer, "Reversal of resistance in microorganisms by help of non-antibiotics", *J. Antimicrob. Chemother.*, vol. 59, no. 6, pp. 1271-1279, 2007. [http://dx.doi.org/10.1093/jac/dkm071] [PMID: 17403708]

[27] L. Amaral, M. Martins, and M. Viveiros, "Phenothiazines as anti-multi-drug resistant tubercular agents", *Infect. Disord. Drug Targets,* vol. 7, no. 3, pp. 257-265, 2007. [http://dx.doi.org/10.2174/187152607782110022] [PMID: 17897062]

[28] A.D. Mosnaim, V.V. Ranade, M.E. Wolf, J. Puente, and M. Antonieta Valenzuela, "Phenothiazine molecule provides the basic chemical structure for various classes of pharmacotherapeutic agents", *Am. J. Ther.,* vol. 13, no. 3, pp. 261-273, 2006. [http://dx.doi.org/10.1097/01.mjt.0000212897.20458.63] [PMID: 16772768]

[29] V. Nastasa, V. Pradines, I.R. Andrei, M. Boni, M.L. Pascu, and R. Miller, "Studies about the generation and characterisation of microdroplets with a controlled content", *Optoelectron. Adv. Mater.-Rapid Commun.,* vol. 4, no. 11, pp. 1916-1919, 2010.

[30] A. Dinache, M. Boni, and M.L. Pascu, "Phenothiazine derivatives interaction with laser radiation", *Rom Rep Phys,* vol. 65, no. 3, pp. 1078-1091, 2013.

[31] T. Alexandru, A. Armada, B. Danko, A. Hunyadi, A. Militaru, M. Boni, V. Nastasa, A. Martins, M. Viveiros, M.L. Pascu, J. Molnar, and L. Amaral, "Biological Evaluation of Products Formed from the Irradiation of Chlorpromazine with a 266 nm Laser Beam", *Biochem. Pharmacol. (Open Access),* vol. 2, no. 1, p. 1000109, 2013.

[32] M.L. Pascu, V. Nastasa, A. Smarandache, A. Militaru, A. Martins, M. Viveiros, M. Boni, I.R. Andrei, A. Pascu, A. Staicu, J. Molnar, S. Fanning, and L. Amaral, "Direct modification of bioactive phenothiazines by exposure to laser radiation", *Recent Pat Antiinfect Drug Discov,* vol. 6, no. 2, pp. 147-157, 2011. [http://dx.doi.org/10.2174/157489111796064515] [PMID: 21517738]

[33] A.M. Armada, T. Alexandru, D. Machado, B. Danko, A. Hunyadi, A. Dinache, V. Nastasa, M. Boni, J. Ramos, M. Viveiros, J. Molnar, M.L. Pascu, and L. Amaral, "The in vitro activity of products formed from exposure of chlorpromazine to a 266 nm laser beam against species of mycobacteria of human interest", *In Vivo,* vol. 27, no. 5, pp. 605-610, 2013. [PMID: 23988894]

[34] I.R. Andrei, T. Tozar, A. Dinache, M. Boni, V. Nastasa, and M.L. Pascu, "Chlorpromazine transformation by exposure to ultraviolet laser beams in droplet and bulk", *Eur. J. Pharm. Sci.,* vol. 81, pp. 27-35, 2016. [http://dx.doi.org/10.1016/j.ejps.2015.09.017] [PMID: 26432595]

[35] G.R. Buettner, R.D. Hall, C.F. Chignell, and A.G. Motten, "The stepwise biphotonic photoionization of chlorpromazine as seen by laser flash photolysis", *Photochem. Photobiol.,* vol. 49, no. 3, pp. 249-256, 1989. [http://dx.doi.org/10.1111/j.1751-1097.1989.tb04103.x] [PMID: 2734362]

[36] S. Navaratnam, B.J. Parsons, G.O. Phillips, and A.K. Davies, "Laser flash photolysis study of the photoionisation of chlorpromazine and promazine in solution", *J. Chem. Soc. Faraday Trans. 1 Phys. Chem. Condens. Phases,* vol. 74, pp. 1811-1819, 1978.

[37] A.K. Davies, E.J. Land, S. Navaratnam, B.J. Parsons, and G.O. Phillips, "Pulse radiolysis study of chlorpromazine and promazine free radicals in aqueous solution", *J. Chem. Soc. Faraday. Trans. 1 Phys. Chem. Condens. Phases,* vol. 75, pp. 22-35, 1979.

[38] M.L. Pascu, B. Danko, A. Martins, N. Jedlinszki, T. Alexandru, V. Nastasa, M. Boni, A. Militaru, I.R. Andrei, A. Staicu, A. Hunyadi, S. Fanning, and L. Amaral, "Exposure of chlorpromazine to 266 nm laser beam generates new species with antibacterial properties: contributions to development of a new process for drug discovery", *PLoS One,* vol. 8, no. 2, p. e55767, 2013. [http://dx.doi.org/10.1371/journal.pone.0055767] [PMID: 23405212]

[39] A. Simon, T. Alexandru, M. Boni, V. Damian, A. Stoicu, V. Dutschk, and M.L. Pascu, "Interaction of solutions containing phenothiazines exposed to laser radiation with materials surfaces, in view of biomedical applications", *Int. J. Pharm.,* vol. 475, no. 1-2, pp. 270-281, 2014. [http://dx.doi.org/10.1016/j.ijpharm.2014.08.031] [PMID: 25148730]

[40] M. Boni, V. Nastasa, A. Militaru, A. Smarandache, I.R. Andrei, A. Staicu, and M.L. Pascu, "Laser beams interaction with liquids in optofluidic experiments", *Rom Rep Phys,* vol. 64, pp. 1179-1194, 2012.

# CHAPTER 10

# Microdroplets of Laser Irradiated Drug Solutions: Surface Tension and Contact Angle

**Ligia Frunza***, **Irina Zgura***, **Valeriu Florin Cotorobai**, **Constantin Paul Ganea** and **Stefan Frunza**

*National Institute of Materials Physics, Magurele, Ilfov, Romania*

**Abstract:** Precision measurements of surface tension and contact angle were performed on some medicines from the phenothiazine class (chlorpromazine, promethazine, and thioridazine) at different temperatures and concentrations in water, using the pending drop and the sessile drop methods. A specialised system DSA 100 (Krüss) was used for these measurements. Precautions were taken for an optimal acquisition of images (thermal isolation and heated windows of the chamber holding the studied droplets). Contact angles either static or dynamic were determined. Density measurements were also performed for the solutions. The same experiments were performed on these solutions after UV laser irradiation. The obtained values were discussed in comparison with the solvent behaviour and with the behaviour of other drugs (of alizarin dye class) as well. The average values of the advancing and receding angle, determined using a home developed program, are here presented for the first time.

**Keywords:** Chlorpromazine, CPZ, Daunorubicin, Density measurement, Doxorubicin, Drop interface, Dynamic contact angle, Laser irradiation, Nd:YAG laser, Pendent drop, Polytetrafluoroethylene, PTFE, Promethazine, PMZ, Sessile drop, Surface tension, Thioridazine, TZ, Ultrapure water, Vancomycin.

## INTRODUCTION

Bubble and drop interfaces are nowadays the fundamental entities on which the operating principle of many smart products for the everyday life is based (see for example, a recent review [1]). Significant improvements of these products can be achieved by using and controlling the specific properties of interfaces between the working liquids and structures holding them. For a better understanding of the interfacial phenomena, very precise values of the surface tension are needed due to the importance of this quantity in characterising the (drug) solutions.

* **Corresponding authors Ligia Frunza & Irina Zgura:** National Institute of Materials Physics, 077125 Magurele, Romania; Tel: 0040 213690185; Fax: 0040 213690177; E-mails: lfrunza@infim.ro, irina.zgura@infim.ro

**Mihail Lucian Pascu (Ed.)**

Several methods have been already developed to measure the surface tension of the liquids [2 - 5]. Among these, the most used are the equilibrium static methods in which still a slow flow of the liquid is able to influence the alignment of the molecules of an anisotropic fluid at the interface and consequently the measured value of the surface tension. The static methods based on the drop profile analysis present advantages as compared to the dynamic ones because they are independent of the contact angle and are based on an exact physical theory.

The pending drop method (used in this study as well) was improved in time by increasing the number of points taken into consideration for numerical computations [6 - 8] and by the technique to find the drop profile [6, 8, 9]. Actually, there are instruments to analyse the profile of a pendant drop assumed to take a shape corresponding to the equilibrium condition given by Young-Laplace equation [10]:

$$\Delta p = \sigma\left(\frac{1}{R_1} + \frac{1}{R_2}\right) \quad \textbf{(1)}$$

where $\sigma$ is the interfacial tension of the two media in contact, $R_1$ and $R_2$ are the main curvature radii of the surface in a point having the coordinate z measured from an arbitrary level; $\Delta p = \Delta\rho \times g \times z$, with $g$ representing the gravitational acceleration and $\Delta\rho = \rho_1 - \rho_2$ is the difference between the densities $\rho_1$ and $\rho_2$ of the two media. The results obtained following this procedure were comparable to those reported in the literature by using standard methods for measuring surface tension [6]. Among the advantages of the method one can mention that the calibration is simple and consists in finding the optical magnification in the captured image using metrological standards. The solid surfaces of the equipment involved in the method do not need special cleaning which is a great advantage as compared to Wilhelmy method where the used plate cleanness is crucial.

Ultraviolet (UV) light was used to control the dynamic surface tensions of mixed surfactant systems, *e.g.* containing sodium dodecylsulfate and an azobenzene derivative [11]. Light influences the dynamic surface tension of these solutions by driving the isomerization of the azobenzene moiety from cis to trans. The values of dynamic surface tension of an illuminated aqueous solution are with up to 25 mN/m lower than that of a solution not previously exposed to UV light. Furthermore, the evolution of changes induced in the molecular structure by exposure to optical radiation was studied. It was expected to induce the photo degradation (depolymerization) of large molecules and the formation of low-molecular-weight fractions [12] as observed in the case of free-falling water drops containing polysaccharides exposed to a $CO_2$ laser beam. The effect increases when the organic material has a significant absorbance in the range of laser lines, because the first step is the absorption of photons by the material [13].

Some of the drugs are photosensitive compounds; when exposed to visible (white) light or to UV radiation they suffer degradation yielding new photoproducts. For example, a laser beam with high energy was applied [14] to chlorpromazine (CPZ) solutions in order to increase antibacterial activity of resulting mixture. The micro/nanodroplets containing antibiotic/antitumour medicine solutions were exposed to laser beams intending to overcome the resistance to treatments with those drugs developed in different kinds of diseases including some types of tumours. Studies on interactions of laser radiation *in vitro* with biological products indicated that the activities of enzymes, gamma globulins and blood-group substances could be altered differentially [15]. Ultraviolet laser irradiation [16] induces modifications of parental compounds [17 - 27], generating photoproducts with enhanced biological activity against bacteria and cancer cells [22, 23].

This contribution presents the results of measurements concerning the surface tension and contact angles as surface properties and the density as bulk property for drug solutions in water of some members of phenothiazine class (promethazine-PMZ, thioridazine-TZ and CPZ) in the initial form, as well as after irradiation with an UV laser beam for different time intervals. In comparison, the same characteristics were determined for drugs from other classes, much more investigated from this point of view (doxorubicin-DOX, daunorubicin-DNR and vancomycin-VCM). Drug concentrations cover the therapy interval, *e.g.* for DOX which is used in ophthalmology and cancer therapy, the concentration interval is $10^{-3}$-$10^{-6}$ M.

# EXPERIMENTAL

## Materials

The studied samples consisted of different aqueous solutions of some drugs from phenothiazine class and alizarin dye (Table **1**).

**Table 1. Structure of the investigated drugs.**

| Drug/Acronym | Structure | Purchased from |
|---|---|---|
| **Promethazine/PMZ** | 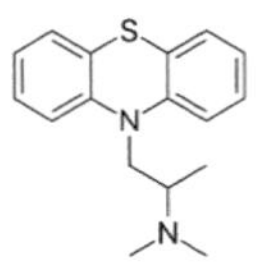 | Sigma Aldrich (Madrid branch) |

*(Table 1) contd.....*

| Drug/Acronym | Structure | Purchased from |
|---|---|---|
| **Chlorpromazine/CPZ** | 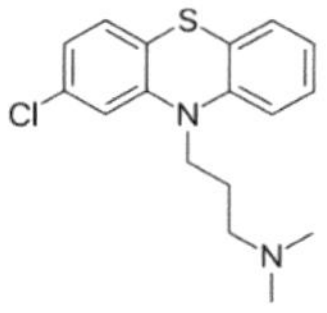 | Sigma Aldrich (Madrid branch) |
| **Thioridazine/TZ** | 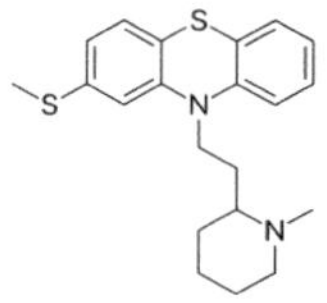 | Sigma Aldrich (Madrid branch) |
| **Doxorubicin/DOX** | 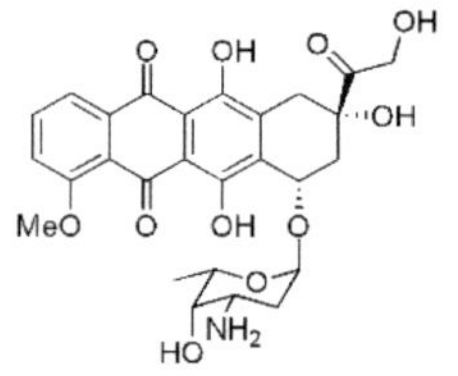 | Sigma-Aldrich, (DE) purity >98% |
| **Daunorubicin/DNR** | 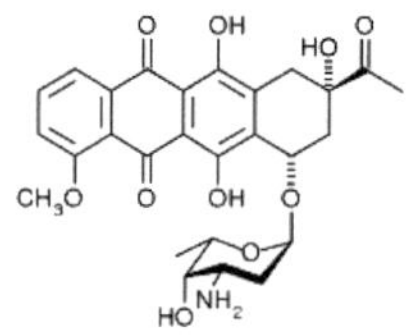 | Sigma-Aldrich, (DE) purity ≥90% |
| **Vancomycin/VCM** | 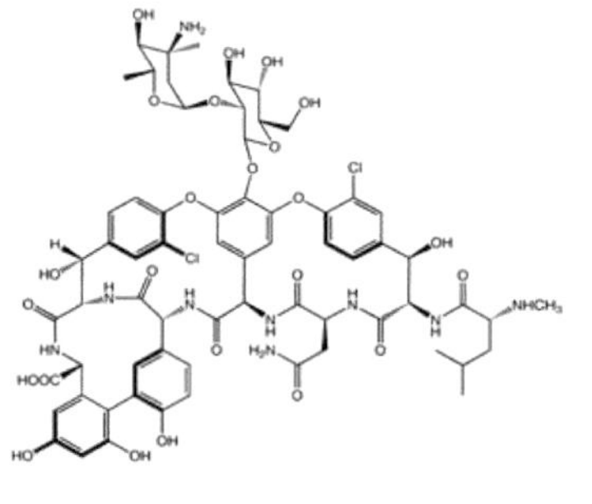 | BioChemica (AppliChem, DE) purity 93% |

Different drug concentrations were prepared. For PMZ-type drugs, the concentration was 20 mg/ml (namely between ~54 and 70 mM as function of the molecular mass of the drug) while for DOX, DNR and VCM, the concentration was mostly in the range $10^{-3}$-$10^{-6}$ mol/l. The behaviour of solutions was compared to that of solvent alone (water). The solvent was either double distilled water or ultrapure (de-ionised) water produced as described in ref [26].

To assure the accuracy of surface tension measurements the solutions were previously filtered using filters made of polytetrafluoroethylene (PTFE) with pores having the diameter of 0.2 μm while the very fine particles were eliminated with quartz filters. The presence of solid particles coming from starting solutions or appearing by precipitation after irradiation or by accident could induce systematic or accidental errors by accumulation of these particles in the pending drops. This accumulation may result in elongated drops which leads to smaller values than the real ones for the corresponding surface tensions.

The presence of dissolved gases in investigated liquids is a problem which can also affect measured values of density and consequently the values of the ratio $\sigma/\rho$ obtained from surface measurements. The degassing of liquid just before the measurement of density was performed in a syringe by creating a low vacuum resulted from displacement of the piston above the liquid and by mildly shaking the syringe.

Medicines solutions were prepared just before exposure to laser radiation. The samples were stored at 2-5°C after exposure to laser beam and kept in the dark.

The support surfaces used in the experiments of contact angle measurements were cleaned by washing (*e.g.* PTFE and glass materials, up to 10 times) in a recipient utilising warm tap water. Titanium surface was cleaned applying a jet of isopropyl alcohol and then drying. Afterwards, the supports were cut into small pieces and placed on a sample holder.

## Laser Irradiation in the UV Spectral Range

Laser irradiation of the drug solutions was performed in the laboratories of National Institute of Laser, Plasma and Radiation Physics. The laser beam was obtained by FHG (fourth harmonic generation) of a Nd:YAG laser at 266 nm. The reason to utilise this $4^{th}$ harmonic of the laser fundamental beam was due to the broad CPZ peak absorbance at 254-256 nm. Selection of beam energy at 6.5 mJ was made since CPZ at a minimum concentration can be rapidly (30 min) and completely transformed at this energy level [17]. Pulse repetition rate was 10 pps. The exposure time of the samples to laser beam varied between 0.5 to 4 h (see details in Chapters 9 and 11).

In order to control and reproduce the exposure conditions (beam energy, alignment with respect to the sample, focusing), the laser radiation was filtered to cut-off the Nd:YAG laser fundamental beam using a Neutral Filter, then collimated with a telescope and focused on sample with a long focal distance lens. The access of laser beam to the sample was controlled *via* computer or a manually acted shutter.

## Surface Tension Measurements

The surface tension was determined using the pending drop method and the specialised system DSA 100 (Krüss) improved to ensure a very good image collection: better thermal isolation, better environment isolation of measuring chamber, heating the reading windows to prevent condensation of vapors in this chamber.

The images of a stable liquid drop pending at the end of a needle were registered and then processed with a convenient computer program. As it is described by Krüss Company, from the equilibrium condition given by the Laplace equation one can express the drop profile in a parametric form by three equations, which give explicitly the link between the surface tension and the drop shape. The theoretical profile of the drop was computed by integration of parametric equations with a Runge-Kutta technique. A second program used for numerical computations was developed in the laboratory [28]. In this case, several steps were followed: (i) the determination of drop coordinates, (ii) a preliminary fit of drop profile, for which both experimental and theoretical results were reduced to the same characteristic dimension and (iii) a final fit from the results obtained in the previous step. Two main assumptions are made when drop shape analysis was performed: pending drop has rotational symmetry around a vertical axis which means that in a horizontal plane the aspect of the drop does not depend on observation direction. The pending drop method is a static one; therefore viscosity and inertia do not play an important role in the determination of drop shape, gravity and surface tension being the only forces which establish the drop shape.

The instrument DSA 100 was able to furnish liquid droplets of the order of picoliters. The syringe needle at the end of which was attached the droplet entered a thermally isolated box with transparent windows where the temperature could be controlled with a precision of 0.01°C from the room temperature up to 400°C. Taking into account that the drug is utilised inside the human body the measurements were made at temperatures close to the physiologic ones.

The drop image was registered with a video camera.

## Contact Angle Measurements

The contact angles (CA) were measured at the liquid-air interface of drug solutions in contact with solid surfaces, yielding wettability of each surface. Sessile drop method was chosen to find the static equilibrium contact angles on three model smooth surfaces (substrates). One of them was made from PTFE, the other one was from titanium either as vacuum deposited thin film on glass plates or as rolled sheet. The third model surface was a hydrophilic one, a soda lime glass slide. Specific processing by the commercial program or a home developed program was applied to put in evidence and to eliminate the artifacts concerning the drop meniscus, image contrast, drop limit under the shadow, *etc* [29 - 31]. Droplet profile was fitted with various functions (2$^{nd}$ degree polynomial, ellipse, circle, Young-Laplace profile *etc*). The first order derivative at left and right contact points with substrate gives the contact angle values. A representative example of a sessile drop used in these studies is given in the following (Fig. **1**): the fitted curves are seen in green, and blue; the tangents to the droplet profile in the points of the contact with the substrate are drawn with red. The left and right CAs have close values.

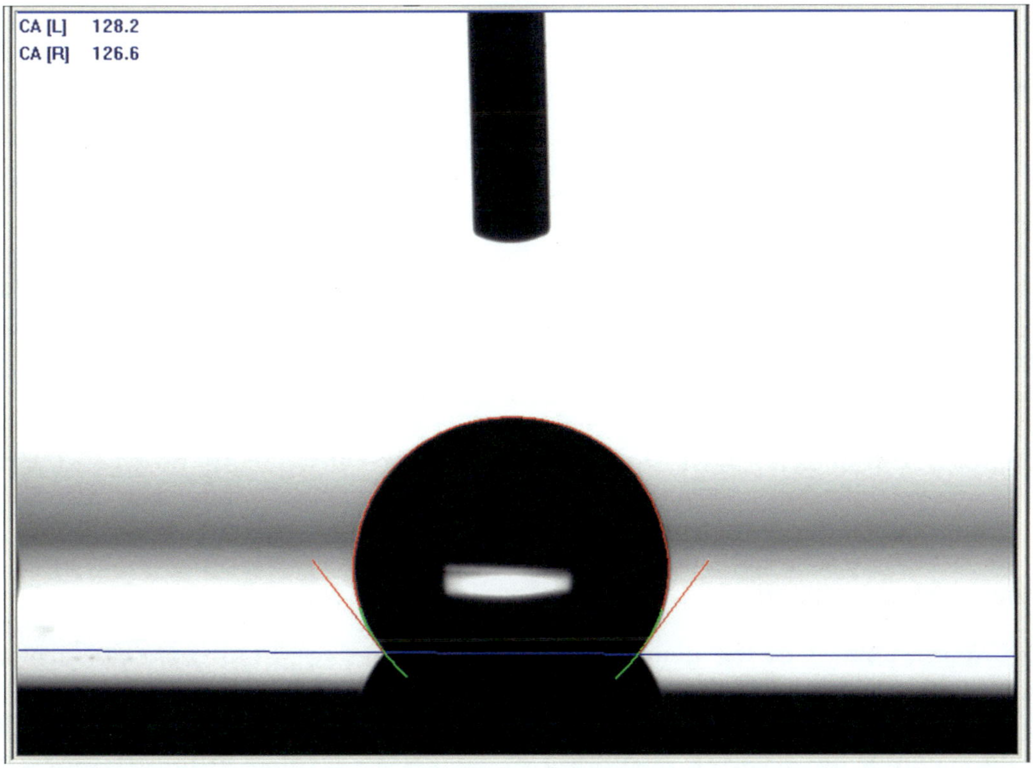


**Fig. (1).** Contact angle for pure water upon silicon wafer. The values expressed in degrees of the left (L) and right (R) CA appear in the image corner.

For a precise investigation of a system containing solvate(s) it is important to associate the equilibrium measurements to the dynamic ones. The obtained results could be analysed on the basis of a known mechanism, *e.g.* diffusion-controlled adsorption.

The dynamic contact angles, namely the advancing angle $\theta_A$ and the recessing angle $\theta_R$, with the well-known relation between them are very important, this relation imposing the equilibrium contact angle $\theta_C$ [32]

$$\theta_C = arccos\frac{r_A cos\theta_A + r_R cos\theta_R}{r_A + r_R} \quad \textbf{(2)}$$

where

$$r_A = \sqrt[3]{\frac{sin^3\theta_A}{2-3cos\theta_A + cos^3\theta_A}} \quad \textbf{(3)}$$

and

$$r_R = \sqrt[3]{\frac{sin^3\theta_R}{2-3cos\theta_R + cos^3\theta_R}} \quad \textbf{(4)}$$

The DSA 100 instrument was programmed to perform a suitable pumping cycle: liquid injection at a rate between 7.5 and 15 μl/min in order to have a droplet growing from 0 to 3 μl or 5 μl followed by suction at a rate 7.5-15 μl/min to have the droplet collapse back until disappearance. For each solution, at least 3 cycles advance/retreat were executed and registered on multimedia files. A typical movie counted 750 frames during 120 s. The movies were subsequently processed with the analysis software coming with DSA 100, and the advancing and receding angles during the pumping cycle were obtained this way.

**Density Measurements**

The measurements of the density of water and drug solutions were performed with a high precision digital densitometer (DMA 4500, produced by Paar) equipped with a new generation stainless cell. Temperature control was made using a Peltier table with a resolution of ±0.001°C, leading to uncertainties in density measurements of ca. $\pm 1 \cdot 10^{-5}$ $g \cdot cm^{-3}$. The instrument displays the apparent weight of the tube filled with the investigated liquid at a fixed temperature; ρ1 is given by:

$$\rho_1 = \frac{W_1^2 - W_a^2}{W_w^2 - W_a^2}(\rho_w - \rho_a) + \rho_a \quad \textbf{(5)}$$

where the indices *a* and *w* refer to air and water respectively (which are the calibration fluids), with known temperature variation of the density. The densitometer was calibrated using de-ionised and double distilled water, their densities being taken from literature [33]. For ultrapure water a value at 25°C of 0.99704 g/cm$^3$ was found, in very good agreement with data from literature.

A degassing procedure was made just in the syringe used to fill the measurement cell of the densitometer. With the syringe kept tight, the piston was displaced so that a low pressure space is created above the liquid. The bubbles developed inside liquid escaped by shaking the syringe *ca.* 15 min.

## RESULTS AND DISCUSSION

As already stated in Introduction, in this chapter are further presented results about surface tension and contact angle measurements for non-irradiated liquids and for laser-irradiated ones. First of all, these measurements concern water, in the forms used as solvent (doubly-distilled and de-ionised or ultrapure), and solutions of drugs from the alizarin class. In the following, it was supposed a small and monotone variation of the density in a narrow temperature interval namely between 25 and 40°C. The density was measured at 25°C; therefore the surface tension at 40°C must be viewed only as a rough value.

Irradiated samples were already analysed by several methods: pH measurements, UV-Vis-NIR absorption spectroscopy, laser induced fluorescence and thin layer chromatography (TLC) [16, 20]. The purpose of these investigations was to study and describe the modified properties of medicines to further investigate their specific interactions with materials.

### Non-irradiated Liquids

The data for the two sorts of water used as solvent and for solutions of drugs from alizarin class are presented.

#### *Water*

The variation of surface tension and sample volume as a function of time and temperature for double distilled and de-ionised water is presented in Fig. (**2**).

The two data sets put in evidence some differences in the behaviour of samples, that could be explained by different oxygen concentrations in the corresponding solutions (oxygen concentration diminishes with rising the temperature and with elapsing the time) and the small, uncontrolled evaporation of water from sample, during the measurement (though the measurement enclosure included also a small reservoir with a free water surface of around 2 cm$^2$ to reduce evaporation from the droplet) (see Fig. **3**). One can see that these measurements agree with literature [34 - 37], that reports a value of 72 mN/m at 25°C which decreases significantly with temperature.(a)

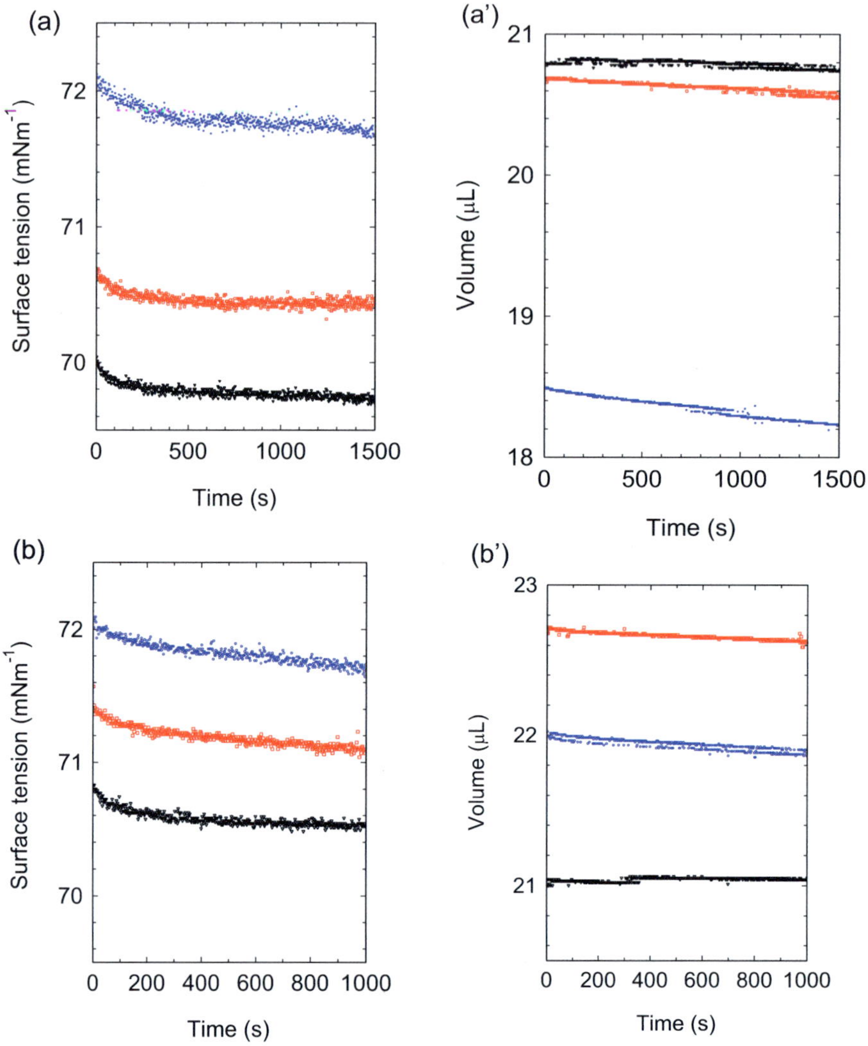


**Fig. (2).** Variation of surface tension and sample volume with time and temperature, for double distilled water (a, b) and ultrapure water (a', b'). The temperatures are as follows: 22.40°C (blue), 30.00°C (red) and 34.85°C (black) for double distilled water and respectively 20.60°C (blue), 25.15°C (red) and 29.90°C (black) for ultrapure water.

The surface tension depends on temperature (Fig. **3**). Generally, surface tension decreases with increasing temperature and reaches the zero value at the critical point.

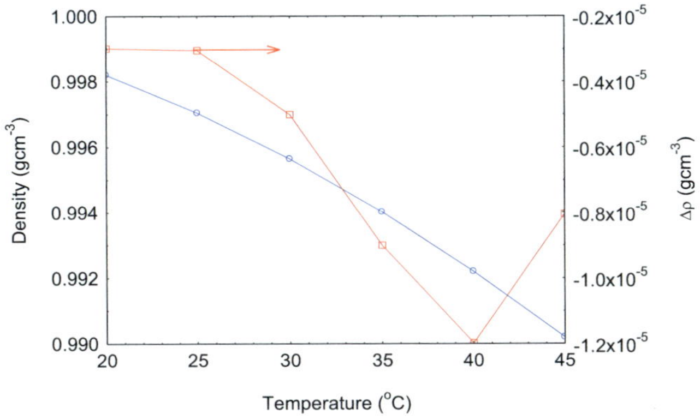


**Fig. (3).** Variation of the density (blue) and its difference (red) against the theory value (for double distilled water) as a function of temperature, for ultrapure water.

Surface tension of double distilled water remains quasi-constant when the droplet size is increasing (Fig. **4**).

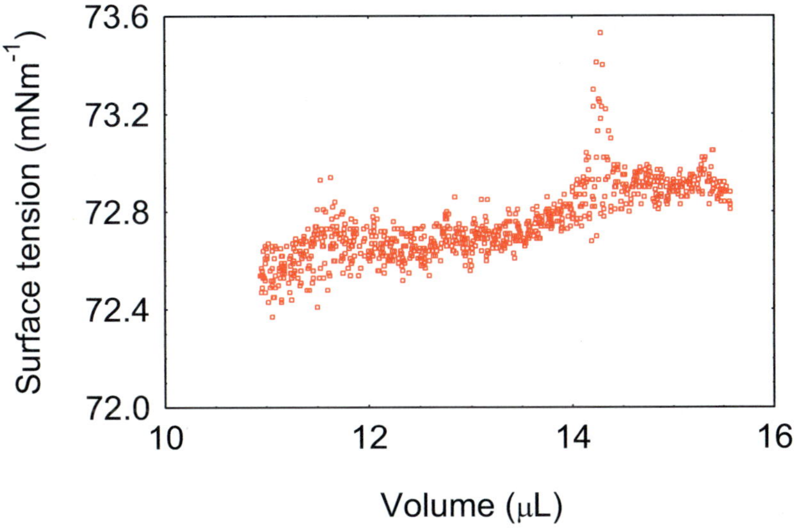


**Fig. (4).** Variation of surface tension for double distilled water with the droplet volume.

The measurement of surface tension, $\sigma$, for ultrapure water provided values comparable to those given by the reference tables, with a weak temporal variation (0.5 mN/m) during the first 10 min since drop generation. This variation might be related to the reduction of drop surface by evaporation.

Depending on their structure, the present solvates may affect the surface tension, for example they decrease the surface tension down to a certain point. Beyond this point the solvate does not have any more effect like in the case of surfactants. The complications with this effect come from the possibility for a solvate to be present in different concentrations both at the drop surface and in its bulk. The difference between surface and bulk concentration varies from a system to another.

### *Drugs Solved in Water*

The solutions of some drugs (DOX, DNR and VCM) in water were prepared at dilutions of $10^{-3}$-$10^{-5}$ M. The results on the density measurements for DOX are presented in Fig. (**5**).

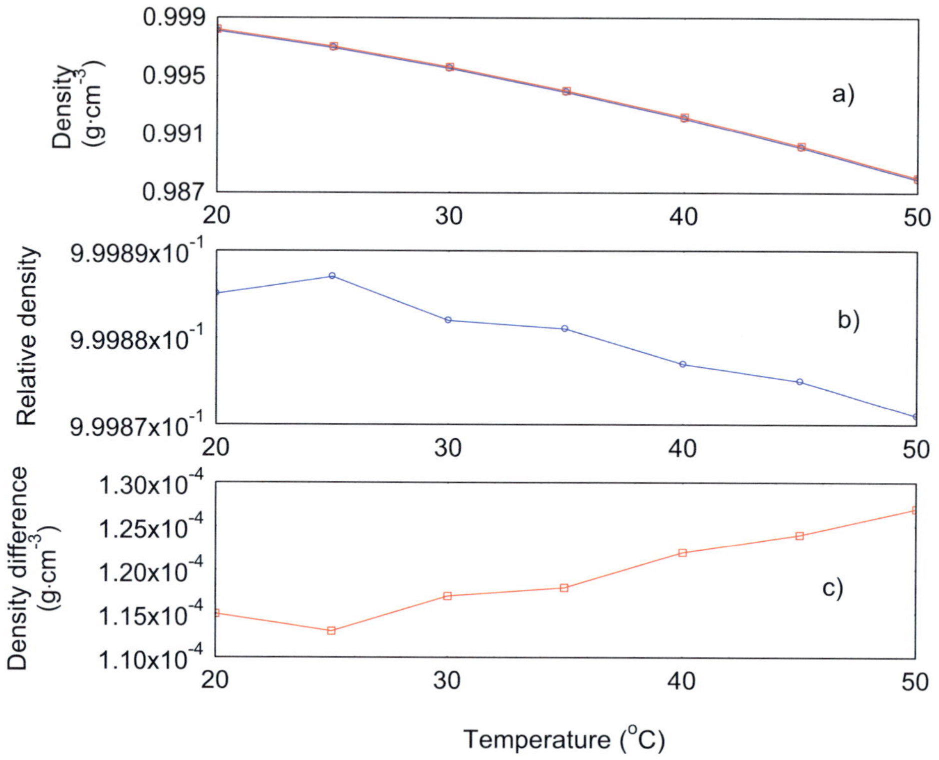


**Fig. (5). a)** Variation of density with the temperature for DOX solution ($10^{-4}$M) (blue curve) and for ultrapure water (red curve). For the ease of comparison, the graph includes the values for **b)** relative density (solution/water) and **c)** density difference (water-solution).

The interaction of the antitumour agent DOX with DNA was investigated in the aqueous solution by UV-resonance Raman spectroscopy and surface enhanced Raman spectroscopy by Lee *et al.* [38]. Thus, they found that the intercalation complexes of DOX with adenine and guanine are formed *via* a hydrogen bond between the N-7 positions of purine bases and the hydroxyl group of DOX. One may suppose that the same OH group of DOX is involved in the bindings made in our experimental solutions, leading to the variation of surface tension (Fig. **6**).

As one can see in Fig. (**6**), the drug solutions have surface tension values which are different from the water value. It is also visible a decrease in time of the surface tension for solutions, that is not found for double distilled water. Since determination of surface tension does not depend on droplet volume, the observed change for drug solutions can only be related to a change of the concentration (by evaporation) or to a possible change in the composition of solution due to exposure to air.

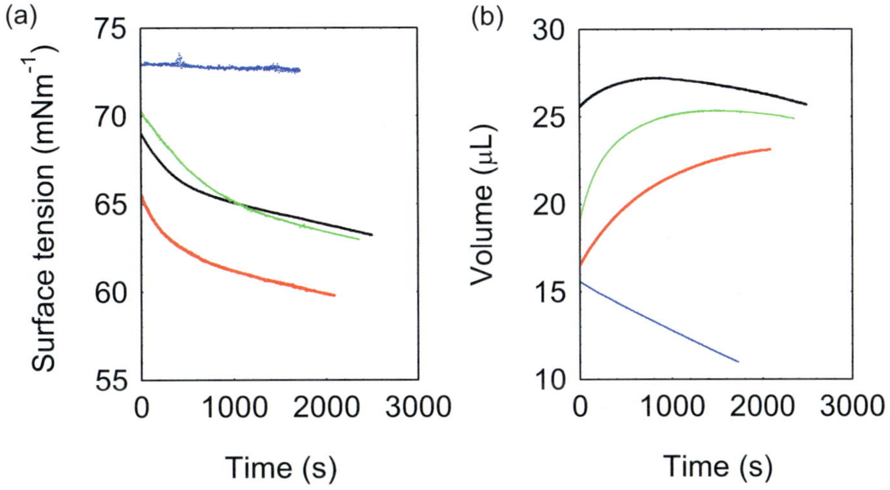


**Fig. (6).** The time evolution of the surface tension a) and sample volume b) for the drug solutions and double distilled water (blue): DNR ($10^{-4}$ M, red), DOX ($10^{-4}$ M, black) and VCM ($10^{-4}$ M, green).

The evolution in time (decrease) of the surface tension can be determined by the dynamical as well as equilibrium conditions taking place at the interface. The observed decrease of surface tension is mainly related to solvate adsorption at the interface [39]. There are relaxation processes involved in adsorption dynamics [40 - 42]. In the case of the soluble surfactants, the diffusion in one or in both bulk phases is a responsible relaxation process. In most cases diffusion is considered the mechanism which controls the whole phenomenon, being the slowest process. Inside the adsorbed layer, other kinetic processes can take place, depending on the nature of the active species and on the liquid bulk phases (like the surface aggregation, surface reactions and conformational changes). A characteristic time corresponds to each of these processes. Even though adsorption dynamics phenomena act at molecular level, they are related to a transient surface tension which can be monitored at macroscopic scale (for example by interfacial tension measurements). The temporal variation of surface tension may give information concerning the processes which control the adsorption (diffusion or other dominant processes).

The classical model of adsorption controlled by diffusion is normally used here for data interpretation. In this approach, the increase of adsorption is supposed to make equal the diffusion flux at the surface. As already mentioned, the coupling of surface tension measurements at equilibrium with dynamical surface tension measurements allows a more precise investigation of the systems containing surfactant(s). The first type of measurements on systems with different surfactant concentrations helps to establish an adequate thermodynamic model and furnishes the parameter of the adsorption isotherm. The second type of measurements may be analysed in the frame of a diffusion-controlled adsorption model and might allow the evaluation of diffusion coefficients.

The proteins have the property to agglomerate and to anchor at the air/water interface, which leads to characteristic evolution of surface tension [43]. Therefore, these aspects might be important to be considered in the case of our drugs.

In addition, there are some cases when an increase of surface tension is observed, for example in the presence of a magnetic field [44]. Thus, for magnetic fields of 10 T an increase of 1.8% was found. To explain such a behaviour it was considered that a stabilisation of hydrogen bonds takes place, which increases Helmholtz free energy of the bulk (by the increase of diamagnetic energy and of entropy) and consequently of the surface energy of the sample, the final result being the increase of surface tension. A similar behaviour might be thought to have certain drugs studied in this chapter.

In another group of measurements, the advancing and receding contact angles were measured for sessile droplets of these drug (DOX, DNR and VCM) solutions. The physics of a sessile droplet of growing volume states that its contact angle with the support is increasing from the initial value (before the pumping start) until it reaches a maximum value (the advancing angle), remaining then constant as long as the pumping continues and the droplet expands. When the liquid is drawn from the sessile droplet, the contact angle starts to decrease until attaining a minimum value (the receding angle) that will remain then constant during suction.

Figs. (**7**-**8**) illustrate representative results for the evolution of advancing and receding angles for droplets of double distilled water or drug solutions in water deposed on glass and PTFE surfaces: Fig. (**7a**) shows the image of a droplet for a $10^{-5}$ M solution of DOX on a hydrophilic substrate (glass) and Fig. (**7b**), on a hydrophobic surface (PTFE). The time evolution of the corresponding CAs is given in Fig. (**8**) for several temperatures.

PTFE coatings are superhydrophobic for common aqueous solutions which are used in biology, biotechnology and chemical sensor applications [45]. Advancing contact angles of 173° for aqueous solutions were measured on the PTFE surface; sliding angles (support inclination angle at which drops start rolling down) of 2° show a very low hysteresis of contact angle [45].

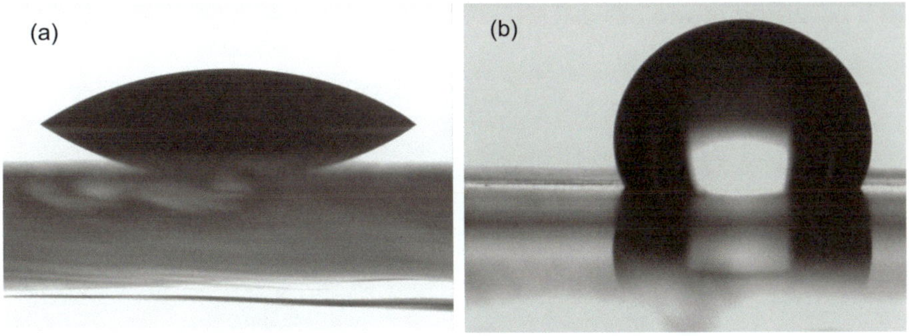


**Fig. (7).** Drop images of DOX (2x10$^{-5}$ M) solution onto different substrates (**a**) glass; (**b**) PTFE.

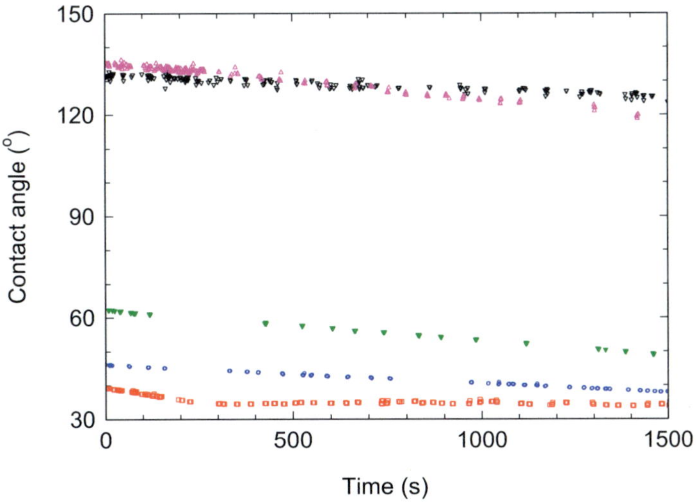


**Fig. (8).** CA for DOX (2×10$^{-5}$ M) as function of time and temperature, upon different substrates: glass, at 25°C (blue circles); glass, at 30°C (red squares); PTFE, at 18.9°C (black dawn triangles); PTFE, at 30°C (magenta up triangles); silicon, at 20°C (green filled circles).

One can see in Fig. (**9**) that, by pumping water, the CA increases up to the largest possible value without increasing the base diameter of the droplet. This maximum angle is the advancing angle. Extracting afterward the liquid, the CA decreases at constant base diameter of the droplet. When the base diameter of the droplet starts to decrease, the contact angle has the smallest possible value and this CA remains constant: it is the receding angle. The same sequence of pumping cycle was applied in all studied cases, but the images for the droplets and for the variation of CA values with droplet diameter were additionally given only for VCM (below, in Fig. **10**) as being representative. These graphs were calculated using the laboratory developed (homemade) program [28] and are shown for the first time. The measurements for double distilled water on glass give an advancing angle of ~42° and a receding angle of ~8°.

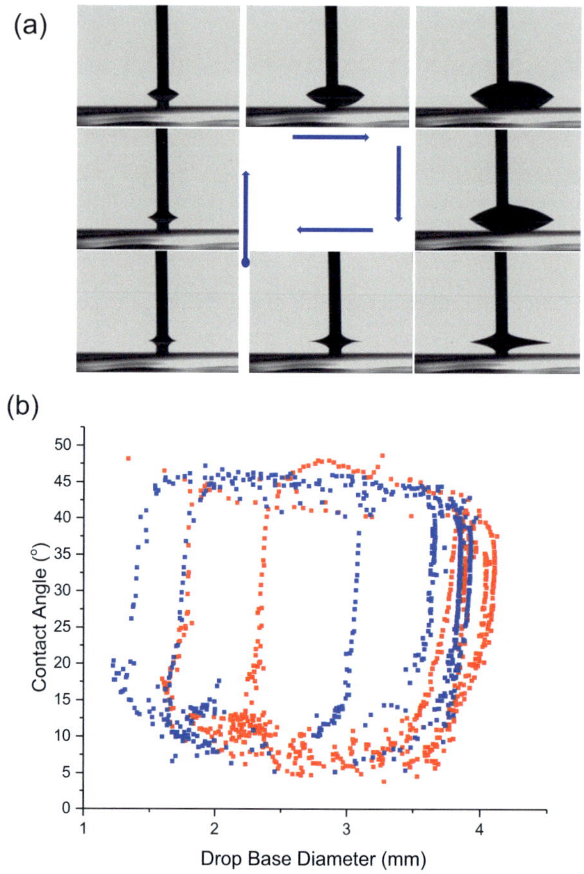


**Fig. (9).** Evolution of advancing and receding angles for a droplet of double distilled water on a hydrophilic surface (glass). **a**) Typical image during a pumping cycle. **b**) Evolution of average angle for two droplets of water (experimental points are in blue and red) deposed on different plate zones.

The pumping cycle is the same as described above for ultrapure water. By pumping VCM solution, the droplet diameter size and the advancing angle increase up to a constant value of CA, followed by an increase of the diameter at constant advancing CA. After that, droplet diameter decreases during liquid extraction; then receding CA decreases until it remains constant.

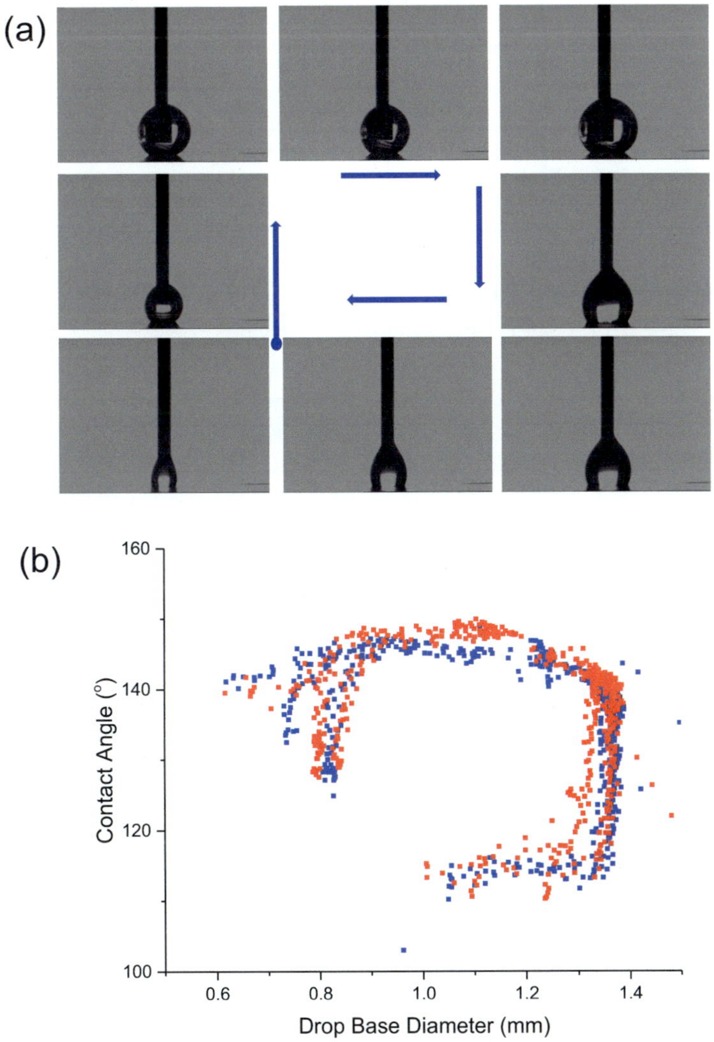


**Fig. (10).** Evolution of advancing and receding angles for a droplet of VCM ($10^{-4}$ M) on a hydrophobic surface (PTFE). **a**) Typical image during a pumping cycle. **b**) Evolution of average angle for two droplets (experimental points are in blue and red) deposed on different support zones.

The measurements for droplets of VCM $10^{-4}$ M solution on PTFE give the advancing contact angle of ~145° and the receding angle ~115°. For the other solutions investigated, the values of contact angles are grouped in the Table **2**.

**Table 2. Dynamic contact angle values [degrees-(°)] of DOX, DNR and VCM for different supports.**

| **Support** **Liquid*** | **Glass** | | **Teflon** | |
|---|---|---|---|---|
| | **Advance** (°) | **Receding** (°) | **Advance** (°) | **Receding** (°) |
| **Double distilled water** | 42 | 8 | 130 | 112 |
| **DNR** ($10^{-4}$ M) | 45 | 10-15(var)** | ~130 | ~65 |
| **DOX** ($10^{-4}$ M) | 43.5 | <7.5 | 125-135 | ~45 |
| **VCM** ($10^{-4}$ M) | ~50 | ~16 | ~145 | 115 |

*concentration of the drug, between parentheses; ** the angle varies with glass treatment.

The investigated drug molecules are amphiphilic, having both apolar and polar parts (with nitrogen and oxygen containing groups).

DOX interacts easily through -C=O and -OH groups, with specific reagents. Also, aggregates of molecules placed one over the other are known [46], mainly due to binding by hydrogen bonds and by $\pi$–$\pi$ bonds. We can talk about the acidic properties of DOX, due to the presence of ionisable phenolic OH groups and about the basic properties due to the atoms with non-participating electron pairs (oxygen, nitrogen) capable of forming coordinative bonds. Saturated cycles can be found in non strained chair shape (in the picture above, there is a saturated cycle) or in the more tense boat shape, in which both ends of the cycle are on the same side of the median plan. The balance between the two forms depends largely on the cycle substituents and their interaction with the rings/neighbouring groups. pH also plays an important role here, as well as radiation or temperature (*i.e.* as function of the way in which energy is transmitted to molecule).

The observed decrease is primarily related to adsorption at the interface of the component solute; then different possible phenomena occur in the adsorbed layer [39]. Thus, when in a liquid-air or liquid-surfactant system, the interface is perturbed (*e.g.* with a contraction or expansion), the system is out of adsorption equilibrium and a number of processes start to contribute to rebalance the system. The whole resulting relaxation process, which brings the system back to equilibrium, can generally be called adsorption dynamics or adsorption kinetics. The phenomena might be as follows: aggregation on the surface, surface reactions, conformational changes. The time variation of the surface tension may give information if adsorption is diffusion-controlled or if there are other dominant processes. The classic model of diffusion-controlled adsorption is normally used for data interpretation.

The data presented in the table lead to the conclusion that drug solutions at concentration $10^{-4}$ M do have a behaviour close to that of double distilled water on the chosen hydrophilic surface (glass). On hydrophobic surfaces (such as PTFE) the receding angles for solutions of DNR $10^{-4}$ M and DOX $10^{-4}$ M are markedly different from those of water or of VCM $10^{-4}$ M solution.

The advancing contact angles for drug solutions on both substrates (glass and PTFE) are close enough to the values measured for double distilled water.

## Laser Irradiated Medicine Solutions

The purpose of the surface tension measurements was to study the active surface characteristics of the three phenothiazine derivatives for different time intervals of exposure. Measurements were performed on aqueous drops at the medicine solution - air interface. The measurements indicate that ultrapure water alone has much higher surface tension (71-72 mN/m) than solutions containing phenothiazines (see further).

Surface tension variation with droplet age is represented in the following, either for original or for irradiated solutions. Contact angle values are shown as function of time during a pumping cycle, when drop diameter increases and then decreases as it is seen above for water and VCM.

### *Promethazine Solution*

The time variation of surface tension of a solution of PMZ is shown in Fig. (**11**).

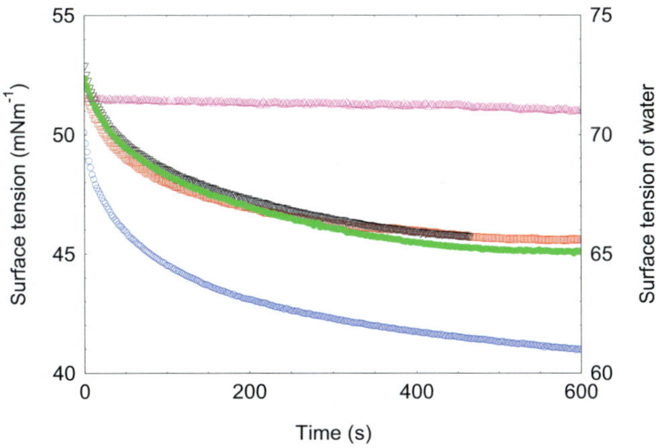


**Fig. (11).** Time variation of surface tension for PMZ solution (20 mg/ml) in ultrapure water after laser irradiation; comparison is made with the solvent (magenta) and the original solution (blue). Density is 1.00091 g/cm$^3$ at 25°C. Irradiation time: 30 min (red), 40 min (black), ~55 min (green).

Laser irradiated solution shows a surface tension decreasing in time, having a value higher than the original solution, but smaller than ultrapure solvent. The molecule of the drug is transformed and broken under the laser light. As it was already established, out of hundreds of photoreaction products, some were identified, namely: promazine (PZ), promazine sulfoxide (PZ-SO), 2-hydroxy promazine (PZ-OH), 2-hydroxy promazine sulfoxide (PZ-OH-SO) and chlorpromazine sulfoxide (CPZ-SO) [16, 23].

Fig. (**12a**) indicates smaller contact angles on PTFE, demonstrating better wetting properties of solutions, compared to water, for concentration starting from 20 mg/ml. On the other hand, the titanium support (Fig. **12b**) is better wetted by the irradiated solution than by the non-irradiated one.

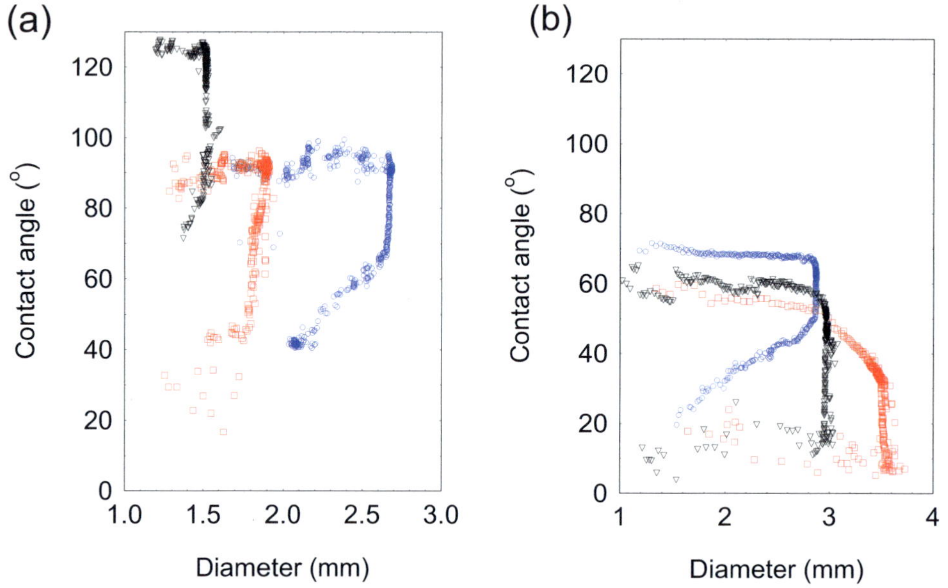


**Fig. (12).** CA values for the laser irradiated PMZ solution (20 mg/ml in ultrapure water) (red symbols) in comparison with the pure solvent (black symbols) and original solution (blue symbols) on **a**) PTFE and **b**) titanium.

PMZ lowers the surface tension at concentrations ranging from $2\times10^{-5}$ M to $5\times10^{-4}$ M [47] as it is the case of solutions considered here.

### *Thioridazine Solutions*

The time variation of the surface tension is represented in Fig. (**13**) while the variation of contact angles is given in Fig. (**14**).

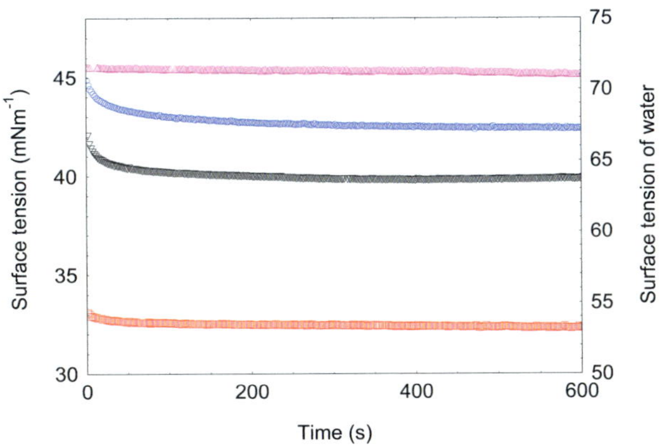


**Fig. (13).** Time variation of surface tension $\sigma$ for TZ solution (20 mg/ml) in ultrapure water after laser irradiation; comparison is made with the solvent (magenta triangles) and the original solution (blue circles). Density is 1.00094 g/cm$^3$ at 25°C. Age after irradiation: red squares-1 h; black down triangles-20 h.

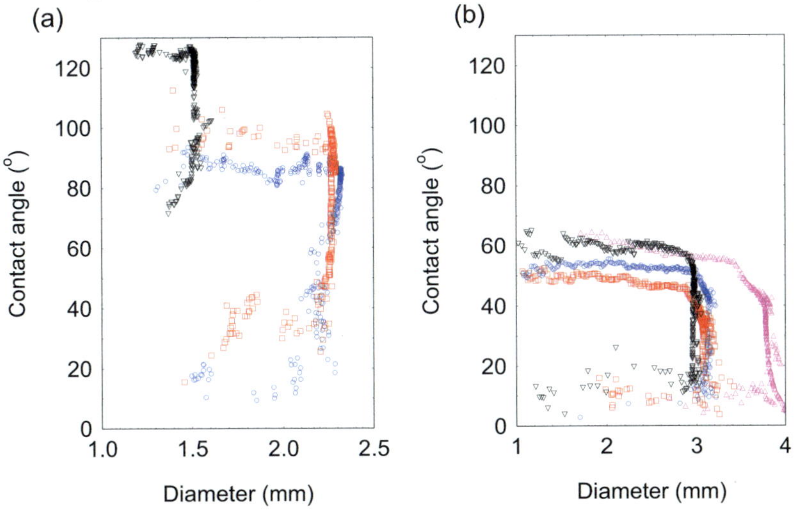


**Fig. (14).** CA values of irradiated TZ solution (20 mg/ml in ultrapure water) in comparison with the behaviour of pure solvent (black down triangles) and original solution (blue circles; red squares). The supports were (**a**) PTFE and (**b**) Ti as sheet (blue symbols) or thin film (red squares).

It is known that TZ hydrochloride is a racemic mixture of two enantiomers. The presence of a thiomethyl radical S-$CH_3$ in the position 2, conventionally occupied by a halogen, is unique in this class of compounds and makes more interesting its study: Thiomethyl group gives to TZ hydrochloride its high toxicity. Therefore, its surface and self-association properties may have a strong influence on its biological activity that makes this compound even more interesting: thus, it has association characteristics that derive from the rigid, tricyclic hydrophobic groups [48]. Aggregation characteristics have been determined by density and surface air/solution interface can be formed at a certain concentration (~0.010 mol/l) which is in reasonable agreement with the value detected by conductivity (~0.012 mol/l). Moreover, a minimum area of 1.50 $nm^2$/molecule was calculated, that is of the order of those reported for other phenothiazine drugs, between 1.32 and 1.54 $nm^2$/molecule. The distance between TZ monomers on the surface is mainly due to the linking of phenyl rings of the structure in the form of a rigid planar nucleus that causes the molecules of this drug to be so separated at interface.

Surface tension is related to the amount of drug at the air/solution interface and after its saturation, it can be considered to be fairly constant in a first approximation, not reflecting changes occurring in the bulk solution [48]. However, the concentrations considered in these studies are much lower than the critical value. Anyway, the self-association of phenothiazine drugs in aqueous solution is known for a long time [49]. An increase of critical concentration was observed between 25 and 45°C, accompanied by an increase of the degree of micellar ionisation, whereas no significant temperature dependence of the aggregation number could be discerned. Moreover, there are differences in the modes of association of the investigated drugs which are related to structural features of drug molecules [50, 51].

### *Chlorpromazine Solutions*

The time variation of surface tension is represented in Fig. (**15**) while the variation of the contact angles is given in Fig. (**16**). From Fig. (**15**) one can deduce that the surface tension of irradiated CPZ solution does not depend on aging time.

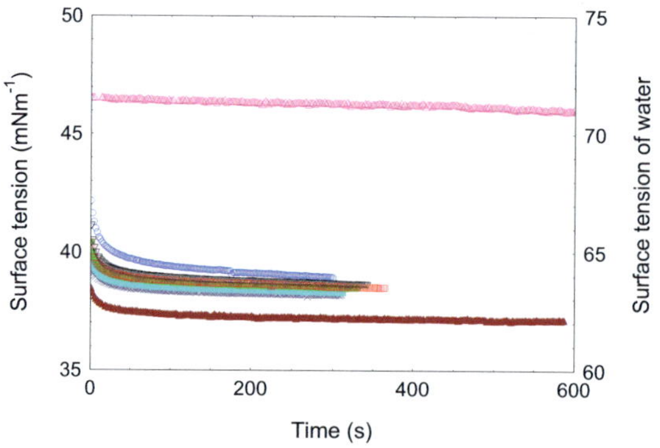


**Fig. (15).** Surface tension values of irradiated CPZ solution (20 mg/ml in ultrapure water) in comparison with pure solvent (magenta) and original solution (brown). The aging time of this irradiated CPZ solution increases from up to down for curves of the following colors: blue, red, black and green. This time covers *ca.* 30 min. Density at 25°C is 1.00162 $g/cm^3$.

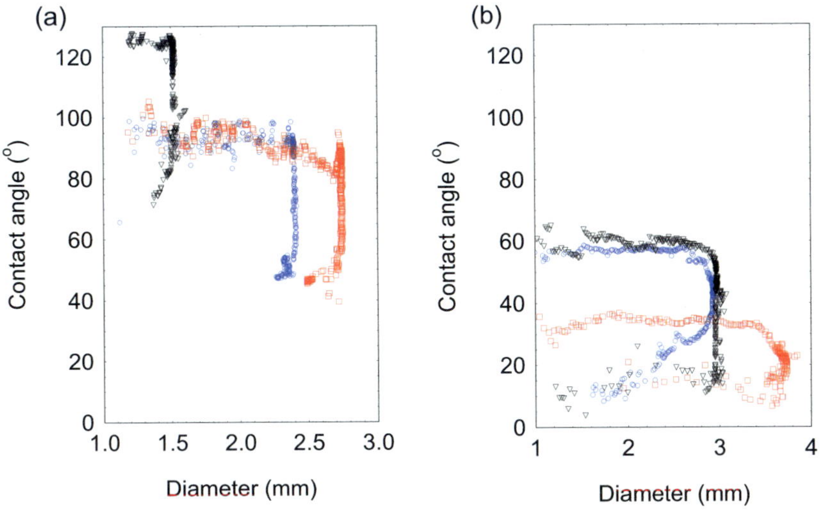


**Fig. (16).** CA values of the irradiated CPZ solution (20 mg/ml in ultrapure water) (red symbols) in comparison with the behaviour of the pure solvent (black down triangles) and original solution (blue circles). The supports were **a**) PTFE; **b**) Ti as sheet. Density at 25°C is 1.00162 g/cm$^3$.

It was already shown that CPZ lowers the surface tension (of a Ringer solution, pH 7) at concentrations between $10^{-7}$ M and $5\times10^{-6}$ M [47]. The surface tension is lowered by the formation of a virtually monomolecular film at the interface, as shown by compression studies. It is interesting that the irradiation product of CPZ, CPZ-SO, barely lowers surface tension at $10^{-3}$ M.

All identified photoproducts are much more polar than CPZ, their polarity increasing from top to bottom. The photoproducts with the highest polarity can be observed in the samples irradiated from 5 up to 120 min. After only 1 min of exposure to UV laser radiation three photoproducts can be visible on TLC plate. PZ is one of the photoproducts. On the other hand, the identified photoproducts using LC-TOF/MS, were: PZ, promazine sulfoxide (PZ-SO), 2-hydroxy promazine (PZ-OH), 2-hydroxy promazine sulfoxide (PZ-OH-SO), chlorpromazine sulfoxide (CPZ-SO), and three other compounds with m/z values of 292, 300, and 308 a.m.u [15].

Contact angle experiments [20] have found that CPZ, PZ and PMZ exhibit better wetting properties comparative to water due to lower surface tension. The contact angles of the three medicine solutions at different exposure times have shown that these angles depend on drug structure and irradiation time. The values of contact angles for the three medicines as watery solutions, before and after UV irradiation are collected in Table **3**.

The arrows in Table **3** suggest the up/down variation of the contact angle by irradiation.

**Table 3. Dynamic contact angle values [degrees-(°)] of medicine of phenothiazine class for different supports.**

| Support | Exposure** (°) | Titan film | | Titan sheet | | Teflon | |
|---|---|---|---|---|---|---|---|
| Liquid* | | Advance (°) | Receding (°) | Advance (°) | Receding (°) | Advance (°) | Receding (°) |
| Ultrapure water | N | | | 60 | 12 | 125 | 80 |
| | I | | | - | - | - | - |
| PMZ | N | | | 67 | 37 | 93 | 50 |
| | I | | | 55 ↓ | 15 ↓ | 91 | 38 ↓ |
| TZ | N | 52 | 10 | 52 | 9 | 85 | 20 |
| | I | | | 59 ↑ | 10 | 98 ↑ | 35 ↑ |
| CPZ | N | | | 63 | 21 | 92 | 48 |
| | I | | | 33 ↓ | 15 ↓ | 92 | 40 ↓ |

*concentration of the drug, 20 mg/ml in ultrapure water.
** N- non irradiated; I- irradiated.

It has been proved that CPZ photoionisation takes place during irradiation of solution with different UV light sources [52 - 58]. The modifications made in the CPZ structure consist in generation of new radicals, resulting in sulfoxides and in the loss of the chloride. Pascu *et al.* [14, 17] observed the almost complete destruction of CPZ molecules after 4 h exposure to 266 nm laser beam at lower concentration (2 mg/ml), while Alexandru *et al.* [16] show that the main photoproducts generated by exposure to laser radiation of 266 nm at concentration of 2 mg/ml are: PZ, promazine sulfoxide (PZ-SO), 2-hydroxy promazine (PZ-OH), 2-hydroxy promazine sulfoxide (PZ-OH), chlorpromazine sulfoxide (CPZ-SO) and other three compounds, chemical paths being proposed for the formation of these.

Simon *et al.* [20] using a CPZ concentration of 20 mg/ml observed that CPZ parent compound does not suffer a full destruction after 4 h of irradiation and the new obtained photoproducts (the same as above) are stable in time [17]. The changes within CPZ structure were identified utilising different measuring techniques. After exposure to UV laser radiation, the pH of the samples declines fast during the 4 h exposure. According to the role of the photo-ionisation processes [54, 57], the pH decrease is due to the increase of hydrogen ion concentration, as explained by CPZ sulfoxidation or by dechlorination [59]. In addition, Simon *et al.* [20] found that CPZ has much shorter triplet state lifetime in comparison with PZ. Excitation of CPZ and PZ molecules at 266 nm results in a monophotonic photoionisation [57], the nitrogen atoms of phenothiazines gain positive charge at reduced pH values [60], justifying the observed pH decrease.

CPZ, PZ and PMZ possess similar structures: the same highly hydrophobic ring scaffold and a different hydrophilic side chain extension at the N-10 position along with a substituent at C-2 position [61]. The differences (Cl as CPZ substituent and propyl for PMZ can explain the absorption and fluorescence properties which differ in function of phenothiazine derivative, as well as the physicochemical conditions [43].

## CONCLUSIONS

This contribution presents the results of the measurements concerning surface tension and contact angles as surface properties and density as bulk property for drug solutions in water of some members of phenothiazine class (PMZ, TZ and CPM) in the initial form, as well as after irradiation with UV laser beam for different time intervals. The reported results show the following:

The drug solutions have a surface tension smaller than that of the ultrapure water, which is, otherwise, an expected result. After irradiation, the values of $\sigma$ evolve in time (which have to be correlated with the results from other measurements) towards values of $\sigma$ from the case of non irradiated solutions.

By irradiation, the value of $\sigma$ changes as compared to the value corresponding to the non irradiated material: a) by increasing, in the case of PMZ and CPZ solutions; b) by decreasing, in the case of TZ solution.

For PMZ solution, the fast variation is more important and covers a larger interval of time than for other two drug solutions under study. These features might be caused by a greater reactivity of PMZ to some of the components of the surrounding atmosphere namely $O_2$, $CO_2$.

Surface tension depends on: dissolved gases (oxygen), chemical modifications of drugs, relative air humidity, drug concentration, outgasing of the solution in the syringe. In the case of DOX (compared to VCM case) it decreases only for "higher" concentration. The contact angle determined on the left side is close to that on the right side. The contact angle increases with concentration decreasing. The contact angle is higher for DOX if compared to VCM.

For comparison, the same characteristics were determined for drugs from other class, which are much more investigated from this point of view: DOX, DNR and VCM. The drug concentrations cover the therapy interval, *e.g.* for DOX which is used in ophthalmology and cancer therapy, the concentration interval is $10^{-3}$- $10^{-6}$ M.

One may suppose that OH groups of DOX are involved in bindings by hydrogen bonds in the solution, leading to the observed variation of the surface tension.

The time variation of the surface tension may give information if adsorption is diffusion-controlled or if there are other dominant processes. The classic model of diffusion-controlled adsorption is normally used for data interpretation.

The data concerning the contact angles lead to the conclusion that drug solutions at concentration $10^{-4}$ M do have a behaviour close to that of double distilled water on the chosen hydrophilic surface (glass). On hydrophobic surfaces (PTFE) receding angles for solutions of DNR $10^{-4}$ M and DOX $10^{-4}$ M are markedly different from those of water or of VCM $10^{-4}$ M solution. The advancing contact angles for drug solutions on both substrates (glass and PTFE) are close enough to the values measured for double distilled water.

## CONFLICT OF INTEREST

The authors confirm that this chapter content has no conflict of interest.

## ACKNOWLEDGEMENTS

The authors thank the Romanian Ministry of Education and Research (by UEFISCDI resort), for the financial support under the 41-018 PALIRT Project and CORE Program, Project PN 16 48/2016.

## REFERENCES

[1] V. Dutschk, T. Karapantsios, L. Liggieri, N. McMillan, R. Miller, and V.M. Starov, "Smart and green interfaces: from single bubbles/drops to industrial environmental and biomedical applications", *Adv. Colloid Interface Sci.*, vol. 209, pp. 109-126, 2014. [http://dx.doi.org/10.1016/j.cis.2014.02.020] [PMID: 24679903]

[2] H.H. Girault, D.J. Schiffrin, and B.D. Smith, "The measurement of interfacial tension of pendant drops using a video image profile digitizer", *J. Colloid Interface Sci.*, vol. 101, no. 1, pp. 257-266, 1984. [http://dx.doi.org/10.1016/0021-9797(84)90026-2]

[3] S.H. Anastasiadis, J.K. Chen, J.T. Koberstein, A.F. Siegel, J.E. Sohn, and J.A. Emerson, "The determination of interfacial tension by video image processing of pendant fluid drops", *J. Colloid Interface Sci.*, vol. 119, no. 1, pp. 55-66, 1987. [http://dx.doi.org/10.1016/0021-9797(87)90244-X]

[4] F.K. Hansen, and G. Rødsrud, "Surface tension by pendant drop", *J. Colloid Interface Sci.*, vol. 141, no. 1, pp. 1-9, 1991. [http://dx.doi.org/10.1016/0021-9797(91)90296-K]

[5] A. Semmler, R. Ferstl, and H-H. Kohler, "New Laser Technique for Automatic Interfacial Tension Measurements: Laser Scanning Drop Shape Analysis (LASDA)", *Langmuir,* vol. 12, no. 17, pp. 4165-4172, 1996.
[http://dx.doi.org/10.1021/la9515433]

[6] A.W. Adamson, *Physical chemistry of surfaces*, 5th ed., John Wiley & Sons: Chichester, 1991.

[7] A.I. Rusanov, and V.A. Prokhorov, *Interfacial tensiometry.*, vol. 3. Elsevier: Amsterdam, New York, 1996.
[http://dx.doi.org/10.1016/S1383-7303(96)80026-3]

[8] A.W. Neumann, J.K. Spelt, Ed., *Applied surface thermodynamics.*, vol. 63. CRC Press: New York, 1996.

[9] J. Jůza, "The pendant drop method of surface tension measurement: equation interpolating the shape factor tables for several selected planes", *Czech. J. Phys.,* vol. 47, no. 3, pp. 351-357, 1997.
[http://dx.doi.org/10.1023/A:1022808017830]

[10] P-G. de Gennes, F. Brochard-Wyart, and D. Quéré, *Capillarity and wetting phenomena: drops, bubbles, pearls, waves*. Springer: New York, 2010.

[11] J.Y. Shin, and N.L. Abbott, "Using Light to Control Dynamic Surface Tensions of Aqueous Solutions of Water Soluble Surfactants", *Langmuir,* vol. 15, no. 13, pp. 4404-4410, 1999.
[http://dx.doi.org/10.1021/la981477f]

[12] G.A. Baranov, A.A. Belyaev, S.B. Onikienko, S.A. Smirnov, and V.V. Khukharev, "Hydrodynamic effects responsible for the modification of polysaccharides in an aqueous solution under laser irradiation", *Tech. Phys. Lett.,* vol. 28, no. 9, pp. 719-721, 2002.
[http://dx.doi.org/10.1134/1.1511764]

[13] J. Wang, H. Niino, and A. Yabe, "Etching of transparent materials by laser ablation of an organic solution", *RIKEN Rev.,* vol. 32, pp. 43-46, 2001.

[14] M.L. Pascu, V. Nastasa, A. Smarandache, A. Militaru, A. Martins, M. Viveiros, M. Boni, I.R. Andrei, A. Pascu, A. Staicu, J. Molnar, S. Fanning, and L. Amaral, "Direct modification of bioactive phenothiazines by exposure to laser radiation", *Recent Pat Antiinfect Drug Discov,* vol. 6, no. 2, pp. 147-157, 2011.
[http://dx.doi.org/10.2174/157489111796064515] [PMID: 21517738]

[15] E. Klein, S. Fine, Y. Laor, M.S. Litwin, J. Donoghue, and L. Simpson, "Laser Irradiation of the Skin", *J. Invest. Dermatol.,* vol. 43, no. 6, pp. 565-570, 1964.
[http://dx.doi.org/10.1038/jid.1964.199] [PMID: 14234863]

[16] T. Alexandru, A. Staicu, A. Pascu, E. Radu, A. Stoicu, V. Nastasa, A. Dinache, M. Boni, L. Amaral, and M.L. Pascu, "Characterization of mixtures of compounds produced in chlorpromazine aqueous solutions by ultraviolet laser irradiation: their applications in antimicrobial assays", *J. Biomed. Opt.,* vol. 20, no. 5, p. 051002, 2015.
[http://dx.doi.org/10.1117/1.JBO.20.5.051002] [PMID: 25365041]

[17] T. Baran, A. Bîrjoianu, and M. Rusu, "Variations of the surface tension coefficient of aqueous solutions of chlorpromazine as a function of the concentration and UV irradiation", *Rev. Med. Chir. Soc. Med. Nat. Iasi,* vol. 76, no. 1, pp. 131-137, 1972.
[PMID: 5016774]

[18] A. Felmeister, and R. Schaubman, "Influence of ultraviolet irradiation on the surface activity of phenothiazine drugs", *J. Pharm. Sci.,* vol. 58, no. 10, pp. 1232-1234, 1969.
[http://dx.doi.org/10.1002/jps.2600581015] [PMID: 5349110]

[19] M.L. Pascu, B. Danko, A. Martins, N. Jedlinszki, T. Alexandru, V. Nastasa, M. Boni, A. Militaru, I.R. Andrei, A. Staicu, A. Hunyadi, S. Fanning, and L. Amaral, "Exposure of chlorpromazine to 266 nm laser beam generates new species with antibacterial properties: contributions to development of a new process for drug discovery", *PLoS One,* vol. 8, no. 2, p. e55767, 2013. [http://dx.doi.org/10.1371/journal.pone.0055767] [PMID: 23405212]

[20] A. Simon, T. Alexandru, M. Boni, V. Damian, A. Stoicu, V. Dutschk, and M.L. Pascu, "Interaction of solutions containing phenothiazines exposed to laser radiation with materials surfaces, in view of biomedical applications", *Int. J. Pharm.,* vol. 475, no. 1-2, pp. 270-281, 2014. [http://dx.doi.org/10.1016/j.ijpharm.2014.08.031] [PMID: 25148730]

[21] A. Smarandache, A. Simon, T. Tozar, V. Nastasa, and M.L. Pascu, "Stability studies on Promethazine unexposed and exposed to UV laser radiation", *Proc SPIE,* 2015p. 954916

[22] A. Dinache, M. Boni, T. Alexandru, E. Radu, A. Stoicu, I.R. Andrei, A. Staicu, L. Liggieri, V. Nastasa, M.L. Pascu, and M. Ferrari, "Surface properties of Vancomycin after interaction with laser beams", *Colloids Surf. Physicochem. Eng. Asp.,* vol. 480, pp. 328-335, 2015. [http://dx.doi.org/10.1016/j.colsurfa.2014.08.023]

[23] T. Alexandru, M.L. Pascu, B. Danko, V. Nastasa, M. Boni, A. Militaru, I.R. Andrei, A. Staicu, A. Hunyadi, A. Armada, M. Viveiros, and L. Amaral, "Generation and biological evaluation of the products formed from the exposure of Phenothiazine to a 266nm laser beam", *Proc SPIE,* 2013p. 88820T [http://dx.doi.org/10.1117/12.2032666]

[24] I.R. Andrei, T. Tozar, A. Dinache, M. Boni, V. Nastasa, and M.L. Pascu, "Chlorpromazine transformation by exposure to ultraviolet laser beams in droplet and bulk", *Eur. J. Pharm. Sci.,* vol. 81, pp. 27-35, 2016. [http://dx.doi.org/10.1016/j.ejps.2015.09.017] [PMID: 26432595]

[25] A. Smarandache, "Laser beams interaction with polidocanol foam: molecular background", *Photomed. Laser Surg.,* vol. 30, no. 5, pp. 262-267, 2012. [http://dx.doi.org/10.1089/pho.2011.3187] [PMID: 22417110]

[26] A. Smarandache, J. Kristiansen, J.B. Christensen, and M-L. Pascu, "Optical Studies of the Spectral Properties of Phenothiazines", *Lett. Drug Des. Discov.,* vol. 9, no. 4, pp. 352-360, 2012. [http://dx.doi.org/10.2174/157018012799859918]

[27] A. Staicu, A. Pascu, M. Boni, M.L. Pascu, and M. Enescu, "Photophysical study of Zn phthalocyanine in binary solvent mixtures", *J. Mol. Struct.,* vol. 1044, pp. 188-193, 2013. [http://dx.doi.org/10.1016/j.molstruc.2012.12.010]

[28] T. Beica, R. Moldovan, I. Zgura, and S. Frunza, "Improved method for determining the temperature dependence of interfacial tension at liquid crystal-isotropic liquid interface", *Proc Rom Acad. Ser A,* vol. 7, pp. 1-6, 2006.

[29] V.F. Cotorobai, "Drop images with reflection spots: improved processing for determination of wetting contact angles", *Rom Rep Phys,* vol. 64, pp. 467-481, 2012.

[30] I. Zgura, T. Beica, I.L. Mitrofan, C.G. Mateias, D. Pirvu, and I. Patrascu, "Assessment of the Impression materials by investigation of the hydrofylicity", *Dig. J. Nanomater. Biostruct.,* vol. 5, pp. 749-755, 2010.

[31] A.C. Popescu, L. Duta, G. Dorcioman, I.N. Mihailescu, G.E. Stan, I. Pasuk, I. Zgura, T. Beica, I. Enculescu, A. Ianculescu, and I. Dumitrescu, "Radical modification of the wetting behavior of textiles coated with ZnO thin films and nanoparticles when changing the ambient pressure in the pulsed laser deposition process", *J. Appl. Phys.,* vol. 110, no. 6, p. 64321, 2011. [http://dx.doi.org/10.1063/1.3639297]

[32] R. Tadmor, "Line energy and the relation between advancing, receding, and young contact angles", *Langmuir,* vol. 20, no. 18, pp. 7659-7664, 2004. [http://dx.doi.org/10.1021/la049410h] [PMID: 15323516]

[33] D.R. Lide, Ed., *Handbook of chemistry and physics*, 76th ed., CRC Press: Boca Raton, 1995.

[34] W.D. Harkins, and F.E. Brown, "The determination of surface tension (free surface energy), and the weight of falling drops: the surface tension of water and benzene by the capillary height method", *J. Am. Chem. Soc.,* vol. 41, no. 4, pp. 499-524, 1919. [http://dx.doi.org/10.1021/ja01461a003]

[35] N.B. Vargaftik, B.N. Volkov, and L.D. Voljak, "International Tables of the Surface Tension of Water", *J. Phys. Chem. Ref. Data,* vol. 12, no. 3, p. 817, 1983. [http://dx.doi.org/10.1063/1.555688]

[36] J. Alejandre, D.J. Tildesley, and G.A. Chapela, "Molecular dynamics simulation of the orthobaric densities and surface tension of water", *J. Chem. Phys.,* vol. 102, no. 11, p. 4574, 1995. [http://dx.doi.org/10.1063/1.469505]

[37] G. Vazquez, E. Alvarez, and J.M. Navaza, "Surface Tension of Alcohol + Water from 20 to 50oC", *J. Chem. Eng. Data,* vol. 40, pp. 611-614, 1995. [http://dx.doi.org/10.1021/je00019a016]

[38] C.J. Lee, J.S. Kang, M.S. Kim, K.P. Lee, and M.S. Lee, "The study of doxorubicin and its complex with DNA by SERS and UV-resonance Raman spectroscopy", *Bull. Korean Chem. Soc.,* vol. 25, pp. 1211-1216, 2004. [http://dx.doi.org/10.5012/bkcs.2004.25.8.1211]

[39] F. Ravera, M. Ferrari, E. Santini, and L. Liggieri, "Influence of surface processes on the dilational visco-elasticity of surfactant solutions", *Adv. Colloid Interface Sci.,* vol. 117, no. 1-3, pp. 75-100, 2005. [http://dx.doi.org/10.1016/j.cis.2005.06.002] [PMID: 16054585]

[40] F. Ravera, M. Ferrari, and L. Liggieri, "Adsorption and partitioning of surfactants in liquid-liquid systems", *Adv. Colloid Interface Sci.,* vol. 88, no. 1-2, pp. 129-177, 2000. [http://dx.doi.org/10.1016/S0001-8686(00)00043-9] [PMID: 11185695]

[41] J. Eastoe, and J.S. Dalton, "Dynamic surface tension and adsorption mechanisms of surfactants at the air-water interface", *Adv. Colloid Interface Sci.,* vol. 85, no. 2-3, pp. 103-144, 2000. [http://dx.doi.org/10.1016/S0001-8686(99)00017-2] [PMID: 10768479]

[42] J.K. Ferri, and K.J. Stebe, "Which surfactants reduce surface tension faster? A scaling argument for diffusion-controlled adsorption", *Adv. Colloid Interface Sci.,* vol. 85, no. 1, pp. 61-97, 2000. [http://dx.doi.org/10.1016/S0001-8686(99)00027-5] [PMID: 10696449]

[43] L. Caseli, D.C. Masui, R.P. Furriel, F.A. Leone, and M.E. Zaniquelli, "Adsorption kinetics and dilatational rheological studies for the soluble and anchored forms of alkaline phosphatase at the air/water interface", *J. Braz. Chem. Soc.,* vol. 16, no. 5, pp. 969-977, 2005. [http://dx.doi.org/10.1590/S0103-50532005000600013]

[44] Y. Fujimura, and M. Iino, "Magnetic Field Increases the Surface Tension of Water", *J. Phys. Conf. Ser.,* vol. 156, no. 012028, 2009.

[45] E. Burkarter, C.K. Saul, F. Thomazi, N.C. Cruz, S.M. Zanata, L.S. Roman, and W.H. Schreiner, "Electrosprayed superhydrophobic PTFE: a non-contaminating surface", *J. Phys. Appl. Phys.,* vol. 40, no. 24, p. 7778, 2007. [http://dx.doi.org/10.1088/0022-3727/40/24/027]

[46] V. Malatesta, F. Morazzoni, L. Pellicciari-Bollini, and R. Scotti, "The interaction between superoxide dismutase and doxorubicin. An electron spin resonance approach", *J. Chem. Soc. Faraday Trans. 1 Phys. Chem. Condens. Phases,* vol. 83, no. 12, p. 3669, 1987.

[47] P.M. Seeman, and H.S. Bialy, "The surface activity of tranquilizers", *Biochem. Pharmacol.,* vol. 12, pp. 1181-1191, 1963.
[http://dx.doi.org/10.1016/0006-2952(63)90093-5] [PMID: 14074118]

[48] M.A. Cheema, S. Barbosa, P. Taboada, E. Castro, M. Siddiq, and V. Mosquera, "A thermodynamic study of the amphiphilic phenothiazine drug thioridazine hydrochloride in water/ethanol solvent", *Chem. Phys.,* vol. 328, no. 1-3, pp. 243-250, 2006.
[http://dx.doi.org/10.1016/j.chemphys.2006.07.003]

[49] D. Attwood, D. Doughty, V. Mosquera, and V. Perez Villar, "Complex aggregation patterns for the self-association of phenothiazine drugs in aqueous solution", *J. Colloid Interface Sci.,* vol. 141, no. 2, pp. 316-321, 1991.
[http://dx.doi.org/10.1016/0021-9797(91)90327-5]

[50] D. Attwood, R. Blundell, V. Mosquera, and M. Garcia, "Association and Surface Properties of Amphiphilic Benzodiazepine and Benzothiazepine Drugs in Aqueous Solution", *J. Colloid Interface Sci.,* vol. 161, no. 1, pp. 19-23, 1993.
[http://dx.doi.org/10.1006/jcis.1993.1434]

[51] D. Attwood, R. Blundell, and V. Mosquera, "Light Scattering Studies on the Association of Phenothiazine Drugs in Aqueous Solutions of Low Ionic Strength", *J. Colloid Interface Sci.,* vol. 157, no. 1, pp. 50-54, 1993.
[http://dx.doi.org/10.1006/jcis.1993.1156]

[52] A. Felmeister, and C.A. Discher, "Photodegradation of Chlorpromazine Hydrochloride", *J. Pharm. Sci.,* vol. 53, no. 7, pp. 756-762, 1964.
[http://dx.doi.org/10.1002/jps.2600530712] [PMID: 14209473]

[53] T. Iwaoka, and M. Kondo, "Mechanistic Studies on the Photooxidation of Chlorpromazine in Water and Ethanol", *Bull. Chem. Soc. Jpn.,* vol. 47, no. 4, pp. 980-986, 1974.
[http://dx.doi.org/10.1246/bcsj.47.980]

[54] A.G. Motten, G.R. Buettner, and C.F. Chignell, "Spectroscopic studies of cutaneous photosensitizing agentsVIII. A spin-trapping study of light induced free radicals from chlorpromazine and promazine", *Photochem. Photobiol.,* vol. 42, no. 1, pp. 9-15, 1985.
[http://dx.doi.org/10.1111/j.1751-1097.1985.tb03540.x] [PMID: 4059358]

[55] C.F. Chignell, A.G. Motten, and G.R. Buettner, "Photoinduced free radicals from chlorpromazine and related phenothiazines: relationship to phenothiazine-induced photosensitization", *Environ. Health Perspect.,* vol. 64, pp. 103-110, 1985.
[http://dx.doi.org/10.1289/ehp.8564103] [PMID: 2869942]

[56] G.R. Buettner, A.G. Motten, R.D. Hall, and C.F. Chignell, "Free radical production by chlorpromazine sulfoxide, an ESR spin-trapping and flash photolysis study", *Photochem. Photobiol.,* vol. 44, no. 1, pp. 5-10, 1986.
[http://dx.doi.org/10.1111/j.1751-1097.1986.tb03556.x] [PMID: 3018808]

[57] G.R. Buettner, R.D. Hall, C.F. Chignell, and A.G. Motten, "The stepwise biphotonic photoionization of chlorpromazine as seen by laser flash photolysis", *Photochem. Photobiol.,* vol. 49, no. 3, pp. 249-256, 1989.
[http://dx.doi.org/10.1111/j.1751-1097.1989.tb04103.x] [PMID: 2734362]

[58] R.D. Hall, G.R. Buettner, and C.F. Chignell, "The biphotonic photoionization of chlorpromazine during conventional flash photolysis: spin trapping results with 5,5-dimethyl-1-pyrroline-N-oxide", *Photochem. Photobiol.,* vol. 54, no. 2, pp. 167-173, 1991.
[http://dx.doi.org/10.1111/j.1751-1097.1991.tb02003.x] [PMID: 1664109]

[59] B. Ljunggren, "Phenothiazine phototoxicity: toxic chlorpromazine photoproducts", *J. Invest. Dermatol.,* vol. 69, no. 4, pp. 383-386, 1977.
[http://dx.doi.org/10.1111/1523-1747.ep12510313] [PMID: 20478]

[60] M.D. Kabir-ud-Din, A. Al-Ahmadi, A.Z. Naqvi, and M. Akram, "Micellar properties of a phenothiazine drug in presence of additives", *Colloid J.,* vol. 71, no. 4, pp. 498-502, 2009. [http://dx.doi.org/10.1134/S1061933X09040115]

[61] S. Mahajan, and R.K. Mahajan, "Interactions of phenothiazine drugs with surfactants: a detailed physicochemical overview", *Adv. Colloid Interface Sci.,* vol. 199-200, pp. 1-14, 2013. [http://dx.doi.org/10.1016/j.cis.2013.06.008] [PMID: 23933135]

## CHAPTER 11

# Interaction of Laser Beams with Medicine Solutions in Bulk

**Angela Staicu[1], Adriana Samarandache[1], Tatiana Tozar[1], Alexandru Stoicu[1], Ruxandra Pirvulescu[2] and Mihail Lucian Pascu[1,3,*]**

[1] *National Institute for Laser, Plasma and Radiation Physics, Magurele, Ilfov, Romania*

[2] *University of Medicine and Pharmacy "Carol Davila", Emergency University Hospital, Ophthalmology Clinic, Bucharest, Romania*

[3] *Faculty of Physics, University of Bucharest, Magurele, Ilfov, Romania*

**Abstract:** The chapter presents a synthesis of results obtained on resonant interaction of laser radiation with medicines, in bulk, such as: cytostatics (methotrexate, 5-fluorouracil), phenothiazines (thioridazine, promethazine), quinazoline derivative (BG 1188) and hydantoin derivative (SZ-2). The effect of laser radiation on the studied samples is evaluated by analytical techniques out of which steady state absorption spectroscopy, laser induced fluorescence and FTIR spectroscopy, phosphorescence of generated photosensitised singlet oxygen and thin layer chromatography (TLC) used for qualitative photoproducts evidencing are approached. The molecular structural changes are suggested and the obtained photoproducts under laser exposure are analysed by spectroscopic means. This kind of data is related to a new method to combat multiple drug resistance acquired by microorganisms and tumours which develop new chemicals with bactericide or antitumour effects by exposing existing medicines that have low effects or are un-efficient, to laser beams.

**Keywords:** Absorption spectroscopy, Cytostatics, 5-fluorouracil, FTIR, Hydantoin derivatives, Laser induced fluorescence, Laser photochemistry, Methotrexate, Phenothiazine, Photosensitive medicines, Promethazine, Quinazoline derivatives, Singlet oxygen phosphorescence, Thin layer chromatography, Thioridazine.

## INTRODUCTION

A problem of high importance at the moment which has to be overcome in some treatments is the acquirement of multiple drug resistance (MDR) by bacteria or, more general, microorganisms. The use of an important number of already known

* **Corresponding author Mihail Lucian Pascu:** Laser Department, National Institute for Laser, Plasma and Radiation Physics, Magurele, Ilfov, Romania; Tel/Fax: 0040 214575739; E-mail: mihai.pascu@inflpr.ro

medicines became in the last decades inefficient in fighting MDR and new medicines as well as new approaches in using existing drugs to overcome MDR have to be explored. Photoactivation of some cytostatics such as 5-fluoruracil (5-FU) and methrotrexate (MTX) in order to be used as photosensitisers was pointed out in [1, 2]. Under UV laser radiation exposure, chlorpromazine (a well-known phenothiazine) exhibits molecular structure modifications in such a way that the photoproducts gain antimicrobial activity [3, 4].

Here is presented a synthesis regarding the use of laser methods to induce modifications of medicines and to obtain new photoproducts with photo-enhanced activity considering the following categories: cytostatics (MTX and 5-FU) [1, 2, 5, 6], phenothiazines (thioridazine-TZ, promethazine-PMZ) [3, 4, 7, 8], quinazoline derivative (BG 1188) [9], hydantoin derivative (SZ-2) [10, 11]. The effect of laser radiation on the studied samples is evaluated by analytical techniques such as steady state absorption and laser induced fluorescence (LIF) spectroscopy, Fourier transformed infrared (FTIR) spectroscopy, phosphorescence of generated photosensitised singlet oxygen and thin layer chromatography (TLC). These methods are used primarily for, at least, qualitative identification of photoproducts. Preclinical tests made on bacteria and tumours of these laser modified drugs are shown in Chapters 13 and 14, respectively.

## MATERIALS AND METHODS

### Materials

The studied compounds were selected based on several criteria: (i) the cytostatics, out of which two of the most "popular" medicines were chosen for which there are extensive applications reports; (ii) the phenothiazines were selected because this class contains photosensitive products and in some cases they have bactericide effects even if at high, toxic concentrations; (iii) the quinazoline derivative was specifically designed to fight MDR by using the mechanisms specific to the efflux pumps of bacteria; (iv) the hydantoin derivative was designed to fight MDR resistant tumours.

#### *Methotrexate (MTX)*

MTX ($C_{20}H_{22}N_8O_5$) belongs to the pterine group (folic acid, aminopterin, *etc.*) and it is an antifolate used, on one hand, in chemotherapy and on the other in treating diseases such as psoriasis and rheumatoid arthritis. MTX stops the DNA synthesis by inhibiting the action of dihydrofolate reductase (DHFR) enzyme which is responsible for the tetrahydrofolate (THF) level within the cell. THF plays an important role in purine nucleoside synthesis and in DNA synthesis.

The difference between MTX structure and that of THF consists in the presence of $NH_2$ radical in the pteridine ring instead of OH. These similar structures make possible the action of MTX on DNA synthesis.

Fig. (**1**) shows the 3D chemical structure of MTX optimised with the DFT method (B3LYP functional; 6-31G(d,p) basis set) using Gaussian09 [12].

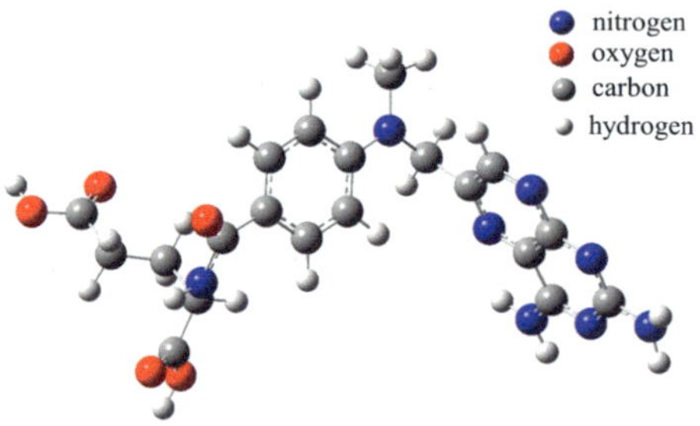


**Fig. (1).** The 3-D optimised structure of MTX for fundamental electronic state.

MTX used in the studied samples was produced by EBEWE Arzneimittel G.m.b.H (DE). The samples were MTX solutions prepared in natural saline at concentrations between $10^{-4}$ M and $10^{-5}$ M.

## *5-Fluorouracil (5-FU)*

5-FU (5-fluoro-2,4(1*H*,3*H*)-pyrimidinedione; $C_4H_3FN_2O_2$) is an antimetabolite used in the chemotherapy of a variety of solid cancer tumours. It is a chemically synthesised fluorinated pyrimidine, administered as a sodium salt; 5-FU (Sindan, ROU) solutions in natural saline (0.9% NaCl) at $10^{-4}$ M concentration M and adjusted to pH = 8.4 by the addition of NaOH were studied.

5-FU presents 2 tautomeric forms: the diketo (lactam) form, which is a major component, more stable and more weakly fluorescent tautomer, and the enol-keto (lactim) form (Fig. **2**), as a minor component but fluorescent tautomer.

**Fig. (2).** The 5-fluorouracil (5-FU) tautomers structure.

## *Phenothiazines*

The common chemical structure of phenothiazines consists of a three-ring structure in which two benzene rings are joined by sulphur and a nitrogen atom is placed at nonadjacent positions.

Phenothiazine derivatives have shown antibacterial activity when they were photosensitised by visible light [13 - 15]. They were used in the so called photodynamic antimicrobial chemotherapy (PACT) [16]. This supposes to deliver the drug to target and then to irradiate the place in order to activate singlet oxygen, reactive oxidative species or free radicals which act on bacteria strains. A particularity, that was found out recently ([3, 4]), is that under UV laser radiation, chlorpromazine (CPZ, that is the best known phenothiazine) undergoes molecular modifications in such a way that the photo-exposed solutions gained activity against a *Staphylococcus aureus* reference strain.

The difference between PACT and the experiments reported here was that in the latter case the drug was irradiated before application on bacteria targets (*i.e.* photoproducts were obtained by exposure of the parent compound to laser beam and then applied on bacteria cultures). Therefore, the potential for lasers to obtain relatively stable compounds that become biologically active comes out to be promising. This approach is expected to yield a plethora of new derivatives of phenothiazines that may be difficult to be prepared *via* conventional chemical manipulation methods. Studies regarding CPZ, TZ and PMZ interaction with laser light are reported in [3, 4, 17]. They indicate spectral changes of investigated samples, suggesting the formation of sulfoxide derivatives, the generation of a radical cation that has absorption bands in the visible spectral range and the formation of stable compounds after several hours of irradiation with a frequency quadrupled pulsed laser beam of an Nd:YAG laser (266 nm). In this chapter, some results obtained on PMZ and TZ are shown.

Thioridazine (TZ) (10-[2-(1-methyl-2-piperidinyl)ethyl]-2-(methyltio)-10*H*-phenothiazine; $C_{21}H_{26}N_2S_2$) (Fig. **3**) is an antipsychotic drug in phenothiazine class. It is used as a neuroleptic substance and in certain countries is considered to be the second line treatment for schizophrenia. Having a phenothiazine nucleus, it affects the central nervous system roughly to the same extent as chlorpomazine. Still, due to accumulated data concerning side-effects such as cardiac failure, angina, cardiomyopathy and severe cardiac arrhythmias, its use in the UK and other countries is not recommended since 2000 [18].

The compound studied here was purchased from Sigma Aldrich (DE) and solutions were prepared in ultrapure water at 20 mg/ml concentration.

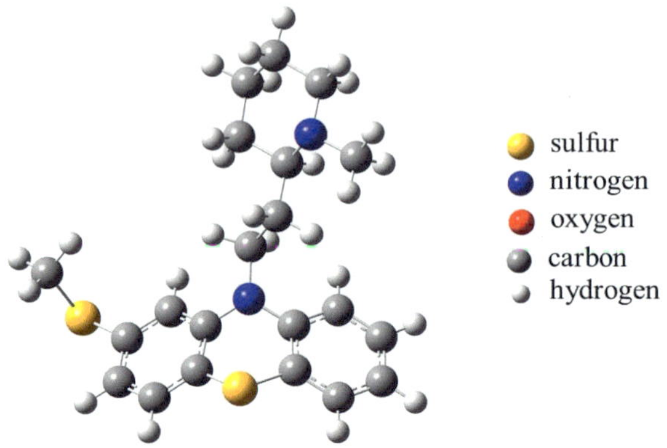


**Fig. (3).** 3D optimised chemical structure of TZ.

Promethazine (PMZ) (*N,N,*α-trimethyl-10*H*-phenothiazine-10-ethanamine; $C_{17}H_{20}N_2S$), with the chemical structure shown in Fig. (**4**) is a histamine receptor antagonist. It is a chiral compound, found as a mixture of enantiomers. Due to its potency and length of action it is widely used for its central sedative effects. The hydrochloride form is an odourless, white or faintly yellow crystalline powder with a bitter taste. It slowly oxidizes in moist environment and turns blue when exposed to air, so it should be stored in airtight, heat-resistant containers [18].

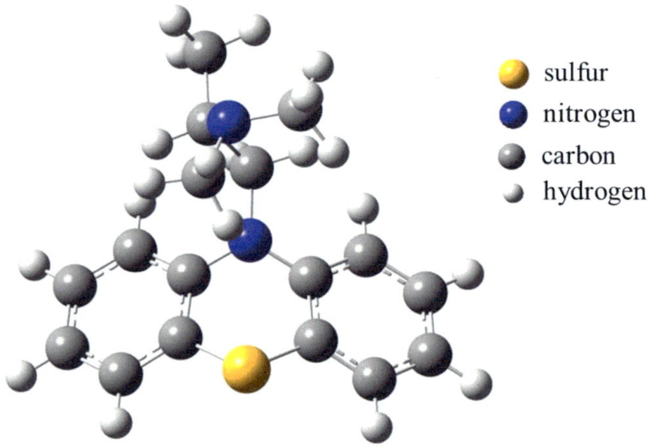


**Fig. (4).** 3D optimised chemical structure of PMZ.

The studied compound was purchased from Sigma Aldrich (DE) and the studied solutions were prepared in ultrapure water at 20 mg/ml concentration.

## *Hydantoin SZ-2*

The ring of hydantoin (2,4-imidazolidinedione) sparingly appears in nature. As a ring system it was first synthesised by Baeyer in 1861, but its structure was correctly assigned in 1870 by Strecker [19]. Hydantoin derivatives enjoy special attention in the medical field due to their pharmacological activity as anticonvulsants [20], antiandrogens [21], antidiabetics [22], antimicrobial [23], anticancer and antimetastatics [24, 25]. Several studies demonstrate radical uptake abilities of imidazolidine derivative compounds [26]. Guanidine and spiroiminodihydantoins are known to mispair with adenine or guanine as a consequence of their structural similarity to nitrogenous bases of DNA [27] that can help in cancer control. Hydantoins like 5-hydroxyhydantoin and 5-methyl-5-hydroxyhydantoin may act as blocking lesion factors for DNA polymerases [28]. Recent studies assert that some hydantoin derivatives inhibit the P-glycoprotein efflux pump of mouse T-lymphoma cells and act synergistically with doxorubicin [29]. In an anti-cancer drug screening program [30], 5-(2-Phenyl-3′-indolal)-2-thiohydantoin has shown inhibitory activity on several cancer lines organised into subpanels representing leukaemia, melanoma and lung, colon, kidney, ovary, breast, prostate and central nervous system cancer; 5-benzyliden--hydantoin is known to interact with DNA *via* intrinsic and extrinsic pathways [31].

The hydantoin derivative generically named SZ-2 ((5Z)-5-(3-chlorobenzylidene) - 2-thioxo-4-imidazolidinone) is part of a set of new synthesised chemical compounds as modulators of cancer efflux pumps [32] at Jagiellonian University in Krakow (Poland). The chemical structure of SZ-2 is shown in Fig. (**5**).

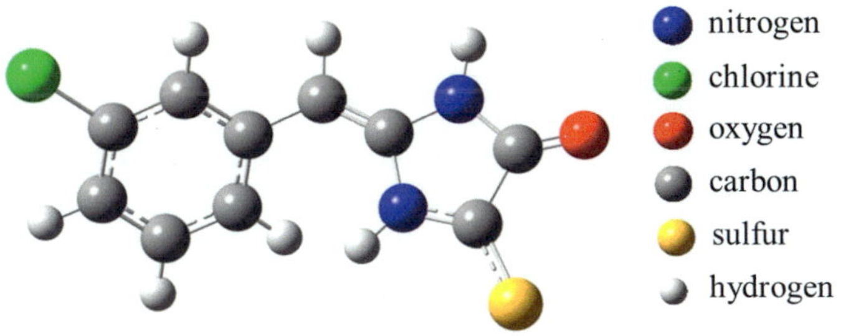


**Fig. (5).** 3D optimised chemical structure of SZ-2.

The compound is presented under the form of light yellow crystals, keeping the appearance of thiohydantoins [33]. It was prepared as solution in dimethyl sulfoxide (DMSO) at different concentrations and it was preserved in dark at 4°C. DMSO shows some advantages in biomedical applications especially as a permeation enhancer to improve drug delivery through tissue's epithelium [34], although side effects could appear [35].

## *Quinazoline BG 1188*

Out of the broad class of heterocyclic compounds, quinazolines are among the most important group. They received special attention in medicinal chemistry, because of their large panel of pharmacological properties, including diuretic [36], antihypertensive, antitumour [37], antiallergic [38], antimicrobial [39, 40], analgesic [41], anti-inflammatory [42] and antifungal [43] effects.

Based on the importance of these molecules, researchers focused their attention towards synthesis, characterisation, and evaluation of novel quinazoline derivatives in order to find more potent biologically active molecules.

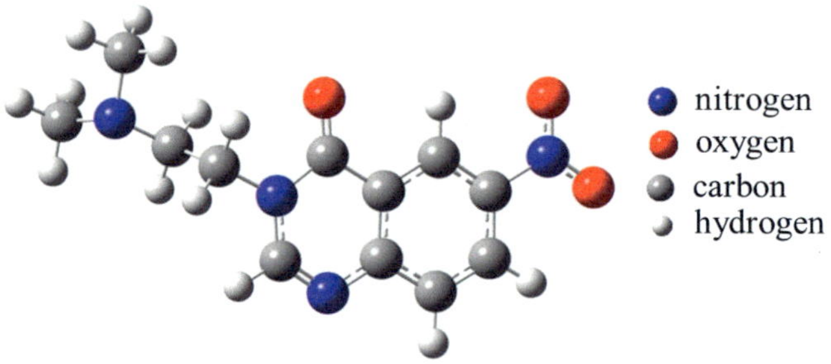


**Fig. (6).** 3D optimised chemical structure of BG 1188.

The pharmaceutical compound belonging to quinazoline derivatives, generically named BG 1188 (Fig. **6**) is part of a series of potential drugs that were synthesised at UMR-MD1 (Faculté de Pharmacie et de Médecine, Marseille Université, France) in order to increase the antibiotic susceptibility of Gram-negative bacteria presenting multidrug-resistant phenotypes [44].

This yellow, solid chemical compound was dissolved in ultrapure de-ionised water (upw) and solutions were prepared in $10^{-5}$-$10^{-3}$ M concentration range. The solvent was delivered *via* a sterile filter (TKA Pacific UP/UPW6 and TKA Genpure ultra-pure water system accessory, Thermo Electron LED GmbH, DE; flow rate -at 15°C: 6 l/h and retention of bacteria and particles: 99%). Its bacterial content is <1 CFU/ml, particle content <1 at a resistivity of 18.2 MΩxcm and 0.055 µS/cm conductivity, at 25°C).

## Methods

The irradiation of the compounds is made with different UV pulsed lasers or with continuous wave (cw) Hg and Xe lamps in some cases. The radiation emitted by optical lamps covered at least one absorption band of the irradiated sample. As for lasers, their beams were chosen so that the wavelengths were emitted in the spectral domain of an absorption band of the samples. Two types of lasers were used. One, was a pulsed nitrogen laser with emission at 337.1 nm, 1 ns pulse time width, pulse repetition rate 10 pps and average beam energy 0.3 mJ. This was in the case of MTX and 5-FU irradiation. Irradiation time was varied between 0 min and 60 min. Analysis of the fluorescence of these compounds was made with an Aminco Bowman spectrofluorimeter.

For exposure of phenothiazines (TZ, PMZ), quinazoline (BG1118) and hydantoin (SZ-2), the beam of a second kind of laser was used, emitted as THG (third harmonic generation, 355 nm) and FHG (fourth harmonic generation, 266 nm) by a pulsed Nd:YAG laser system. Pulse repetition rate was 10 pps at 5 ns pulse time width and the average beam energy was varied between 6 mJ and 38 mJ. The exposure time of samples to laser beams was up to 4 h. LIF emitted by the sample was collected with an optical fibre and sent to a high-resolution spectrometer (Ocean Optics HR4000CG-UV-NIR or Spectra Pro 2750, Acton Research - spectral resolution: 0.75 nm and 0.8 nm, respectively). In these experiments modification of the content of samples exposed to laser radiation was evidenced in real time by LIF which was measured during samples irradiation.

The scheme of the experimental set-up utilised for irradiation and LIF studies is depicted in Fig. (**7**) where all the other associated techniques used in samples evaluation are illustrated.

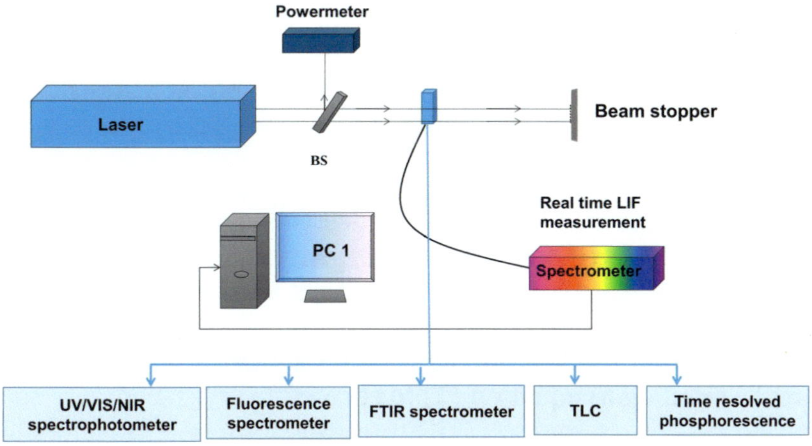


**Fig. (7).** The irradiation/LIF set-up and the investigation methods applied on exposed samples.

Before and after exposure, the absorption spectra of the solutions were measured with a Perkin Elmer Lambda 950 UV-Vis-NIR spectrometer in order to detect modifications of the molecular structures.

For analysing structural changes under UV irradiation, IR absorption spectra of samples were recorded using a FTIR spectrometer (Nicolet iS50, with 4 $cm^{-1}$ resolution). Samples were prepared as KBr pellets or solutions were dried on the surface of a KRS-5 window. The analysis was made in the 4,000-400 $cm^{-1}$ range.

The experimental set-up for measuring time-resolved phosphorescence signal of the photosensitised singlet oxygen at $\lambda$ = 1,270 nm is described in detail elsewhere [45] and is illustrated in Fig. (**8**).

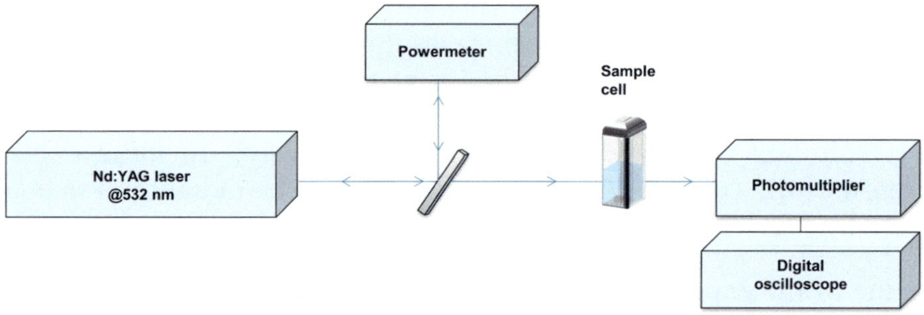


**Fig. (8).** Singlet oxygen phosphorescence kinetics measurement set-up.

The phosphorescence was detected by a near infrared (NIR) photomultiplier (Hamamatsu H10330 NIR-PMT Module) and recorded with a digital oscilloscope (Tektronix DPO-7254). The lifetime of singlet oxygen as a transient species is obtained by fitting the experimental phosphorescence decay curve with a mono-exponential function.

Data regarding time resolved phosphorescence of singlet oxygen allows finding the quantum yield (QY) of the singlet oxygen generation ($\Phi$). To do this, a relative method to compare a reference standard with known quantum yield of singlet oxygen with the experimental data was used. The reference and the probe samples were measured in the same experimental conditions and quantum yield was calculated according to equation (1) [45]:

$$\Phi = \Phi_{ref} \frac{I}{I_{ref}} \frac{A_{ref}}{A} \frac{n^2}{n_{ref}^2} \frac{\tau}{\tau_{ref}} \qquad \textbf{(1)}$$

where Φ is singlet oxygen QY, *I* is singlet oxygen phosphorescence intensity at the moment of laser pulse input ($t$ = 0) in measured sample, *A* is optical absorption of the sample at laser excitation wavelength (355 nm or 532 nm), $\tau$ is lifetime of singlet oxygen in measured sample and *n* is solvent's refractive index. Subscript *ref* refers to values for standard sample. The phosphorescence intensity at $t$=0 was obtained by extrapolating to zero the curve of the mono-exponential fitting of experimental data.

The interaction of singlet oxygen with MTX was demonstrated by the determination of singlet oxygen quenching rates [45]. The lifetime of singlet oxygen phosphorescence was measured for solution mixtures of MTX (with varied concentration between zero to 5 mM) and a known photosensitiser (verteporfin) kept at constant concentration (50 μM) in N-dimethylformamide (DMF) as solvent. The rate constant for singlet oxygen quenching by MTX was determined from the Stern-Volmer plots of $1/\tau$ *versus* MTX concentration using the equation (2):

$$\frac{1}{\tau}=\frac{1}{\tau_0}+k[c] \qquad \textbf{(2)}$$

where τ is the singlet oxygen excited state lifetime in the presence of quencher, $\tau_o$ is the excited state lifetime in the absence of it, *k* is the $^1O_2$ quenching rate constant and [*c*] is the concentration of the tested compound.

A qualitative analysis of photoproducts generated by irradiating 20 mg/ml TZ and PMZ was performed using thin layer chromatography (TLC). This technique is explained in detail in Chapter 9. The stationary phase was an aluminium plate coated with 0.25 mm silica gel $60F_{254}$ (DC-Alufolien Kieselgel 60F254; Merck) and the mobile phase was a mixture of acetone: methanol: 25% ammonia (50:50:1, V:V:V). Quantities of 1 μl of unirradiated, irradiated 60 min, 120 min and 240 min TZ and PMZ were applied, respectively, on the plate at 1 cm from the margins and 1 cm distance between each sample spot. After solvent evaporation, TLC plate was introduced in a developing chamber that contained the mobile phase and developed at room temperature. Separation of compounds from mixture is based on their polarity. After removal, the plate is dried and the chromatogram is visualised under UV light at 254 nm and 365 nm using an UV Cabinet Overview Chromo-Vue® Cabinet C-65 (UVP). Finally, the plates were photographed for both visualising wavelengths and analysed, as well.

## RESULTS AND DISCUSSIONS

### Absorption Spectroscopy

#### *Methotrexate*

The absorption spectra for MTX solution in natural saline at $2.5\times10^{-5}$ M for exposure times between 0 min and 60 min to the radiation emitted by a cw Hg lamp are shown in Fig. (**9**). A variation of absorbance of the main bands with the irradiation dose takes place. The absorption band with the peak at 303 nm has a decrease in intensity and the one with the maximum at 371 nm is enhanced when irradiation time is increased. It should be noted that the spectrum of a control solution kept in the same conditions except irradiation is not changing in time, so that the effect mentioned above is due to irradiation process, indeed. These facts suggest that MTX molecules undergo structural changes upon UV irradiation and the main resulted photoproducts have a structure related to the parent MTX.

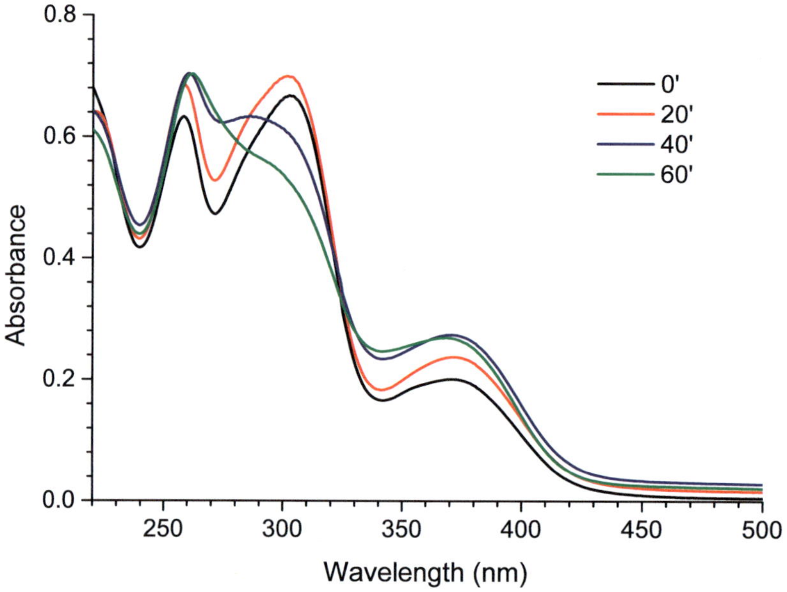


**Fig. (9).** Absorption spectrum of MTX solution in natural saline at $2.5\times10^{-5}$ M, for UV Hg lamp irradiation; exposure time between 0 min and 60 min.

The absorption spectra of MTX solutions at $2.5\times10^{-5}$ M for irradiation with $N_2$ laser during time intervals between 0 min and 30 min are shown in Fig. (**10**). The exposure to the monochromatic laser radiation at 337.1 nm induced spectral modifications which differ from those produced by exposure to polychromatic UV lamp radiation.

In the case of lamp irradiation (Xe - data not shown - or Hg lamp), the MTX spectra have the same behaviour, *i.e.* the absorption peak at 303 nm is decreasing in intensity with irradiation time while for the one at 371 nm, the absorbance is increasing with exposure duration. For $N_2$ laser irradiation, there is another spectral behaviour: the absorption band with maximum located at 303 nm does not exhibit significant changes with irradiation, but an attenuation of the absorption band with the peak at 371 nm is observed. This can be due to the fact that 337.1 nm radiation is resonant with the first absorption band corresponding to the electronic transition from fundamental to first excited singlet state of MTX, while the lamp irradiation with wavelengths starting from 300 nm, excites MTX molecules to higher excited electronic levels as well.

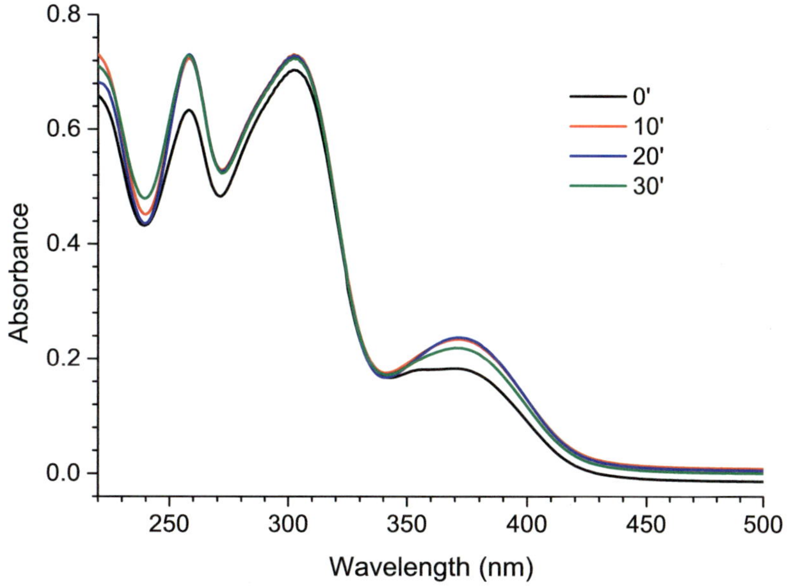


**Fig. (10).** Absorption spectra of MTX solutions (at $2.5\times10^{-5}$ M in natural saline) at several irradiation times (0 min, 10 min, 20 min and respectively 30 min) with $N_2$ laser beam.

## *5-Fluorouracil*

5-FU solutions in natural saline (0.9% NaCl) at $10^{-4}$ M, adjusted to pH=8.4 by addition of NaOH, were studied. The light source was a $N_2$ pulsed laser, emitting at 337.1 nm with an average beam energy of 350 μJ.

The absorption spectra of 5-FU solutions exposed to $N_2$ laser beam for time intervals between zero and 5 min, are shown in Fig. (**11**).

They exhibit two spectral bands in the range 250-350 nm. The main absorption band wavelength is 279 nm. 5-FU spectra did not show any significant changes and this suggests that under exposure to 337.1 nm laser radiation, 5-FU molecule does not change.

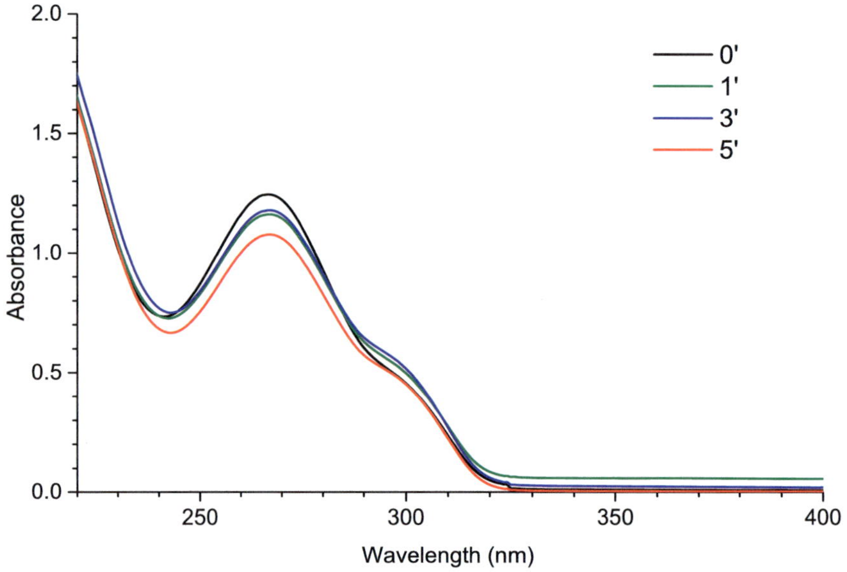


**Fig. (11).** The absorption spectra of 5-FU solutions ($10^{-4}$ M) irradiated with $N_2$ laser beam.

### *Phenothiazines*

The absorption spectra modifications under exposure to 266 nm for solutions of two phenothiazine derivatives (PMZ and TZ) at 20 mg/ml in ultrapure water were investigated. Spectra are saturated between 400-200 nm due to the high values of the extinction coefficient. It is necessary to analyse a dilution of 0.2 mg/ml for both unirradiated and irradiated samples in order to observe the spectral modifications resulted from the resonant interaction of 266 nm laser beam with phenothiazines' molecules.

The absorption spectrum of the dilution of 0.2 mg/ml from 20 mg/ml unirradiated TZ (Fig. **12**) shows three bands with peaks at 230 nm, 262 nm and 315 nm. In absorption spectra, after irradiation of the sample with 266 nm laser beam, several modifications were observed that include hypochromic and hypsochromic shifts.

The band at 262 nm is subject of a hypochromic shift of 6.1% if the sample is irradiated 60 min and of 10.2% for 240 min exposure time. The band with the peak at 315 nm of the unirradiated sample suffers a hypsochromic shift of 2 nm by the end of exposure time interval and becomes broader.

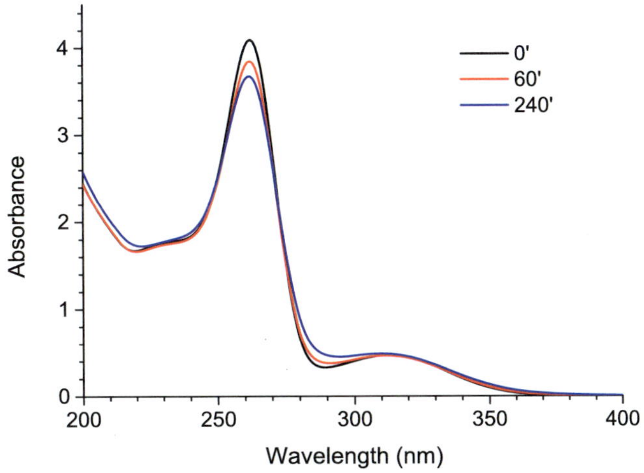


**Fig. (12).** UV-Vis absorption spectra of 20 mg/ml unirradiated and irradiated TZ recorded for diluted solution at 0.2 mg/ml.

The UV-Vis absorption spectrum of 0.2 mg/ml diluted from 20 mg/ml unirradiated PMZ (Fig. **13**) presents 2 bands with peaks at 250 nm and 300 nm, respectively. The $\pi \rightarrow \pi^*$ transitions occur for peaks located at lower wavelengths and for the rest, the transitions are $n \rightarrow \pi^*$. The second peak is due to the presence of sulphur lone-electron pairs in phenothiazine ring structure. Also, (i) the hypsochromic shift of the peak situated between 300-320 nm, observed when polar solvents are used and (ii) the low value of the extinction coefficient compared to that of the other peak, confirm the $n \rightarrow \pi^*$ transition [46].

After 60 min exposure to UV laser radiation the band at 250 nm shows a 2 nm hypsochromic shift and a hypochromic shift of 10.5%, as well, with respect to the values of unirradiated sample.

Exposure of 20 mg/ml PMZ solution to 266 nm laser beam for a time interval of 240 min produces in the absorption spectra a 3 nm hypsochromic shift and also a hypochromic one of 13.2% in comparison with that of the unirradiated solution. These changes in absorption spectra are accompanied by the appearance of another band with maximum at 551 nm for the solution irradiated 60 min and two bands with peaks at 554 nm and 671 nm for the sample exposed to UV laser during 240 min (inset Fig. **13**).

The changes observed in the absorption spectra of TZ and PMZ indicate photodegradation of parental compound and formation of photoproducts, out of which some absorb in the same spectral range as parental compound and others in the visible spectral range.

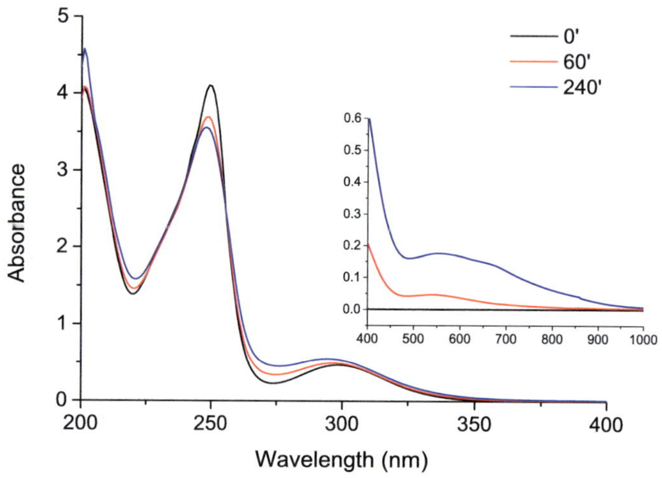


**Fig. (13).** UV-Vis absorption spectra of 20 mg/ml unirradiated and irradiated PMZ recorded for diluted solution at 0.2 mg/ml and undiluted solution (inset).

## *Hydantoin SZ-2*

UV-Vis absorption spectra of all imidazolidines, whose structures are very close to uracil framework exhibit absorption peaks in 200-250 nm and 300-400 nm spectral ranges. The first absorption band is attributed to $\pi\rightarrow\pi^*$ transition of N-C=O chromophore and the second to the corresponding transition of carbonyl group [47]. The difference in peak positions for investigated compounds is due to the effect of attached substituents. These spectral characteristics are due most likely to $\pi\rightarrow\pi^*$ transitions of the carbonyl group of imidazolidine ring rather than $n\rightarrow\pi^*$ transitions of it. This is because the benzene ring connected to the hydantoin core causes extensive delocalisation in these structures and determines $\pi\rightarrow\pi^*$ transitions [48, 49].

The UV-Vis absorption spectrum of SZ-2 is characterised by several broad bands due to overlapping of vibronic transitions [50]. The higher magnitude of peaks located in UVA (at 364 nm and 379 nm) could be observed. They are assigned to carbonyl chromophore in position 4 of the imidazolidine ring due to greater delocalisation [47].

SZ-2 solution samples were exposed to pulsed Nd:YAG laser beam in two experimental sessions. In first, the sample was irradiated with THG radiation emitted at 355 nm up to 30 min. In the second experiment, was involved FHG radiation emitted at 266 nm up to 60 min.

For laser exposed SZ-2 solution sample at 355 nm, the UV-Vis absorption spectra reveal that the intensity of the absorption band placed in 350-400 nm spectral range constantly decreases with the irradiation time along with the blue shift and the decrease in intensity of the peak at 293 nm. After 10 min irradiation with pulsed laser beam, these bands disappear completely, as is observable in Fig. (**14**) and only the 254 nm peak increases with the time length of laser beam exposure.

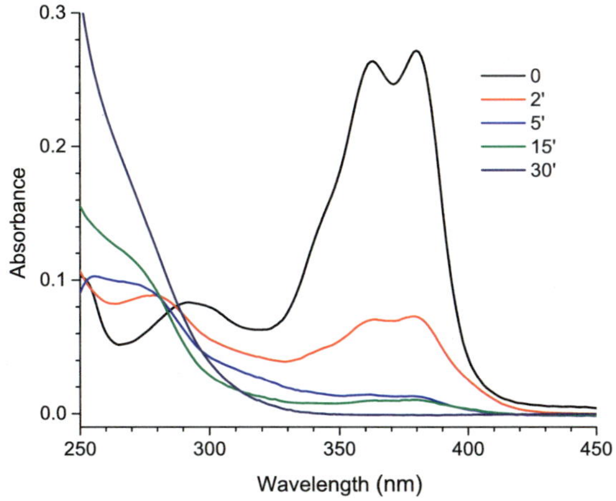


**Fig. (14).** UV-Vis absorption spectra evolution with 355 nm laser exposure (38 mJ, irradiation up to 30 min) for SZ-2 solution in DMSO at concentration $10^{-4}$ M.

For FHG laser exposure of SZ-2 solution, a continuous flattening of the absorption spectra with time exposure is noted, accompanied by a hypsochromic shift of the peak at 293 nm. Nevertheless, the absorption at 254 nm peak does not exhibit a so drastic change and a hyperchromic (blue shift) occurs (Fig. **15**).

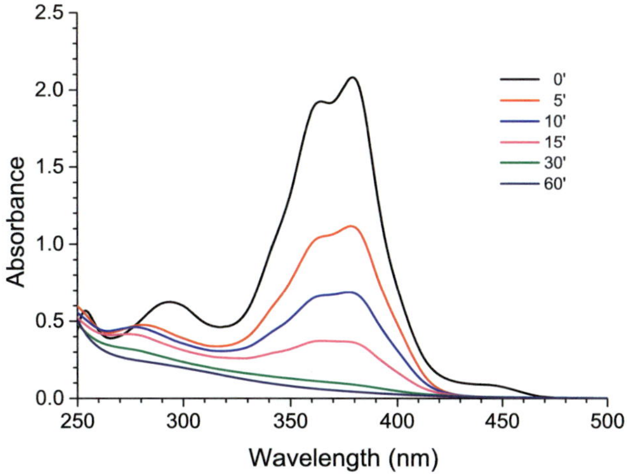


**Fig. (15).** The UV-Vis absorption spectra evolution under 266 nm laser irradiation (6.5 mJ) for SZ-2 solution in DMSO at concentration $10^{-3}$ M.

## *Quinazoline BG 1188*

As for the UV-Vis absorption properties of the quinazoline derivative BG 1188 solution in upw, its spectrum shows three absorption bands with peak intensities located at 207 nm, 224 nm and 313 nm (Fig. **16**).

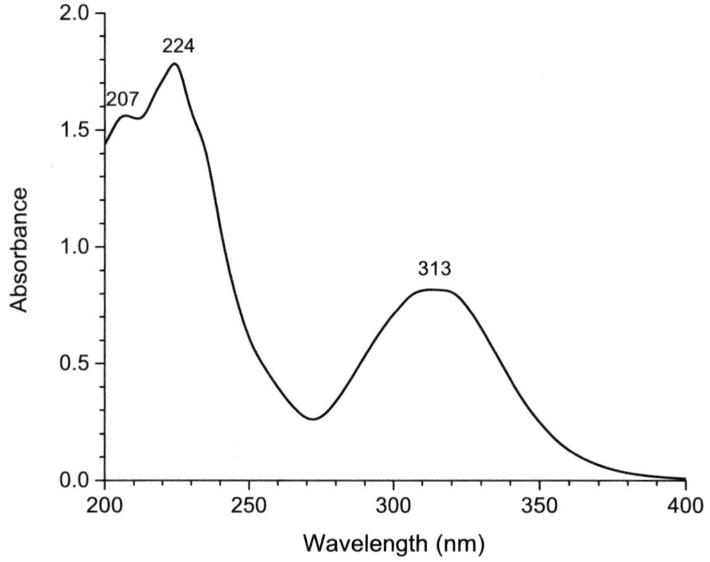


**Fig. (16).** UV-Vis absorption spectrum of BG 1188 solution in ultrapure water (upw).

The bands with absorption maxima at 224 nm and 313 nm are typical to quinazoline derivatives [51]. The band looking like a shoulder with maximum at 207 nm could be due to the influence of atmospheric oxygen.

The peak at 224 nm having the highest molar extinction coefficient might be assigned to a $\pi \rightarrow \pi^*$ (1B) transition of benzene ring. In the same manner, the band with maximum at 314 nm is characterised by a minor molar extinction coefficient, and it might be due to $\pi \rightarrow \pi^*$ (1La si 1Lb) transitions. The bandwidth can probably be attributed to the overlap of both $\pi \rightarrow \pi^*$ and $n \rightarrow \pi^*$ transition [51].

Concerning BG 1188 exposed to Nd:YAG pulsed laser beams, the spectra highlight three isosbestic points, as shown in Fig. (**17a**). Considering the first 5 min of irradiation, the three intersection points maintain their position at respectively 250 nm, 286 nm, and 354 nm, which allows to consider the following two hypotheses: (i) a photochemical reaction is in progress in the exposed solution during irradiation, in which case there are two molecular species having the same molar absorption coefficients and these two absorbing components are a reactant and a photoproduct, or (ii) the two absorbing components are in equilibrium and their relative proportions are controlled by the concentration of other component, typically the concentration of hydrogen ions.

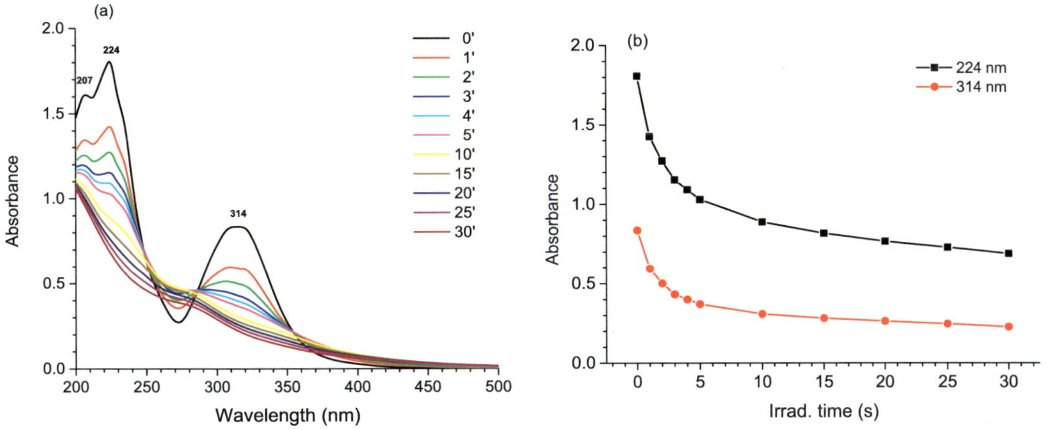


**Fig. (17).** (**a**) the UV-Vis absorption spectra of unirradiated/THG laser exposed BG 1188 solution in upw; (**b**) time evolution of absorption peaks intensities at 224 nm and 314 nm during irradiation experiment.

Starting from irradiation times higher than 10 min, the evolution of the isosbestic points can be explained either by formation of a reaction intermediate in irradiated solution, or by involvement of a third absorption species, in equilibrium. There may be involved some photochemical interactions between both absorbing species present in irradiated solutions, as well [52].

The exponential decrease of the peak arising at 224 nm (Fig. **17b**) has its origin in the aromatic ring of the compound [51] and shows that the exposure to pulsed laser radiation in the given conditions affects BG 1188 quinazoline moiety. In the same manner, the absorption band at 314 nm undergoes an exponential decrease, suggesting molecular changes.

## Fluorescence Spectroscopy

### *Methotrexate*

The excitation spectrum of an MTX solution in natural saline at $2.5\times10^{-5}$ M and its evolution when exposed to UV irradiation with Hg lamp is shown in Fig. (**18**). The emission band was centred on 470 nm. The exposure time was varied between 0 and 60 min. Spectra show two excitation bands with peaks at 300 nm and respectively 400 nm. Furthermore, the intensity of these fluorescence excitation bands is increasing with irradiation time.

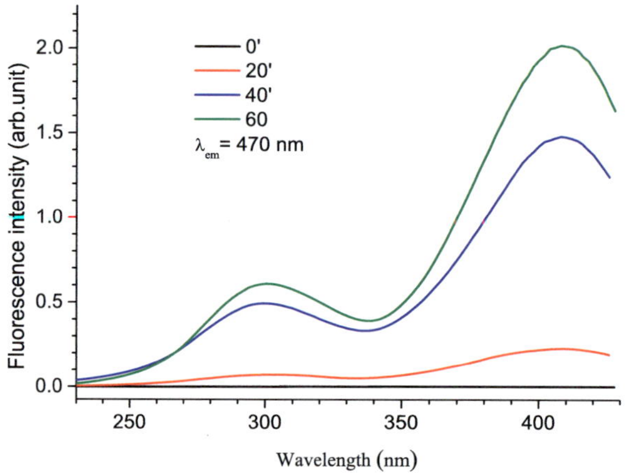


**Fig. (18).** The excitation spectrum of a MTX solution in natural saline at $2.5\times10^{-5}$ M, UV irradiation (Hg lamp), exposure time between 0 min and 60 min (470 nm emission wavelength).

The same behaviour under cw Hg lamp irradiation is shown by emission spectra of MTX solution at $2.5\times10^{-5}$ M in natural saline (Fig. **19**). The excitation is made at 340 nm and the intensity of fluorescence emission bands is increasing with irradiation time.

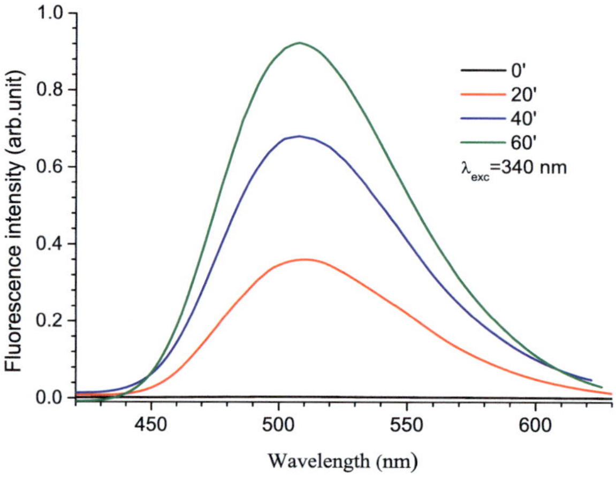


**Fig. (19).** The emission spectra of a MTX solution in natural saline at $2.5\times10^{-5}$ M, for UV irradiation (Hg lamp), exposure time between 0 min and 60 min (excitation at 340 nm).

The excitation and emission spectra of MTX solutions, at the same concentrations when irradiation is made with pulsed $N_2$ laser beam (337.1nm) show the same characteristics as the ones for lamp irradiation, except that fluorescence intensity is 10-15 times lower than in the case of irradiation with Xe or Hg lamps, at the same exposure times. For $N_2$ laser irradiation at 0.5 mJ energy per pulse, the emitted fluorescence intensity is also increasing with irradiation, but the values of intensity are about 10-15 times lower than in the case of lamp irradiation. This is due to the smaller radiation dose applied on the sample at laser exposure.

The absorption and fluorescence spectra of MTX show that major modifications of MTX molecules when irradiated with UV-Vis light (emitted by a lamp or pulsed laser radiation) take place.

MTX belongs to the pterine group (folic acid, aminopterin *etc*) which contains photosensitive compounds and under UV radiation decomposes in 6-formylpterin derivatives and p-aminobenzoylglutamic acid [53]. Studies about the folic acid photolysis [54, 55] indicate that singlet oxygen is implied in the photochemical process. The folic acid molecule undergoes cleavage and oxidation yielding the 6-formylpterin and p-aminobenzoylglutamic acid. This suggests that MTX decomposes under UV irradiation in 2,4-diamino-formylpteridine and the same acid. For alkaline media (which is the case of the present study), a deprotonation of the amide group in pterin group can take place as well. The interaction of singlet oxygen with MTX is evaluated below in this chapter.

### *5-Fluorouracil*

The fluorescence excitation spectra of 5-FU solutions were recorded for 440 nm emission wavelength. They are shown in Fig. (**20**), where one can see the relative intensity of the excitation fluorescence as function of irradiation time. The irradiation was made for time intervals between 1 min and 5 min. and the radiation source was the $N_2$ laser at 0.3 mJ energy/pulse.

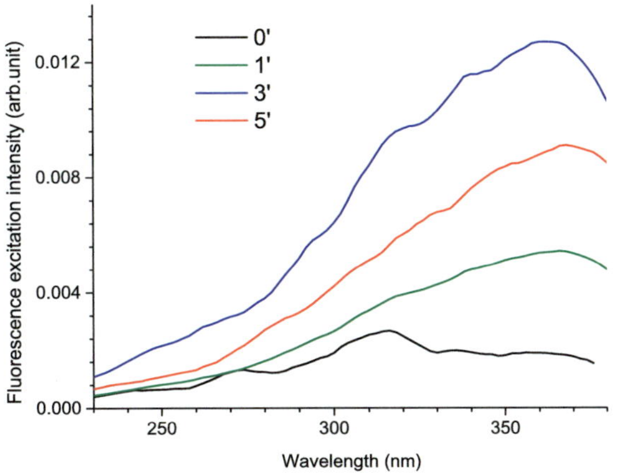


**Fig. (20).** The fluorescence excitation spectra of 5-FU solution at irradiation time between 0 min and 5 min.

The peak of fluorescence excitation band is placed at 360 nm and its intensity increases with the irradiation dose up to 3 min, when starts to decrease.

The dispersed fluorescence spectra excited at 360 nm shown by 5-FU solutions as function of exposure time to $N_2$ laser radiation, are presented in Fig. (**21**).

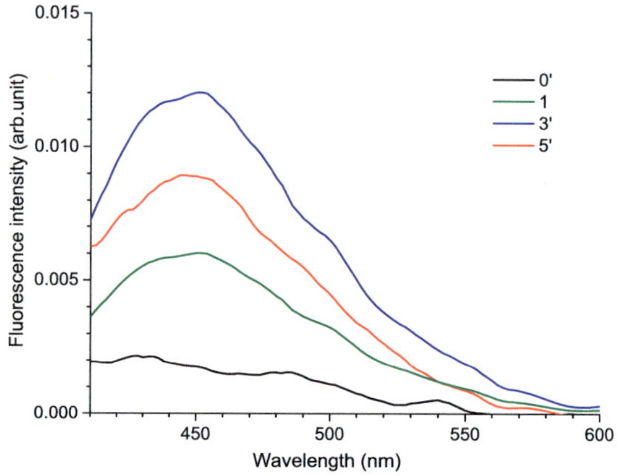


**Fig. (21).** The fluorescence emission spectra of 5-FU solutions exposed to $N_2$ laser radiation.

This fluorescence emission band is placed in the 400-600 nm spectral range with an intensity peak observed at 450 nm. It is obvious that 5-FU solution became fluorescent under 337.1 nm irradiation.

It can be observed (Fig. **22**) that after irradiation with $N_2$ laser beams, the dispersed fluorescence intensity of 5-FU solution increases more than six times up to 3 min irradiation and then decreases with further exposure.

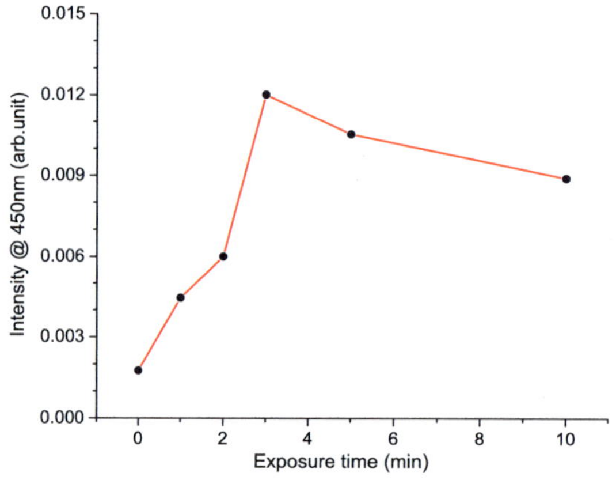


**Fig. (22).** Relative intensity of fluorescence emission peak at 450 nm *versus* the exposure time to 337.1 nm laser radiation.

5-FU belongs to pyrimidines' group which can have different tautomeric forms. For these type of species, it was shown that excitation spectrum is different than the mirror form of absorption spectrum when the solution contains two distinct tautomer forms [56].

The results shown here suggest that under UV laser irradiation the diketo (lactam) form of 5-FU, which is a major component for unirradiated solution, more stable and more weakly fluorescent tautomer transforms into the enol-keto (lactim) form (Fig. **2**) which has fluorescence properties.

### *Phenothiazines*

The emission spectra of phenothiazine derivatives that have substituents at the 2nd and 10th position of the ring structure are characterised by only one emission band with the peak between 440 nm and 470 nm if acetonitrile, methanol or phosphate buffer are used as solvents. If at the 2nd position -$SCH_3$ or -$CF_3$ groups are present, the shift to higher wavelength is significantly greater ($\Delta\lambda$ = 10-25 nm). Also, the emission spectra strongly depend on the used solvent and with increasing polarity, the spectrum is shifted to longer wavelengths [46].

LIF spectrum of 20 mg/ml TZ irradiated 240 min was recorded in real time and one may observe only a band at 470 nm. The fluorescence intensity is increasing for irradiation time intervals up to 40 s, after which it starts to decrease (Fig. **23**). Increasing the exposure time, one can observe a red shift of 9 nm by the end of irradiation process and also a hypochromic shift of 10% compared to spectrum obtained after 40 s exposure.

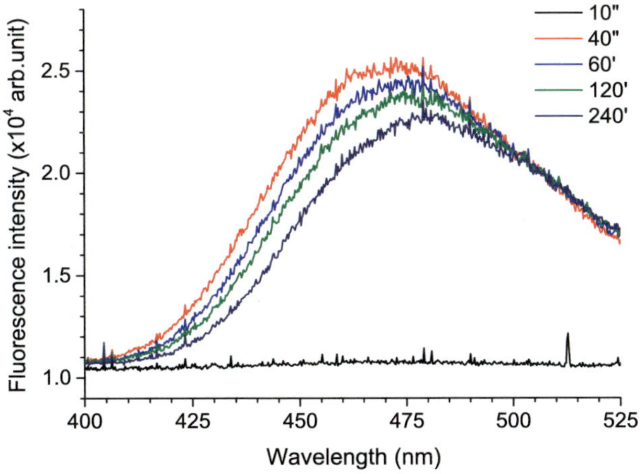


**Fig. (23).** LIF spectra of 20 mg/ml TZ solution irradiated up to 240 min.

LIF spectra of 20 mg/ml PMZ irradiated up to 240 min (Fig. **24**) exhibit a maximum of fluorescence intensity after 2 min of solution exposure to 266 nm laser beam and are centred on two emission wavelengths: at 445 nm and 492 nm.

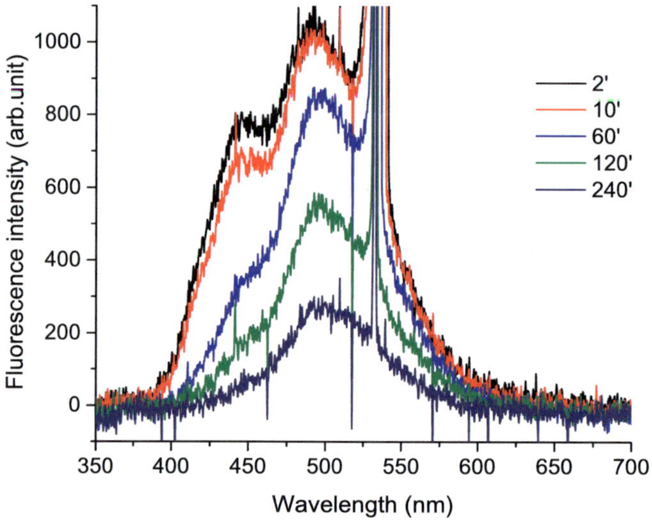


**Fig. (24).** LIF spectra of 20 mg/ml PMZ irradiated up to 240 min.

After 10 min exposure, a slight modification of the emission spectra can be observed, where the first band present a hypochromic shift of 14.5% compared to unirradiated sample. During the irradiation process the fluorescence intensity decreases and the first band disappears by the end of exposure to UV laser radiation. After 60 min irradiation, the intensity of fluorescence for the band at 445 nm exhibits a hypochromic shift of 58.1% and after 120 min it decreases with 80.5%. In the same way, the band at 492 shows a decrease in intensity up to 74.6% and a red shift of 8 nm by the end of the irradiation time. So, only one broad band with maximum at 500 nm is present after 240 min exposure.

The fluorescence intensity peaks displacement to longer wavelength that was measured for emission spectra of irradiated TZ and PMZ can be assigned to the modification of the substituent from the molecular ring structure of the phenothiazine [46]. It was shown that LIF spectrum is not influenced by N-alkyl chain [57] the emission peaks being located at the same wavelength for parental compound and for 10-alkilated derivatives.

## *Hydantoin SZ-2*

The photosensitivity of SZ-2 hydantoin derivative solution in DMSO is highlighted by its LIF spectra behaviour as well [10]. LIF spectra of bulk sample of SZ-2 at $10^{-3}$ M in DMSO exhibit two emission bands that undergo significant changes upon 266 nm laser irradiation (Fig. **25**).

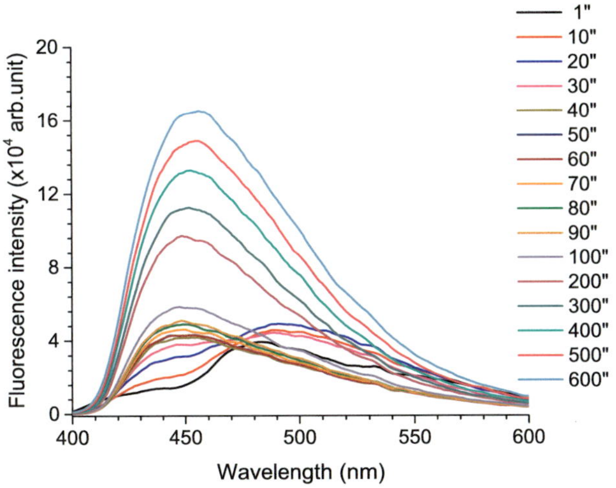


**Fig. (25).** LIF spectra of bulk samples of SZ-2 solutions at $10^{-3}$ M in DMSO *versus* irradiation time.

The first band arises at 435 nm, its intensity increases and its maximum shows a red shift (up to 20 nm) with exposure time (Fig. **26a**). The second band has a peak placed at 485 nm, exhibits an increase in intensity and has a 15 nm red shift in the first 20 s of laser exposure, followed further by a linear decrease with irradiation dose up to 50 s when it is vanishing (Fig. **26b**).

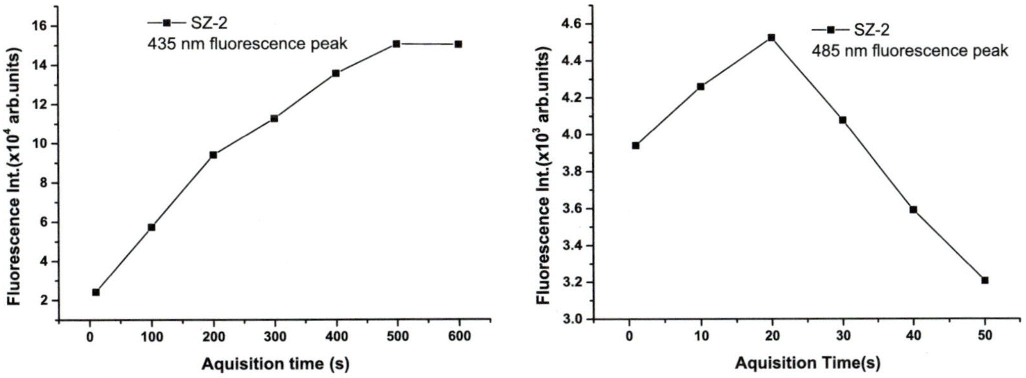


**Fig. (26).** SZ-2: (**a**) the evolution of 435 nm peak of fluorescence intensity; (**b**) the evolution of 485 nm peak of fluorescence intensity.

## *Quinazoline BG 1188*

The evolution of the BG 1188 $10^{-3}$ M solution LIF spectra (Fig. **27**) highlights an increase of the emission peak and a shift of it with 6 nm towards longer wavelengths for an irradiation time between 20 s to 5 min.

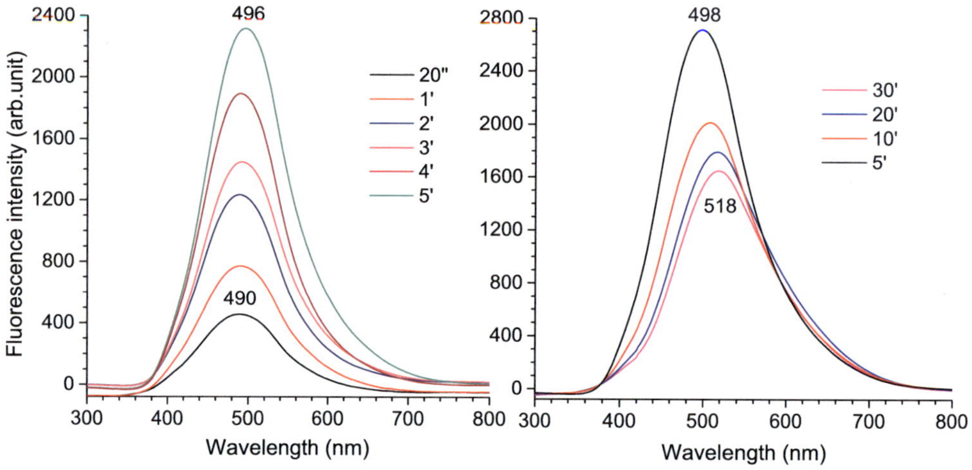


**Fig. (27).** LIF spectra of BG 1188 $10^{-3}$ M solution excited with 355 nm laser radiation.

At exposure times of 5-30 min, the peak of fluorescence intensity decreases and shifts with 20 nm towards higher wavelengths, too. This evolution suggests more than one photoinduced process during irradiation, with at least one intermediary product generation.

## Singlet Oxygen Generation

The interaction of singlet oxygen with MTX was demonstrated by measuring the singlet oxygen quenching rate [58]. In these studies, the lifetime of singlet oxygen phosphorescence was measured for solution mixtures of MTX (with concentrations varied between 0 mM and 5 mM) and a known photosensitiser (verteporfin) kept at constant concentration (50 μm) in DMF.

The rate constants for singlet oxygen quenching by MTX were determined from the Stern-Volmer plots of $1/\tau$ *versus* the MTX concentration shown in Fig. (**28**).

From the fitting line slope, it resulted a singlet oxygen quenching rate constant of $1\times10^{7}$ $M^{-1}s^{-1}$. This proves an interaction of singlet oxygen with MTX molecules and, consequently, an oxidation process that occurs during MTX irradiation which is a process that was suggested in other studies as well.

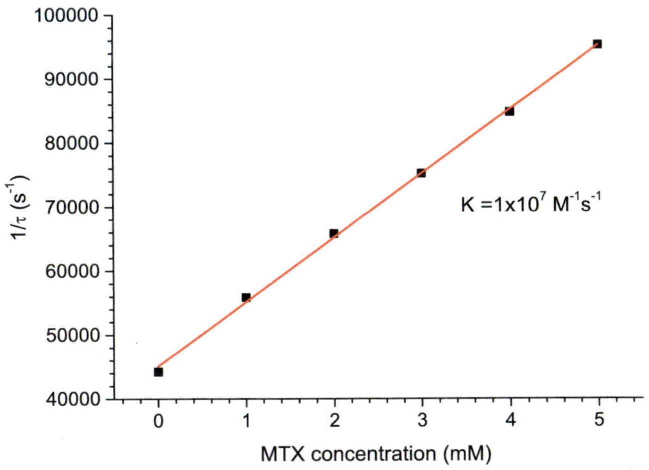


**Fig. (28).** Stern Volmer plot for singlet oxygen interaction with MTX.

## *Hydantoin SZ-2*

In Fig. (**29**) absorption spectra for solutions prepared in DMSO for SZ-2 at $10^{-4}$ M concentration and the standard ZnPc at $5\times10^{-5}$ M are shown.

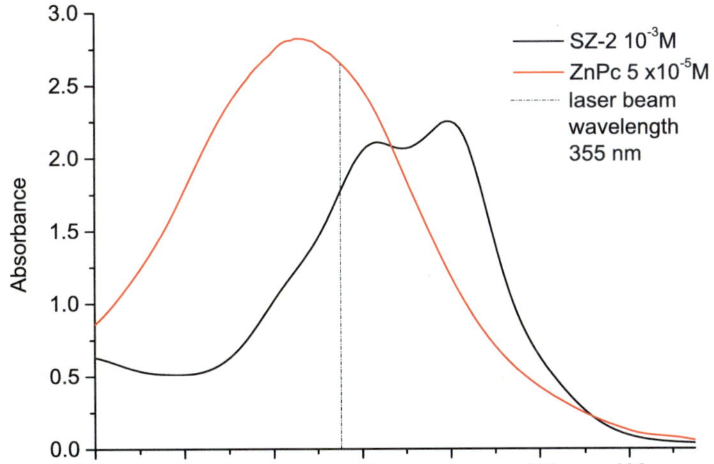


**Fig. (29).** Absorption spectra of SZ-2 $10^{-4}$ M and of standard ZnPc $5\times10^{-5}$ M solutions in DMSO.

The phosphorescence of the singlet oxygen generated by SZ-2 sample is given in Fig. (**30a**). A typical singlet oxygen lifetime of about 6 μs is obtained when DMSO is used as solvent. In Fig. (**30b**), the signal is shown in comparison with the one of the singlet oxygen generated by ZnPc.

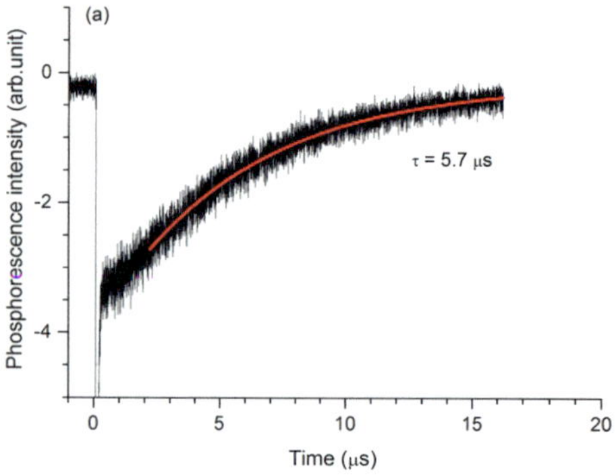


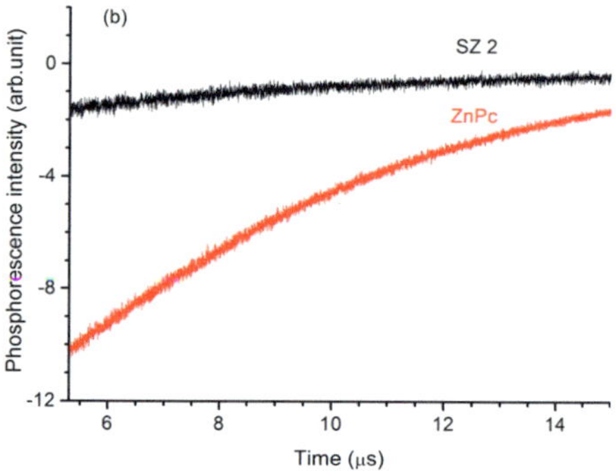


**Fig. (30).** (**a**) the phosphorescence signal for singlet oxygen generated by SZ-2 sample; (**b**) the comparison of the SZ-2 signal to the ZnPc standard.

Using equation (1) and the signals measured in the same conditions for SZ-2 and ZnPc samples (Fig. **30b**), a quantum yield value of 0.12 for SZ-2 singlet oxygen generation was obtained.

## FTIR

FTIR spectroscopy is a non-destructive analytical method with which the molecules are investigated and it is a suitable technique for the identification of characteristic vibrations of functional groups in an asymmetric molecule. The method is based on the study of the interaction between molecular structures and infrared radiation, between 14,000 $cm^{-1}$ and 100 $cm^{-1}$.

When a molecule absorbs IR radiation, the associated energy is transformed into different types of vibrational and rotational modes. Vibration modes are divided into stretching vibrations (symmetrical and asymmetrical) and deformation vibrations (scissoring, rocking, twisting and wagging) that are attributed to characteristic wavenumbers [59].

### *5-FU*

Since no structural modifications were evidenced from the UV-Vis absorption spectra, FTIR investigation was used to identify the change of the molecule's geometry produced as a result of UV irradiation [60]. If pyrimidines have a hydroxy group placed orto or para relative to a nitrogen atom position, they usually transform into the keto form, with a C=O band near 1,700 $cm^{-1}$. FTIR spectrum of non-irradiated 5-FU solution at $5\times10^{-4}$ M and pH=8.6÷9.4 is shown in Fig. (**31**), evidencing typical bands in the range 1,700-1,650 $cm^{-1}$, assigned to conjugated (1,670-1,655 $cm^{-1}$) and nonconjugated (1,695 $cm^{-1}$) C=O stretch.

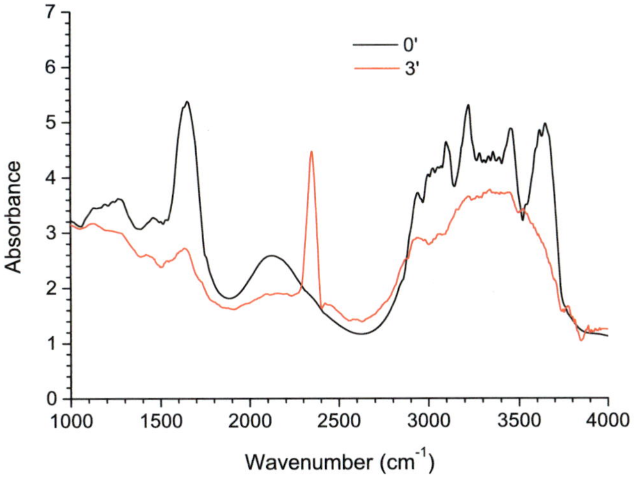


**Fig. (31).** 5-FU FTIR spectrum for unirradiated compared to 3 min irradiated solution at 337.1 nm.

This spectrum seems to be characteristic to the major component diketo (lactam) form, in equilibrium with the enol-keto (lactim) one.

The FTIR spectrum of 5-FU solution, irradiated 3 min with pulsed nitrogen laser (Fig. **31**) shows the 5-FU molecule's geometry change after irradiation. Except the very strong peak at 2,284 cm$^{-1}$ due to $\nu_3$ antisymmetric stretch mode of carbon dioxide, the peaks attributed to conjugated and non-conjugated C=O stretch are strongly diminished, as well as the quadrant (1,590-1,555 cm$^{-1}$ and 1,565-1,562 cm$^{-1}$) and semicircle (1,480-1,400 cm$^{-1}$ and 1,410-1,375 cm$^{-1}$) stretch bands, assigned to keto form.

A signal of only one N-H bond at 3,500-3,300 cm$^{-1}$ and the superimposed wide band due to OH groups (~3,500 cm$^{-1}$) bonded to pyrimidine ring could suggest that after irradiation, the equilibrium was shifted to enol-keto tautomer (lactim form), which explains fluorescence enhancement.

All these observations lead to the conclusion that, by UV irradiation, 5-FU undergoes a tautomerisation process which induces a fluorescence increase for the irradiation exposure time of 3 min. Under this irradiation dose, 5-FU is transformed into the tautomeric keto-enol (lactim) form that is fluorescent.

### *Phenothiazines*

The recorded FTIR spectra were compared in order to identify the molecular modifications obtained after the exposure of phenothiazine solution to laserradiation. For each unirradiated TZ and PMZ is presented a table with the characteristic vibrations of the molecular bonds.

FTIR spectrum of unirradiated 20 mg/ml TZ is presented in Fig. (**32**) and in Table **1** are listed the characteristic vibrations of TZ molecular bonds. Due to the high absorption of water in IR region, the evaporation of it on the KRS-5 support was necessary and the obtained information is only qualitative.

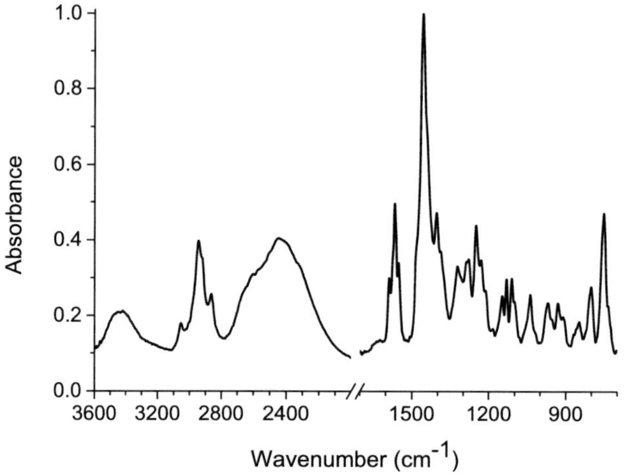


**Fig. (32).** FTIR spectrum of unirradiated 20 mg/ml TZ solution recorded between 3,600-700 $cm^{-1}$.

**Table 1. Characteristic vibrations of the molecular bonds of the TZ molecule identified in FTIR spectrum.**

| TYPE OF VIBRATION | ν ($cm^{-1}$) |
|---|---|
| =C-H $sp^2$ hybridised | 3,050 |
| =C-H $sp^3$ hybridised | 2,950 |
| $N^+H$ hydrate vibration | 2,672-2,470 |
| stretching vibration of the aromatic ring C=C-C | 1,566 and 1,552 |
| deformation vibrations-scissoring C-H from -$(CH_2)_2$- | 1,459 |
| in plane deformation vibrations-scissoring - C-H from ( $N - CH_3$ ) | 1,405 |
| symmetrical stretching vibration S-C din S-$CH_3$ | 1,322-1,280 |
| =C-H in-plane deformation vibrations of 4-adiacent H atoms of the ring system symmetrical stretching vibration S-C din S-$CH_3$ | 1,250 |
| in-plane deformation vibrations of =C-H (aromatic ring) | 1,148 |
| =C-H in-plane deformation vibrations of 4-adiacent H atoms of the ring system | 1,039 |
| in-plane deformation vibrations - scissoring of C-H from S-$CH_3$ | 969 |
| =C-H out of plane deformation vibrations of 4-adiacent H atoms of the ring system/of 1 isolated H atom | 931 |
| =C-H out of plane deformation vibrations of the ring system | 860 |

*(Table 1) contd.....*

| TYPE OF VIBRATION | ν ($cm^{-1}$) |
|---|---|
| =C-H out of plane deformation vibrations of 4-adiacent H atoms of the ring system | 850-753 |
| stretching vibration of C-H from -$(CH_2)_2$-/<br>=C-H out of plane deformation vibrations of the ring system | 736 |

The FTIR spectrum is characterised by specific vibrations of aromatic rings like those at 1,566 $cm^{-1}$ and 1,552 $cm^{-1}$ that are responsible for the stretching vibration of the aromatic ring C=C-C or at 1,148 $cm^{-1}$ for in-plane deformation vibrations of =C-H (aromatic ring). Vibrations characteristic to TZ molecule are present, like symmetrical stretching vibration S-C in S-$CH_3$ (1,322-1,280 $cm^{-1}$) or in-plane deformation vibrations-scissoring of C-H from S-$CH_3$ (969 $cm^{-1}$).

FTIR spectra recorded on 20 mg/ml TZ, for unirradiated and irradiated 60 min, 120 min and 240 min samples with 266 nm laser beam are presented in Fig. (**33**). From TLC analysis (next sub-chapter) it was observed that the compound that has the highest concentration in the sample is TZ, so that all FTIR spectra presented in Fig. (**33**) show only TZ contribution. This claim is also confirmed by the fact that there is no difference between the spectra of samples unirradiated and irradiated 60 min, 120 min and 240 min.

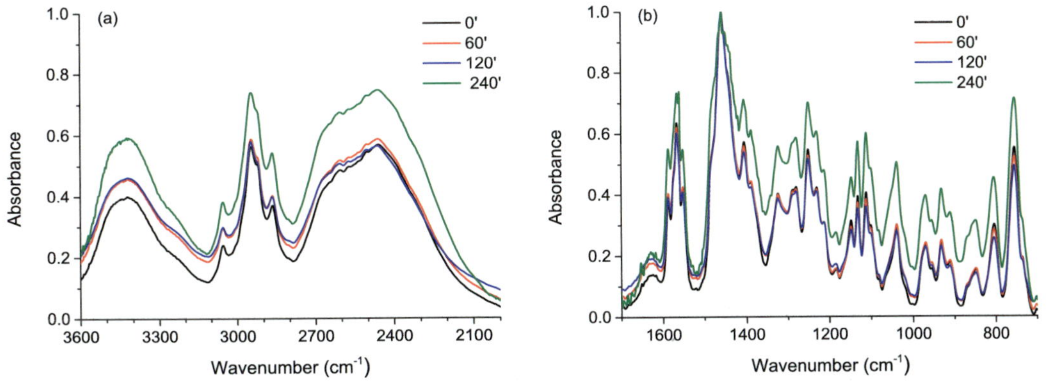


**Fig. (33).** FTIR spectra of 20 mg/ml TZ solutions unirradiated and irradiated 60 min, 120 min and 240 min, between **a**) 3,600-2,000 $cm^{-1}$ and **b**) 1,700-700 $cm^{-1}$.

The FTIR spectrum of unirradiated 20 mg/ml PMZ is presented in Fig. (**34**) and in Table **2** are listed the characteristic vibrations of PMZ molecular bonds. This is characterised by specific vibrations of aromatic rings like those at 1,591 $cm^{-1}$ and 1,569 $cm^{-1}$ that are responsible for the stretching vibration of the ring C=C-C or at 1,250-1,000 $cm^{-1}$ for aromatic =C-H in-plane deformation vibrations-wagging. There are also presented vibrations characteristic for PMZ molecule like in-plane deformation vibrations scissoring a C-H from the functional group (>N – $CH_3$) (1,150-950 $cm^{-1}$).

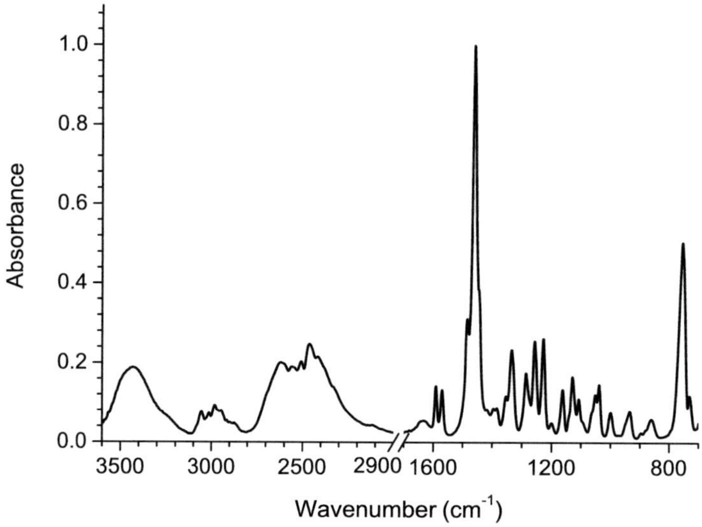


**Fig. (34).** FTIR spectra of unirradiated 20 mg/ml PMZ solution recorded between 3,600-700 $cm^{-1}$.

**Table 2. Characteristic vibrations of the molecular bonds of the PMZ molecule identified in FTIR spectrum.**

| Type of vibration | ν ($cm^{-1}$) |
|---|---|
| =C-H $sp^2$ hybridised | 3,050 and 3,010 |
| =C-H $sp^3$ hybridised | 2,940 and 2,850 |
| $N^+H$ hydrate vibration | 2,672-2,470 |
| stretching vibration of the aromatic ring C=C-C | 1,591 and 1,569 |
| stretching vibration of C-H from >$CH_2$ | 1,459 |
| C-H deformation vibration >N—CH | 1,350-1,315 |
| C-H symetrical deformation vibration >N—$CH_3$ | 1,440-1,390 |
| =C-H in-plane deformation vibrations of 4-adiacent H atoms of the ring system | 1,286 |
| aromatic =C-H in-plane deformation vibrations -wagging | 1,250-1,000 |
| in-plane deformation vibrations-scissoring a C-H from >N – $CH_3$ | 1,150-950 |
| C-N stretching vibration | 1,180-860 |
| out of plane bending vibration of 4-adiacent H atoms of the ring system | 755 |

FTIR spectra recorded on 20 mg/ml PMZ for unirradiated and irradiated 60 min, 120 min and 240 min solutions with a UV laser beam are shown in Fig. (**35**).

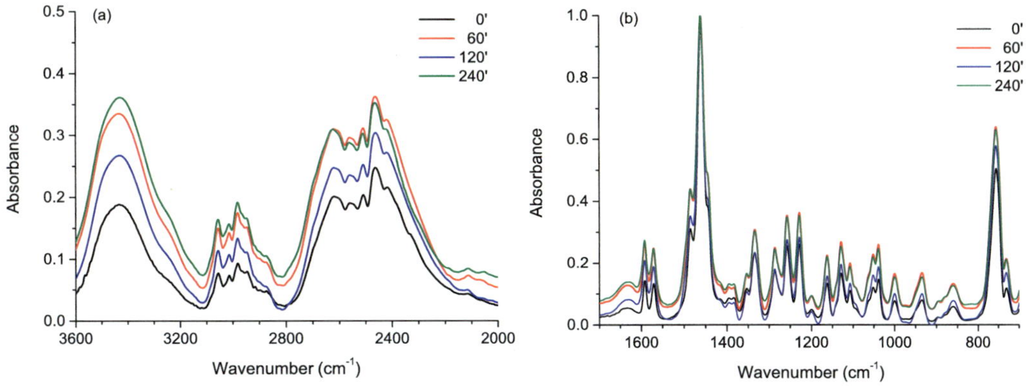


**Fig. (35).** FTIR spectra of 20 mg/ml PMZ solutions unirradiated and irradiated 60, 120 and 240 min, between (**a**) 3,600-2,000 $cm^{-1}$ and (**b**) 1,700-700 $cm^{-1}$.

From TLC analysis, it was observed that the compound that has the highest concentration in the sample is PMZ, so that all FTIR spectra shown in Fig. (**35**) present only PMZ contribution. This is also confirmed by the fact there is no difference between the spectra of the samples unirradiated and irradiated 60 min, 120 min and 240 min.

## *Hydantoin SZ-2*

The FTIR spectrum of SZ-2 solution in DMSO (Fig. **36**) reveals a typical peak for C=O (bending) and C=S (in plane stretching) vibrations at 517 $cm^{-1}$. Regarding the vibrational modes of C-H coupling, although the position of its stretching and bending frequencies remains nearly constant in hydrocarbons, the attachment of $CH_3$ or $CH_2$ to atoms other than carbon, or to a carbonyl group or aromatic ring may result in a significant shift of the C-H stretching and bending frequencies [13, 14]. So, absorption arising from C-H stretching may be observed in the interval 3,200-2,900 $cm^{-1}$, while asymmetrical bending vibration occurs near 1,450 $cm^{-1}$. The out-of-plane C-H bending vibration between 1,000-650 $cm^{-1}$ is assigned to alkene ring and some absorption peaks in 1,563-700 $cm^{-1}$ belong to vibrations involving interaction between C=S stretching and C-N stretching. The peaks at 690 $cm^{-1}$ and 704 $cm^{-1}$ may be assigned to C-Cl in plane stretching vibrations, C=S out-of-plane deformations and torsion of the aromatic ring.

At 1,660 $cm^{-1}$ it occurs the absorption fingerprint of cyclo-alkene due to the C=C stretch vibration coupled with C-C stretching of the adjacent bond. It is shifted compared to the unconjugated alkene typical absorption (~1,640 $cm^{-1}$) probably because the angles between C-C and C=C become smaller and the interaction of these two bonds decreases. Also, between 1,870 $cm^{-1}$ and 1,540 $cm^{-1}$ it can be observed the C=O stretching absorption band. Within its range, the position of C=O is determined by the following factors: the physical state, electronic and mass effects of neighbouring substitutes, conjugation, hydrogen bonding and ring strain. The resonance effect of -NH increases the C=O bond length and reduces the frequency of absorption comparing to the typical ketone fingerprint [13 - 15].

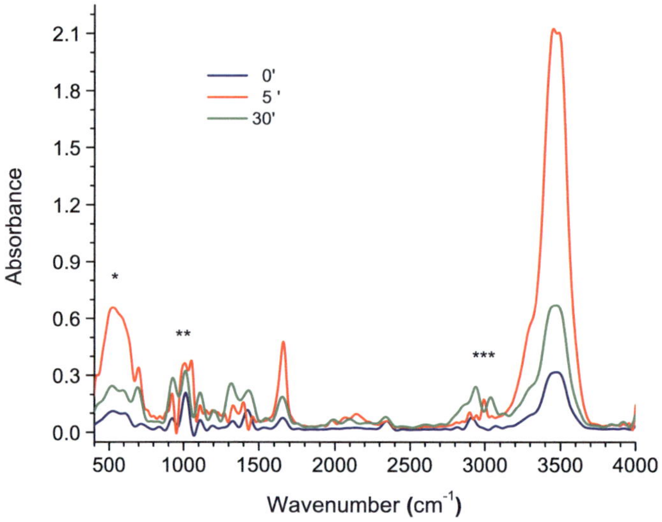


**Fig. (36).** FTIR spectra of SZ-2 solution in DMSO ($10^{-4}$ M) exposed to pulsed Nd:YAG laser beam emitted at 355 nm (THG; 38 mJ) up to 30 min.

This thiohydantoin derivative has two N-H bonds in its molecule. One is adjacent to a carbonyl radical, and the other is surrounded by both carbonyl and thiocarbonyl radicals. Since the two N-H bonds are not equivalent, they show different acid-base properties. Also, addition of DMSO as solvent pointed out different acidic properties of the two N-H bonds [16]. Absorption peaks located in 3,500-3,400 $cm^{-1}$ interval could be assigned to the N-H stretching vibration occurred in diluted solutions. In the meantime, absorption in 900-700 $cm^{-1}$ domain arises from N-H wagging of secondary amine [13]. The two N-H bonds of SZ-2 solution in DMSO give two N-H stretching signals, at 3,448 $cm^{-1}$ and 3,483 $cm^{-1}$. One may suppose that the higher wavenumber band corresponds to the N-H stretching vibration of the N-H bond adjacent to carbonyl radical and the lower wavenumber band to the N-H bond surrounded by the carbonyl and thiocarbonyl radicals [17].

The characteristic absorption frequencies of SZ-2 are briefly structured in Table **3**, where the following items are specified: ν-in plane stretching; δ-bending; *oop*-out of plane deformation; τ-torsion. The calculated frequencies are obtained by molecular modelling using Density Functional Theory (DFT:B3LYP, 6-311G basis) using Gaussian09 [12].

**Table 3. The characteristic absorption frequencies of SZ-2 in DMSO.**

| **Assignment/ Comments** | **Experimental frequencies ($cm^{-1}$)** | **Calculated frequencies ($cm^{-1}$)** |
|---|---|---|
| Aromatic ring/$\delta$, $v$ C=O/$\delta$; C=S/$v$ | 517 | 560 |
| C-Cl/$v$; C=S/*oop* Aromatic ring/$\tau$ | 690; 706 | 688; 704 |
| Aromatic ring/$\delta$, $v$ C=S/$v$; C=O/*oop* | 1,000-930 | 920; 984 |
| Aromatic ring/$v$ N-H/$\delta$; C=S/$v$ | 1,041 | 1,096; 1,112 |
| N-H/$v$ | 3,503 | 3,648 |
| Aromatic ring/$v$, $\delta$ N-H/δ | 1,315; 1,393 | 1,312; 1,384 |
| Aromatic ring/$v$ | 1,450-1,400 | 1,440; 1,472; 1,504 |
| C=O/$v$; N-H/$\delta$ | 1,663 | 1,688 |
| CvH/$v$ | 2,905; 2,912; 2,990 | 3,200 |
| C=C (Chlor-benzene ring)/v | 1,331 | 1,303 |

The resonant interaction of the laser beam with bulk solution was outlined also by FTIR measurements. FTIR spectra of laser irradiated SZ-2 solution in DMSO (Fig. **36**) highlight changes mainly of the C=O bending, C=S in plane stretching (*) and the in plane stretching and bending of the aromatic ring (**). Also, the C-H stretching vibrations (***) are affected.

## *Quinazoline BG 1188*

As for the quinazoline derivative BG 1188, the characteristic absorption frequencies are briefly structured in Table **4**.

**Table 4. The characteristic absorption frequencies of BG 1188 in upw.**

| **Assignment/ Comments** | **Experimental frequencies ($cm^{-1}$)** | **Calculated frequencies ($cm^{-1}$)** |
|---|---|---|
| Aromatic ring/$\delta$, $v$ C=O/$\delta$; $NO_2$/$\delta$ | 568 | 560 |
| Aromatic ring/$\delta$ N-C=O/$\delta$; C-H/*oop* | 635-605<br>630-570 | 600<br>576 |

*(Table 4) contd.....*

| **Assignment/ Comments** | **Experimental frequencies ($cm^{-1}$)** | **Calculated frequencies ($cm^{-1}$)** |
|---|---|---|
| Aromatic ring/*δ*, *v* C=S/*v*; C=O/*oop* | 1,000-930 | 888; 960 |
| Aromatic ring/*v* $CH=CH_2$/*oop* | 1,036 | 1,096 |
| N-H/*v* | 3,428 | 3,480 |
| C-N/v $NO_2$/ *v* | 1,280-1,180 1,343 | 1,136; 1,160 1,224; 1,264; 1,384 |
| Aromatic ring/*v* | 1,450-1,400 | 1,456 |
| C=O/*v*; N-H/*δ* | 1,684 | 1,656; 1,856 |
| C-H/*v* | 2,947; 3,019 | 3,168; 3,296 |
| C-H overtones/δ | 2,750-2,650 | - |

FTIR spectrum of BG 1188 highlights C-H (belonging to aromatic ring) stretching at 3,070 $cm^{-1}$. An overlapping in 3,000-2,800 $cm^{-1}$ range might be due to the stretching of the CH belonging to $N-CH_3$ group and in the range 1,700 $cm^{-1}$ and 1,800 $cm^{-1}$ could be assigned to C=O stretching, while the stretching of the non-conjugated C=C bond may give rise to the maximum at 1,654 $cm^{-1}$ and asymmetric stretching of N-O bonds are responsible for the appearance of band with maximum at 1,556 $cm^{-1}$.

The bands arising at 1,506 $cm^{-1}$ and 1,418 $cm^{-1}$ could be the result of C-C bond belonging to aromatic ring stretching. The symmetric stretching of N-O bond might be responsible for the band at 1,306 $cm^{-1}$ and stretching of aliphatic C-N bond could give rise to peaks in the range 1,250 $cm^{-1}$ and 1,020 $cm^{-1}$.

The overlapped bands in the range 900-675 $cm^{-1}$ might be assigned to *oop* bending vibrations of the C-H, to *oop* bending of aromatic C-C bonds, and to N-H bond wagging vibrations.

Subsequent to laser irradiation of BG 1188 aqueous solution, one can notice changes in the corresponding FTIR spectra (Fig. **37**), as follows: the increase of the peaks arising at (i) 616 $cm^{-1}$, 668 $cm^{-1}$, (ii) in the 1,004-927 $cm^{-1}$ range, (iii) at 1,089 $cm^{-1}$, (iv) at 1,318 $cm^{-1}$ and (v) in the 3,330-2,970 $cm^{-1}$ spectral range. Also, a decrease of the peaks intensities at 853 $cm^{-1}$ and 2,850 $cm^{-1}$ may be observed.

All changes suggest a breaking of C=O and C-N bonds, followed by C-O-H and N-H bonds occurrence. Also, a possible breaking of the benzene ring of the compound is suggested by the decrease of the band at 853 $cm^{-1}$, which is sustained also by the UV-Vis spectra (Fig. **17**) modifications. At the same time, this fact could increase the vibrations amplitude of the pyrimidinic ring.

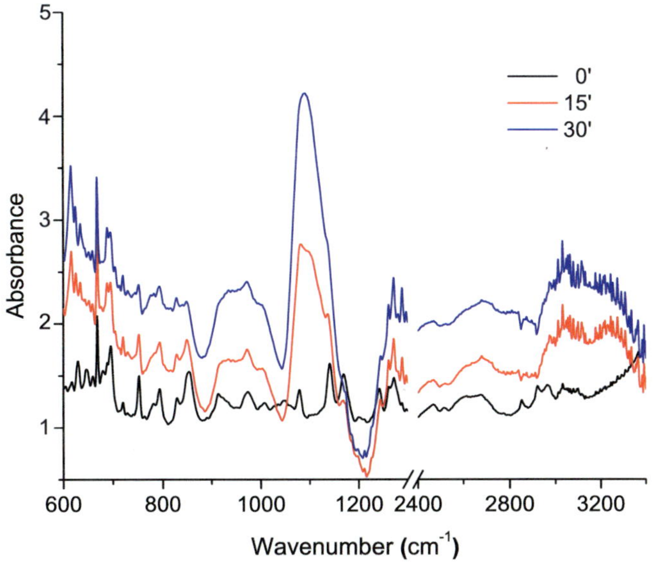


**Fig. (37).** FTIR spectra of the BG 1188 solution in upw ($10^{-3}$ M) exposed to pulsed Nd:YAG laser beam emitted at 266 nm (10 mJ) up to 30 min.

The raising of the bands appeared in the interval 1,318-1,270 cm$^{-1}$ could be due to the splitting of the chain attached to quinazoline core ring, which leads to N-O bond appearance.

In the 3,330-2,970 cm$^{-1}$ vibrational range occur overlapped bands. These bands might be attributed to asymmetric stretching of C-H bonds of R-$CH_3$ belonging to molecules, coupled with stretching of O-H bonds which are raising subsequently C=O double bond breaking. Also, in the same vibrational range, arise absorption bands due to symmetrical and asymmetrical stretching vibrations of N-H bonds [50].

## TLC

TLC analysis allows the visualisation and the direct comparison of photoproducts obtained after exposure to 266 nm laser beam of 20 mg/ml TZ and PMZ solutions (Fig. **38**). Each column corresponds to a specific irradiation time: first column represents the unirradiated solution and, from left to right, are shown the samples irradiated 60 min, 120 min and 240 min. The dotted line from the bottom of TLC plate represents the spots were samples were applied and the one from upper part represents the front line of the solvent.

The unirradiated sample represents the control, it is used to visualise the amount of TZ left in the irradiated solution and it is marked with red dotted line. In the TLC plate visualised at 254 nm it can be observed that all formed photoproducts are more polar than TZ, their polarity increasing from top to bottom. Also, the main compound in the sample after 240 min irradiation is TZ and only four photoproducts or classes of photoproducts having the same polarity are present. Their intensity is direct proportional with the exposure to UV laser radiation and the intensity of photoproducts spot is increasing with prolonged irradiation. Visualisation of TLC plate at 365 nm shows the existence of two new photoproducts/classes of photoproducts that do not absorb at 254 nm, one of them being more polar than TZ (marked with red arrow) and the other less polar that TZ (marked with dotted rectangle).

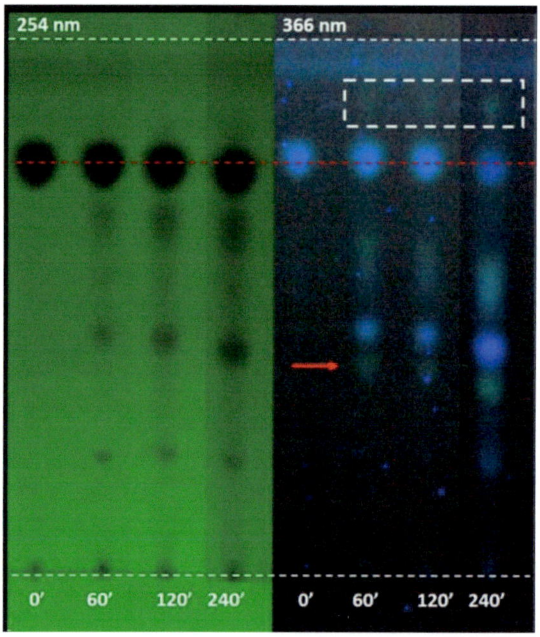


**Fig. (38).** TLC analysis of the 20 mg/ml unirradiated and irradiated TZ samples examined under 254 nm and 365 nm light.

As in TZ case, unirradiated PMZ represents the control in TLC plate and red dotted line represents the amount of PMZ left in sample after each exposure to 266 nm laser radiation. Visualising the plate at 254 nm it can be observed that the major compound is PMZ and a few photoproducts are present (Fig. **39**). Photoproducts are both more polar and less polar than PMZ and their intensity increases with longer periods of exposure to UV laser radiation.

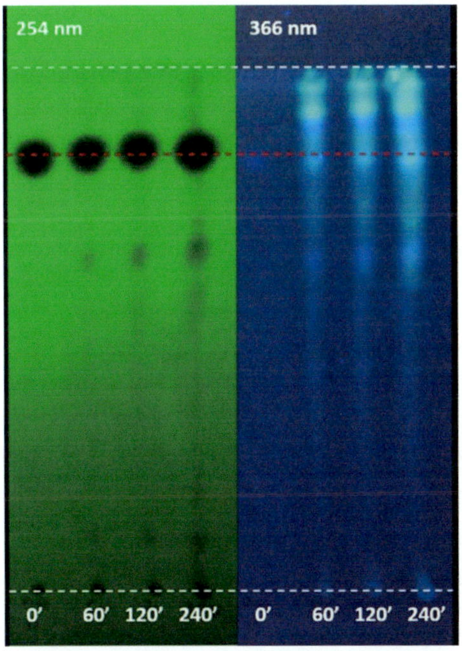


**Fig. (39).** TLC analysis of the 20 mg/ml unirradiated and irradiated PMZ samples examined under 254 nm and 365 nm light.

If the plate is visualised at 365 nm, the photoproducts that are less polar than PMZ are better depicted but those that have the highest polarity are not present in the plate.

## CONCLUSIONS

In this chapter, a synthesis regarding the resonant interaction of several types of medicines with laser radiation is made. Spectroscopic studies are used to emphasise the structural modifications of the drugs under laser irradiation. This investigation can be a powerful tool in establishing the way of action of photosensitive drugs and in defining new strategies to overcome MDR. The potential of lasers to obtain new compounds that become biologically active is promising. This approach is expected to yield a plethora of new compounds that may be difficult to be prepared *via* conventional chemical manipulation methods.

## CONFLICT OF INTEREST

The authors confirm that this chapter content has no conflict of interest.

## ACKNOWLEDGEMENTS

This work has been financed by the Romanian National Authority for Research and Innovation in the frame of Nucleus programme-contract 4N/2016 and the project PN-II-ID-PCE-2011-3-0922.

## REFERENCES

[1] M.L. Pascu, A. Staicu, L. Voicu, M. Brezeanu, B. Carstocea, R. Pascu, and D. Gazdaru, "Methotrexate as a photosensitiser", *Anticancer Res.,* vol. 24, no. 5A, pp. 2925-2930, 2004. [PMID: 15517898]

[2] M.L. Pascu, M. Brezeanu, L. Voicu, A. Staicu, B. Carstocea, and R.A. Pascu, "5-fluorouracil as a phosensitiser", *In Vivo,* vol. 19, no. 1, pp. 215-220, 2005. [PMID: 15796177]

[3] M.L. Pascu, V. Nastasa, A. Smarandache, A. Militaru, A. Martins, M. Viveiros, M. Boni, I.R. Andrei, A. Pascu, A. Staicu, J. Molnar, S. Fanning, and L. Amaral, "Direct modification of bioactive phenothiazines by exposure to laser radiation", *Recent Pat Antiinfect Drug Discov,* vol. 6, no. 2, pp. 147-157, 2011. [http://dx.doi.org/10.2174/157489111796064515] [PMID: 21517738]

[4] M.L. Pascu, B. Danko, A. Martins, N. Jedlinszki, T. Alexandru, V. Nastasa, M. Boni, A. Militaru, I.R. Andrei, A. Staicu, A. Hunyadi, S. Fanning, and L. Amaral, "Exposure of chlorpromazine to 266 nm laser beam generates new species with antibacterial properties: contributions to development of a new process for drug discovery", *PLoS One,* vol. 8, no. 2, p. e55767, 2013. [http://dx.doi.org/10.1371/journal.pone.0055767] [PMID: 23405212]

[5] M.L. Pascu, B.D. Carstocea, M. Brezeanu, L. Voicu, A. Staicu, D.M. Gazdaru, and R.A. Pascu, "Studies on activated cytostatic fluorouracil as photosensitizer: to use in eye tumor treatment", In: *Laser Florence 2003: A Window on the Laser Medicine World,* 2004, pp. 87-95. [http://dx.doi.org/10.1117/12.584388]

[6] M.L. Pascu, L. Danaila, B.D. Carstocea, A. Staicu, S. Truica, and D.M. Gazdaru, "Studies on cytostatics used as photosensitizing material in photodynamic therapy", In: *Laser Florence 2001: a Window on the Laser Medicine World*, 2002, pp. 16-28. [http://dx.doi.org/10.1117/12.486634]

[7] T. Alexandru, M.L. Pascu, B. Danko, V. Nastasa, M. Boni, A. Militaru, I.R. Andrei, A. Staicu, A. Hunyadi, and A. Armada, "Generation and biological evaluation of the products formed from the exposure of Phenothiazine to a 266nm laser beam", *ROMOPTO International Conference on Micro-to Nano-Photonics III,* 2013, p. 88820T [http://dx.doi.org/10.1117/12.2032666]

[8] T. Alexandru, A. Staicu, A. Pascu, E. Radu, A. Stoicu, V. Nastasa, A. Dinache, M. Boni, L. Amaral, and M.L. Pascu, "Characterization of mixtures of compounds produced in chlorpromazine aqueous solutions by ultraviolet laser irradiation: their applications in antimicrobial assays", *J. Biomed. Opt.,* vol. 20, no. 5, p. 051002, 2015. [http://dx.doi.org/10.1117/1.JBO.20.5.051002] [PMID: 25365041]

[9] R.A. Pascu, M. Trifu, M. Dumitrescu, A. Mahamoud, A. Staicu, M. Dicu, B. Carstocea, and M.L. Pascu, "In vivo studies of the effects of alkyl substituted benzo [b] pyridinium compounds exposed to optical radiation", *AIP Conf. Proc.,* vol. 1142, pp. 8-14, 2009. [http://dx.doi.org/10.1063/1.4757152]

[10] A. Smarandache, V. Nastasa, M. Boni, A. Staicu, J. Handzlik, K. Kiec-Kononowicz, L. Amaral, and M.L. Pascu, "Laser beam resonant interaction of new hydantoin derivatives droplets for possible biomedical applications", *Colloid Surface A,* vol. 505, pp. 37-46, 2015. [http://dx.doi.org/10.1016/j.colsurfa.2015.12.002]

[11] A. Smarandache, A. Pascu, I.R. Andrei, J. Handzlik, K. Kiec-Kononowicz, A. Staicu, and M.L. Pascu, "Study of the optical properties of 2-thiohydantoin derivatives", *Rom Rep Phys,* vol. 68, no. 2, pp. 673-683, 2016.

[12] M.J. Frisch, G.W. Trucks, H.B. Schlegel, G.E. Scuseria, M.A. Robb, J.R. Cheeseman, G. Scalmani, V. Barone, B. Mennucci, G.A. Petersson, H. Nakatsuji, M. Caricato, X. Li, H.P. Hratchian, A.F. Izmaylov, J. Bloino, G. Zheng, J.L. Sonnenberg, M. Hada, M. Ehara, K. Toyota, R. Fukuda, J. Hasegawa, M. Ishida, T. Nakajima, Y. Honda, O. Kitao, H. Nakai, T. Vreven, J.A. Montgomery Jr, J.E. Peralta, F. Ogliaro, M. Bearpark, J.J. Heyd, E. Brothers, K.N. Kudin, V.N. Staroverov, R. Kobayashi, J. Normand, K. Raghavachari, A. Rendell, J.C. Burant, S.S. Iyengar, J. Tomasi, M. Cossi, N. Rega, J.M. Millam, M. Klene, J.E. Knox, J.B. Cross, V. Bakken, C. Adamo, J. Jaramillo, R. Gomperts, R.E. Stratmann, O. Yazyev, A.J. Austin, R. Cammi, C. Pomelli, J.W. Ochterski, R.L. Martin, K. Morokuma, V.G. Zakrzewski, G.A. Voth, P. Salvador, J.J. Dannenberg, S. Dapprich, A.D. Daniels, Ö. Farkas, J.B. Foresman, J.V. Ortiz, J. Cioslowski, and D.J. Fox, *Gaussian 09, Revision D.01*Gaussian, Inc.: Wallingford, CT, 2009.

[13] D.A. Phoenix, and F. Harris, "Phenothiazinium-based photosensitizers: antibacterials of the future?", *Trends Mol. Med.,* vol. 9, no. 7, pp. 283-285, 2003.
[http://dx.doi.org/10.1016/S1471-4914(03)00115-1] [PMID: 12900214]

[14] F. Harris, Z. Sayed, S. Hussain, and D.A. Phoenix, "An investigation into the potential of phenothiazinium-based photo-sensitisers to act as PDT agents", *Photodiagn. Photodyn. Ther.,* vol. 1, no. 3, pp. 231-239, 2004.
[http://dx.doi.org/10.1016/S1572-1000(04)00046-8] [PMID: 25048337]

[15] M. Wainwright, D.A. Phoenix, M. Gaskell, and B. Marshall, "Photobactericidal activity of methylene blue derivatives against vancomycin-resistant Enterococcus spp", *J. Antimicrob. Chemother.,* vol. 44, no. 6, pp. 823-825, 1999.
[http://dx.doi.org/10.1093/jac/44.6.823] [PMID: 10590285]

[16] M. Wainwright, "Photodynamic antimicrobial chemotherapy (PACT)", *J. Antimicrob. Chemother.,* vol. 42, no. 1, pp. 13-28, 1998.
[http://dx.doi.org/10.1093/jac/42.1.13] [PMID: 9700525]

[17] A. Smarandache, J. Kristiansen, J.B. Christensen, and M.L. Pascu, "Optical Studies of the Spectral Properties of Phenothiazines", *Lett. Drug Des. Discov.,* vol. 9, no. 4, pp. 352-360, 2012.
[http://dx.doi.org/10.2174/157018012799859918]

[18] National Health Services (NHS), *Shared Care Agreement - Thioridazine for the treatment of adults,* 2014.

[19] C. Avendaño, and J.C. Menendez, "Hydantoin and Its Derivatives", In: *Kirk-Othmer Encyclopedia of Chemical Technology*, John Wiley & Sons, Inc., 2000.
[http://dx.doi.org/10.1002/0471238961.0825040101220514.a01]

[20] K.E. Tiedje, and D.F. Weaver, "Deducing the bioactive face of hydantoin anticonvulsant drugs using NMR spectroscopy", *Can. J. Neurol. Sci.,* vol. 35, no. 2, pp. 232-236, 2008.
[http://dx.doi.org/10.1017/S0317167100008696] [PMID: 18574940]

[21] C. Tran, S. Ouk, N.J. Clegg, Y. Chen, P.A. Watson, V. Arora, J. Wongvipat, P.M. Smith-Jones, D. Yoo, A. Kwon, T. Wasielewska, D. Welsbie, C.D. Chen, C.S. Higano, T.M. Beer, D.T. Hung, H.I. Scher, M.E. Jung, and C.L. Sawyers, "Development of a second-generation antiandrogen for treatment of advanced prostate cancer", *Science,* vol. 324, no. 5928, pp. 787-790, 2009.
[http://dx.doi.org/10.1126/science.1168175] [PMID: 19359544]

[22] Z. Iqbal, S. Ali, J. Iqbal, Q. Abbas, I.Z. Qureshi, and S. Hameed, "Dual action spirobicycloimidazolidine-2,4-diones: antidiabetic agents and inhibitors of aldose reductase-an enzyme involved in diabetic complications", *Bioorg. Med. Chem. Lett.,* vol. 23, no. 2, pp. 488-491, 2013.
[http://dx.doi.org/10.1016/j.bmcl.2012.11.039] [PMID: 23246355]

[23] E. Szymańska, K. Kieć-Kononowicz, A. Białecka, and A. Kasprowicz, "Antimicrobial activity of 5-arylidene aromatic derivatives of hydantoin. Part 2", In: *Farm. Soc. Chim. Ital. 1989,* vol. 57. 2002, no. 1, pp. 39-44.

[24] I. Subtelna, D. Atamanyuk, E. Szymańska, K. Kieć-Kononowicz, B. Zimenkovsky, O. Vasylenko, A. Gzella, and R. Lesyk, "Synthesis of 5-arylidene-2-amino-4-azolones and evaluation of their anticancer activity", *Bioorg. Med. Chem.,* vol. 18, no. 14, pp. 5090-5102, 2010. [http://dx.doi.org/10.1016/j.bmc.2010.05.073] [PMID: 20594860]

[25] M. Mudit, M. Khanfar, A. Muralidharan, S. Thomas, G.V. Shah, R.W. van Soest, and K.A. El Sayed, "Discovery, design, and synthesis of anti-metastatic lead phenylmethylene hydantoins inspired by marine natural products", *Bioorg. Med. Chem.,* vol. 17, no. 4, pp. 1731-1738, 2009. [http://dx.doi.org/10.1016/j.bmc.2008.12.053] [PMID: 19195897]

[26] O. Bozdağ-Dündar, R. Ertan, H.Y. Aboul-Enein, A. Kładna, and I. Kruk, "Free radical scavenging abilities of flavonyl-thiazolidine-2,4-dione compounds", *Luminescence,* vol. 26, no. 1, pp. 10-16, 2011. [http://dx.doi.org/10.1002/bio.1176] [PMID: 19924676]

[27] T.K. Hazra, J.G. Muller, R.C. Manuel, C.J. Burrows, R.S. Lloyd, and S. Mitra, "Repair of hydantoins, one electron oxidation product of 8-oxoguanine, by DNA glycosylases of Escherichia coli", *Nucleic Acids Res.,* vol. 29, no. 9, pp. 1967-1974, 2001. [http://dx.doi.org/10.1093/nar/29.9.1967] [PMID: 11328881]

[28] M. Redrejo-Rodríguez, C. Saint-Pierre, S. Couve, A. Mazouzi, A.A. Ishchenko, D. Gasparutto, and M. Saparbaev, "New insights in the removal of the hydantoins, oxidation product of pyrimidines, via the base excision and nucleotide incision repair pathways", *PLoS One,* vol. 6, no. 7, p. e21039, 2011. [http://dx.doi.org/10.1371/journal.pone.0021039] [PMID: 21799731]

[29] G. Spengler, M. Evaristo, J. Handzlik, J. Serly, J. Molnár, M. Viveiros, K. Kiéc-Kononowicz, and L. Amaral, "Biological activity of hydantoin derivatives on P-glycoprotein (ABCB1) of mouse lymphoma cells", *Anticancer Res.,* vol. 30, no. 12, pp. 4867-4871, 2010. [PMID: 21187464]

[30] S. Suzen, and E. Buyukbingol, "Anti-cancer activity studies of indolalthiohydantoin (PIT) on certain cancer cell lines", In: *Farm. Soc. Chim. Ital. 1989*, vol. 55. 2000, no. 4, pp. 246-248.

[31] A. Cavazzoni, R.R. Alfieri, C. Carmi, V. Zuliani, M. Galetti, C. Fumarola, R. Frazzi, M. Bonelli, F. Bordi, A. Lodola, M. Mor, and P.G. Petronini, "Dual mechanisms of action of the 5-benzyliden--hydantoin UPR1024 on lung cancer cell lines", *Mol. Cancer Ther.,* vol. 7, no. 2, pp. 361-370, 2008. [http://dx.doi.org/10.1158/1535-7163.MCT-07-0477] [PMID: 18281519]

[32] J. Handzlik, G. Spengler, B. Mastek, A. Dela, J. Molnar, L. Amaral, and K. Kieć-Kononowicz, "5-arylidene(thio)hydantoin derivatives as modulators of cancer efflux pump", *Acta Pol. Pharm.,* vol. 69, no. 1, pp. 149-156, 2012. [PMID: 22574520]

[33] S. Jha, J.D. Silversides, R.W. Boyle, and S.J. Archibald, "Hydrogen bonded dimers vs. one-dimensional chains in 2-thiooxoimidazolidin-4-one (thiohydantoin) drug derivatives", *CrystEngComm,* vol. 12, no. 6, pp. 1730-1739, 2010. [http://dx.doi.org/10.1039/b924683e]

[34] J. Baeten, A. Suresh, A. Johnson, K. Patel, M. Kuriakose, A. Flynn, and D. Kademani, "Molecular imaging of oral premalignant and malignant lesions using fluorescently labeled lectins", *Transl. Oncol.,* vol. 7, no. 2, pp. 213-220, 2014. [http://dx.doi.org/10.1016/j.tranon.2014.02.006] [PMID: 24913673]

[35] B.N. Swanson, "Medical use of dimethyl sulfoxide (DMSO)", *Rev. Clin. Basic Pharm.,* vol. 5, no. 1-2, pp. 1-33, 1985. [PMID: 3916302]

[36] H.M. Eisa, M.B. el-Ashmawy, M.M. Tayel, S.A. el-Magd, and H.A. el-Kashef, "Fused pyrimidines. Synthesis of new derivatives of potential diuretic activity", *Boll. Chim. Farm.,* vol. 135, no. 10, pp. 585-590, 1996.
[PMID: 9048447]

[37] A.M. Alafeefy, "Some new quinazolin-4(3H)-one derivatives, synthesis and antitumor activity", *J. Saudi Chem. Soc.,* vol. 15, no. 4, pp. 337-343, 2011.
[http://dx.doi.org/10.1016/j.jscs.2011.06.019]

[38] C.F. Schwender, B.R. Sunday, J.J. Kerbleski, and D.J. Herzig, "Antiallergy activity of 10-oxo-10- H-pyridazino[6,1-b]quinazoline-2-carboxylic acids", *J. Med. Chem.,* vol. 23, no. 8, pp. 964-966, 1980.
[http://dx.doi.org/10.1021/jm00182a030] [PMID: 7401122]

[39] R. Rohini, K. Shanker, P.M. Reddy, V.C. Sekhar, and V. Ravinder, "6-substituted indolo[1,2-c]quinazolines as new antimicrobial agents", *Arch. Pharm. (Weinheim),* vol. 342, no. 9, pp. 533-540, 2009.
[http://dx.doi.org/10.1002/ardp.200900068] [PMID: 19598289]

[40] J. Chevalier, A. Mahamoud, M. Baitiche, E. Adam, M. Viveiros, A. Smarandache, A. Militaru, M.L. Pascu, L. Amaral, and J-M. Pagès, "Quinazoline derivatives are efficient chemosensitizers of antibiotic activity in Enterobacter aerogenes, Klebsiella pneumoniae and Pseudomonas aeruginosa resistant strains", *Int. J. Antimicrob. Agents,* vol. 36, no. 2, pp. 164-168, 2010.
[http://dx.doi.org/10.1016/j.ijantimicag.2010.03.027] [PMID: 20494558]

[41] S. Rádl, P. Hezky, J. Proska, and I. Krejcí, "Synthesis and analgesic activity of some quinazoline analogs of anpirtoline", *Arch. Pharm. (Weinheim),* vol. 333, no. 11, pp. 381-386, 2000.
[http://dx.doi.org/10.1002/1521-4184(200011)333:11<381::AID-ARDP381>3.0.CO;2-A] [PMID: 11129980]

[42] V. Alagarsamy, V. Raja Solomon, R.V. Sheorey, and R. Jayakumar, "3-(3-ethylphenyl)-2-substituted hydrazino-3H-quinazolin-4-one derivatives: new class of analgesic and anti-inflammatory agents", *Chem. Biol. Drug Des.,* vol. 73, no. 4, pp. 471-479, 2009.
[http://dx.doi.org/10.1111/j.1747-0285.2009.00794.x] [PMID: 19291107]

[43] G. Grover, and S.G. Kini, "Synthesis and evaluation of new quinazolone derivatives of nalidixic acid as potential antibacterial and antifungal agents", *Eur. J. Med. Chem.,* vol. 41, no. 2, pp. 256-262, 2006.
[http://dx.doi.org/10.1016/j.ejmech.2005.09.002] [PMID: 16260068]

[44] A. Militaru, A. Smarandache, A. Mahamoud, V. Damian, P. Ganea, S. Alibert, J.M. Pagès, and M.L. Pascu, "Stability Characterization of Quinazoline Derivative BG1188 by Optical Methods", *AIP Conf. Proc.,* vol. 1364, pp. 13-23, 2011.
[http://dx.doi.org/10.1063/1.3626907]

[45] A. Staicu, A. Pascu, M. Boni, M.L. Pascu, and M. Enescu, "Photophysical study of Zn phthalocyanine in binary solvent mixtures", *J. Mol. Struct.,* vol. 1044, pp. 188-193, 2013.
[http://dx.doi.org/10.1016/j.molstruc.2012.12.010]

[46] J.J. Aaron, M.D. Seye, S. Trajkovska, and N. Motohashi, "Bioactive Phenothiazines and Benzo[a]phenothiazines: Spectroscopic Studies, and Biological and Biomedical Properties and Applications", In: *Bioactive Heterocycles VII*Springer: Berlin Heidelberg, 2008, pp. 153-231.
[http://dx.doi.org/10.1007/7081_2008_125]

[47] A. Shah, E. Nosheen, S. Munir, A. Badshah, R. Qureshi, Z.U. Rehman, N. Muhammad, and H. Hussain, "Characterization and DNA binding studies of unexplored imidazolidines by electronic absorption spectroscopy and cyclic voltammetry", *J. Photochem. Photobiol. B,* vol. 120, pp. 90-97, 2013.
[http://dx.doi.org/10.1016/j.jphotobiol.2012.12.015] [PMID: 23474470]

[48] N. Banjac, G. Ušćumlić, N. Valentić, and D. Mijin, "Solvent Effects on the Structure-Activity Relationship of Pharmacological Active 3-Substituted-5,5-Diphenylhydantoins", *J. Solution Chem.*, vol. 36, no. 7, pp. 869-878, 2007.
[http://dx.doi.org/10.1007/s10953-007-9153-2]

[49] H. Moustafa, M.F. Shibl, and R. Hilal, "Electronic Absorption Spectra of Some 2-Thiouracil Derivatives", *Phosphorus Sulfur Silicon Relat. Elem.*, vol. 180, no. 2, pp. 459-478, 2005.
[http://dx.doi.org/10.1080/10426509051718 1]

[50] A. Smarandache, A. Dinache, A. Staicu, J. M. Pages, and M. L. Pascu, *Photosensitivity of new medicines to enhance the antibiotic activity*. to be submitted.

[51] A.T. Aaron, "Effects of solvent on the electronic absorption and fluorescence spectra of quinazolines, and determination of their ground and excited singlet-state dipole moments", In: *Spectrochim. Acta Part Mol. Spectrosc*, vol. 47. 1991, no. s 3–4, pp. 419-430.

[52] A. Smarandache, A. Militaru, H. Goker, A. Pascu, and M.L. Pascu, "Laser methods for pharmaceutical pollutants removal", *Proc. SPIE*, vol. 8411, p. 84111E, 2012.
[http://dx.doi.org/10.1117/12.966414]

[53] K. Robinson, T.F. Pilot, and J.E. Meany, "Time dependent inhibition of xanthine oxidase in irradiated solutions of folic acid, aminopterin and methotrexate", *Physiol. Chem. Phys. Med. NMR*, vol. 22, no. 2, pp. 95-103, 1990.
[PMID: 2100009]

[54] A. Thomas, F.G. Einschlag, M.R. Féliz, and A.L. Capparelli, "First steps in the photochemistry of folate in alkaline medium", *J. Photochem. Photobiol. Chem.*, vol. 116, no. 3, pp. 187-190, 1998.
[http://dx.doi.org/10.1016/S1010-6030(98)00304-9]

[55] A.H. Thomas, G. Suarez, and F.M. Cabrerizo, "R. Martinol, and A. L. Capparellil, "Study of the photolysis of folic acid and 6-formylpterin in acid aqueous solutions", *J. Photochem. Photobiol. Chem.*, vol. 135, no. 2, pp. 147-154, 2000.
[http://dx.doi.org/10.1016/S1010-6030(00)00304-X]

[56] M. Daniels, "Tautomerism of uracil and thymine in aqueous solution: spectroscopic evidence", *Proc. Natl. Acad. Sci. USA*, vol. 69, no. 9, pp. 2488-2491, 1972.
[http://dx.doi.org/10.1073/pnas.69.9.2488] [PMID: 4506769]

[57] M. Barra, G.S. Calabrese, M.T. Allen, R.W. Redmond, R. Sinta, A.A. Lamola, R.D. Small, and J.C. Scaiano, "Photophysical and photochemical studies of phenothiazine and some derivatives: exploratory studies of novel photosensitizers for photoresist technology", *Chem. Mater.*, vol. 3, no. 4, pp. 610-616, 1991.
[http://dx.doi.org/10.1021/cm00016a010]

[58] T. Alexandru, A. Staicu, A. Pascu, A. Dinache, A.B. Abdeladhim, M. Enescu, A. Khatyr, and M.L. Pascu, "Light cleavage of some potential linkers for drug carriers", *Rom Rep Phys*, vol. 68, no. 2, pp. 684-692, 2016.

[59] A. Alvarez-Ordóñez, and M. Prieto, *Fourier Transform Infrared Spectroscopy in Food Microbiology*Springer: Boston, 2012.
[http://dx.doi.org/10.1007/978-1-4614-3813-7]

[60] G. Alagona, C. Ghio, and S. Monti, "Ab initio modeling of competitive drug-drug interactions: 5-fluorouracil dimers in the gas phase and in solution", *Int. J. Quantum Chem.*, vol. 83, no. 3-4, pp. 128-142, 2001.
[http://dx.doi.org/10.1002/qua.1205]

CHAPTER 12

# Lasers in Foams and Emulsions Studies

**Viorel Nastasa[1,2], Mihai Boni[1,3], Alexandru Stoicu[1], Andra Dinache[1], Adriana Smarandache[1] and Mihail Lucian Pascu[1,3,*]**

[1] *National Institute for Laser, Plasma and Radiation Physics, Magurele, Ilfov, Romania*

[2] *ELI-NP, "Horia Hulubei" National Institute for Physics and Nuclear Eng., Magurele, Ilfov, Romania*

[3] *Faculty of Physics, University of Bucharest, Magurele, Ilfov, Romania*

**Abstract:** This chapter presents results regarding production of emulsions and foams and their interaction with laser beams. Foams and emulsions are considered mainly collection of bubbles, or drops of different kind with particular properties, respectively. Here, emulsions of oily vitamin A and water in which various surfactants were introduced are described outlining the conditions in which they have a longer time stability. Laser induced fluorescence emitted by microdroplets that contain emulsions of rhodamine 6G in water and oily vitamin A is shown and the spectral distribution of the fluorescence radiation is described evidencing the enhancement of fluorescence emitted by droplets with respect to bulk. Lasing conditions in droplets are discussed and the role of reflections on foam drop inner structural components is outlined. Foams based on water solutions of vancomycin, produced by droplets interaction with laser beams are described in correlation with biomedical applications. Foams of polidocanol in water produced by Tessari method are described and their use in varicose vein treatments is introduced outlining the role of their exposure in tissues to infrared Nd:YAG laser beams in connection with the more rapid positive treatment effects. Polidocanol foam stability function of several surfactants such as tween 80, glycerin and xanthan gum is presented and discussed.

**Keywords:** Aetoxisclerol, Colloids, Emulsion, Foam, Hanging droplet, Laser, Nanoparticle, Pendant droplet, Polidocanol, Rhodamine 6G, Sclerosing foam, Sclerotherapy, Sessile droplet, Suspended droplet, Tessari technique, Vancomycin, Varicose veins, Vitamin A.

## INTRODUCTION

Droplets with submicron dimensions are of high interest for applications in various domains (pharmaceutical, cosmetic, food industry, *etc.*). The multiple applications of nano-/micro-droplets in emulsions and the need of smaller droplets

[*] **Corresponding author Mihail Lucian Pascu:** Laser Department, National Institute for Laser, Plasma and Radiation Physics, Magurele, Ilfov, Romania; Tel/Fax: 0040 21 4575739; E-mails: mihai.pascu@inflpr.ro

**Mihail Lucian Pascu (Ed.)**

dimensions pushed the efforts to develop better emulsification technologies. Droplets with dimensions smaller than 1 μm are suitable for biomedical applications as transport vectors for medicines to specific targets. The systems using droplets allow substances poorly soluble in water, but oil-soluble, such as vitamins or drugs to be incorporated in a lipophilic phase so that the increase of their local bioavailability takes place. This also stabilises components sensitive to enzymatic degradation, allows a slower, controlled release of components to targets over a prolonged period of time and reduces the side effects of drugs [1]. Based on small particles use in systemic treatments, one predicts that micro-/nano-emulsions uptake improves the efficiency of lipophilic substances The dispersed phase is present in the form of droplets existing in continuous phase [1].

## Emulsions

Emulsions are defined as heterogeneous systems composed of an immiscible or partially miscible liquid denominated as dispersed phase, in another liquid (continuous phase) as droplets with diameters ranging from nanometres to 100 μm. A particular case is represented by oil and water mixtures, the continuous and dispersed phases being determined by several factors such as the percentage of each liquid, the type of surfactant used, *etc.* Multiple emulsions are also possible (*e.g.* O/W/O, W/O/W - where O means oil and W means water) to produce. In order to stabilise the emulsion (avoiding droplet breakage and coalescence) surfactants are used.

The two liquid phases are selected to be fully immiscible and chemically non-reactive which makes their mixture thermodynamically unstable [2]. In the case of a mixture of two immiscible liquids, a molecular exchange between dispersed phase and continuous phase is not possible, and therefore the Ostwald ripening effects are avoided.

One of the characteristics of an emulsion is the mean size and the size distribution of droplets. By selecting the proper emulsification method and/or device(s) these characteristics can be controlled [3].

Emulsions are widespread and have applications in many domains with an impact on daily life, such as: food, pharmaceutical and cosmetic industry. Applications of compound droplets are reported in the literature and are widely used in various domains. One example is the use of droplets (diameter~2.7 mm) that contain emulsions of oil in water to cool a moving steel strip and at the same time lubricate the sprayed areas [4]. By using coaxial electrified jets, stratified capsules with micro- and sub-micrometre dimensions and emulsions composed of immiscible liquids were produced (0.15 μm capsules composed of a water core of 0.1 μm and a surrounding layer of olive oil of 0.025 μm diameter) [5].

The emulsification process implies the use of energy to mix the immiscible liquids. This can be made by mechanical or pseudo-mechanical means, such as rotor-stator systems (high speed homogenisers), high pressure homogenisers, ultrasonic probes, micropore size membranes, microfluidisers, phase inversion systems, *etc.* The emulsification is based on two opposite processes: the drop breakage that produces smaller droplets out of a parent drop, and droplet-droplet coalescence [6]. The evolution of drop-size distribution during emulsification is determined by the competition between these two processes [1].

The stability of an emulsion depends on various factors such as physical properties of interfacial film, rheological properties of continuous phase, composition of the mixture (surface active substances addition). The type of added surfactants/nanoparticles, the presence of electrostatic or steric barriers on droplets, the droplet size distribution, the dispersed/continuous phase ratio, and the temperature are also factors to consider [7 - 8]. By adding surfactants to one or both phases, droplet size as well as droplet-droplet interactions are influenced. Addition of surface active substances decreases the surface tension due to surfactant molecules that migrate to interface, which fact influences droplets sizes and modifies their visco-elastic properties; therefore, this delays droplets coalescence, which leads to an increased emulsion stability [9]. In the case of mechanical methods used to mix the liquids, droplets dimensions are determined by the energy used to apply the procedure (rotating speed, pressure difference *etc.*). Under these conditions the evolution of drop-size distribution in an emulsion is mainly determined by drop breakup process. Function of surfactant concentration in bulk, the newly formed interfaces may be fully or partially covered with surfactants. Due to the function of surfactant concentration in bulk or type of surface active substance one can control drop coalescence effects, drop breakage due to the employed emulsification method and therefore the stability of the emulsion after finishing the emulsification process.

A widely accepted technique for emulsification consists in using a rotor-stator system that can mix substances in controlled conditions (rotating speed, type of rotor, *etc.*). Another emulsification method may use high pressure homogenisers that can mix the immiscible components under high pressure differences. Fluids are pushed at high speed through a small orifice and droplet breakage is produced. Each emulsification method requires energy consumption by the system. The energy may be provided in various ways: trituration, homogenisation, heat, agitation, ultrasound *etc.* There are also emulsification methods that do not require mechanical energy to be dissipated; these emulsions are generated by condensation techniques, are mainly determined by thermodynamic principles and are usually performed in the presence of emulsifying or swelling adjuvants.

The selection of the surfactant for a specific emulsion is made based on HLB (hydrophilic-lipophilic balance), while its concentration is selected function of its CMC (critical micelle concentration). The HLB is a measure of the hydrophilicity or lipophilicity of the surfactant and it is determined by calculating the ratio of hydrophobic to hydrophilic portion of the molecule [2]. This method was developed by Griffin in 1954 for non-ionic surfactants. The HLB for non-ionic surfactants can be obtained from:

$$HLB = 20 \times \frac{M_h}{M} = \frac{weight\,\%\,hydrophile}{5} \quad \textbf{(1)}$$

where $M_h$ is the molecular mass of the hydrophilic part of the molecule and $M$ is the molecular mass of the entire molecule. Depending on HLB value, one can estimate the hydrophilic/lipophilic character of the surfactant (*e.g.* HLB from 10 to 20 means water soluble and HLB<10 water insoluble). The soluble character of a surfactant in water is given by the polarity of the head group (*e.g.* carboxylate, amine, quaternary ammonium *etc.*) which is pH sensitive. To calculate the HLB values for ionic surfactants, one can use the Davies's formula:

$$HLB = 7 + \sum_{i=1}^{m} H_i - n \times 0.475 \quad \textbf{(2)}$$

where $m$ is the number of hydrophilic groups of the molecule, $H_i$ is the value of the $i$-th hydrophilic group and $n$ is the number of lipophilic group in molecule. Each oil/water system has a specific HLB value which is function of the nature of oil and the of surfactant needed by the intended application.

A surfactant is a compound that has two parts, a hydrophilic head (water soluble/water loving part) and a hydrophobic hydrocarbon tail (water insoluble part). These amphiphilic molecules may act as detergents, emulsifiers, foaming agents, *etc.* by lowering liquid/air surface tension, interfacial tension between two liquids, and/or interfacial tension between a liquid and a solid.

At water/oil interface the surfactant molecules tend to migrate at the interface between the two liquids so that the hydrophilic head remains in water phase, while the hydrophobic tail is in oil phase. Depending on different factors, one of them being surfactant concentration, these molecules may also form other energetically favourable structures in water, such as micelles (Fig. **1**).

One of the most important characteristics of a surfactant is the critical micelle concentration (CMC) defined as the threshold concentration of surfactant in bulk, above which micellar structures (micellar aggregates) start to form in the liquid. Any increase of surfactant concentration above this value will lead to more micellar structures. At low surfactant concentration, the amphiphilic molecules will be adsorbed at interface (surface) which in fact will lead to the decrease of interfacial tension (energy of interface is lowered; it is calculated as area times surface tension). As the interface is getting saturated with these molecules, they will start forming micelles to protect the hydrophobic part by removing the hydrophobic tails from the contact with water. After reaching CMC, surface tension remains relatively constant. CMC values depend on various factors such as molecular composition of the surfactant, temperature, pressure and composition of the solution in which the surfactant is present (*e.g.* presence of other surfactants, electrolytes *etc.*). Formation of micelles strongly depends on temperature and micellar structures are formed only above the critical micelle temperature (Krafft temperature) [12]. Below this value, the surfactant is in a crystalline form, and therefore not solubilising in solution unless another surface-active substance is added or the temperature is increased above threshold value.

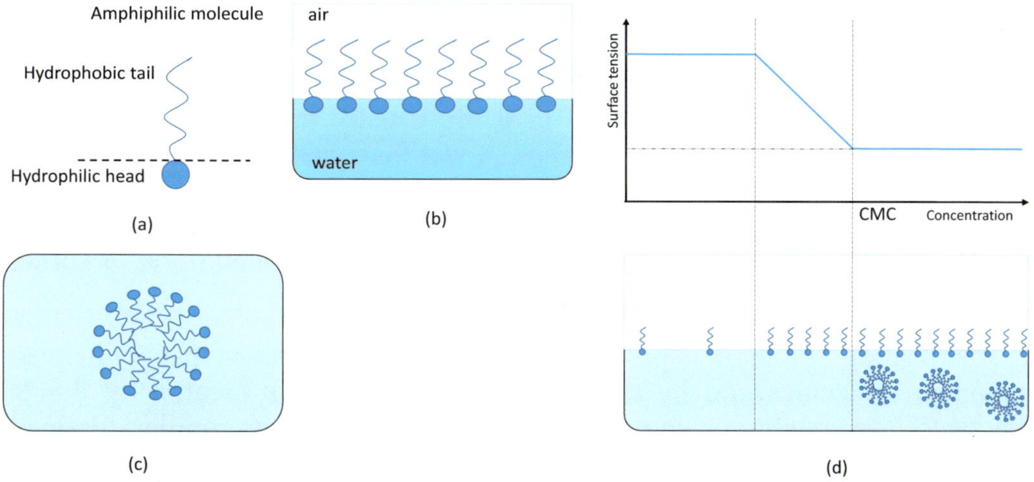


**Fig. (1).** Micelles formation for amphiphilic substances. (a) amphiphilic molecule structure; (b) amphiphilic molecule at air-water interface; (c) micelle; (d) surface tension distribution, function of surfactant concentration (logarithmic scale) [10, 11].

The knowledge of these parameters is very important when using surfactants. In many cases, CMC is the limiting concentration for meaningful use, since adding more surfactants will not influence solution's surface/interfacial tension. If the formation of micelles is the main goal, the value of CMC is a measure of surfactant efficiency (detergency) - a lower CMC means that less surfactant is needed to saturate the interface and form micelles.

The CMC is used extensively in industry. One way to remove surfactants from solutions is to use air bubbles that will pull out surfactant molecules from the bulk to the top; as they rise to surface, surfactant bulk concentration decreases. This will lead to formation of foams. A similar effect can be obtained when using emulsions (decrease of surfactant concentration below CMC). CMC plays an important role in oily soil removal by modifying the contact angles and in extraction of oil as part of an emulsion. In oil industry, it is of high importance to add surfactant solutions with concentrations above CMC, since below this value, the interfacial tension between oil and water phase is very difficult to reduce.

**Foams**

A foam is defined as a dispersion of gas bubbles separated by thin films of liquid or solid phases. Foams have multiple applications in various domains such as medicine, food, cosmetics, petrochemical industry, defence, construction industry, *etc.* They are encountered in daily life, a bath sponge being an example of a solid foam, while the top of a coffee cup is an example of a liquid foam.

Foams can be classified as wet or dry depending on the air/liquid ratio. The foams with a higher percentage of liquid (wet foams) are usually characterised by a fast drainage, unless the liquid is considerably viscous. Since a wet foam has a high liquid content, the bubbles enclosed in it are usually mobile, and have small interbubble contacts and spherical shapes. A wet foam can drain and become dry foam, if it remains stable enough during the draining process. A dry foam is characterised by low liquid content, close bubble packing and a polyhedral internal structure [13 - 16]. Like emulsions, foams also need addition of surface active substances to be stable.

The stabilisation of a foam is produced by van der Waals forces that act between component molecules and by the electrical double layers created by dipolar surfactants. The stability is the result of two competing forces: capillary drainage of liquid towards Plateau borders (Fig. **2**) and disjoining pressure ($\Pi$) [15]. The added surface active substances influence not only the surface tension of the liquid but also rheological (shear and dilatational) properties of the adsorbing layers at gas/liquid interfaces. One of the processes that destabilise a foam is the liquid drainage. Foam decay may also take place due to gas diffusion across films from small to large bubbles (Ostwald ripening) due to their different internal pressures which induce modified gas solubility. A foam is also affected by bubble coalescence, thermal fluctuations, external vibrations, ionising radiation, spontaneous vapour nucleation, instability of interbubble films, *etc.* [16 - 18]. Before bubble rupture, an interbubble film stretches and becomes thinner producing a variation of surfactant concentration at film surface and an increase in surface tension which acts as a restoring force to the lamellae; this is termed the Gibbs-Marangoni effect [19 - 22].

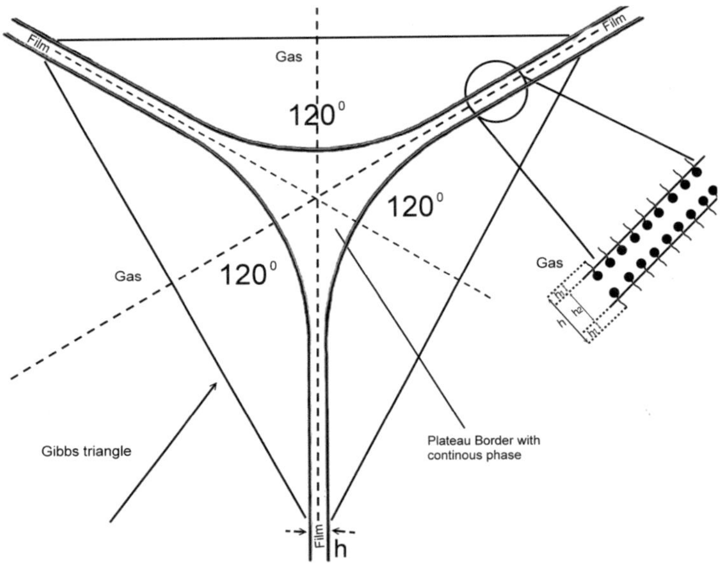


**Fig. (2).** Schematic presentation of a foam Plateau border* [15].

One of the biomedical applications of foams is popular in the treatment of varicose vein disease. The varicosities treatment is challenging due to variable sizes, depths, flow patterns and vessel thicknesses of leg veins. There are several treatment approaches, more or less invasive. The sclerotherapy assisted by laser irradiation proves to be a suitable approach especially for micro-veins with less than 1.5 mm diameter [23]. In sclerotherapy, a detergent like drug is injected into the vein, where produces chemical irritation and endothelial inflammation. A local thrombus is generated that is attached to vessel interior walls and transforms it into a fibrotic cord [24]. Liquid, as well as foamed sclerosant drugs are employed nowadays. Injecting of foamed medicines has advantages, such as for instance the improvement of active agent's contact with vessel wall by forcing out intravascular blood. This leads to a minimum dilution and a deactivation of drug by blood components.

On the other hand, less active substance concentration is needed for therapy. The use of foams allows to better control drug delivery and to pursuit the effect on the treated tissue. Overall, the efficacy of foam sclerotherapy depends on foam quality.

A large number of sclerosants are employed in varicose veins treatment. One of the most used is polidocanol (POL), a mild nonionic surface active substance.

Surfactants reduce the surface tension (ST) of liquids by dispersing their molecules among the liquid surface molecules. As they have polar hydrophilic head and non-polar hydrophobic tail, they are amphiphilic compounds [25].

Many techniques have been developed to generate foams for clinical use. Most of them produce foams by cyclical mechanical agitation of the liquid substance in the presence of a gas, usually air (79% $N_2$ and 21% $O_2$), at gas/liquid ratios ranging from 1:1 to 8:1. One of the most popular and affordable methods to generate foams is the Tessari technique, in which a double syringe system is used connected through a three ways connector [26].

A different treatment approach, especially in cases of smaller veins calibres or when there is a resistance to microfoam sclerotherapy is laser therapy [27]. Leon Goldman started dermatologic application of laser technology [28] and afterwards some authours introduced the selective photothermolysis [29, 30].

Individually considered, these two techniques prove their advantages and drawbacks. When varicose veins are widespread in both legs, a high number of sclerotherapy and maintenance sessions is required [31, 32]. As for laser therapy, different types of lasers and associated devices have been used in the treatment of varicosities in the past decades, with positive, though variable results. Studies comparing these interventions with microfoam sclerotherapy, which is considered the "gold standard", conclude that only Nd:YAG laser application has shown effectiveness [33, 34]. Nevertheless, limitations caused by vessels calibre (for veins with diameters larger than 4 mm), deepness of penetration, pain at high fluences and occurrence of relapses were reported [35].

Fifteen years ago, it was observed that exposing the tissue immediately after injection of POL microfoam, to Nd:YAG laser radiation, venous lesions cleared almost completely and the obtained effects were permanent. Several trials were performed so far in order to determine the optimal procedure, which showed that results were very good after a single session and even better after a second follow-up session observed three weeks later [36]. Considering these encouraging clinical results, the protocol of combining POL microfoam injections with exposure of affected tissues to Nd:YAG laser was recently added to the daily practice of some medical centres being recommended as the first step treatment of large areas of varicose veins.

Spectral investigations of liquid and foamed POL, including subsequent pulsed Nd:YAG laser beam exposure were performed to better understand the physical processes involved in evolution of foaming POL as well as laser beam influence on this sclerotic agent.

## MATERIALS AND METHODS

The substances dealt with in this chapter are: vancomycin (VCM), vitamin A, polidocanol, xanthan gum, tween 80, glycerin, and Rhodamine 6G (Rh6G).

### Materials

#### *Vancomycin*

VCM is a glycopeptide antibiotic used mainly to treat multiple drug resistance (MDR) acquired by Gram-positive bacteria [37]. It is a powerful antibiotic with some severe side effects which makes it usable as ultimate solution in well-defined cases (*e.g.* the treatment of methicillin or penicillin-resistant staphylococcal infections). The molecular structure of VCM is shown in Fig. (**3**). It was optimised with the DFT method (B3LYP functional; 6-31G(d,p) basis set) using Gaussian09 [38]. Due to the high toxicity of VCM, it is of interest to find new ways of using this drug as efficiently as possible, without any/or fewer side effects on human patients, either by improving its effect, or by improving the delivery method, which would lower the used doses. Several methods are introduced in literature and two are described here: (i) incorporation of VCM as active agent in capsules/layered droplets and their use as transport vectors and/or (ii) VCM exposure to UV light to induce structural modifications of molecules that could lead to generation of new photoproducts with higher antibacterial properties and lower toxicity if compared to parent compound [37].

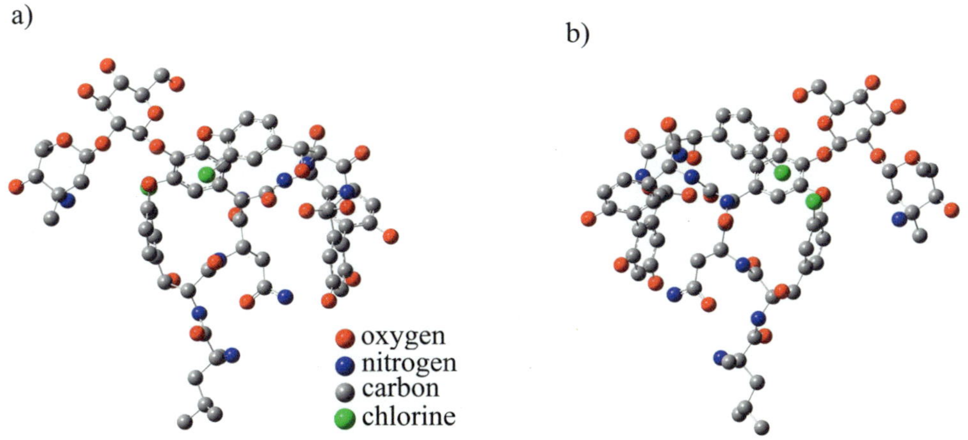


**Fig. (3).** 3D optimised molecular structure of VCM for fundamental electronic state: (**a**) image in an aleatory initial position and (**b**) image rotated anticlockwise with respect to (**a**) position.

## *Vitamin A*

The oily part of the obtained emulsions was vitamin A (retinol diluted in sunflower oil) that is a commercial product found in pharmacies and is commonly used for ophthalmic applications. Vitamin A (retinaldehyde) is necessary in the retina to form the rhodopsin molecule. It has also an important role in the growth of the epithelial cells (retinoic acid - an oxidised form of retinol) [39 - 41]. The 3D chemical structure of vitamin A molecule is shown in Fig. (**4**).

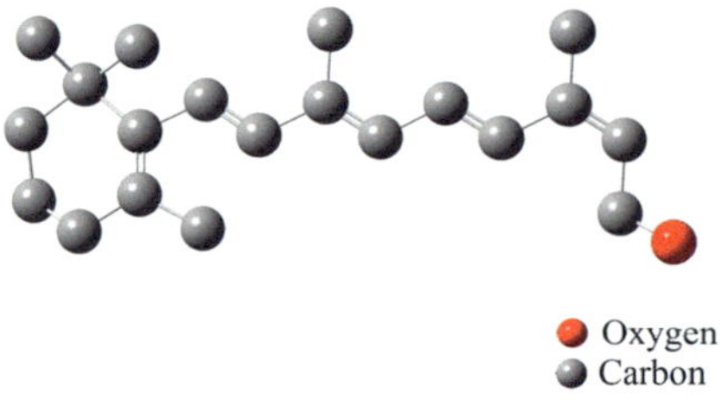


**Fig. (4).** 3D molecular structure of vitamin A.

Vitamin A was used in the studies reported in this chapter because: (i) it is an important nutrient of eye tissues; in Chapter 14 the use of exposed medicines to UV laser beams is described outlining effects on pseudotumour tissues produced on rabbit eyes; (ii) it has antioxidant properties being used in cosmetics and dentistry; (iii) it is soluble in oils which are normally biocompatible and emulsions of it may be easily prepared [39 - 41].

## *Rhodamine 6G*

Rh6G is a fluorescent dye with broad applications in spectroscopy. It is also known as a laser dye or high gain medium in dye lasers with an absorption peak close to 532 nm (second harmonic generated from a Nd:YAG laser fundamental beam) and emission between 555 nm and 585 nm [42]. A more detailed description of this substance is made in Chapter 9.

## *Tween 80*

Tween 80 is a non-ionic surfactant: (polysorbate 80-polyoxyethylene sorbitan monooleate) mostly used in cosmetic and food industry. It has pharmaceutical applications, being used to stabilise aqueous formulations, or as emulsifier. The 3D chemical structure of tween 80 molecule is shown in Fig. (**5**).

Tween 80 was utilised in the experiments reported here as an emulsifier able to prolong the lifetime of emulsions in order to make them available larger time intervals for investigations and applications.

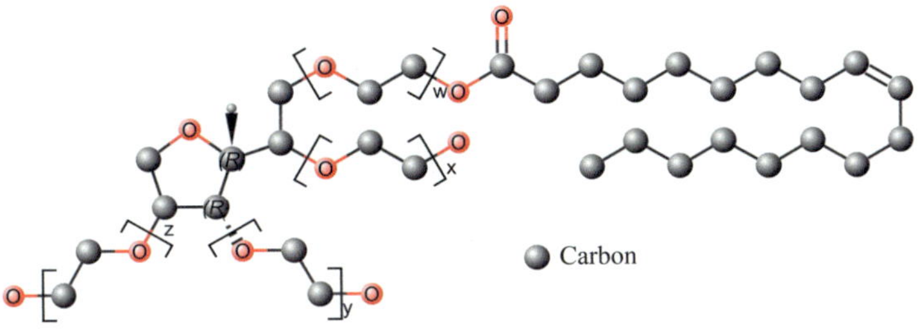


**Fig. (5).** 3D molecular structure of tween 80 [43]. R, x, y, z, w are subjects of patent coverage.

## *Xanthan gum*

Xanthan gum ($C_{115}H_{184}O_{59}P_2^{-4}$) is an extracellular polysaccharide secreted by *Xanthomonas campestris.* It is obtained from the fermentation of glucose, sucrose or lactose and is mainly used as a food additive, emulsion stabiliser, and rheology modifier due to its non-Newtonian behaviour (strong shear thinning). It is also used in pharmaceutics as suspending and stabilising agent in oral and topical forms, for production of sustained release matrix tablets or for proved mucoadhesive properties [44]. Its 3D molecular structure is shown in Fig. (**6**).

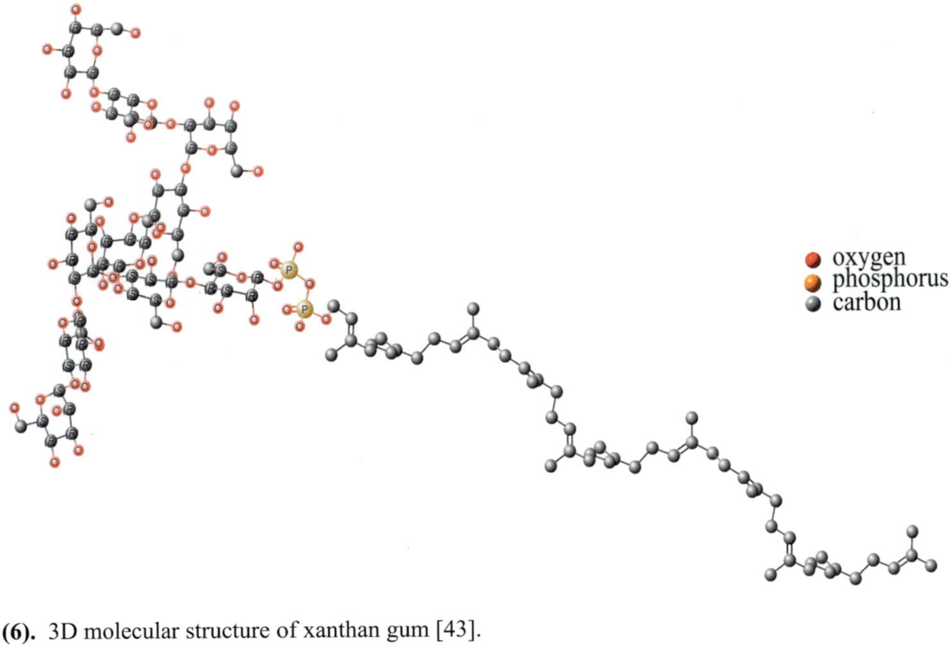


**Fig. (6).** 3D molecular structure of xanthan gum [43].

## *Glycerin*

Glycerin is a colourless, odorless, and viscous liquid with low toxicity, used in pharmaceutics (humectant or lubricant) and food industry (humectant, solvent, sweetener, thickening agent). It makes the stability of foams and emulsions bigger by increasing the viscosity of the compounds used to produce them. Its 3D molecular structure is given in Fig. (**7**).

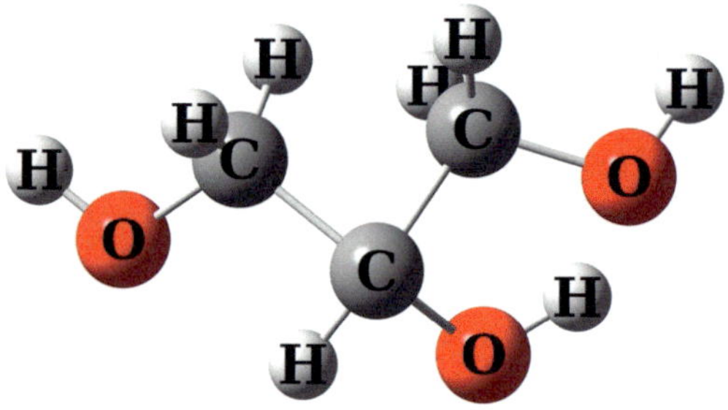

**Fig. (7).** 3D molecular structure of glycerin.

## *Polidocanol*

POL, also known as aethoxyscklerol or lauromacrogol is a synthetic fatty alcohol, an alkyl polyglycol ether of lauryl alcohol; it is a nonionic surfactant composed of an alkyl chain of 12 carbon atoms and an ethylene oxide chain of 9 ethylene oxide units ($C_{30}H_{62}O_{10}$; 3D optimised molecular structure given in Fig. **8**).

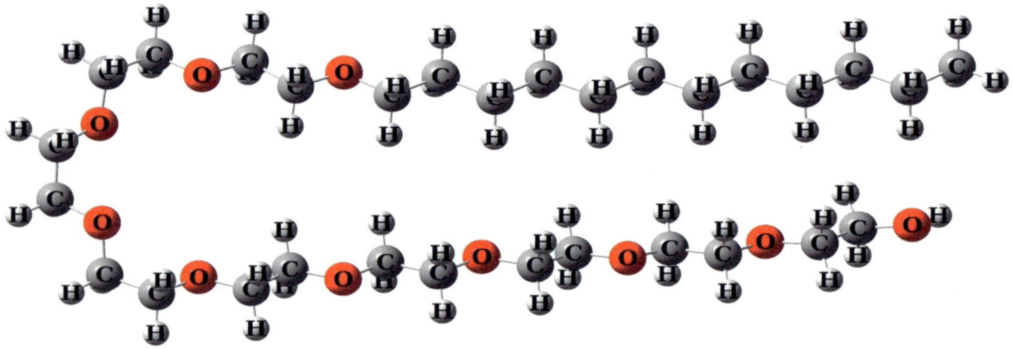

**Fig. (8).** 3D optimised (DFT 6-31G) molecular structure of POL [38].

It has multiple applications in various domains such as cosmetics (hair-conditioner, shampoo, body lotions and facial cream) and medicine. One of the most important operations with this surfactant is performed in sclerotherapy for treatment of varicose vein disease. Upon injection in the vein, it generates a lipid bilayer that produces disruption of cell surface membrane; the process depends on POL concentration. It also activates calcium signalling and nitric oxide pathways and produces endothelial cell death [45].

## Methods

First, standard pharmaceutical form of POL 2% (Kreussler & Co. GmbH, DE) was used in an experimental study as an ampoule of 2 ml injectable solution. Its composition is a mixture of the following substances: lauromacrogol 400 (POL) (active) - 40 mg, sodium hydrogen phosphate (buffer-inactive), potassium dihydrogen phosphate (buffer - inactive), ethyl alcohol (inactive) - 0.10 ml and pure water (inactive) - 2 ml. This is why, when the solution is exposed to laser radiation, one should take into account the interaction of all above mentioned compounds that exist in the commercially available product.

POL foam was obtained using the Tessari method which employs two disposable 5 ml syringes (Luer Slip) connected to a three-way stopcock [46]. A mixture of drug solution and atmospheric air (1:4 ratio) was passed 40 times between the two syringes and the resulting hand-made foam preserved its characteristics for approximately 5 minutes.

Another experimental session implied POL (Sigma-Aldrich Chemie GmbH, DE), which was prepared as solution of 0.5% in a mixture of ultra pure water (upw)

delivered *via* a sterile filter (TKA Pacific UP/UPW6, Thermo Electron LED GmbH, DE), and ethanol (Merck Millipore, DE) (ratio 1:1).

The optical properties of investigated solutions were acquired using a UV-Vis-NIR spectrophotometer (Lambda 950, Perkin Elmer, USA). The spectral resolution was 0.05 nm for UV-Vis and 0.2 nm for NIR. Quartz cuvettes of 1 mm optical path were employed. The overall measuring error of the equipment was ±0.004% [47].

FTIR was also used to characterise POL and a FTIR spectrometer (Nicolet IS50, Thermo Scientific, USA) was operated with Omnic 9 Standard software at 8 $cm^{-1}$ experimental spectral resolution.

The first experimental session made use of POL 2% commercially available solution in quartz cell of 10 mm thickness, which was irradiated with the fundamental beam of the pulsed Nd:YAG laser (Excel Technology, model Surelite II, Continuum, USA) emitted at 1,064 nm. Average beam energy sent on sample was 50 mJ and the maximum irradiation time was 30 min.

In another experiment, 120 µl POL solution 0.5% in upw-ethanol was exposed in quartz cell of 1 mm optical path in successive sessions up to 30 min to 266 nm irradiation utilising the 4th harmonic (FHG) of the Nd:YAG laser fundamental beam. The average beam energy on the sample was 6.5 mJ. The experimental arrangement is schematically shown in Fig. (**9**).

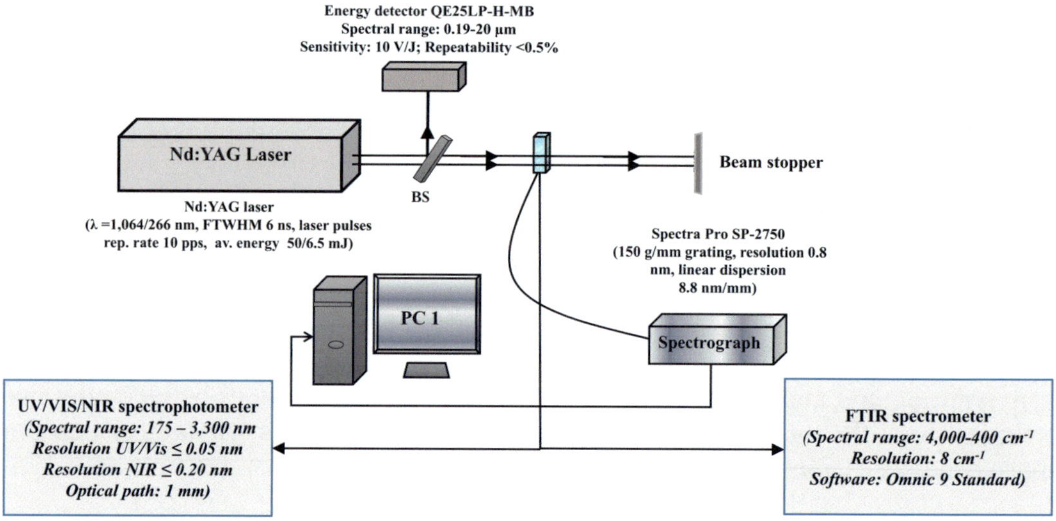


**Fig. (9).** Laser exposure experimental set-up and the used survey devices.

Laser beam was sent on sample cell after crossing an UG5 optical filter used to cut-off residual 532 nm radiation. In irradiation sequence with FHG laser beam, 10% of it was directed to a powermeter (QE25 Gentec, CND) to monitor the beam energy delivered to sample cell. Laser beam characteristics were: λ=266 nm, pulse repetition rate 10 pps and 6 ns - full time width at half maximum (FTWHM). After each exposure session, the effects on solution samples in bulk were evaluated based on their UV-Vis-NIR and FTIR absorption spectra.

Also, foam and solution samples that contain POL were investigated by measuring Raman spectra harvested with a home-made set-up which is shown in Fig. (**10**). The arrangement includes a Nd:YAG laser that excites Raman spectra by collecting Raman emission at 90° with respect to laser excitation beam direction. A spectrograph (Spectra PRO 750, Acton Research, Princeton Instruments, NJ, USA) that is coupled with an ICCD camera (PI-MAX: 1,024, Princeton Instruments, NJ, USA) is used to record and evaluate Raman signals using dedicated software. The laser radiation in this experiment is the second harmonic (SHG) of the pulsed Nd:YAG laser emitted at 532 nm, at a 250 mJ average beam energy.

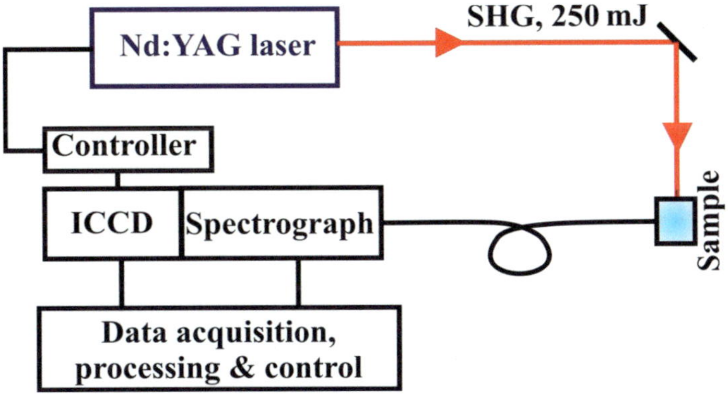


**Fig. (10).** The Raman spectra measuring system.

The dynamic surface tension (DST) was measured with a Drop Profile Tensiometer (PAT1, Sinterface, DE). The experiments were performed on pendant, simple or layered droplets. The same device was used to evaluate rheological properties of the solutions involved in emulsification process. A "surface disturbance-relaxation experiment" was made by exploring the harmonic perturbations followed by Fourier analysis of the obtained data after the surface tension value reached a *plateau* (constant value, within experimental errors). The oscillation amplitude of the droplet's volume is ±1 $mm^3$ added/subtracted to/from the initial drop volume and the utilised frequencies of oscillation were 0.005, 0.008, 0.01, 0.02, 0.04, 0.05, 0.08, 0.1, 0.16 and 0.2 Hz.

Droplet/bubble size distribution was explored with an optical microscope and with a particle sizer that approximates particle dimensions by light scattering. The utilised microscope was an Axiostar Plus Carl Zeiss Microscope system, equipped with a Cannon camera that was set to record samples' pictures. Each sample was scanned to cover the entire statistically recommended dimensions' interval. At the same time, samples were analysed by light scattering measurements made with a commercial equipment (Malvern Mastersizer) which has ±1% measuring error.

The viscosity of liquids whose foams were prepared was measured with a Physica MCR 301 Anton Paar rheometer that assesses the viscosity in function of the shear rate [15].

The immiscible solutions were mixed according to several methods such as rotor-stator (low or high speed/Ultra turrax), high pressure homogeniser (HPH) or double syringe system (DSS). These methods are presented, comparatively and in detail elsewhere [1, 3, 48]. The viscosity values of the studied immiscible liquids without any additives are: $\eta_{oil}$=49.2 mPa s for oily vitamin A solution and $\eta_{aqueous}$=1.08 mPa·s for aqueous VCM solution. As regards emulsification, the relatively high viscosity of oily phase restricts droplet breakage which leads to emulsions with larger droplets. With respect to the stability, the low viscosity of continuous aqueous phase indicates an easy and, consequently, fast phase separation. This holds in particular for mixtures with low volume percentage of the dispersed oil phase. At the same time, for emulsions composed of oily phase in proportions higher than 35%, dispersed droplets are in contact with each other forming a network of interconnected droplets inside the aqueous phase. This delays the phase separation since shearing makes some droplets slide and squeeze among other droplets. This process is amplified in the case of droplets covered by surfactants that not only resist coalescence but give interfacial rigidity to droplets, as well. For emulsions containing dispersed phase at such high volume percentage the emulsion stability is dictated by the viscosity of emulsion and not the viscosity of immiscible phases. The above comments need to be adjusted when additives that modify drastically viscosity values are used [3].

To assess the droplets size distribution, three statistical mean diameters are measured: $D_{[4,3]}$ (mean volume, also called De Brouckere mean diameter), $D_{[3,2]}$ (mean surface area, or Sauter mean diameter), and $D_{[1,0]}$ (arithmetic, or number-length mean) which are given by equation (3):

$$D_{[4,3]} = \frac{\Sigma d^4}{\Sigma d^3}; \quad D_{[3,2]} = \frac{\Sigma d^3}{\Sigma d^2}; \quad D_{[1,0]} = \frac{\Sigma d}{\Sigma n} \tag{3}$$

where $d$ is the individual droplets diameter and $n$ is the number of measured droplets [49].

The description of the three mean diameters shows that $D_{[1,0]}$ is weighted towards the droplet population with the largest number of droplets in emulsion which is usually the case of the smallest droplets. $D_{[4,3]}$ is weighted towards the droplet population which occupies the largest volume in emulsion which is often that of the largest droplets. $D_{[3,2]}$ is weighted towards the droplet population with the highest interfacial area in emulsion which is usually obtained for an effective (not actual) droplet size from the combination of droplets with different sizes and proportions in the mixture. For low volume fractions of the dispersed phase, emulsion stability is better described by $D_{[4,3]}$ since large droplets are usually more affected by buoyancy and therefore are easier freely separated. For high volume fractions of the dispersed phase, phase separation is described best by $D_{[1,0]}$ since small droplets can move between large droplets and they rise slower due to their small size. In most other cases, $D_{[3,2]}$ are the recommended indicators of emulsion stability [3].

Next, some results regarding emulsion and foam generation with biomedical applications are presented.

## RESULTS

Table **1** presents the droplet mean size corresponding to various types of emulsifying method used to mix VCM solutions in water and oily vitamin A. It becomes obvious that as the rotor speed increases, the droplets' mean diameter decreases. This effect is the result of a higher amount of energy dissipated by the homogeniser to the sample which leads to droplet disruption. Ultra Turrax use may lead to the production of generations of droplets whose sizes are in nanometer range. This holds also for HPH.

**Table 1. Statistical mean droplet diameters for different types of emulsifiers and processing intensity levels for 10% oil volume in an emulsion composed of VCM solution in water and oily vitamin A [3].**

| Method | Conditions | $D_{[4,3]}$ (µm) | $D_{[3,2]}$ (µm) | $D_{[1,0]}$ (µm) |
|---|---|---|---|---|
| Stirred tank | 300 rpm | 445 | 404 | 110 |
| | 350 rpm | 351 | 253 | 62.3 |
| | 400 rpm | 310 | 218 | 53.3 |
| | 450 rpm | 227 | 149 | 32 |
| | 600 rpm | 190 | 121 | 22.4 |
| High speed homogeniser (HSH) | 10,000 rpm | 0.423 | 0.162 | 0.070 |
| | 20,000 rpm | 0.326 | 0.135 | 0.065 |
| High pressure homogeniser (HPH) | 800 bar | 1.241 | 0.337 | 0.072 |
| HSH followed by HPH | 20,000 rpm+800 bar | 0.285 | 0.133 | 0.063 |
| DSS | 50 cycles | 25 | 13.6 | 0.969 |

In practical terms, the combination of Ultra Turrax with HPH does not reduce further the droplet size, since droplets with around 60-70 µm diameters are already produced. The emulsion produced with DSS technique is composed of droplets with dimensions smaller than those obtained with the stirring system. On the other hand, with DSS much larger droplets are produced: $D_{[4,3]}$ and $D_{[3,2]}$ are almost two orders of magnitude larger and $D_{[1,0]}$ is only one order of magnitude larger than in emulsions produced with Ultra Turrax and HPH systems. Yet, the smaller droplets obtained with DSS have about 1 µm diameter which, in general, can ensure the stability of the obtained emulsion.

Emulsions produced by HPH and DSS also show an increased stability: DSS generated emulsion containing tween 80 at 50 ppm, presents no phase separation during 2.5 h after the stop of mixing. Emulsion produced by Ultra turrax followed by HPH containing 3,200 ppm tween 80 is stable more than two weeks [1].

Droplet dimensions in emulsions dDroplet dimensions in emulsions depend on the energy used for mixing; Fig. (**11**) shows the evolution of droplet diameters function of mixing system for emulsions that contain water, vitamin A and tween 80 as a stabilising surfactant.

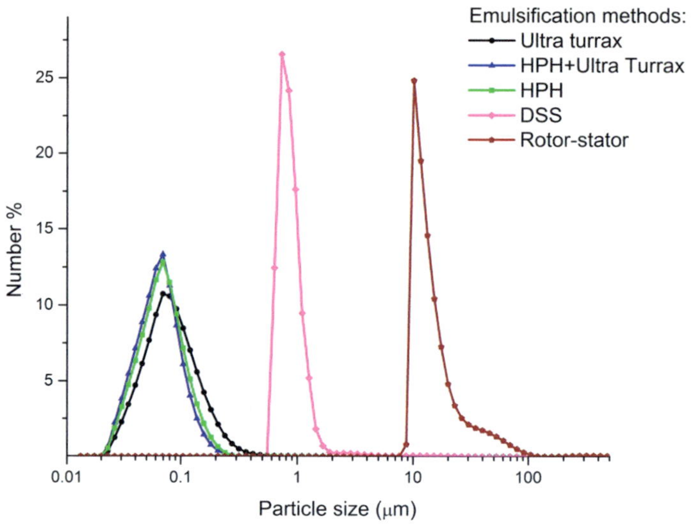


**Fig. (11).** Distributions of droplets' sizes when produced by different emulsification methods. Emulsion of water, vitamin A and tween 80.**

In Fig. (**12**) viscosity results for 10% oil emulsions with VCM in water are introduced. The emulsions with smaller droplets dimensions (stable homogenous emulsions) have lower viscosities and are less dependent on shear rate than emulsions obtained in the stirred tank (composed of larger droplets). For stirred tank method, large droplets make viscosity measurements predisposed to errors due to phase separation inside the rheometer test cell. This is suggested also by the viscosity value at 10 Hz shear rate which is close to the value of pure oily vitamin A: $\eta_{oil}$ = 49.2 mPa·s. Moreover, DSS emulsion presents a similar shear-thinning behaviour to Ultra Turrax and HPH emulsions (similar slope of viscosity *versus* shear rate). All in all, a good stability against phase separation for the cases of DSS emulsions is expected.

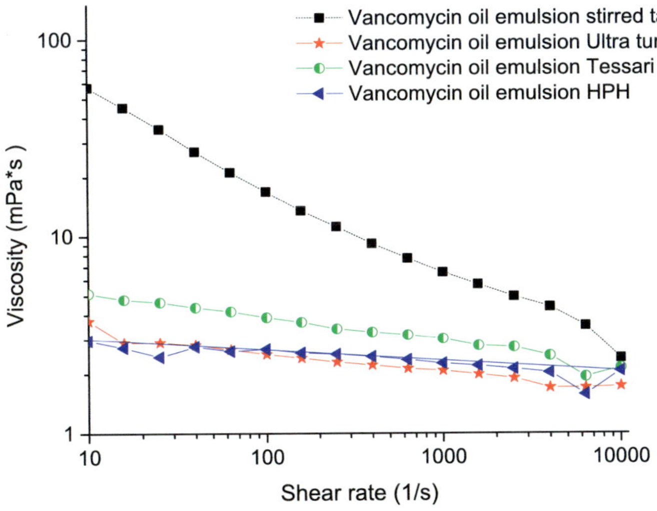


**Fig. (12).** Viscosity measurements on emulsions of VCM with oily vitamin A, function of homogenising method.**

For improving the stability of emulsions produced by DSS, one can add various quantities of certain additives that can act as stabilisers. Table **2** presents the mean droplets' size when different substances are added to a liquid/liquid mixture.

**Table 2. Statistical mean droplet diameters for DSS emulsion (VCM in water and vitamin A in oil) after addition of certain emulsion stabilisers - 10% oil in emulsion [3].**

| **Stabiliser (% of total volume)** | **$D_{[4,3]}$ (μm)** | **$D_{[3,2]}$ (μm)** | **$D_{[1,0]}$ (μm)** |
|---|---|---|---|
| DSS emulsion with no additives | 25 | 13.6 | 0.969 |
| $C_{30}H_{62}O_{10}$ 0.5% | 19.0 | 10.1 | 1.00 |
| Xanthan gum 0.03% | 27.6 | 15.2 | 0.97 |
| Xanthan gum 0.15% | 25.5 | 11.8 | 0.99 |
| Tween 80 3% | 25.0 | 13.6 | 0.97 |
| $C_{30}H_{62}O_{10}$ 0.5%+tween 80 0.1% | 25.2 | 13.4 | 1.00 |
| $C_{30}H_{62}O_{10}$ 0.5%+tween 80 1% | 25.9 | 9.17 | 1.14 |
| $C_{30}H_{62}O_{10}$ 0.5%+tween 80 10% | 29.1 | 16.8 | 1.11 |
| $C_{30}H_{62}O_{10}$ 0.5%+glycerin 0.1% | 21.7 | 11.6 | 0.98 |
| $C_{30}H_{62}O_{10}$ 0.5%+glycerin 1% | 24.7 | 18.8 | 1.07 |
| $C_{30}H_{62}O_{10}$ 0.5%+glycerin 10% | 20.8 | 11.6 | 1.01 |
| $C_{30}H_{62}O_{10}$ 0.5%+glycerin 0.1%+xanthan gum 0.1% | 26.8 | 13.9 | 0.98 |
| $C_{30}H_{62}O_{10}$ 0.5%+xanthan gum 0.02% | 21.2 | 10.4 | 1.01 |

*(Table 2) contd.....*

| Stabiliser (% of total volume) | $D_{[4,3]}$ (μm) | $D_{[3,2]}$ (μm) | $D_{[1,0]}$ (μm) |
|---|---|---|---|
| $C_{30}H_{62}O_{10}$ 0.5%+xanthan gum 0.07% | 23.7 | 11.0 | 1.00 |
| $C_{30}H_{62}O_{10}$ 0.5%+xanthan gum 0.3% | 10.0 | 6.2 | 1.02 |

The addition of stabilisers in the specified concentration range has, generally, no strong influence on droplets dimensions. Some additives have even minor effects on statistical mean diameters with no clear trend whatsoever.

For a VCM solution exposed to laser radiation, a structural modification can be observed. These modifications may lead to generation of new photoproducts which coexist with the effect of generation of foams. Out of the experimental conditions, the parameter that has the most important effect on foam generation is the beam energy.

The minimum value of laser beam energy for which foam was obtained is 40 mJ for a 0.5 ml and 7 mJ for 5 μl. The experimental conditions for foam generation in a VCM solution at $2\times10^{-3}$ M and a volume of 0.5 ml were: irradiation wavelengths 266 nm and 532 nm, average beam intensity 40 mJ, optical path 5 mm. VCM is not surface active substance and surface tension measurements performed on VCM solutions in water show ST values close to water ST [50].

The VCM exposure in bulk for one hour produces a colour change of the solution from colourless (the unirradiated sample) to yellow with appearance of foam at the liquid surface. After 2 h exposure, the colour of the solution becomes orange with an increased quantity of foam at the liquid surface. The obtained foam is stable for several minutes after laser exposure is stopped.

VCM solutions were also exposed as one single droplet with V = 5 μl, at 266 nm and 532 nm (unseparated fourth and second harmonic of Nd:YAG laser beam, respectively) and at beam energy of 7 mJ. The beam waist was tailored to cover the entire surface (cross section) of the droplet.

In Fig. (**13**), it can be seen that while exposure time is increased, VCM solution droplet gradually turns into foam. In approximately 40 min, the 5 μl droplet is converted into foam. Foam generation depends on both beam energy and laser pulse duration. The mechanism involved in foam generation is laser ablation in liquids. Studies have shown that short pulses lead to photomechanic ablation, together with cavitation beneath the surface due to a transient thermoelastic pressure, while for pulses longer than the acoustic relaxation time, energy accumulation leads to an increase of local temperature that produces smaller droplets that are expelled from the volume of solution [51].

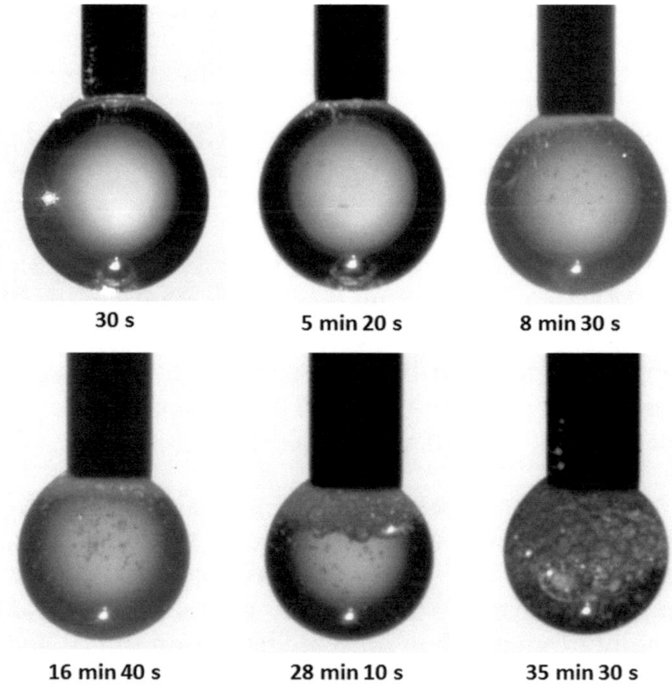


**Fig. (13).** Foam generation in VCM droplets in water during laser irradiation.

Bubbles are formed due to acoustic wave which accompanies the laser pulse that produces nucleation into the liquid. The temperature of the solution was also monitored and, although there is a small increase of it, this is not enough to explain bubble generation, a much stronger effect being the one of the acoustic wave [52]. This process is called "cold ablation" since it is not obtained due to thermal expansion but to "tensile field" generated by pressure containment [53].

In the following, are presented measurements of laser induced fluorescence (LIF) in pendant droplets of emulsion which contains Rh6G in distilled water and oily vitamin A in comparison with pendant droplets that contain only Rh6G water solution. The laser dye Rh6G was chosen due to its well-known properties: absorption spectra, high quantum yield and fluorescence lifetime. Emulsion was prepared by Tessari method [46, 54] and the solvent was water since its evaporation rate is lower than for other Rh6G possible solvents (*e.g.* ethyl alcohol).

The use of Rh6G in an emulsion is of interest due to the possibilities it opens along several directions. One, is the introduction in the vein of a mixture (containing medicine, laser dye at very low concentration and a surfactant) and the local application of the laser beam, to excite the emulsion and improve the drug treatment action. Another direction could be the enhancement of optical signals (LIF, Raman) emitted by R6G content if solution is mixed in an emulsion.

The emission spectra are recorded in both cases for several concentrations of Rh6G, variable energies of the pumping laser beam and different time intervals for irradiation. A 7 μl droplet volume was used for LIF measurements.

For solutions, five different concentrations of laser dye ($10^{-5}$ M, $5\times10^{-5}$ M, $10^{-4}$ M, $5\times10^{-4}$ M and $10^{-3}$ M) were prepared by dissolving Rh6G in distilled water. The Rh6G was selected for this study due to its high quantum yield in water (>90%) and for its acceptable biocompatibility. As for oily phase used for emulsion preparation, it was chosed vitamin A in oil. The surfactant tween 80 was added at a concentration below CMC, in order to increase the emulsions stability.

The absorption spectra were measured for 2 ml solution of Rh6G in water at different concentrations. Rh6G, tween 80, and oily vitamin A absorption spectra are presented in Fig. (**14**). The presence of tween 80 and oily vitamin A in emulsion's composition does not influence LIF characteristics and measurements from the optical point of view. They present absorption in UV and no absorption in the visible spectral range, where LIF pumping wavelength is used to excite dye molecules. Moreover, when the emulsion is exposed to 532 nm, molecular structures of tween 80 and oily vitamin A are not modified. Also, Rh6G is not soluble in oil and therefore Rh6G molecules do not migrate at oil-water interface. Each emulsion component remains in its own solvent, oil and water respectively, because they are immiscible, even when emulsion is exposed to laser irradiation.

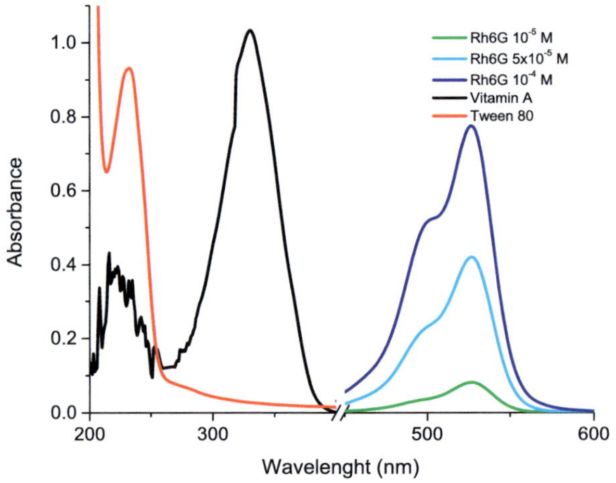


**Fig. (14).** Absorption spectra of Rh6G solution in water at different concentrations and of tween 80 in water and vitamin A in oil.

For the case of dye, if concentration is modified from $10^{-5}$ M to $10^{-4}$ M, as Lambert-Beer law predicts [55], absorption is exponentially increased. At the lowest concentration of laser dye, $10^{-5}$ M, one obtains the specific absorption band, with the peak corresponding to free monomers absorption. By increasing the concentration, a short-wavelength shoulder arises with respect to the first peak as previously reported [55, 56]. It is known that in particular conditions, for example by increasing concentration, two or more dye molecules could interact and form aggregates.

Function of their structure, the aggregates are divided into two categories: those which present head-to-tail structures, named J-type dimer aggregates and those with face-to-face stacked H-type dimer aggregates. These newly formed species could enhance or diminish the fluorescence emission of the dye. Usually, the difference between these types of aggregates can be made by observation of the spectral shift of the second maximum with respect to monomer absorption peak. In this case, H-type aggregates are formed and they imply a diminishing of fluorescence and a bathochromic shift of the absorption band.

Along these lines, LIF measured on pendant droplets of solution or emulsion containing Rh6G, for a pumping energy of 2.2 mJ is presented. LIF spectra are averaged on 10 pulses (1 s), for both samples. As can be observed in Figs. (**15-17**), fluorescence emission of droplets that contain emulsion shows the typical fluorescence broadband of Rh6G. In the case of droplets that contain solution, LIF emission spectra show two peaks: one specific to fluorescence emission as in the case of emulsion and the other obtained at longer wavelength showing an amplification of emitted radiation.

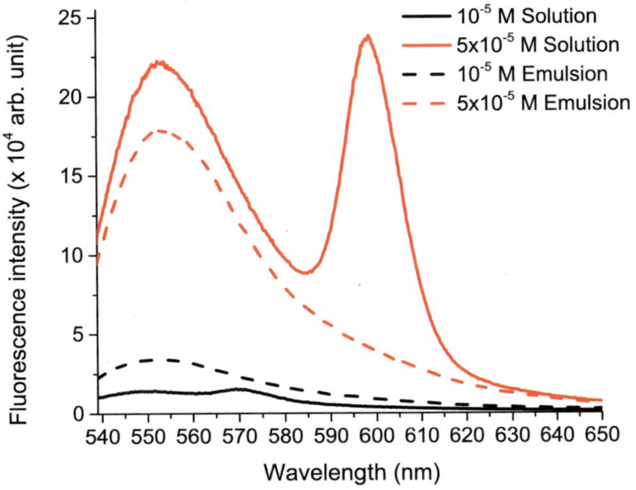


**Fig. (15).** LIF emitted by pendant droplets containing solution of Rh6G in water or emulsion with Rh6G in water at $10^{-5}$ M and $5\times10^{-5}$ M concentrations with oily vitamin A.

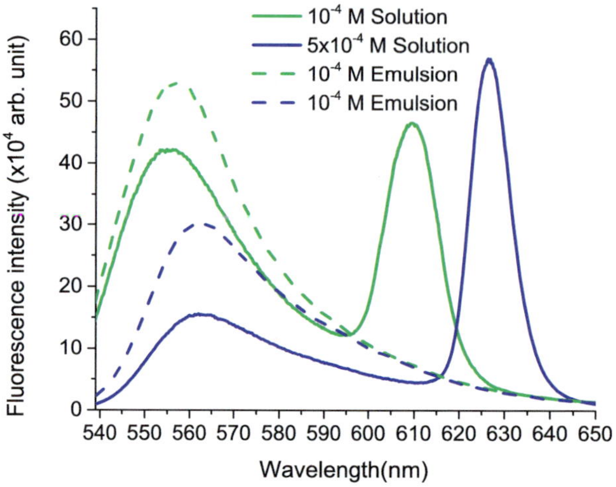


**Fig. (16).** Laser induced fluorescence emitted by pendant droplets containing solution or emulsions (prepared as in Fig. **15**) of Rh6G at $10^{-4}$ M and $5\times10^{-4}$ M concentrations.

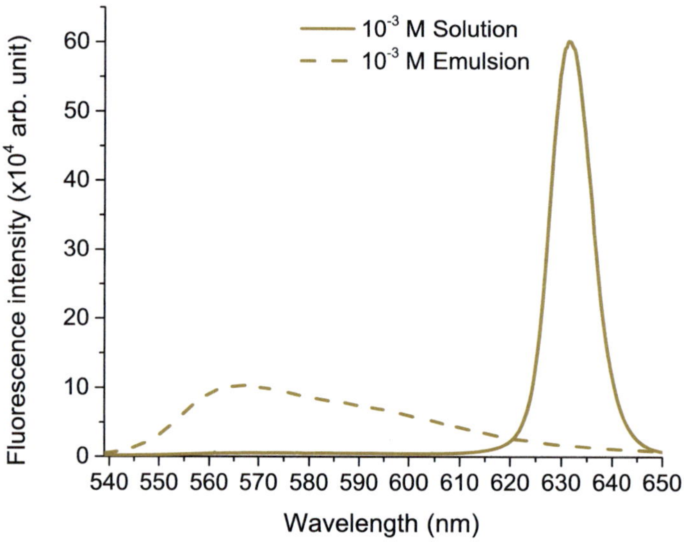


**Fig. (17).** LIF emitted by pendant droplets containing Rh6G solution in water and emulsion (prepared as in Fig. **15**) of Rh6G at $10^{-3}$ M concentration.

In the case of solution as well as emulsion, by increasing Rh6G concentration from $10^{-5}$ M to $10^{-4}$ M (Figs. **15** and **16**) the fluorescence emission is increased. When concentration exceeds $10^{-4}$ M, the fluorescence signal decreases. This phenomenon is known to be due to reabsorption of emitted signal by fluorophores in a large number [54]. Xanthene dyes class, which includes rhodamines, has a pronounced reabsorption effect due to the broad overlap of absorption and emission spectra. As a result, the quantum yield obtained for a diluted solution of Rh6G is higher than in the latter case (*i.e.* at higher dye concentrations). Another indicative of reabsorption is the emission maximum that is red shifted with increasing concentration of the dye. Furthermore, absorption spectra presented in Fig. (**14**) proved the formation of H-aggregates [55] when concentration is increased. It is known that when H-aggregates are excited, their relaxation occurs with a suppressed signal due to a low fluorescence quantum yield. The presence of H-aggregates at higher concentrations explains the strong Stokes shift of fluorescence emission for both types of droplets.

A pendant droplet containing emulsion of Rh6G solution and oily vitamin A exposed to a laser beam presents only the broadband fluorescence if compared to a pendant droplet containing solution of Rh6G in water. The difference could be explained by the fact that the pumping beam energy of 2.2 mJ is too low in the former case to produce an amplification of emitted radiation because light scattering in emulsion is significantly higher than in simple solution. When pumping beam energy is increased to 10 mJ, the second peak of emission spectra appears in the case of emulsion as well. The changes of emission spectra when the pumping energy is increased are illustrated in Fig. (**18**) for emulsion and in Fig. (**19**) for solution of Rh6G.

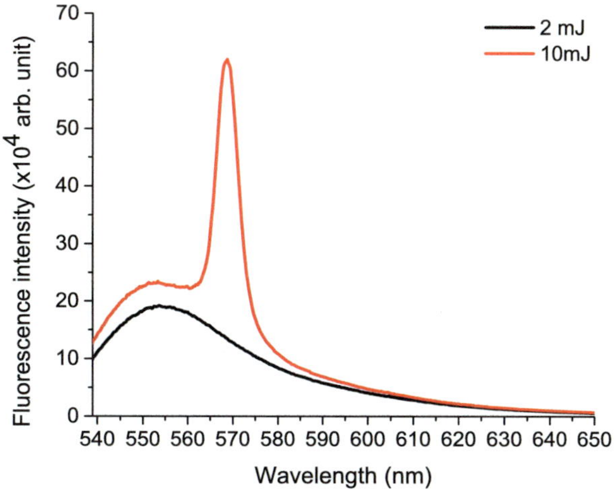


**Fig. (18).** LIF emitted by pendant droplets containing emulsion of Rh6G at different pumping energies.

The amplified peak is obtained at 570 nm for emulsion and at 580 nm for solution. The blue shift of amplified peak obtained in emulsion compared to solution is explained by the modification of the polarity of the interfacial medium.

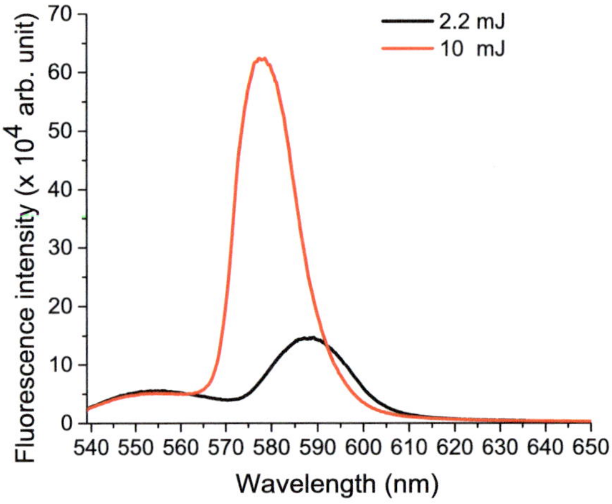


**Fig. (19).** LIF emitted by pendant droplet containing Rh6G aqueous solution with laser dye at different pumping energies. Droplet volume 7µl.

In Figs. (**20** and **21**), LIF spectra measured for irradiation time up to 1 min are presented for emulsion and respectively solution in order to estimate LIF evolution in time. For this study Rh6G concentration of $5\times10^{-5}$ M and pumping energy of 10 mJ were chosen. The droplet volume was kept 7 µl.

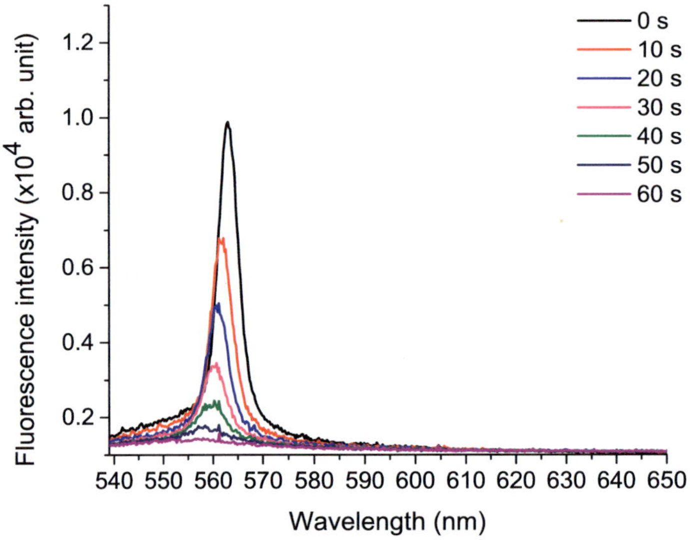


**Fig. (20).** Time evolution of emission spectra obtained for pendant droplets (7 µl) containing emulsion of Rh6G solution in water ($5\times10^{-5}$ M) and oily vitamin A. Droplets were exposed 60 s (600 pumping pulses) at 10 mJ pumping beam energy. Each spectrum is an average of fluorescence emission measured for 10 pulses.

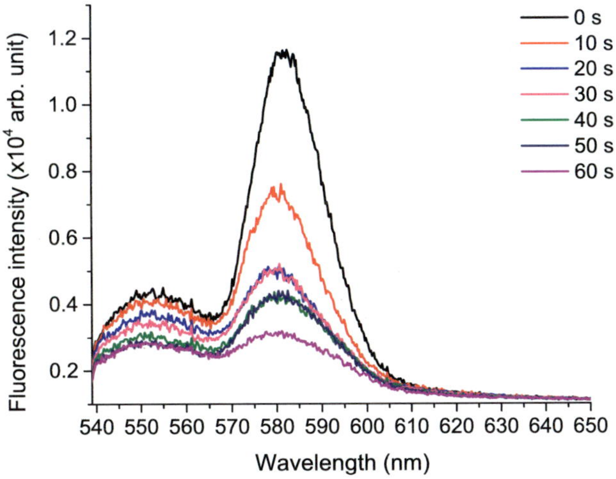


**Fig. (21).** Time evolution of emission spectra obtained for pendant droplets (7 μl) containing solution of Rh6G in distilled water ($5\times10^{-5}$ M) and oily vitamin A. The droplets were exposed 60 s (600 pumping pulses) at 10 mJ pumping beam energy. Each spectrum is an average of fluorescence emission acquired for 10 pulses.

The acquisition time is equivalent to the exposure time of droplet to the pumping beam. The decrease of LIF intensity in time is due to a longer exposure to radiation which leads to the destruction of fluorophores by photo bleaching processes [57].

Looking at emission spectra and their evolution in time it can be observed that for emulsion, the amplified peak is 5 times narrower than in solution. Also, the intensity of emitted radiation is more intense in emulsion than in solution, even if in former case the number of dye molecules is smaller due to the ratio oil/solution that is 10/90. The amplification of LIF signal in micro cavities is treated from theoretical point of view in [58].

Besides the geometry of the system, a modification of environment of fluorophores could lead to enhancement or quenching of fluorescence signal. In the case of emulsion, the viscosity is increased due to the presence of oily vitamin A in aqueous Rh6G solution. Thereby, molecules of Rh6G have an inhibited motion inside the droplet, resulting in decrease of internal conversion of absorbed radiation and consequently in higher fluorescence quantum yield. This behaviour is similar to results reported in the literature for alcohol solvents [59, 60].

In emulsion, the existence of reflecting surfaces of oil micro droplets, has an important effect in fluorescence enhancement, similar to optical antenna phenomena [61, 62]. The reflecting surfaces induce modification of local electromagnetic (EM) field intensity and produce a large number of Rh6G excited molecules between oil microspheres.

The ratio $I/\Delta\lambda$ (where $I$ is fluorescence peak intensity and $\Delta\lambda$ is the full width at the half maximum of fluorescence band-FWHM) is presented function of acquisition time for emulsion and for solution in Fig. (**22**). The ratio $I/\Delta\lambda$ is in average 5 times greater in emulsion than in solution. The pendant droplets emission decreases exponentially in the first 20 s for both cases; afterwards the decreasing is lower in solution than in emulsion. This shows the important role that solvent plays in fluorescence quenching due to its viscosity and its polarity.

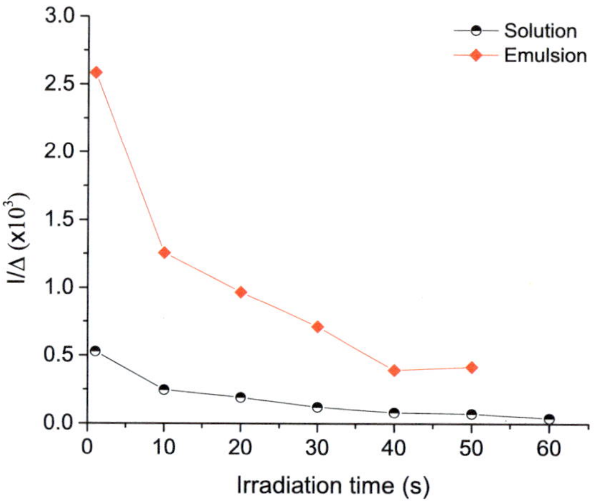


**Fig. (22).** Amplified peak intensity *versus* amplified peak bandwidth ratio function of data acquisition time for pendant droplets of solution of Rh6G in water and emulsion containing Rh6G and oily vitamin A. The droplets were exposed 60 s (600 pumping pulses) at the pumping energy of 10 mJ.

A narrow peak is revealed in fluorescence spectra of pendant droplets corresponding to signal amplification. To obtain this amplification in emulsion, the pumping energy has to be increased, compared to solution. The peak value is higher in emulsion, even if the quantity of fluorophores is lower than in solution. Moreover, the amplification peak is narrower in emulsion and it is blue shifted [63, 64] with respect to the peak obtained in solution. Even if the same laser dye is used, Rh6G, emission properties are different, this effect being due to the modification of viscosity and to the presence of reflecting oil surfaces in emulsion.

Foam stability is of great importance for applications. It may be a wished or an unwished effect and, depending of applications, there are ways to increase or to decrease the stability of a foam. Next, some results regarding the stability of a foam with biomedical applications are presented. Table **3** shows the bubble mean size for foams containing different concentrations of POL and atmospheric air. By increasing the concentration of the sclerosing agent, a small variation on bubble size is observed.

**Table 3. Mean bubble size D $_{[1,0]}$ for POL concentrations used in this study.**

| Concentration of POL | $D_{[1,0]}$ (µm) |
|---|---|
| 0.5%~(8×10$^{-3}$ M) | 47.1 |
| 1%~(1.6×10$^{-2}$ M) | 54.9 |
| 3%~(4.8×10$^{-2}$ M) | 64.1 |

In Fig. (**23**), the evolution of bubbles average dimensions and foam half time function of concentration of surfactant are shown; following addition of 10% tween 80 (10,000 ppm), it is obtained a 2.5 times increase of POL foam stability.

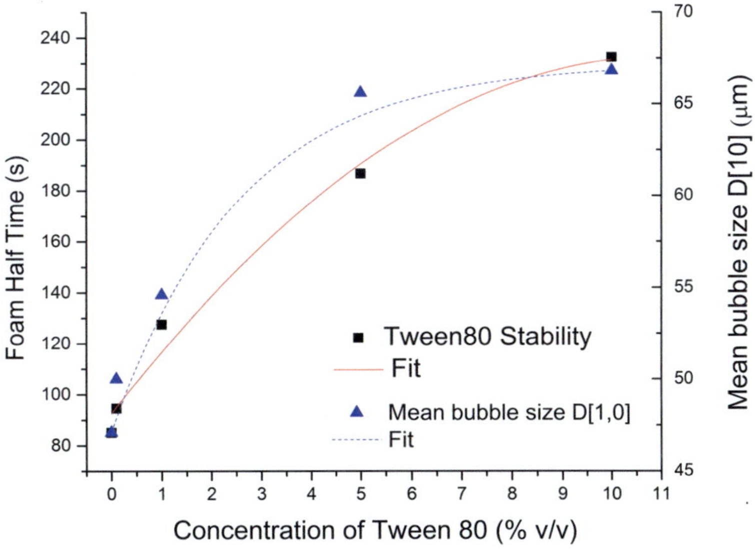


**Fig. (23).** Effect of the concentration of tween 80 on POL 0.5% foam stability.*

Fig. (**24**) shows the effects of adding glycerin to an air/water mixture that contains POL 0.5%. The increase of glycerin concentration produces a linear increase of foam stability mainly due to its contribution to the viscosity of the mixture. Glycerin is a viscous fluid and its presence increases significantly the viscosity of liquid mixture delaying the drainage through bubble interstices. This holds, although the increased viscosity leads to an increased mean bubble size. One should consider that high viscosity inhibits bubble breakage into smaller size bubbles during pumping by syringes. Presence of glycerin also makes films between bubbles more rigid, leading to increased foam stability as well [15].

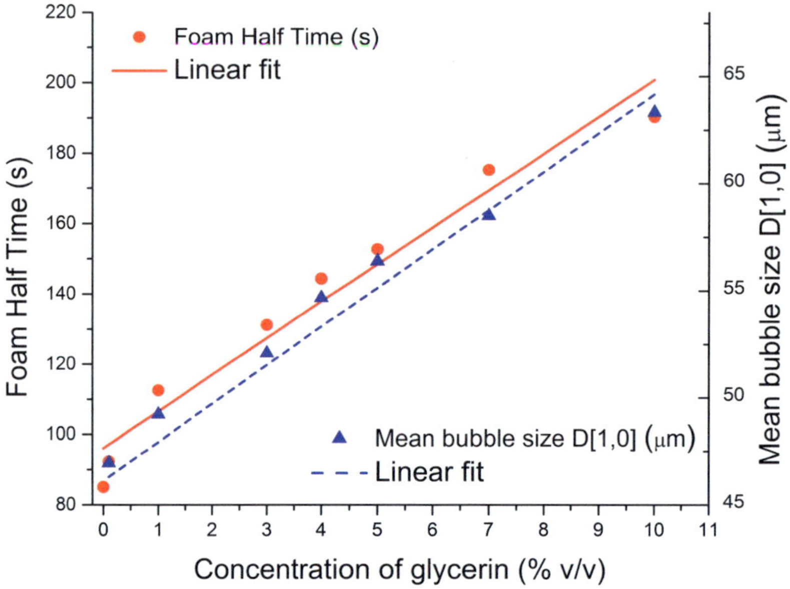


**Fig. (24).** Foam half time and mean bubble size distribution after adding glycerin at several concentrations.*

The foam half time for tween 80 and glycerin at concentrations of up to 10% v/v is approximately 200 s. However, increasing the concentrations of glycerin, the stability of foam produced by POL/glycerin mixture decreases (at 50% glycerine, the foam is destabilised almost instantly). At the same time, for tween 80, at 20% concentration, FHT is 481.2 s, which is a much higher value than the one obtained for the same concentration of glycerin which is 80.86 s (smaller that obtained for 10% v/v). This is opposite from what expects based on the different rheological properties of the two substances (glycerin is more viscous than tween 80) [15].

The effects of adding a non-Newtonian liquid in the mixture (xanthan gum) are presented in Fig. (**25**). Both FHT and mean bubble size are increased with addition of various quantities of xanthan gum. However, comparing to results obtained on tween 80 and glycerin, the foam FHT for a 0.42% w/v xanthan gum is longer than one hour and 47 times higher than the stability of POL foam without any added substances. A 0.1% (1000 ppm) of xanthan gum added to an air/POL mixture produces an increase in FHT of 6.5 times (~550 s).

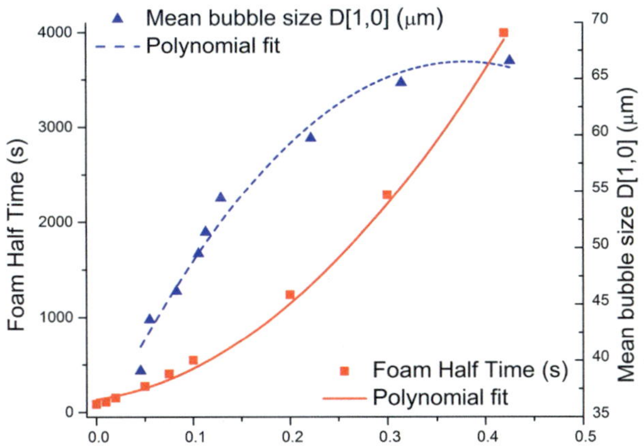


**Fig. (25).** Evolution of foam half time and of mean bubble size distribution with addition of xanthan gum.*

At the same time, the mean bubble size variation is small, the values obtained being close to those of POL 0.5% without any added substance. Measurements evidenced that less than 5% of bubbles have diameters larger than 200 μm, most of them having dimensions around 50 μm; the average bubble size is 66.6 μm.

Measurements made on mixtures of xanthan gum (0.08% w/v) and glycerin or tween 80 (3% v/v) showed no variation in bubble mean size distribution with an increased stability, if compared to results obtained on mixtures without xanthan gum (*e.g.* FHT: POL 0.5%+glycerin 3%+xanthan gum 0.08%=459.3 s; bubble mean size 42.8 μm and FHT: POL 0.5%+tween 80 3%+xanthan gum 0.08%=487.3 s; bubble mean size 37.6 μm).

The physicochemical stability of gas/liquid films for the examined substances is evaluated by DST measurements performed on pendant droplets. The interfacial dilational viscoelasticity of gas/liquid films is assessed by Fourier transform method to evaluate the effect of induced harmonic perturbations on surface tension. All the performed measurements used a single concentration for each added substance to POL 0.5% (*e.g.* 0.1% xanthan gum, 3% tween 80 and 3% glycerin). The concentrations were selected function of their influence on foam stability; a low value of viscosity was necessary to avoid the increase of bubble diameters.

In Table **4**, are introduced ST values at equilibrium (a plateau is reached after 4,000 s) for several POL concentrations and POL 0.5% mixtures with the selected compounds. The ST values of mixtures at equilibrium are very close to POL ST figures. The data also show no differences between glucerin, tween 80 and xanthan gum, ST values at equilibrium being quite close. A difference may be noticed in the data obtained by Fourier analysis of the harmonic perturbations induced in hanging droplet. Visco-elastic modulus (|E|) and phase lag (Φ) for all used substances and mixtures are computed from the $E_{Im}$ and $E_{Re}$ obtained through Fourier analysis of the rheological data produced by the induced harmonic perturbations [65, 66].

**Table 4. Surface tension values at equilibrium, for POL mixed with glycerin, tween 80 or xanthan gum at different concentrations.**

| Solution | ST (mN/m) |
|---|---|
| POL 0.5% | 30.7±0.3 |
| POL 1% | 30.5±0.4 |
| POL 3% | 30.1±0.3 |
| Glycerin 3% | 71.4±0.3 |
| Glycerin 3%+POL 0.5% | 31.7±0.4 |
| Tween 80 3% | 34.9±0.4 |
| Tween 80 3%+POL 0.5% | 32.8±0.2 |
| Xanthan gum 0.1% | 71.9±0.3 |
| Xanthan gum 0.1%+POL 0.5% | 31.4±0.5 |

$$|E| = \sqrt{E_{re}{}^2 + E_{im}{}^2} = \frac{\Delta\sigma_m}{\Delta A_m/A_0} = ratio\ of\ amplitudes \quad \textbf{(4)}$$

$$E_{re} = |E| \times cos\phi, E_{im} = |E| \times sin\phi \quad \textbf{(5)}$$

For the studied substances, the values of $E_{re}$ are bigger than $E_{im}$ which is an indicator that the gas/liquid interfaces composed by the selected substances show a predominantly rather viscous than elastic character. This conclusion is confirmed by the visco-elastic phase lag results (for pure elastic behaviour Φ should be zero) [15]. Xanthan gum and its mixture with POL present the lowest viscoelasticity figures, whereas tween 80 and its mixture show the highest viscoelasticity values. For both substances and for low frequencies, for xanthan gum in particular, the viscoelasticity values of the pure substances are smaller than those of mixtures. Following the mixture of selected substances with POL, |E| value decreases significantly. This is an effect which is not observed on tween 80. One explanation is that POL and tween 80 have a similar value of |E| and addition of tween 80 does not modify the visco-elastic behaviour of the mixture.

The modification of |E| values could be one of the reasons for the increased stability of POL/xanthan gum foam. A similar stability is not obtained for glycerine, because of its increased viscosity and of its independence with respect to the shear rate that makes mixing procedures with the used techniques more difficult. For xanthan gum/POL and tween 80/POL mixtures visco-elastic phase values are smaller than for pure substances for all used oscillation frequencies. In the case of glycerin/POL mixture, this occurs for oscillation frequencies larger than 0.1 Hz.

The absorption and scattering processes jointly govern the propagation of light in turbid biological media. The study of optical absorption and of light scattering properties of marketed sclerosing agents becomes necessary to clarify the mechanisms involved in combined sclerotherapy and light therapy used for varicose veins treatment. The absorption spectra of commercially available POL show no significant absorption in UV-Vis and exhibit very weak peaks in NIR, as shown in Fig. (**26**). They are the result of superposed absorption of all the compounds contained in the pharmaceutically available POL 2% [67].

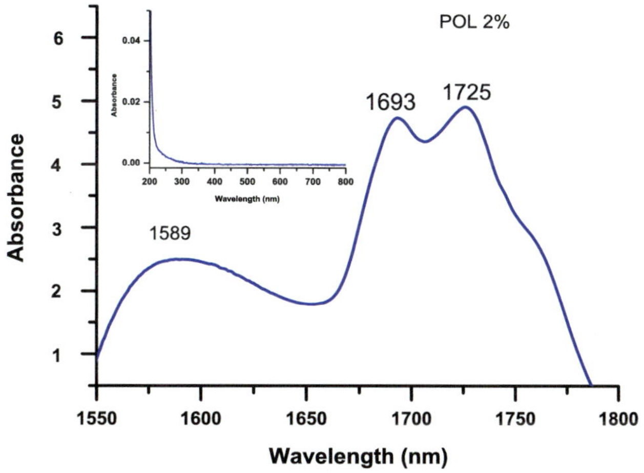


**Fig. (26).** UV-Vis-NIR absorption spectrum of pharmaceutical available POL 2%.

The optical characterisation of commercially available POL, based on the absorption spectrum registered in UV-Vis-NIR spectral ranges, highlights absorption peaks in NIR, as results from Fig. (**27**) [47]. The investigated solution totally transmits in UV-Vis. These spectral features are due to the superposed absorption of all the compounds included in solution. As for solutions of POL 0.5% prepared in a mixture of ultra pure water and ethanol, the spectral behaviour is shown in Fig. (**28**). It emphasises also absorption peaks centered in NIR spectral range.

Many reports show that under controlled exposure to laser beams, one may induce structural molecular changes of the parent compounds and one can generate new photoproducts that could confer to the irradiated mixtures enhanced or new pharmaceutical properties [68 - 70].

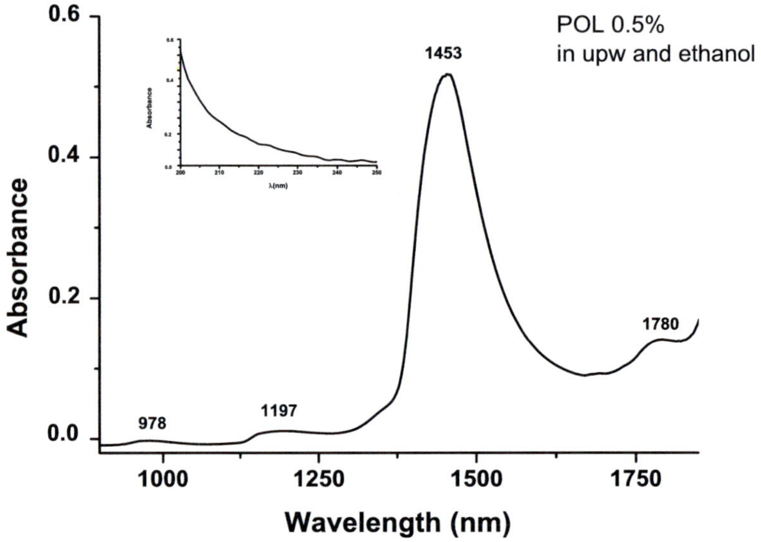


**Fig. (27).** The absorption spectrum of POL solution prepared in a mixture of water/ethanol, in NIR spectral range.

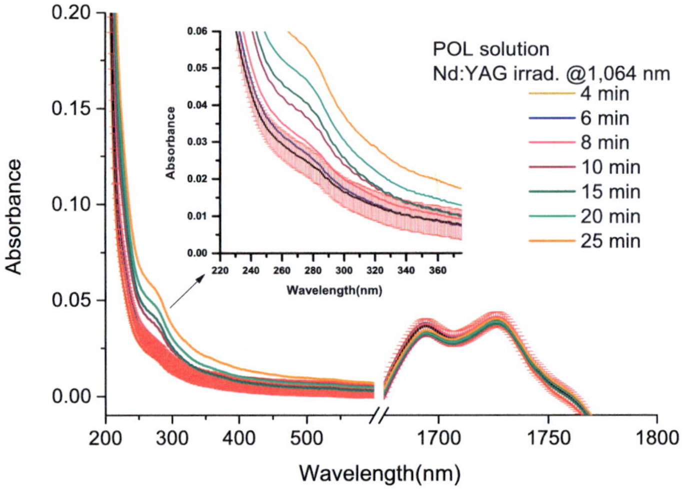


**Fig. (28).** Absorption spectra of pharmaceutically available POL 2%, exposed to 1,064 nm laser radiation.

After exposure of POL solutions at pulsed 1,064 nm radiation with specific irradiation doses, absorption spectra (Fig. **28**) at wavelengths higher than 500 nm remain unmodified within the measuring errors limits. However, in the spectral range 250-285 nm changes appear that, as a trend, are larger than error margins. For exposure time at 4, 6 or 8 min, absorption curves remain within the error limits of the measuring system. This behaviour could be an effect of a nonlinear process, namely of the absorption of 4 photons at 1,064 nm, which corresponds to a single photon absorption at 266 nm [67].

Following these observations, POL 0.5% solutions were exposed to FHG of Nd:YAG laser, as previously described. After each exposure session, FTIR spectra were recorded in view of evidencing structural modifications.

Furthermore, POL microfoam was produced from solution previously exposed to laser radiation. Its optical characteristics were also investigated by measuring FTIR spectra. The reference FTIR spectrum of unirradiated POL solution contains characteristic functional groups. The C-H out-of-plane bending vibrations occur when POL is foamed. Also, C-O-H bending vibrations are affected in this case (Fig. **29**).

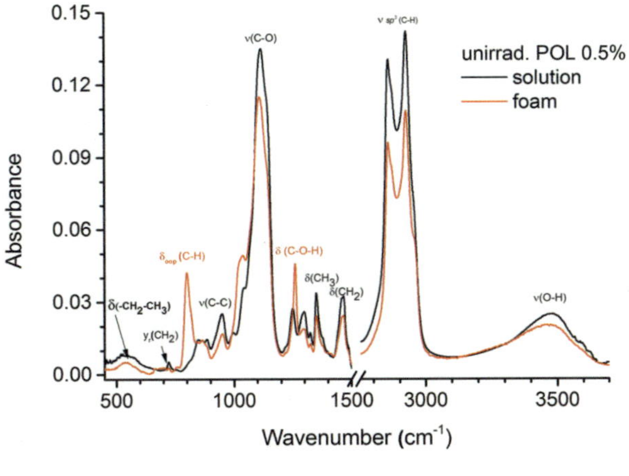


**Fig. (29).** FTIR spectra of unirradiated POL.

The FTIR spectrum of POL microfoam prepared after 10 min of laser beam exposure at 266 nm exhibits almost the same modifications as unirradiated samples, bringing up the long chain bending; C-O bond shows the changes of $sp^3$ C-H stretching vibrations (Fig. **30**). No significant differences are recorded between FTIR data of unirradiated and irradiated POL solution, or the corresponding foam [47].

The experimental measurements suggest that the use of foamed POL leads to the increase of the optical path of laser radiation in foam sample by light scattering. This causes an increase of total absorption which, according to Lambert-Beer's law, is proportional with the product between absorption coefficient, optical path length in the sample and absorbent concentration. The absorption of beam's energy in the foam can be boosted by the multiplication of the impacts of the photons by collisions with gas bubbles. Moreover, in these conditions even the number of changed POL molecules could increase.

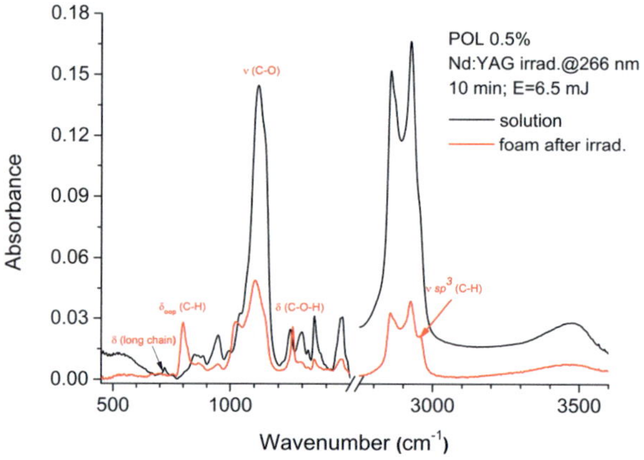


**Fig. (30).** FTIR spectra of POL 0.5% solution exposed 10 min to pulsed laser beam at 266 nm.

In an attempt to characterise the increase of efficacy of the laser beam if sclerosing drug is used as foam, solution and foam samples of POL were investigated using Raman measurements. The intensities of Raman signals were higher for foam than for simple solution samples (Fig. **31**).

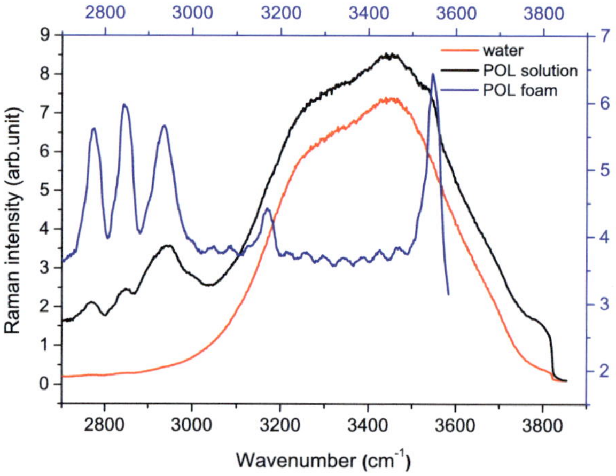


**Fig. (31).** Raman spectra of POL in solution and foam presentations.

This laser light scattering enhancement is due to a longer optical path of the beam in the foam sample. More, Raman vibrational lines obtained from the foam sample are more structured and stronger than those of the liquid specimens.

The broad band of Raman POL solution spectrum in the 3,500-3,200 $cm^{-1}$ interval is due to the OH stretching produced in water solutions and in alcohols.

As for Raman spectra of foam sample, signals below 3,000 $cm^{-1}$, particularly those centred on 2,850 $cm^{-1}$ and 2,935 $cm^{-1}$ are specific to asymmetric/symmetric C-H stretching of long linear aliphatic chain. This is the case of the chemical structure of the active compound in commercially available drug.

The characteristic band profiles in the 3,600-3,200 $cm^{-1}$ range indicate hydroxyl groups that linked with weak Raman signals between 1,650 $cm^{-1}$ and 1,450 $cm^{-1}$ are typical for OH stretch signature.

Raman signals were also measured at different time moments after preparation of foam samples (Fig. **32**). Raman spectrum changes relatively fast and after 5 min from foam production its lines are 10 times less intense than the initial values measured immediately after preparation.

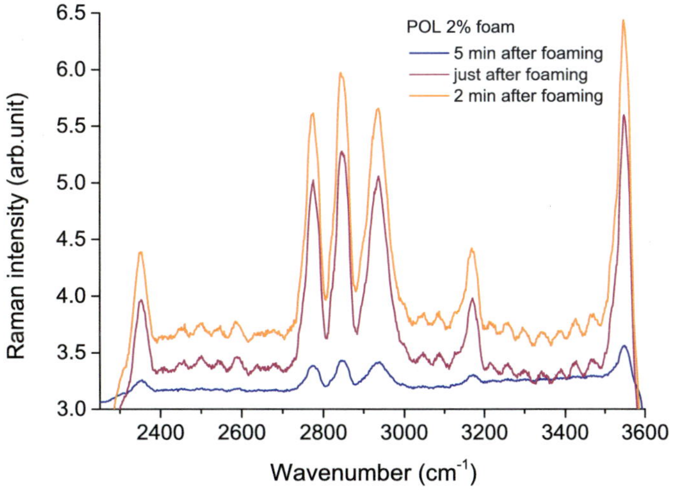


**Fig. (32).** The influence of foam separation in liquid and remaining foam on Raman spectra of POL sample.

The results indicate that for foam application some parameters must be taken into account related mainly to foam cohesion, such as bubble dimensions. It becomes important to chose the time moment of exposure of the varicose veins injected with foamed drug to laser radiation [71].

The time moment should be chosen with respect to the time when preparation of foam was finalised.

## CONCLUSIONS

Various methods of emulsification and the effect of surface active agents and/or rheology modifier substances were studied. The average droplet dimensions are proportional with the energy dissipated by the selected method while the stability of obtained emulsion is influenced by the used emulsifier. Moderately stable emulsions may be obtained by using a double syringe method with characteristics that recommend them for medical applications.

With the variation of the dye concentration in liquid droplets (*e.g.* Rh6G dissolved in water), a fluorescence signal accompanied by a lasing peak can be obtained. At the same dye concentration in emulsions, fluorescence signal may be enhanced especially due to the reflecting surfaces of oil microdroplets within emulsion. These results have applications in optical imaging as biocompatible fluorescent markers. A lasing effect is also observed in emulsion when the pumping energy is increased. In both systems, the maximum of lasing signal depends on droplet content and this can lead to generation of tunable microlasers that could be used in medical and/or technological applications.

The generation of foams starting from VCM solution by irradiation with a laser beam is possible and accompanies in specific conditions modifications induced to the VCM molecule and generation of new surface active photoproducts. Following these modifications, an antibacterial activity of the new photoproducts is expected. The exposure of VCM to laser radiation accompanied by foam generation has the potential to enhance its biological activity.

The stability of foams containing agents used in sclerotherapy (such as polidocanol) is of high importance since the efficacy of the treatment depends on it. Addition of surfactants or subtances that in certain conditions influence the rheological properties of the mixture represents a method to finely tune foam stability and prepare it for a particular, intended application.

Clinical results show that foam sclerotherapy used in combination with photothermolysis increases the efficiency of varicose veins treatment [21, 34]. These practical observations are sustained by experimental measurements, which show that it is recommended to make a compromise between bubbles dimension and foam stability in time, so that the ST value is high enough to produce sclerosis of the target tissue and the foam does not self-destroys and dilutes too rapidly in the blood stream before acting on vein.

Comparing spectroscopic data with clinical experimental results, one may conclude that improvement of POL 2% action on varicose vein by exposure of impregnated tissues to 1,064 nm laser beam is possible due to the following, most important mechanisms:

*for wavelengths around 260 nm, laser radiation may be absorbed by active chemical(s) in the pharmaceutically available drug solution. The interaction mechanisms between veins tissues and medicine in the presence of laser radiation are not completely clarified yet. It is possible that nonlinear absorption effects take places in the vein tissues, such as absorption of 4 photons at 1,064 nm (this corresponds to a transition produced by absorption of one photon at 266 nm), which may be responsible for further effects in them;

*at $\lambda$=1,064 nm the absorption may be produced by ethyl alcohol and by the main chromophores in the target such as hemoglobin, methemoglobin and melanin; this absorption may contribute to the sclerosis of veins in exposed area but it remains to clarify the possible mechanisms which lead to this effect. Also, modifications of the molecular structure of POL may be produced by Nd:YAG laser radiation at 1,064 nm once this drug is introduced in the tissue.

The effect of laser beam may be amplified if the drug is injected in the tissue as a foam since light scattering in it becomes more important and absorption becomes larger. So, one may have a larger number of modified drug molecules which may interact with the tissue. On the other side, Raman spectra are more intense for foam than for simple solution samples. The laser light scattering enhancement is, most probably, due to a longer optical path of laser beam in foam sample and vibrational lines that correspond to foam sample are more structured and stronger.

As for the most recommended moment to expose the varicose vein injected in advance with foamed POL to laser radiation, the results of Raman study, performed at different moments during foam lifetime, indicate that some parameters must be considered at applications, the bubble dimensions referred to the foam cohesion being one of the most important. On this line, in the utilised experimental conditions, the suggested time to apply the laser treatment on the previously foam injected tissue is at 2 min after foaming. This remark takes into account the enhanced scattering signal obtained at this moment, which allows both, a better penetration of laser radiation in the tissue and a more efficient modification of drug molecules.

By analysing FTIR spectra of solution and POL 0.5% microfoam in the two investigated circumstances (unirradiated and Nd:YAG laser exposed samples), molecular structure changes of foam comparing with liquid POL are found.

## NOTES

* Reprinted from V. Nastasa, K. Samaras, Ch. Ampatzidis, T.D. Karapantsios, M.A. Trelles, J. Moreno-Moraga, A. Smarandache, M.L. Pascu, "Properties of polidocanol foam in view of its use in sclerotherapy", *Int. J. Pharm.*, vol. 478, pp. 588-596, 2014.

** Reprinted from V. Nastasa, K. Samaras, M. L. Pascu, T. D. Karapantsios "Moderately stable emulsions produced by a double syringe method" *Colloid. Surface A,* vol. 460, pp. 321-326, 2014.

## CONFLICT OF INTEREST

The authors confirm that this chapter content has no conflict of interest.

## ACKNOWLEDGEMENTS

This work has been financed by the National Authority for Research and Innovation in the frame of Nucleus programme-contract 4N/2016 and the projects PN-II-ID-PCE-2011-3-0922 and PN-II-PT-PCCA-2011-3.1-1350, PN-III-P2-2.1-PED-2016-0420.

## REFERENCES

[1] V. Nastasa, K. Samaras, I.R. Andrei, M.L. Pascu, and T. Karapantsios, "Study of the formation of micro and nano-droplets containing immiscible solutions", *Colloid Surface A,* vol. 382, no. 1, pp. 246-250, 2011.
[http://dx.doi.org/10.1016/j.colsurfa.2011.01.016]

[2] W.C. Griffin, "Calculation of HLB Values of Non-Ionic Surfactants", *J. Soc. Cosmet. Chem.,* vol. 5, no. 4, pp. 249-256, 1954.

[3] V. Nastasa, K. Samaras, M.L. Pascu, and T.D. Karapantsios, "Moderately stable emulsions produced by a double syringe method", *Colloid Surface A,* vol. 460, pp. 321-326, 2014.
[http://dx.doi.org/10.1016/j.colsurfa.2014.01.044]

[4] B. Prunet-Foch, F. Legay, M. Vignes-Adler, and C. Delmotte, "Impacting Emulsion Drop on a Steel Plate: Influence of the Solid Substrate", *J. Colloid Interface Sci.,* vol. 199, no. 2, pp. 151-168, 1998.
[http://dx.doi.org/10.1006/jcis.1997.5339]

[5] I.G. Loscertales, A. Barrero, I. Guerrero, R. Cortijo, M. Marquez, and A.M. Gañán-Calvo, "Micro/nano encapsulation via electrified coaxial liquid jets", *Science,* vol. 295, no. 5560, pp. 1695-1698, 2002.
[http://dx.doi.org/10.1126/science.1067595] [PMID: 11872835]

[6] D. Schuster, *Encyclopedia of Emulsion Technology*, CRC Press, 1996.

[7] E. Staples, J. Penfold, and I. Tucker, "Adsorption of Mixed Surfactants at the Oil-Water Interface", *J. Phys. Chem. B,* vol. 104, no. 3, pp. 606-614, 2000.
[http://dx.doi.org/10.1021/jp992955d]

[8] T. Cosgrove, J.S. Phipps, and R.M. Richardson, "Neutron reflection from a liquid/liquid interface", *Colloids Surf.,* vol. 62, no. 3, pp. 199-206, 1992.
[http://dx.doi.org/10.1016/0166-6622(92)80002-J]

[9] E. Dickinson, "Food emulsions and foams: Stabilization by particles", *Curr. Opin. Colloid Interface Sci.,* vol. 15, no. 1-2, pp. 40-49, 2010. [http://dx.doi.org/10.1016/j.cocis.2009.11.001]

[10] "Attension Applications", *Biolin Scientific.* Available at: http://www.biolinscientific.com/

[11] S.F. Burlatsky, V.V. Atrazhev, D.V. Dmitriev, V.I. Sultanov, E.N. Timokhina, E.A. Ugolkova, S. Tulyani, and A. Vincitore, "Surface tension model for surfactant solutions at the critical micelle concentration", *J. Colloid Interface Sci.,* vol. 393, pp. 151-160, 2013. [http://dx.doi.org/10.1016/j.jcis.2012.10.020] [PMID: 23153677]

[12] N.A. Malik, and A. Ali, "Krafft temperature and thermodynamic study of interaction of glycine, diglycine, and triglycine with hexadecylpyridinium chloride and hexadecylpyridinium bromide: A conductometric approach", *J. Mol. Liq.,* vol. 213, pp. 213-220, 2016. [http://dx.doi.org/10.1016/j.molliq.2015.11.017]

[13] L.W. Schwartz, and R.V. Roy, "A mathematical model for an expanding foam", *J. Colloid Interface Sci.,* vol. 264, no. 1, pp. 237-249, 2003. [http://dx.doi.org/10.1016/S0021-9797(03)00425-9] [PMID: 12885541]

[14] C. Stubenrauch, and B. Rippner Blomqvist, "Foam Films, Foams and Surface Rheology of Non-Ionic Surfactants: Amphiphilic Block Copolymers Compared with Low Molecular Weight Surfactants", In: *Colloid Stability,* Wiley-VCH Verlag GmbH & Co. KGaA, 2006. [http://dx.doi.org/10.1002/9783527631070.ch11]

[15] V. Nastasa, K. Samaras, Ch. Ampatzidis, T.D. Karapantsios, M.A. Trelles, J. Moreno-Moraga, A. Smarandache, and M.L. Pascu, "Properties of polidocanol foam in view of its use in sclerotherapy", *Int. J. Pharm.,* vol. 478, no. 2, pp. 588-596, 2015. [http://dx.doi.org/10.1016/j.ijpharm.2014.11.056] [PMID: 25433198]

[16] M. Kostoglou, E. Georgiou, and T.D. Karapantsios, "A new device for assessing film stability in foams: Experiment and theory", *Colloid Surface A,* vol. 382, no. 1-3, pp. 64-73, 2011. [http://dx.doi.org/10.1016/j.colsurfa.2010.12.028]

[17] J. Krägel, and S.R. Derkatch, "Interfacial shear rheology", *Curr. Opin. Colloid Interface Sci.,* vol. 15, no. 4, pp. 246-255, 2010. [http://dx.doi.org/10.1016/j.cocis.2010.02.001]

[18] A. Britan, M. Liverts, G. Ben-Dor, S.A. Koehler, and N. Bennani, "The effect of fine particles on the drainage and coarsening of foam", *Colloid Surface A,* vol. 344, no. 1-3, pp. 15-23, 2009. [http://dx.doi.org/10.1016/j.colsurfa.2009.03.011]

[19] I.B. Ivanov, *Thin Liquid Films.*, 1st ed. CRC Press: New York, 1988.

[20] J. Maldonado-Valderrama, and J.M. Patino, "Interfacial rheology of protein-surfactant mixtures", *Curr. Opin. Colloid Interface Sci.,* vol. 15, no. 4, pp. 271-282, 2010. [http://dx.doi.org/10.1016/j.cocis.2009.12.004]

[21] R. Pal, "Influence of interfacial rheology on the viscosity of concentrated emulsions", *J. Colloid Interface Sci.,* vol. 356, no. 1, pp. 118-122, 2011. [http://dx.doi.org/10.1016/j.jcis.2010.12.068] [PMID: 21257179]

[22] N.C. Mishra, H-J. Müller, H. Möhwald, and R. Krastev, "Interactions and stability of foam films from pentaethyleneglycol monodecyl ether", *Colloid Surface A,* vol. 282-283, pp. 92-98, 2006. [http://dx.doi.org/10.1016/j.colsurfa.2005.12.014]

[23] J. Moreno-Moraga, E. Hernández, J. Royo, J. Alcolea, M.J. Isarría, M.L. Pascu, A. Smarandache, and M. Trelles, "Optimal and safe treatment of spider leg veins measuring less than 1.5 mm on skin type IV patients, using repeated low-fluence Nd:YAG laser pulses after polidocanol injection", *Lasers Med. Sci.,* vol. 28, no. 3, pp. 925-933, 2013. [http://dx.doi.org/10.1007/s10103-012-1180-6] [PMID: 22886138]

[24] P. Redondo, and J. Cabrera, "Microfoam sclerotherapy", *Semin. Cutan. Med. Surg.,* vol. 24, no. 4, pp. 175-183, 2005.
[http://dx.doi.org/10.1016/j.sder.2005.10.005] [PMID: 16387261]

[25] L.Y. Zakharova, and A.I. Konovalov, "Supramolecular systems based on cationic surfactants and amphiphilic macrocycles", *Colloid J.,* vol. 74, no. 2, pp. 194-206, 2012.
[http://dx.doi.org/10.1134/S1061933X12020147]

[26] A. Cavezzi, and L. Tessari, "Foam sclerotherapy techniques: different gases and methods of preparation, catheter versus direct injection", *Phlebology,* vol. 24, no. 6, pp. 247-251, 2009.
[http://dx.doi.org/10.1258/phleb.2009.009061] [PMID: 19952380]

[27] M.A. Trelles, I. Allones, X. Álvarez, M. Vélez, C. Buil, R. Luna, and O. Trelles, "Long-pulsed Nd:YAG 1064 nm in the treatment of leg veins: Check up of results at 6 months in 100 patients", *Med. Laser Appl.,* vol. 20, no. 4, pp. 255-266, 2005.
[http://dx.doi.org/10.1016/j.mla.2005.09.001]

[28] L. Goldman, D.J. Blaney, D.J. Kindel Jr, M.F. Richfield, P. Owens, and E.L. Homan, "Effect of the Laser Beam on the Skin", *J. Invest. Dermatol.,* vol. 42, no. 3, pp. 247-251, 1964.
[http://dx.doi.org/10.1038/jid.1964.56] [PMID: 14130640]

[29] R.R. Anderson, and J.A. Parrish, "Selective photothermolysis: precise microsurgery by selective absorption of pulsed radiation", *Science,* vol. 220, no. 4596, pp. 524-527, 1983.
[http://dx.doi.org/10.1126/science.6836297] [PMID: 6836297]

[30] J.M. Garden, L.L. Polla, and O.T. Tan, "The treatment of port-wine stains by the pulsed dye laser. Analysis of pulse duration and long-term therapy", *Arch. Dermatol.,* vol. 124, no. 6, pp. 889-896, 1988.
[http://dx.doi.org/10.1001/archderm.1988.01670060035012] [PMID: 3377518]

[31] J. Alòs, P. Carreño, J.A. López, B. Estadella, M. Serra-Prat, and J. Marinel-Lo, "Efficacy and safety of sclerotherapy using polidocanol foam: a controlled clinical trial", *Eur. J. Vasc. Endovasc. Surg.,* vol. 31, no. 1, pp. 101-107, 2006.
[http://dx.doi.org/10.1016/j.ejvs.2005.08.018] [PMID: 16226901]

[32] E. Rabe, F.X. Breu, A. Cavezzi, P. Coleridge Smith, A. Frullini, J.L. Gillet, J.J. Guex, C. Hamel-Desnos, P. Kern, B. Partsch, A.A. Ramelet, L. Tessari, and F. Pannier, Guideline Group, "European guidelines for sclerotherapy in chronic venous disorders", *Phlebology,* vol. 29, no. 6, pp. 338-354, 2014.
[http://dx.doi.org/10.1177/0268355513483280] [PMID: 23559590]

[33] M.A. Munia, N. Wolosker, C.G. Munia, W.S. Chao, and P. Puech-Leão, "Comparison of laser *versus* sclerotherapy in the treatment of lower extremity telangiectases: a prospective study", In: *Dermatol. Surg. Off. Publ. Am. Soc. Dermatol. Surg. Al,* vol. 38. 2012, no. 4, pp. 635-639.

[34] J.L. Levy, C. Elbahr, E. Jouve, and S. Mordon, "Comparison and sequential study of long pulsed Nd:YAG 1,064 nm laser and sclerotherapy in leg telangiectasias treatment", *Lasers Surg. Med.,* vol. 34, no. 3, pp. 273-276, 2004.
[http://dx.doi.org/10.1002/lsm.20010] [PMID: 15022257]

[35] N.S. Sadick, "Long-term results with a multiple synchronized-pulse 1064 nm Nd:YAG laser for the treatment of leg venulectasias and reticular veins", *Dermatol. Surg.,* vol. 27, no. 4, pp. 365-369, 2001. [PMID: 11298708]

[36] J. Moreno-Moraga, A. Smarandache, M.L. Pascu, J. Royo, and M.A. Trelles, "1064 nm Nd:YAG long pulse laser after polidocanol microfoam injection dramatically improves the result of leg vein treatment: a randomized controlled trial on 517 legs with a three-year follow-up", *Phlebology,* vol. 29, no. 10, pp. 658-666, 2014.
[http://dx.doi.org/10.1177/0268355513502786] [PMID: 23989971]

[37] A. Dinache, M. Boni, T. Alexandru, E. Radu, A. Stoicu, I.R. Andrei, A. Staicu, L. Liggieri, V. Nastasa, M.L. Pascu, and M. Ferrari, "Surface properties of Vancomycin after interaction with laser beams", *Colloid Surface A,* vol. 480, pp. 328-335, 2015. [http://dx.doi.org/10.1016/j.colsurfa.2014.08.023]

[38] M.J. Frisch, G.W. Trucks, H.B. Schlegel, G.E. Scuseria, M.A. Robb, J.R. Cheeseman, G. Scalmani, V. Barone, B. Mennucci, G.A. Petersson, H. Nakatsuji, M. Caricato, X. Li, H.P. Hratchian, A.F. Izmaylov, J. Bloino, G. Zheng, J.L. Sonnenberg, M. Hada, M. Ehara, K. Toyota, R. Fukuda, J. Hasegawa, M. Ishida, T. Nakajima, Y. Honda, O. Kitao, H. Nakai, T. Vreven, J.A. Montgomery Jr, J.E. Peralta, F. Ogliaro, M. Bearpark, J.J. Heyd, E. Brothers, K.N. Kudin, V.N. Staroverov, R. Kobayashi, J. Normand, K. Raghavachari, A. Rendell, J.C. Burant, S.S. Iyengar, J. Tomasi, M. Cossi, N. Rega, J.M. Millam, M. Klene, J.E. Knox, J.B. Cross, V. Bakken, C. Adamo, J. Jaramillo, R. Gomperts, R.E. Stratmann, O. Yazyev, A.J. Austin, R. Cammi, C. Pomelli, J.W. Ochterski, R.L. Martin, K. Morokuma, V.G. Zakrzewski, G.A. Voth, P. Salvador, J.J. Dannenberg, S. Dapprich, A.D. Daniels, Ö. Farkas, J.B. Foresman, J.V. Ortiz, J. Cioslowski, and D.J. Fox, *Gaussian 09, Revision D.01*. Gaussian, Inc.: Wallingford, CT, 2009.

[39] S.A. Tanumihardjo, "Vitamin A: biomarkers of nutrition for development", *Am. J. Clin. Nutr.,* vol. 94, no. 2, pp. 658S-665S, 2011. [http://dx.doi.org/10.3945/ajcn.110.005777] [PMID: 21715511]

[40] G. Wolf, "The discovery of the visual function of vitamin A", *J. Nutr.,* vol. 131, no. 6, pp. 1647-1650, 2001. [PMID: 11385047]

[41] *Office of Dietary Supplements - Vitamin A*. Available at: https://ods.od.nih.gov/factsheets/VitaminA-HealthProfessional/

[42] F. J. Comes, F. P. Schäfer (Ed.): Dye Lasers, 3rd enlarged and revised edition, Vol. 1 aus der Reihe: *Topics in Applied Physics*, Springer-Verlag, Berlin 1990.

[43] *The PubChem Project*. Available at: https://pubchem.ncbi.nlm.nih.gov/

[44] E. Bouyer, G. Mekhloufi, V. Rosilio, J-L. Grossiord, and F. Agnely, "Proteins, polysaccharides, and their complexes used as stabilizers for emulsions: alternatives to synthetic surfactants in the pharmaceutical field?", *Int. J. Pharm.,* vol. 436, no. 1-2, pp. 359-378, 2012. [http://dx.doi.org/10.1016/j.ijpharm.2012.06.052] [PMID: 22759644]

[45] D.M. Eckmann, "Polidocanol for endovenous microfoam sclerosant therapy", *Expert Opin. Investig. Drugs,* vol. 18, no. 12, pp. 1919-1927, 2009. [http://dx.doi.org/10.1517/13543780903376163] [PMID: 19912070]

[46] L. Tessari, "Nouvelle technique d'obtention de la sclero-mousse", *Phlebologie,* vol. 53, no. 1, pp. 129-133, 2000.

[47] A. Smarandache, A. Staicu, V. Nastasa, J. Moreno-Moraga, J. Royo de la Torre, M. Trelles, R. Pirvulescu, and M. L. Pascu, "Physical properties of laser irradiated sclerosing foams", *Rom Rep Phys,* vol. 67, no. 4, pp. 1480-1490, 2015.

[48] M. Boni, V. Nastasa, I.R. Andrei, A. Staicu, and M.L. Pascu, "Enhanced fluorescence emitted by microdroplets containing organic dye emulsions", *Biomicrofluidics,* vol. 9, no. 1, p. 014126, 2015. [http://dx.doi.org/10.1063/1.4913648] [PMID: 25784965]

[49] D. Lee Black, M.Q. McQuay, and M.P. Bonin, "Laser-based techniques for particle-size measurement: A review of sizing methods and their industrial applications", *Pror. Energy Combust. Sci.,* vol. 22, no. 3, pp. 267-306, 1996. [http://dx.doi.org/10.1016/S0360-1285(96)00008-1]

[50] V. Nastasa, V. Pradines, I.R. Andrei, M. Boni, M.L. Pascu, and R. Miller, "Studies about the generation and characterisation of microdroplets with a controlled content", *Optoelectron. Adv. Mater. Rapid Commun.,* vol. 4, no. 11, pp. 1916-1919, 2010.

[51] G. Paltauf, and P.E. Dyer, "Photomechanical processes and effects in ablation", *Chem. Rev.,* vol. 103, no. 2, pp. 487-518, 2003.
[http://dx.doi.org/10.1021/cr010436c] [PMID: 12580640]

[52] S. Lazare, V. Tokarev, A. Sionkowska, and M. Wiśniewski, "Surface foaming of collagen, chitosan and other biopolymer films by KrF excimer laser ablation in the photomechanical regime", *Appl. Phys., A Mater. Sci. Process.,* vol. 81, no. 3, pp. 465-470, 2005.
[http://dx.doi.org/10.1007/s00339-005-3260-y]

[53] A.A. Oraevsky, S.L. Jacques, and F.K. Tittel, "Mechanism of laser ablation for aqueous media irradiated under confined-stress conditions", *J. Appl. Phys.,* vol. 78, no. 2, p. 1281, 1995.
[http://dx.doi.org/10.1063/1.360370]

[54] L. Tessari, A. Cavezzi, and A. Frullini, "Preliminary experience with a new sclerosing foam in the treatment of varicose veins", *Dermatol. Surg.,* vol. 27, no. 1, pp. 58-60, 2001.
[PMID: 11231246]

[55] M. Sauer, J. Hofkens, and J. Enderlein, *Handbook of fluorescence spectroscopy and imaging: from single molecules to ensembles,* Wiley-VCH: Weinheim, 2011.
[http://dx.doi.org/10.1002/9783527633500]

[56] C. Würth, M. Grabolle, J. Pauli, M. Spieles, and U. Resch-Genger, "Relative and absolute determination of fluorescence quantum yields of transparent samples", *Nat. Protoc.,* vol. 8, no. 8, pp. 1535-1550, 2013.
[http://dx.doi.org/10.1038/nprot.2013.087] [PMID: 23868072]

[57] M. Boni, V. Nastasa, I.R. Andrei, A. Staicu, and M.L. Pascu, "Enhanced fluorescence emitted by microdroplets containing organic dye emulsions", *Biomicrofluidics,* vol. 9, no. 1, p. 014126, 2015.
[http://dx.doi.org/10.1063/1.4913648] [PMID: 25784965]

[58] C.G. Garrett, W. Kaiser, and W.L. Bond, "Stimulated Emission into Optical Whispering Modes of Spheres", *Phys. Rev.,* vol. 124, no. 6, pp. 1807-1809, 1961.
[http://dx.doi.org/10.1103/PhysRev.124.1807]

[59] D. Magde, G.E. Rojas, and P.G. Seybold, "Solvent Dependence of the Fluorescence Lifetimes of Xanthene Dyes", *Photochem. Photobiol.,* vol. 70, no. 5, pp. 737-744, 1999.
[http://dx.doi.org/10.1111/j.1751-1097.1999.tb08277.x]

[60] J.R. Sánchez-Valencia, J. Toudert, L. González-García, A.R. González-Elipe, and A. Barranco, "Excitation transfer mechanism along the visible to the Near-IR in rhodamine J-heteroaggregates", *Chem. Commun. (Camb.),* vol. 46, no. 24, pp. 4372-4374, 2010.
[http://dx.doi.org/10.1039/c0cc00087f] [PMID: 20473447]

[61] S. Kühn, U. Håkanson, L. Rogobete, and V. Sandoghdar, "Enhancement of single-molecule fluorescence using a gold nanoparticle as an optical nanoantenna", *Phys. Rev. Lett.,* vol. 97, no. 1, p. 017402, 2006.
[http://dx.doi.org/10.1103/PhysRevLett.97.017402] [PMID: 16907406]

[62] A.I. Dragan, and C.D. Geddes, "Metal-enhanced fluorescence: The role of quantum yield, Q0, in enhanced fluorescence", *Appl. Phys. Lett.,* vol. 100, no. 9, p. 93115, 2012.
[http://dx.doi.org/10.1063/1.3692105]

[63] A.S. Klymchenko, E. Roger, N. Anton, H. Anton, I. Shulov, J. Vermot, Y. Mely, and T.F. Vandamme, "Highly lipophilic fluorescent dyes in nano-emulsions: towards bright non-leaking nano-droplets", *RSC Advances,* vol. 2, no. 31, p. 11876, 2012.
[http://dx.doi.org/10.1039/c2ra21544f]

[64] J. Brujić, C. Song, P. Wang, C. Briscoe, G. Marty, and H.A. Makse, "Measuring the coordination number and entropy of a 3D jammed emulsion packing by confocal microscopy", *Phys. Rev. Lett.,* vol. 98, no. 24, p. 248001, 2007.
[http://dx.doi.org/10.1103/PhysRevLett.98.248001] [PMID: 17677997]

[65] V.G. Babak, J. Desbrières, and V.E. Tikhonov, "Dynamic surface tension and dilational viscoelasticity of adsorption layers of a hydrophobically modified chitosan", *Colloid Surface A,* vol. 255, no. 1-3, pp. 119-130, 2005.
[http://dx.doi.org/10.1016/j.colsurfa.2004.12.029]

[66] V.B. Fainerman, S.V. Lylyk, E.V. Aksenenko, N.M. Kovalchuk, V.I. Kovalchuk, J.T. Petkov, and R. Miller, "Effect of water hardness on surface tension and dilational visco-elasticity of sodium dodecyl sulphate solutions", *J. Colloid Interface Sci.,* vol. 377, no. 1, pp. 1-6, 2012.
[http://dx.doi.org/10.1016/j.jcis.2012.03.030] [PMID: 22484167]

[67] A. Smarandache, M. Trelles, and M.L. Pascu, "Measurement of the modifications of Polidocanol absorption spectra after exposure to NIR laser radiation", *J. Optoelectron. Adv. Mater.,* vol. 12, no. 9, pp. 1942-1945, 2010.

[68] I.R. Andrei, T. Tozar, A. Dinache, M. Boni, V. Nastasa, and M.L. Pascu, "Chlorpromazine transformation by exposure to ultraviolet laser beams in droplet and bulk", *Eur. J. Pharm. Sci.,* vol. 81, pp. 27-35, 2016.
[http://dx.doi.org/10.1016/j.ejps.2015.09.017] [PMID: 26432595]

[69] T. Alexandru, A. Staicu, A. Pascu, E. Radu, A. Stoicu, V. Nastasa, A. Dinache, M. Boni, L. Amaral, and M.L. Pascu, "Characterization of mixtures of compounds produced in chlorpromazine aqueous solutions by ultraviolet laser irradiation: their applications in antimicrobial assays", *J. Biomed. Opt.,* vol. 20, no. 5, p. 051002, 2015.
[http://dx.doi.org/10.1117/1.JBO.20.5.051002] [PMID: 25365041]

[70] A. Smarandache, J. Kristiansen, J.B. Christensen, and M.L. Pascu, "Optical Studies of the Spectral Properties of Phenothiazines", *Lett Drug Des Discov,* vol. 9, no. 4, pp. 352-360, 2012.
[http://dx.doi.org/10.2174/157018012799859918]

[71] A. Smarandache, "Laser beams interaction with polidocanol foam: molecular background", *Photomed. Laser Surg.,* vol. 30, no. 5, pp. 262-267, 2012.
[http://dx.doi.org/10.1089/pho.2011.3187] [PMID: 22417110]

## CHAPTER 13

# Application of Laser Modified Medicines in Fighting Multiple Drug Resistance Acquired by Microorganisms

**Tatiana Tozar**[1], **Alexandru Stoicu**[1], **Viorel Nastasa**[1], **Marcela Popa**[2,3], **Adriana Smarandache**[1], **Marieta Costache**[2], **Mariana Carmen Chifiriuc**[2,3] and **Mihail Lucian Pascu**[1,4,*]

[1] *National Institute for Lasers, Plasma and Radiation Physics, Magurele, Ilfov, Romania*

[2] *Faculty of Biology, University of Bucharest, Bucharest, Romania*

[3] *Research Institute of the University of Bucharest - ICUB, University of Bucharest, Bucharest, Romania*

[4] *Faculty of Physics, University of Bucharest, Bucharest, Romania*

**Abstract:** The development of new medicines and alternatives to existing antimicrobial agents represents a high priority in treating infections caused by multi-drug resistant bacteria. This is reported mainly, as multiple resistance acquired by bacteria at treatment with antibiotics. The use of laser radiation to photo-generate new compounds from known medicines can be an effective method to obtain a better antimicrobial activity against infections with bacteria and fungi. The irradiation with a laser beam emitted at 266 nm of 2 mg/ml aqueous chlorpromazine and thioridazine solutions for periods of time varying from 1 to 240 min leads to photochemical changes in the molecular structure of the parental compounds and to generation of new photoproducts with enhanced antimicrobial, antifungal and antibiofilm activity. The susceptibility of broad panels of Gram-negative and Gram-positive bacteria and fungi in planktonic and biofilm state, to the unirradiated and irradiated CPZ and TZ were performed in order to highlight the possible use of these substances for the development of novel antimicrobial agents. The antimicrobial activity was evaluated by quantitative methods, *i.e.* minimum inhibitory concentration and minimum biofilm eradication concentration assays. Both CPZ and TZ irradiated solutions presented, as cocktails of medicines obtained after laser irradiation, enhanced antimicrobial, antifungal and antibiofilm activity when compared to the unirradiated samples.

**Keywords:** *ABC 230*, Antimicrobial activity, Bacteria, Biofilm, *Candida albicans*, Chlorpromazine, *Cryptococcus neoformans*, Fungi, *Klebsiella pneumoniae*, Laser, MBEC, MDR, MIC, Multiple drug resistance,

* **Corresponding author Mihail Lucian Pascu:** Laser Department, National Institute for Laser, Plasma and Radiation Physics, Magurele, Ilfov, Romania; Tel/Fax: 0040 21 4575739; E-mail: mihai.pascu@inflpr.ro

*Pseudomonas aeruginosa*, Susceptibility, Thioridazine, *Staphylococcus aureus*, UV laser radiation.

## INTRODUCTION

In 1883, Bernthsen published the first paper describing the synthesis of phenothiazine molecules [1]. Between 1883 and 1950, intense research was dedicated to synthesising biologically and medicinally active phenothiazine derivatives, which resulted in the discovery of the 10-dialkyl derivatives prepared by Charpentier [2]. These derivatives came into wide clinical use due to their multiple therapeutic properties. In 1988 it was reported that over four thousand phenothiazine derivatives were synthesised and investigated for a wide range of biological and pharmacological activities. Phenothiazine derivatives were found to have the following clinical effects: antihistaminic, sedative, antiemetic, tranquilising, antipruritic, antipsychotic, anthelmintic, anticholinergic, analgesic, antitussive, anti-inflammatory, antimutagenic, *etc.* [3].

One of the most important and widely used phenothiazine derivatives is chlorpromazine (CPZ), a drug mostly prescribed for its antipsychotic effects. The description and CPZ molecular geometry is shown in Chapter 9. Another important phenothiazine derivative is thioridazine (TZ), a piperidine phenothiazine derivative prescribed for the treatment of schizophrenia and psychosis. The description and molecular geometry of thioridazine can be found in Chapter 11.

Within the wide range of biological and medical effects, phenothiazine derivatives have been proven to possess cancer chemo-preventive effects, such as inhibition of the P-gp transport function, inhibition of PKC activity, decrease of cell proliferation, and anti-CaM activity [4 - 6]. Literature data support the idea that (i) substituents at C-2 position of the tricyclic core of phenothiazine and (ii) length of alkyl side chain connected to nitrogen at position 10 (N-10) of the thiazine ring, are influencing compounds activity against cancer cells. The type of substituents present in phenothiazine core structure plays a larger role than the length and nature of side chain [7].

Also, phenothiazine compounds are considered to increase cellular sensitivity to cytostatic drugs by reversing the multi-drug resistance (MDR) of neoplastic cells by a strong inhibiting effect on P-gp - dependent mechanism of MDR [4]. Phenothiazine C-2 position substituents that have been proven to reverse MDR include the following functional groups: $-COCH_3$, $-COCH_2CH_3$, $-COCH_2CH_2CH_3$, $-SOCH_3$, $-SO_2CH_3$, $-SO_2N(CH_3)_2$ [8].

The nitrogen atom at position 10 exerts a pyramidalisation effect on the structure of phenothiazine compounds. A high pyramidal character has been linked to increased biological activity, especially the antitumour effects on AIDS-related leukaemia and lymphomas, since the molecule becomes more reactive with the increase of these characteristics [9].

A relatively new approach in the fight against MDR is the study of laser modified phenothiazine derivatives. Most compounds that have a phenothiazine core structure are highly sensitive to light in general, showing visible changes such as colour modification of drug powder or solution when exposed to light. The colours can vary from light green, blue or yellow to dark brown. The change in colour depends mostly on the nature of the substituents at C-2 and N-10 position. By exposing these drugs to certain laser beams having particular wavelengths, it has been proven that the process of photodegradation can be directed so that the mixture that results from this exposure has antibacterial properties [10 - 13]. As for cytotoxic effects of irradiated solutions, a lower toxicity is observed compared with the unirradiated sample when tested against Murine Swiss albino fibroblast, 3T3 and human epithelial carcinoma HeLa cell lines [14].

In this chapter, photochemical modifications produced by the interaction of laser radiation at 266 nm with CPZ and TZ aqueous solutions are shown. Promising results are obtained regarding the antimicrobial, antifungal and antibiofilm activity of irradiated CPZ and TZ samples against a panel of pathogens.

## MATERIALS AND METHODS

The studied compounds are CPZ and TZ (98.9%, Sigma Aldrich, DE). Solutions of CPZ and TZ having a concentration of 2 mg/ml were irradiated with a 266 nm pulsed laser beam (fourth harmonic of the Nd:YAG laser, Excel Technology, Surelite II model) having an average pulse energy of 6.5 mJ. The characteristics of Nd:YAG laser are the following: 6 ns full time width at half maximum (FTWHM), 10 Hz pulse repetition rate. The optical path length was 1 cm, the beam cross section at interface of the spectrophotometric cell with air 0.38 $cm^2$ and the fluence of the laser beam 17.1 $J/cm^2$.

The irradiation was carried out for different exposure time intervals of 1, 5, 15, 30, 120, and 240 min for CPZ and 1, 15, 60, 120 and 240 min for TZ. For these studies, 2 ml of aqueous solutions of CPZ and TZ were used, which were continuously mixed/agitated at 700 rpm with a magnetic stir bar to homogenise the samples and to prevent formation of precipitates.

The susceptibility of microorganisms to unirradiated and irradiated samples was tested *in vitro*, using as methods the *minimum inhibitory concentration* (MIC) and *minimum biofilm eradication concentration* (MBEC). Non-antibiotics solutions, such as CPZ and TZ solutions, exposed to UV laser radiation were tested on various types of bacteria and fungi. The tested microorganisms were (i) Gram-positive bacteria: *Staphylococcus aureus (S. aureus)* 6538, methicillin-susceptible *S. aureus* (MSSA), methicillin resistant *S. aureus* (MRSA) and (ii) Gram-negative bacteria: *Acinetobacter baumannii "complex"* (ABC) 230, *Pseudomonas aeruginosa (P. aeruginosa)* 27833, *Klebsiella pneumoniae (K. pneumoniae)* 40, and the fungal strains *Cryptococcus neoformans (C. neoformans)* 204092 and *Candida albicans (C. albicans)* 26790.

The MIC values were determined by microdilution assay, using a 96-well polystyrene microtiter plate in which the antimicrobial agent was diluted in a volume of 0.1 ml culture medium by serial binary dilution. In the wells containing different concentrations of antimicrobial agent was then added a fixed volume of pathogen suspension and after proper incubation, MIC is determined by the lowest concentration where no visible growth of pathogen is observed [15]. Two out of the 12 wells for each row represented the positive growth control (culture medium and inoculum) and the negative control (sterile culture medium only) in order to validate the microbial growth and to exclude the contamination of culture medium.

The inoculum volume, culture medium, incubation period, temperature and pH can be varied from one experiment to another, so that the results of assays can be different, if MIC values are compared. A difference of two-fold dilution can be accepted and the result can be considered reproducible [15].

The steps involved in performing a MIC assay are:

1. *Selection of microbial culture for the inoculum:* isolated colonies from an agar plate containing fresh microbial cultures of 18-24 h.
2. *Preparation of microbial suspension*-two available methods:
   *2.1 Growth method:* from an agar plate one selects three to five isolated colonies of the same morphological type of cells with a wire loop and one transfers them to a tube containing 4-5 ml broth medium. The next step is to incubate the suspension at 35°C until a turbidity of 0.5 McFarland standard is obtained [16].
   *2.2 Direct suspension method:* as an alternative method, the inoculum can be prepared by the direct transfer of the isolated colonies in a tube containing culture medium or saline suspension and adjusting it to a turbidity corresponding to 0.5 McFarland standard [16].
3. *Addition of culture medium to the wells:* adding the desired volume of culture medium to each well, depending on the initial chosen concentration of the

antimicrobial agent.

4. *Addition of the antimicrobial agent:* volumetric two-fold serial dilution of the antimicrobial agent are made in culture medium found in microtiter plate.
5. *Addition of bacteria from the suspension:* the desired amount of microbial suspension is added in each well with the exception of the negative control.
6. *Incubation of the microtiter plate:* the plate is incubated at 37°C for 18-24 h.
7. *Reading the* MIC *results:* after incubation, the microtiter plate is placed in a photometric device, where absorbance of each well is recorded. MIC determined value for the drug is the lowest concentration that induced the absence of a clear growth of the pathogen, provided that the positive control contains one. The pathogen growth can be seen as a turbidity or a pellet on the bottom of the well [17]. Absorbance is recorded with using a microplate reader at 630 nm.

Results obtained in recent years have shown that bacteria which grow on surfaces have the tendency to form heterogeneous communities containing various species that can live together, interact and probably exchange genetic and chemical information. These aggregate structures are called biofilms and constitute an important defence mechanism against immune effectors, antimicrobial substances and other stressors. The organisms within the biofilms can have substantially different properties compared to the same organisms in their respective planktonic states. Bacteria within the biofilm are able to transfer horizontally genetic material and to exchange information regarding population size and metabolic states [18].

Both artificial and natural growth conditions indicate the formation of communities amongst surface dependent populations, as well as an apparent systematic/non-random evolution in these communities. This is a challenge in understanding the way in which a relatively primitive organism can develop and withstand in complex, stressful and even hostile environments. Heterogeneity refers to the distribution of biomass, as well as the distribution of activities. On one hand, it is clear that biomass distribution is caused by one or more biological activities such as growth of some organisms at the expense of others, cell transfer *via* community environment, bacterial motility coupled with chemotaxis, *etc*. On the other hand, cell activity reflects interactions between cells and their environment, involving either nutrients from external environment or metabolic interactions between community members. The understanding of relationship between structure and function and the investigation of these biofilms represent a major aim of modern microbial ecology [19].

The inhibition of biofilm development cannot be investigated by standard assays used in evaluation of biocides, disinfectants, antibiotics or antimicrobials against microorganisms, due to the fact that response of bacteria in biofilm is different from that in planktonic form. Some of the important differences are the following:

1. Gene expression in biofilms is different than in suspension, the adherent bacteria physiological state making them more resistant to antimicrobial agents than the bacteria from the suspension.
2. Biofilm cannot be totally penetrated by antimicrobial agents.
3. Biofilms are more resistant to antimicrobials due to their heterogeneity (*e.g.*, selection of persisters population, low oxygen pressure in the depth of the biofilm *etc.*).
4. Contaminants from the environment or antibiotic inactivating or modifying enzymes can be accumulated in biofilm thus affecting the activity of antimicrobial agent [20].

The method for testing MBEC is suitable for the screening of new antimicrobial agents and for the determination of MIC and MBEC values in clinical cases. This method consists in the use of the same microtitration plates used to determine the MIC value. After 24 h of incubation at 37°C, the content of each well is removed by decantation and washed three times with 300 ml of sterile saline solution [21] or sterile distilled water. Attached bacteria remaining in the well are fixed by their exposure to hot air at 60°C for 60 min in an incubator [21], or by adding an amount of ethanol or methanol and maintained in the wells for 5 to 10 min, followed by removal of the content and addition of 130 ml of crystal violet dye (1%) to each well. After 15 min, the excess is removed and the plate washed. Finally, 130 ml of acetic acid is added to each well and the optical density is recorded at 490 nm using a microtiter plate reader.

In order to determine MIC for the specific case of the substances reported here, the direct method was used in order to prepare a cell suspension for inoculation, using colonies from a microbial culture incubated for 24 h. In each well from the first column, were added 150 μl of medium culture (Mueller-Hinton) and 30 μl of sample, followed by two-fold serial dilutions. Thus, concentrations of test compounds ranged between 330 μg/ml to 0.64 μg/ml. Finally, 20 ml of suspended bacteria were added to all wells, except the positive control. The microtitration plates containing culture medium, the samples and the bacteria were incubated at 37°C for 24 h, and then absorbance of each well was recorded at 630 nm using a Multi-Mode Microplate Reader Synergy™ HTX.

The MBEC values were determined after reading the absorbance of each well at 630 nm. The content of the plate was discarded and washed three times with distilled water. After drying, 130 ml of ethanol were added and after 5-10 min (the time interval in which the biofilm is fixed on the wall surface of the well) it was discarded. Further, 130 ml of 1% crystal violet were added and following another 15 min the content was discarded and the excess dye was removed. The next step involved the addition of 130 ml of 33% acetic acid per well in order to re-suspend the stained biofilm. The final step consisted in recording the absorbance at 490 nm of each well using the same Multi-Mode Microplate Reader Synergy™ HTX.

## RESULTS

The *in vitro* activity of CPZ in the case of Gram-positive bacteria *S. aureus* 6538, *S. aureus* MSSA and *S. aureus* MRSA is shown in Fig. (**1**). In the graphs below, for each MIC and MBEC value the standard deviation is presented, if different from zero.

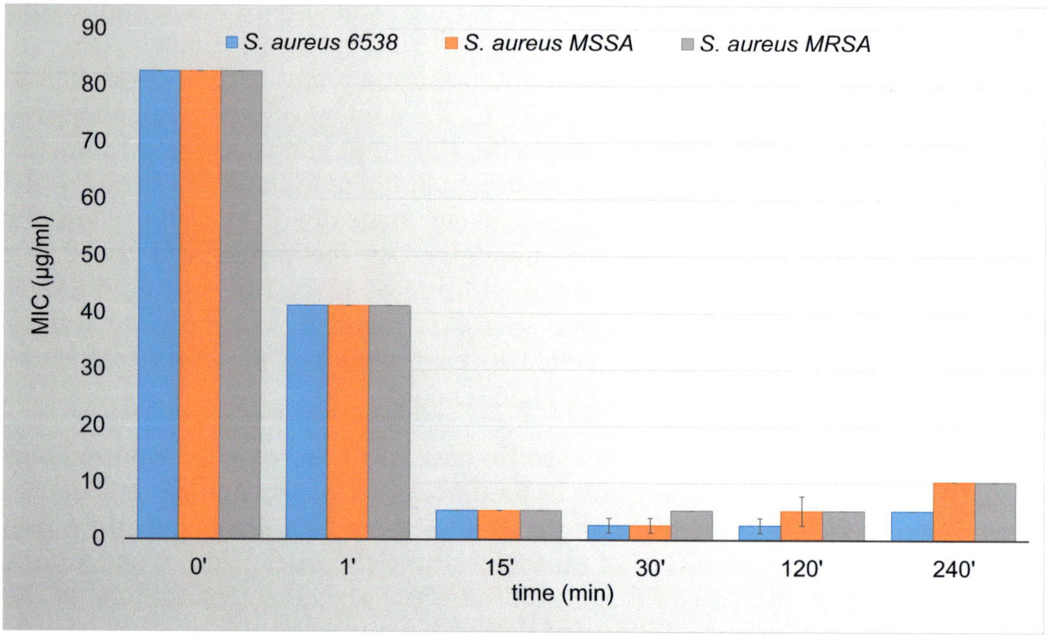


**Fig. (1).** MIC values obtained by testing the antimicrobial activity of unirradiated CPZ and of irradiated CPZ for time intervals between 1-240 min against *S. aureus* 6538, *S. aureus* MSSA and *S. aureus* MRSA.

The analysis shows that in the case of *S. aureus* 6538, for unirradiated CPZ is obtained a MIC value of 82.5 µg/ml and this is reduced by half, 41.3 µg/ml, in case of CPZ exposed only 1 min to UV laser radiation. For solutions irradiated up to 15 and 240 min, MIC is 5.16 µg/ml and the lowest MIC of 2.58 µg/ml is obtained for CPZ exposed to UV radiation for 30 and 120 min. In this case, a significant improvement of antimicrobial activity of irradiated CPZ is observed compared to the unirradiated one, requiring only 30 min irradiation with a laser beam emitted at 266 nm; MIC=2.58 μg/ml results.

For each MIC and MBEC value obtained in the case of bacteria and fungi, the standard deviation was calculated and is represented, if different from zero, in each graph. In statistics and probability theory, standard deviation is a measure of the variability of a set of data [22].

For *S. aureus* MSSA it was observed the same MIC evolution, where for unirradiated CPZ was obtained 82.5 μg/ml and for 1 min irradiated CPZ it resulted 41.2 μg/ml. The lowest MIC, *i.e.* the best antibacterial activity, was of 2.58 μg/ml for 30 min irradiated CPZ. Therefore, these results suggest that it is enough to irradiate CPZ only 30 min to obtain an improved antimicrobial activity up to 32 times higher compared to unirradiated CPZ.

Susceptibility of *S. aureus* MRSA to unirradiated and irradiated CPZ was tested and a MIC=82.5 μg/ml was obtained for unirradiated CPZ. After only 1 min exposure, it is reduced by half. For an exposure to 266 nm radiation between 15 and 120 min it resulted the same MIC value of 5.16 μg/ml, followed by an increase of MIC for 240 min irradiation of CPZ, respectively 10.3 μg/ml. Accordingly, the optimum irradiation time for which the best antibacterial effect was obtained was 15 min.

The same method was applied for testing the susceptibility of Gram-negative bacteria to unirradiated and irradiated CPZ. Tested Gram-negative bacteria were: *ABC* 230, *P. aeruginosa* 27833 and *K. pneumoniae* 40. The obtained MIC values are presented in Fig. (**2**). The difference between Gram-positive and Gram-negative bacteria is that the latter has, in addition to the first, a second barrier represented by the outer membrane, which results in its ability of being less susceptible to antibiotics than Gram-positive bacteria [23].

The susceptibility of Gram-negative bacteria to unirradiated and irradiated CPZ was lower if compared to the susceptibility obtained for Gram-positive bacteria. For *ABC* 230, the MIC obtained for unirradiated and one min irradiated CPZ was 41.3 μg/ml. The lower MIC is obtained for 15 and 30 min irradiation time intervals, at 5.16 μg/ml. If exposure duration increases, MIC increases to 10.30 μg/ml for 120 min and to 20.6 μg/ml for 240 min. As a result, one may consider that after 120 min irradiation the potency of compound decreases, the optimum irradiation time being 15 min.

For *P. aeruginosa* 27833 strain no improvement of antimicrobial activity was observed when unirradiated CPZ was compared to irradiated CPZ up to 240 min, the same MIC value of 330 μg/ml being obtained for all tested solutions. Comparatively, *K. pneumoniae* 40 proved to be less resistant to both unirradiated and irradiated CPZ solution. The lowest MIC obtained was 82.5 μg/ml for CPZ irradiated 1 min, two times lower than that obtained for unirradiated CPZ. As irradiation time increases, the parent compound seems to be photodegraded and antimicrobial activity of the sample decreases, suggesting that generated photoproducts are not effective in killing *K. pneumoniae* 40 strain.

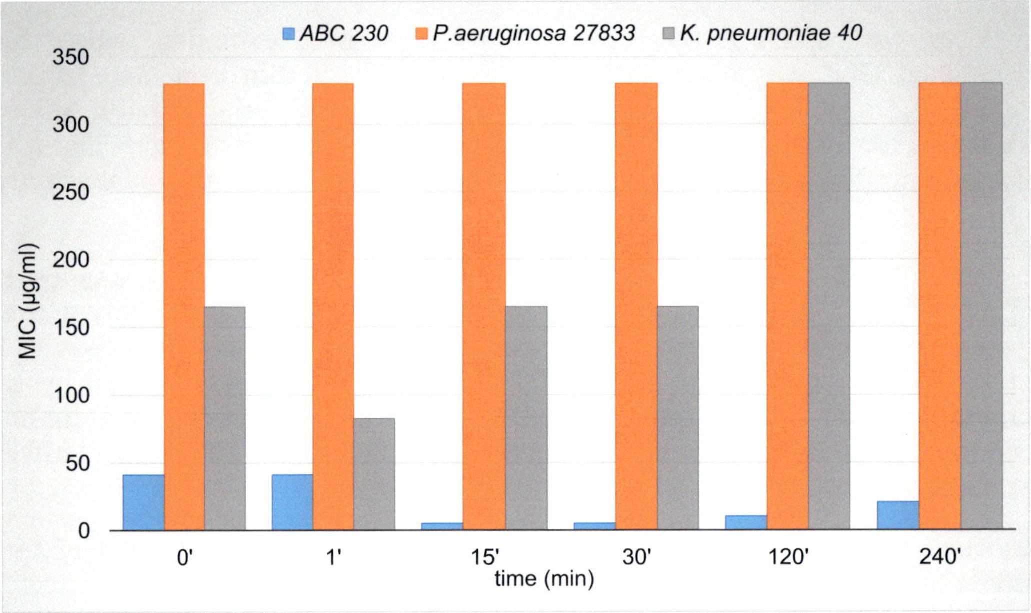


**Fig. (2).** MIC values obtained by testing the antimicrobial activity of unirradiated CPZ and of irradiated CPZ for time intervals between 1-240 min against *ABC* 230, *P. aeruginosa* 27833 and *K. pneumoniae* 40.

In addition to test the Gram-positive and Gram-negative bacteria to unirradiated and irradiated CPZ for various periods of time, the same susceptibility assays were performed for the fungal strains *C. neoformans* 204092 and *C. albicans* 26790 and MIC values are shown in Fig. (**3**).The antifungal activity of CPZ against *C. neoformans* 204092 suggests an improvement in antimicrobial activity if CPZ is irradiated. For unirradiated CPZ the obtained MIC was 10.31 μg/ml.

The lowest MIC value, 5.16 μg/ml, was obtained for 30 min irradiation of CPZ aqueous solution. In this case, the optimum irradiation time is 30 min.

*C. albicans* 26790 proved to be more resistant to unirradiated and irradiated CPZ compared to results obtained for *C. neoformans* 204092. MIC for unirradiated CPZ was 41.3 μg/ml. The lower MIC for irradiated samples was obtained for CPZ exposed 15 min to 266 nm laser beam, followed by an increase of its value with prolonged exposure, arriving at 330 µg/ml for 240 min irradiation. Thus, optimum irradiation time is 15 min.

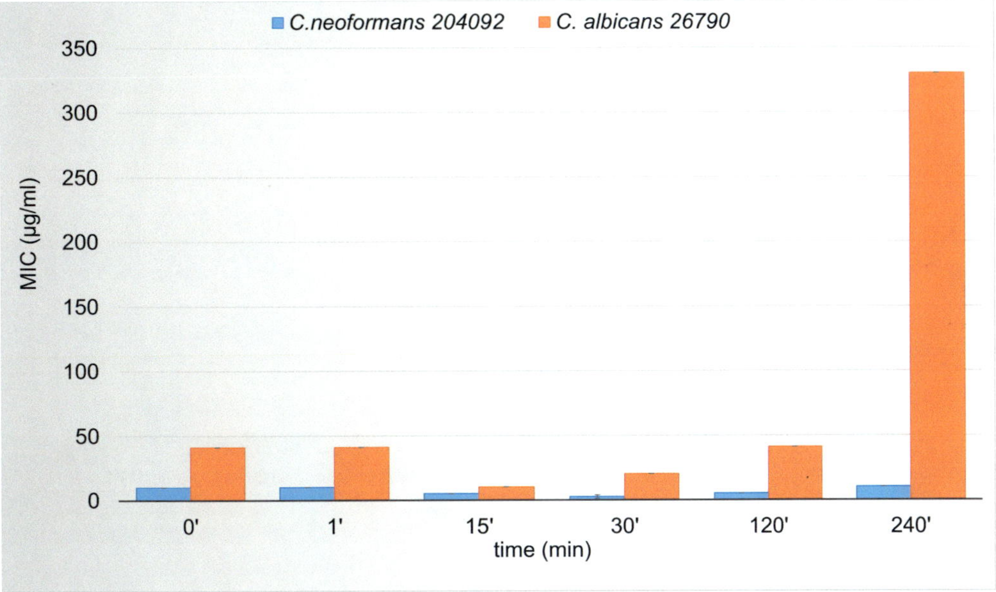


**Fig. (3).** MIC values obtained by testing the antimicrobial activity of unirradiated CPZ and of irradiated CPZ for time intervals between 1-240 min against *C. neoformans* 204092 and *C. albicans* 26790.

In order to determine the susceptibility of microorganisms grown in biofilm state to 2 mg/ml unirradiated and irradiated CPZ between 1-240 min with a laser beam at 266 nm, the MBEC assay was performed. For this type of analysis there were studied the bacteria that formed the most well developed biofilms at interaction with the surface of wells. Therefore, assays included Gram-positive bacteria: *S. aureus* 6538, *S. aureus* MSSA and *S. aureus* MRSA, as well as Gram-negative: *ABC* 230 and *P. aeruginosa* 27833.

Fig. (**4**) shows the results obtained for MBEC assay for Gram-positive bacteria *S. aureus* 6538, *S. aureus* MSSA and *S. aureus* MRSA. For unirradiated CPZ and one min irradiated CPZ it was obtained the same MBEC value for all Gram-positive bacteria, namely 165 µg/ml, respectively 41.30 µg/ml.

When exposure times were 15 min and 30 min, in the case of *S. aureus* MSSA, and *S. aureus* MRSA, it resulted a MBEC value of 5.16 µg/ml and in the case of *S. aureus* 6538, 2.58 µg/ml. With longer exposure to 266 nm laser beam, more exactly above 120 min, the effectiveness of test solutions is decreasing as MBEC values increase.

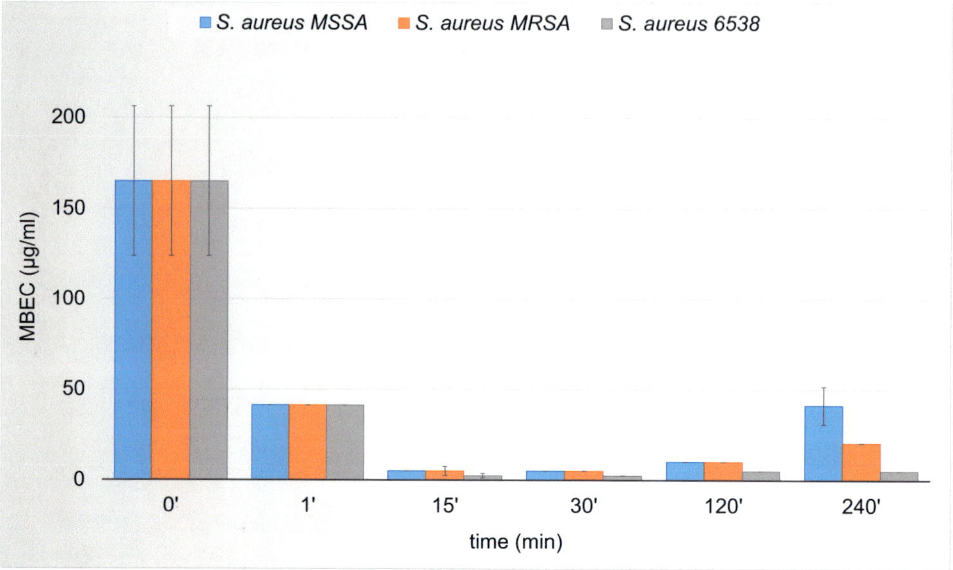


**Fig. (4).** MBEC values obtained by testing the antimicrobial activity of unirradiated CPZ and of irradiated CPZ for time intervals between 1-240 min against biofilm formation of *S. aureus* 6538, *S. aureus* MSSA and *S. aureus* MRSA.

Comparing the results of MIC and MBEC assays on Gram-positive bacteria it can be admitted that the solutions of CPZ irradiated 15 and 30 min efficiently destroy bacteria in both planktonic and biofilm forms. In the case of *S. aureus* 6538 strain the lowest MIC and MBEC values were 2.58 µg/ml resulted from 30 min irradiation of CPZ solution and for *S. aureus* MRSA this was 5.16 µg/ml for 15 min exposure of CPZ.

MBEC assays were determined for Gram-negative *ABC* 230 and *P. aeruginosa* 27833 bacteria and are shown in Fig. (**5**). For *P. aeruginosa* 27833 strain there is no difference regarding the antibiofilm activity of unirradiated CPZ and irradiated CPZ between 1 min and 240 min.

For *ABC* 230 strain, one can observe an improvement of antibiofilm activity of irradiated CPZ compared to unirradiated CPZ. The best MBEC value, 5.16 µg/ml, was obtained for 30 min irradiation of CPZ, being 32 times lower than that obtained for irradiated solution.

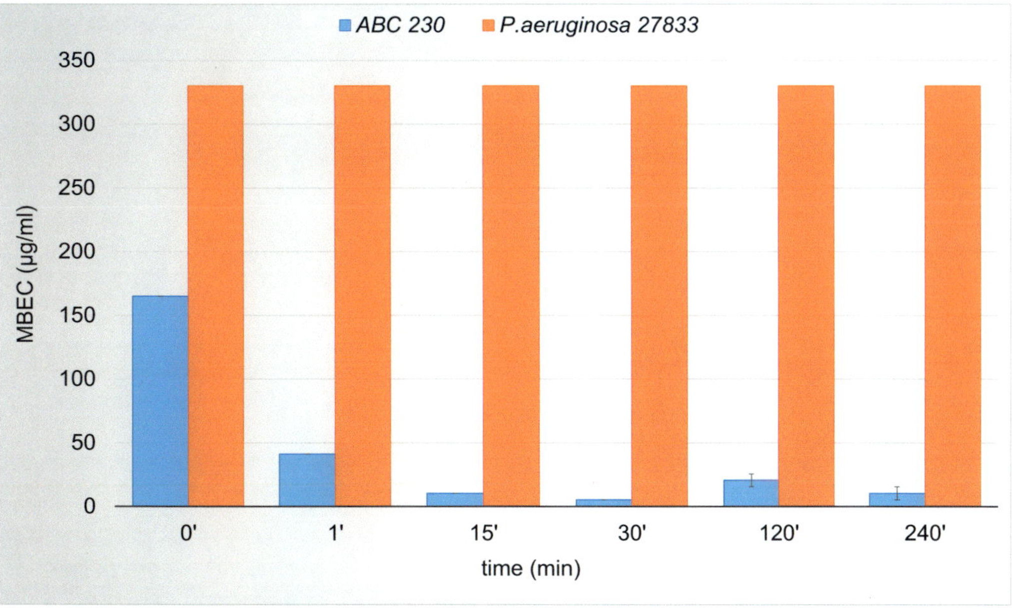


**Fig. (5).** MBEC values obtained by testing the antimicrobial activity of unirradiated CPZ and of irradiated CPZ for time intervals between 1-240 min against biofilm formation of *ABC* 230 and *P. aeruginosa* 27833.

One may consider that the optimum irradiation time of 2 mg/ml CPZ is 30 min, effectively eliminating bacteria both in planktonic and in biofilm forms at the same concentrations and irradiation times. These results suggest that 2 mg/ml CPZ can be used successfully as an antimicrobial, antifungal and antibiofilm agent in treating bacterial and fungus infections if it is exposed to UV laser radiation between 15 and 30 min.

The *in vitro* antimicrobial activity of unirradiated and irradiated TZ between 1min and 240 min against Gram-positive bacteria *S. aureus* 6538, *S. aureus* MSSA and *S. aureus* MRSA is shown in Fig. (**6**). For *S. aureus* 6538, MIC value of unirradiated TZ is 20.63 µg/ml, and for the sample irradiated 1 min, its value decreases by half, at 10.31 µg/ml. With prolonged TZ exposure to UV beam, MIC continues to decrease as follows: 60 min-5.16 µg/ml, 120 min and 240 min - 2.58 µg/ml. To obtain an improvement of antibacterial activity about 10 times higher between unirradiated and irradiated TZ, from 20.63 µg/ml to 2.58 µg/ml, it is necessary a 120 min exposure time to 266 nm laser beam.

For *S. aureus* MSSA the same MIC dependence on the irradiation time is observed, where for unirradiated TZ the obtained MIC was 20.63 µg/ml and for 15 min exposed TZ, the value was 10.31 µg/ml. The lowest MIC of 2.58 µg/ml, and so the best antibacterial activity, was obtained for 120 min and 240 min irradiated TZ. Therefore, an irradiation of TZ for 120 min is enough to obtain an improved antimicrobial activity up to 10 times higher compared to that of unirradiated TZ.

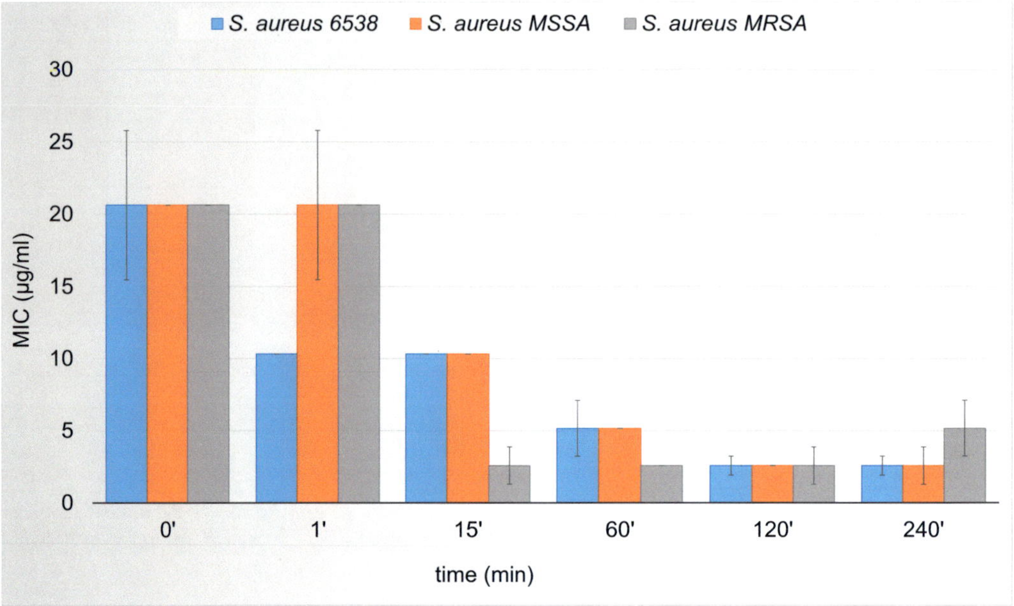


**Fig. (6).** MIC values obtained by testing the antimicrobial activity of unirradiated TZ and of irradiated TZ for time intervals between 1-240 min against *S. aureus* 6538, *S. aureus* MSSA and *S. aureus* MRSA.

When testing the susceptibility of *S. aureus* MRSA to unirradiated and irradiated TZ, it was observed a lower MIC for irradiated samples compared to previous strains. Therefore, for unirradiated and 1 min irradiated sample MIC is the same, respectively 20.63 µg/ml. The antimicrobial activity of samples irradiated between 15 and 120 min is the same and MIC is 2.58 µg/ml. Accordingly, the optimum irradiation time of TZ to obtain the best antimicrobial activity in the shortest exposure time, is 15 min.

The *in vitro* antimicrobial activity of unirradiated and irradiated TZ was tested for Gram-negative bacteria *ABC* 230, *P. aeruginosa* 27883 and *K. pneumoniae* 40 and MIC values are presented in Fig. (**7**). In the case of *ABC* 230 strain, MIC obtained for unirradiated TZ was 41.25 µg/ml and remains the same for all exposure time intervals to UV radiation. Therefore, for this strain, there is no improvement of antibacterial activity of the irradiated samples.

In the case of *P. aeruginosa* 27833, the activity of unirradiated TZ is lower than for *ABC 230*, the MIC value being 165 μg/ml. For 1 and 15 min irradiated TZ, the same MIC is obtained as for unirradiated TZ. The best result is obtained for 60 min irradiation of TZ for which a MIC of 82.5 μg/ml was obtained. At 240 min exposure to UV laser radiation, it was obtained a higher MIC value compared to unirradiated TZ, namely, 330 μg/ml. Thus, it can be said that the optimum irradiation time is 60 min.

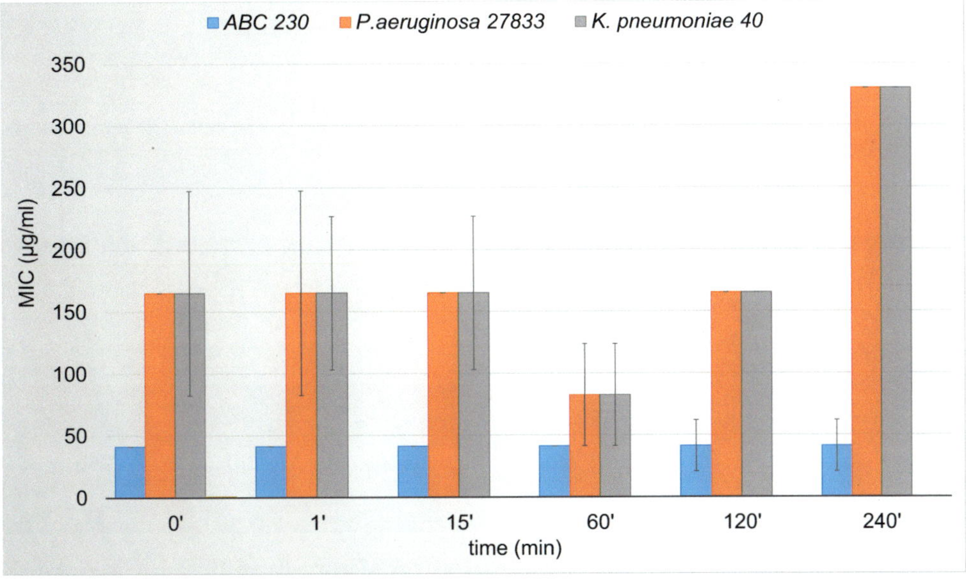


**Fig. (7).** MIC values obtained by testing the antimicrobial activity of unirradiated TZ and of irradiated TZ for time intervals between 1-240 min against *ABC* 230, *P. aeruginosa* 27833 and *K. pneumoniae* 40.

For *K. pneumoniae* 40 it was observed the same dependence of MIC on irradiation time as in the case of *P. aeruginosa* 27833. Thus, for the unirradiated sample it has been obtained 165 μg/ml, and for irradiated solution the lower MIC value was achieved for 60 min irradiation, respectively 82.5 μg/ml. For 240 min irradiation time were obtained MIC values higher than for unirradiated TZ, that is 330 μg/ml. Thus, one may consider that the optimum irradiation time with 266 nm laser beam is in this case 60 min.

Besides determining the susceptibility of Gram-positive and Gram-negative bacteria to unirradiated and irradiated TZ for different time intervals, the same assay was performed for fungi *C. neoformans* 204092 and *C. albicans* 26790, for which MIC values are shown in Fig. (**8**).

The antibacterial activity of TZ against *C. neoformans* 204092, represented by MIC value suggests an improvement from 5.16 μg/ml for the unirradiated TZ and

1 and 15 min irradiated TZ to 1.29 μg/ml for TZ exposed to laser radiation between 60 and 240 min. These results suggest that the sample irradiated 60 min is four times more efficient than unirradiated TZ. In this case, the optimum irradiation time is 60 min.

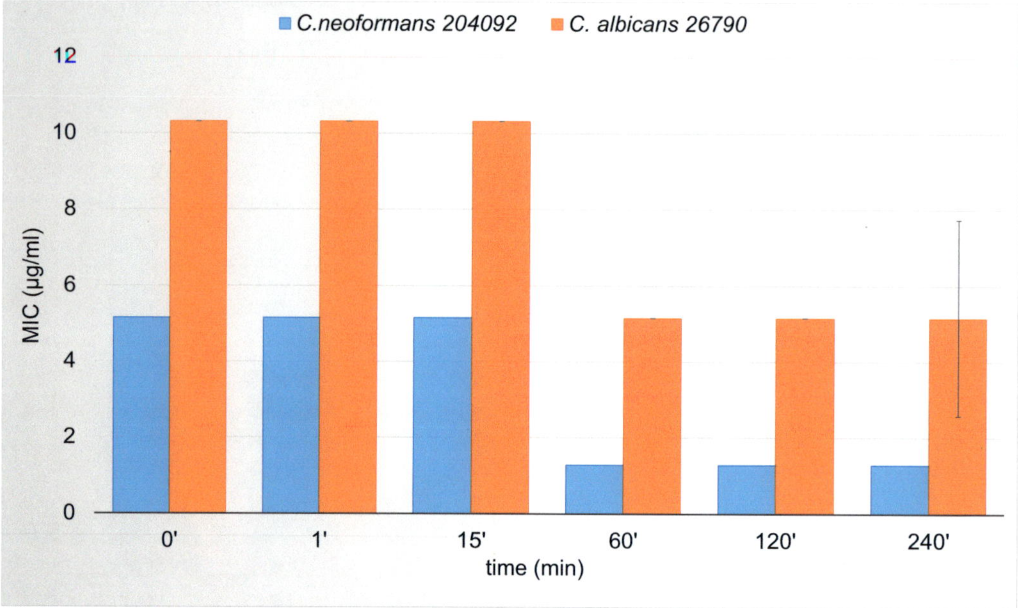


**Fig. (8).** MIC values obtained by testing the antimicrobial activity of unirradiated TZ and of irradiated TZ for time intervals between 1-240 min against *C. neoformans* 204092 and *C. albicans* 26790.

*C. albicans* 26790 has shown to be more resistant to unirradiated and irradiated TZ than *C. neoformans* 204092. The MIC obtained for unirradiated TZ and irradiated TZ for 1 min and 15 min was the same, respectively 10.31 μg/ml. Solutions irradiated between 60 and 240 min presented an enhanced antimicrobial activity than the rest of samples, resulting in a MIC of 5.16 μg/ml. In this case, the optimum irradiation time is 60 min.

The property of 2 mg/ml TZ irradiated periods of time between 1 to 240 min to eradicate the biofilm formation, both for Gram-positive and Gram-negative bacteria and for fungi is depicted in Figs. (**9** and **10**) by determining the MBEC values. For this assay, bacterial biofilms were investigated. Thus, bacteria from Gram-positive class, *S. aureus* 6538 and *S. aureus* MSSA and from Gram-negative class, *K. pneumoniae* 40, were those who developed biofilm in the positive control (culture medium and bacteria).

MBEC values determined for Gram-positive bacteria are presented in Fig. (**9**); in the case of *S. aureus* 6538, for unirradiated TZ the MIC was 10.31 μg/ml. The antibiofilm activity of irradiated TZ is enhanced compared to unirradiated TZ. By exposing TZ solution to 266 nm laser beam, periods of time ranging between 1 and 60 min, it was obtained a MBEC value of 2.58 µg/ml. With prolonged irradiation, the antibiofilm activity is increasing, being 1.29 µg/ml for TZ exposure times of 120 and 240 min. As in the case of MIC assay, the optimal irradiation time to destroy the bacteria in the biofilm is 120 min.

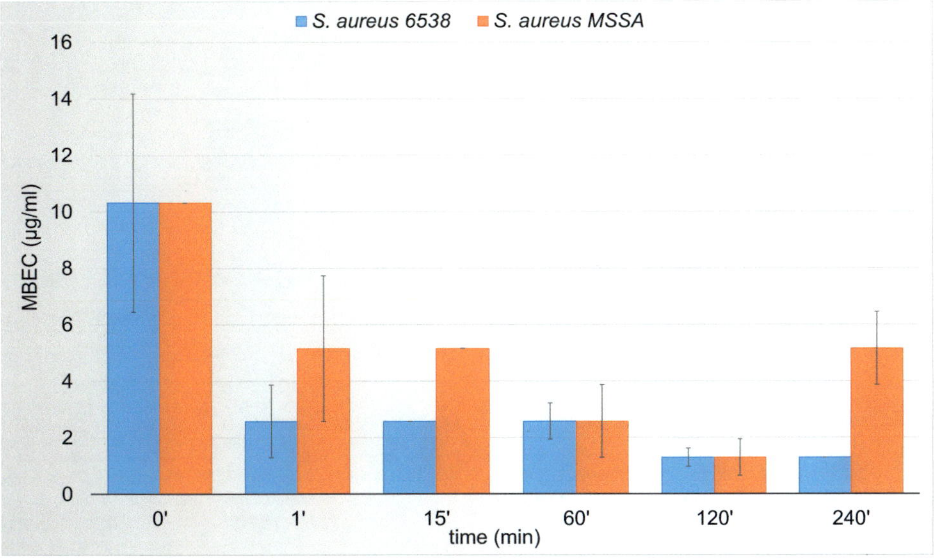


**Fig. (9).** MBEC values obtained by testing the antimicrobial activity of unirradiated TZ and of irradiated TZ for time intervals between 1-240 min against biofilm formation of *S. aureus* 6538 and *S. aureus* MSSA.

*S. aureus* MSSA strain shows for unirradiated TZ, 10.31 µg/ml MBEC, and after one minute of irradiation this becomes two times smaller. The antibiofilm activity increases with TZ prolonged exposure to 266 nm laser beam and it was enhanced 8 times if the solution was irradiated 120 min. For the sample irradiated 240 min, MBEC value increases to 5.16 µg/ml suggesting that photoproducts which are formed are not effective in eradicating the biofilm.

The MBEC values determined for Gram-negative bacteria are shown in Fig. (**10**), where the results for *K. pneumoniae* 40 suggest that there is no difference in terms of antibiofilm activity of unirradiated TZ and irradiated TZ from 1 to 240 min.

The MIC values obtained for unirradiated and irradiated TZ solutions between 1 and 240 min show that the optimum irradiation time of samples with a laser beam emitted at 266 nm varies for each bacterium. Thus, for the Gram- positive bacteria *S. aureus* MRSA, the optimum irradiation time is 15 min; for Gram-negative bacteria *K. pneumoniae* 40 and *P. aeruginosa* 27833 and for fungus

*C. neoformans* 204092 and *C. albicans* 26790 the optimum irradiation time is 60 min and for Gram-positive bacteria *S. aureus* 6538 and *S. aureus* MSSA is 120 min. For Gram-negative bacteria *ABC* 230, irradiated TZ presents no enhancement of antibacterial activity compared to the unirradiated sample.

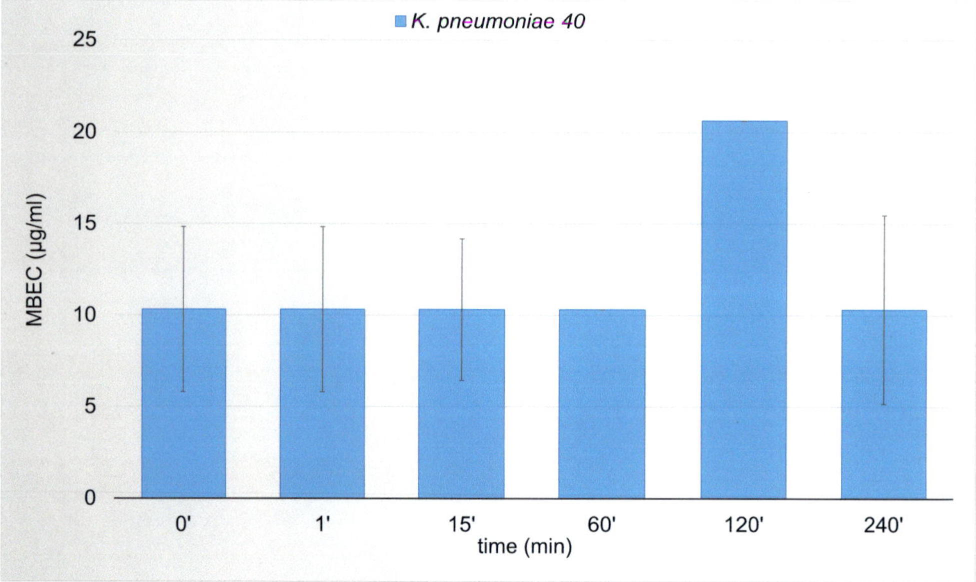


**Fig. (10).** MBEC values obtained by testing the antimicrobial activity of unirradiated TZ and of irradiated TZ for time intervals between 1-240 min against biofilm formation of *K. pneumoniae* 40.

As in CPZ medicine, for all studied microorganisms, antibacterial activity of irradiated TZ samples is more pronounced than in the case of unirradiated samples, with the exception of *ABC* 230, where there was no change of MIC. The differences between MIC values of unirradiated and irradiated TZ at time intervals between 1 and 240 min were not as pronounced as in the CPZ case.

## DISCUSSIONS

### Phenothiazine Antibacterial Activity

Although Gram-positive and Gram-negative bacteria are similarly organised, some essential differences still exist between them. These differences influence the antimicrobial agents' capacity to inhibit the growth of bacteria. However, some antibiotics are active when dealing with both types of bacteria and are often referred to, as wide spectrum antibiotics [24]. Antibiotics are usually small molecules, synthesised by bacteria or fungi, able to kill other bacteria without affecting the person or animal being treated. The synthesised compounds are sometimes referred to as antibacterial agents. In general, all chemical compounds which are not antibiotics but manifest moderate or aggressive antimicrobial activity are part of a group called "non-antibiotics" [25]; phenothiazine compounds successfully fall into this category.

Large scale use of antibiotics may lead to microbial antibiotic resistance, which is fastened by horizontal gene transfers between different strains or species. Resistance genes are transferred by several mechanisms: transformation, conjugation or transduction [26].

Bacterial resistance to antibiotics may occur due to production of beta-lactam enzymes that deactivate the antibiotic by cleaving the ring core structure of penicillin or by attaching a chemical group to antibiotic side-chain that prevents its binding to the ribosome as in the case of streptomycin.

An important antibiotic drug resistance mechanism occurs through the modification of the target binding site, with two possible means: either through a gene mutation that changes the antibiotic binding site or through an enzyme that chemically modifies antibiotic binding site.

Another wide-spread resistance mechanism relies on efflux pumps which are protein structures in the cell wall that act as pumps. These elements require energy in order to hastily remove the antibiotic from cell, faster than drugs' ability to enter the cell. In some cases, efflux pumps have multiple purposes, and so, a single trans-membrane protein is able to remove multiple types of drugs [26].

Pores are bacterial cell structures in charge of importing and exporting various compounds between the intracellular and extracellular environment and often act as portals of entry for antibiotics. One strategy for bacterial resistance is the expression of smaller pores, thus excluding the entrance of a large number of antibiotics.

The remarkable photo-physical and photo-chemical properties of phenothiazine derivatives offer a variety of biological activities and medical applications. The presence of nitrogen and sulphur atoms between the phenothiazine rings, plays an important role in its various interactions with biological systems. Moreover, chemical synthesis of new molecules containing the phenothiazine core structure has generated fields of research aimed at investigating their biological and therapeutic properties. In many cases, new technological breakthroughs have allowed to shed light on the mechanisms of interaction between these new compounds and tissues and cells structures. Because this type of research constitutes an important part of pre-clinic research, it is necessary to determine with accuracy the positive, negative and/or unwanted medical effects of phenothiazine compounds [27].

Phenothiazine and a lot of its derivatives manifest diverse biological activities and have been proved to possess significant properties as insecticides, antifungal, antibacterial and anthelmintic agents [27]. The mechanism of phenothiazine antibacterial activity has been investigated on many biological systems, including *E. coli* and mice skin. For example, Komatsu *et al.* [28] have tested 17 different phenothiazine species using the disk diffusion method and demonstrated that besides a slight inhibition of growth manifested by CPZ, these compounds have no antibacterial effect on *E. coli*. CPZ and TZ are able to reduce oxacillin sensitivity in *S. aureus* MRSA strains but they do not have the same effect on *S. aureus* MSSA [27].

Sharma *et al.* [29] have investigated the efficiency and mechanism of action of CPZ and trifluoperazine on *C. albicans*. The anticandidal effect of these compounds may be due to the interaction with $Ca^{2+}$/calmodulin, which depends on protein phosphorylation. Also, phenothiazine compounds have determined a significant drop in cellular lipid and phospholipid content. Thus, CPZ and trifluoperazine may intervene in cellular cycle during DNA synthesis and cell division stages by affecting calmodulin and phosphorylation of calmodulin-dependent protein. Moreover, the results of these studies have had an important role in the development of new anti-*Candida* medicines [27].

Besides these properties, it has been shown that CPZ acts as an efflux pump inhibitor, showing an important role in the fight against multiple drug resistance in bacteria [30] and also behaving as a stabiliser when it interacts with *S. aureus* and *Vibrio cholera* [31]. CPZ lowers cell membrane permeability with respect to water, as well as to potassium and sodium ions and alters glucose membrane transport which explains its antimicrobial property [31].

An antimicrobial agent, that has a weight lower than 600 Da and is soluble in water, presents a high probability of passing the pores channels filled with water of cell's outer membrane in order to reach a specific target responsible for bacterial resistance. Both CPZ and TZ are small molecules having a molecular weight of 318.86 Da and respectively 370.56 Da, are soluble in water and are lipophilic. One possible mode of action could be the binding to cytoplasmic membrane or to cytoplasm, where its start to exert their effects, after penetrating outer membrane through pores. The antimicrobial agent effects can be of microbiostatic or microbicidal nature [32]. The microbicidal effect refers to irreversible destruction of the microorganism (bactericidal and fungicidal) and microbiostatic effect is the reversible, temporary inhibition of bacterial multiplication (bacteriostatic, fungistatic) [33].

The processes involving the antimicrobial agents like the transport to the cell surface, adsorption, penetration, diffusion, and the interaction with target site, are not instantaneous, the time of action being different from one microorganism to another. Differences depend on the mode of action and the chemical structure and physicochemical properties of the agent [32]. The different values resulted for the optimum irradiation times of CPZ and TZ solutions in order to obtain the best antimicrobial effect against microorganisms suggest that there are necessary different photoproducts or concentrations of photoproducts. For MIC assay, in Gram-positive bacteria were obtained values between 1.29 μg/ml and 10.3 μg/ml for both CPZ and TZ. The optimum exposure time intervals to laser radiation for TZ were higher than for CPZ, respectively 60 and 120 min.

Gram-negative bacteria presented a smaller susceptibility to antimicrobial agents than Gram-positive bacteria, where MIC values for irradiated CPZ and TZ ranged from 5.16 μg/ml to 330 μg/ml. Also, in this case, irradiated TZ is less effective than CPZ, where the recorded MIC values were 82.5 μg/ml for all tested bacteria, with the exception of *ABC* 230 for which no difference was observed between irradiated and unirradiated samples. Furthermore, for fungi, MIC values were obtained between 1.29 μg/ml and 10.31 μg/ml.

The MIC and MBEC assays for CPZ and TZ suggest that they can be considered antimicrobial, antibiofilm and antifungal agents, but their mechanism(s) of antimicrobial action is still unknown. The possible effects which may be mediated by irradiated solution can be: direct inhibition of microbial chromosome replication [34], reduction of antimicrobial resistance through increased efflux of drug [35], inhibition of the motility of bacteria [36], elimination of plasmids [37], or inhibition of efflux pumps [38].

## Applications of Phenothiazines Modified by Exposure to UV Laser Radiation in Biomedicine

One of the most important research direction in biomedical science and applications is to find ways to combat failure of infectious diseases and of cancer treatment, when MDR is acquired by the target. Fighting MDR involves drug discovery and development or activity up-grading of existing drugs. The concept of optically induced structure modification of the available drugs molecules could be a strong pathway to achieve this purpose.

MDR has become a factor seriously limiting the treatment of various diseases, including bacterial [39, 40] and fungal [41] infections and cancer [42, 43].

Worldwide, large efforts are directed to design new drugs or to enhance the potency of existing ones in order to combat MDR. In this respect, recently, scientific community's attention was directed to evaluate antimicrobial activity of the so called non-antibiotics. The term 'non-antibiotics' is taken to include a variety of compounds which are employed in the management of pathological conditions of a non-infectious aetiology, but which modify cell permeability and have been shown to exhibit broad-spectrum antimicrobial activity [44]. Despite belonging to antipsychotic drug class, phenothiazines have been shown to have activity against a wide range of bacteria, including those that are resistant to representatives from two or more classes of antibiotics [44 - 49]. The initially observed effects of phenothiazines in humans, acting on dopamine $D_2$ receptor of neurons, has led to their widespread use for the treatment of various mental disorders, such as anxiety, depression and psychosis. However, during their long history of medicinal use, phenothiazines' pharmaceutical action has been successively completed by clinical observations and experimental studies showing that some of these agents also possess non-psychotropic effects by interfering with molecular pathways playing pivotal roles in cell survival, proliferation, tumour genesis and metastasis. Phenothiazines have been shown to have activity against a wide gamut of bacteria [50] including those resistant to representatives from two or more classes of antibiotics [44 - 46, 48 - 50]. Thus, drugs of the phenotiazine class have been found to interact with ion channels and efflux pumps such as P-glycoprotein (P-gp) and multidrug resistance associated protein 1 (MRP-1), augmenting cytoplasmic retention of chemotherapeutic agents [51]. Phenothiazines have also been shown to selectively inhibit various signalling pathways involving 3'-Phosphoinositide-dependent kinase (PDK1)/Akt [52] and phosphatidylinositol-3 kinase (PI3K)/Akt [53] or calmodulin (CaM) [54], to induce apoptosis of cells derived from a variety of cancers such as that of the ovary [53], colon [55], lung, kidney, prostate and breast [56], leukaemia and glioblastoma, while showing a lower cytotoxic effect on non-malignant blood lymphocytes [57] or normal cerebral tissue [58]. Moreover, direct interaction of phenothiazines with lipid bilayers adds yet another element to their plethora of pharmacological effects responsible for interaction with fundamental cellular processes [59]. Hence, phenothiazines may be considered as lead compounds for development of new and effective drugs used in MDR treatments of bacteria.

Chemical manipulation of a compound for the creation of new derivatives is limited by existing organic chemistry, is time consuming and allows a quite small percent yield of an active product. According to literature reports, exposure of a phenothiazine to UV-Vis yields a product that is more biologically active than un-exposed parent [60, 61]. However, because the product is highly unstable, the demonstration of increased activity was usually possible only when exposure of phenothiazine to incoherent UV radiation takes place in presence of the biological target, for instance bacteria [62, 63]. Moreover, because the source of incoherent UV radiation emits a wide spectrum of wavelengths, little is known of the response of a given phenothiazine to a much narrower spectrum, such as that provided by a laser emitter. Lasers have been used in recent years for localised effects on tissues when the tissue and the phenothiazine are simultaneously irradiated [64]. The difference between exposures of a phenothiazine to an UV laser, versus an incoherent UV radiation is that laser produces a specific wavelength ($\lambda$) that may be, practically, as narrow as $10^{-3}$ nm or even lower. In laser case, an energy level at the given wavelength is $10^{3}$-$10^{5}$ fold greater, at the same $\lambda$, than that produced by the UV incoherent source employed in biological systems [65].

Because of the specificity of wavelength and beam energy emitted by a laser, the exposure of a phenothiazine to given laser beams yields rearrangements of molecule(s) whose bioactivity may be studied without concurrent exposure of biological target to the laser, that is, provided the rearranged compound(s) is sufficiently stable for study. Moreover, if exposure of phenothiazine to laser radiation is conducted in the presence of a reactive compound, entirely new compounds can be quickly created within hours if not minutes that may be difficult, if not impossible to create *via* conventional chemical manipulation.

The results reported in [66] demonstrate that, indeed, exposure of phenothiazines that have antibacterial activity to lasers alters their absorption and fluorescence spectra and causes development of coloured precipitates which are reversible with time. Because altered spectra remain stable for about 4 days and colour is transient, one may conclude that at least two distinct molecular species result from exposure to laser beams.

The fast modification of the molecular structure of non-antibiotics by exposure to laser radiation is seldom reported, at the moment. Also, the use of nano-/micro-droplets produced by a larger diameter droplet interaction with laser beams in view to transport modified medicines to eye tissues targets infected with bacteria is an approach that is for the first time proposed in [10 - 12, 14, 67 - 71].

## CONCLUSIONS

The susceptibility of microorganisms to unirradiated and irradiated water solutions of phenothiazine derivatives, CPZ and TZ were investigated *in vitro* by analysing the minimum inhibitory concentration and the minimum biofilm eradication concentration. The samples were irradiated with a laser beam emitted at 266 nm by a pulsed Nd:YAG laser for time intervals between 1 and 240 min.

The results showed a strong antifungal activity of both irradiated CPZ and irradiated TZ, where the best MIC for both *C. neoformans* 204092 and *C. albicans* 26790 were obtained for 30 min irradiated TZ.

The tests on Gram-positive bacteria showed a strong antimicrobial activity of irradiated CPZ and TZ, where for CPZ were obtained better results from the point of view of the optimum irradiation time.

Gram-negative bacteria are less susceptible to antimicrobial agents than Gram-positive bacteria. In this case, unirradiated and irradiated TZ have a moderate action, smaller than that of CPZ. For *ABC* 230 strain there is no difference between MIC values of unirradiated and irradiated TZ solution for 1 and 240 min. The same applies for *P. aeruginosa* when unirradiated and irradiated CPZ solutions were tested.

The MBEC assay performed on *S. aureus* 6538 and on *S. aureus* MSSA shows that irradiated TZ is more effective in preventing the adherence of bacteria on the wall of the well, when compared with the results obtained for irradiated CPZ.

Therefore, the use of UV laser radiation emitted at 266 nm for exposing non-antibiotic solutions for different periods of time in order to generate photoproducts with antimicrobial properties can be considered a new method that can be used to fight microbial resistance.

## CONFLICT OF INTEREST

The authors confirm that this chapter content has no conflict of interest.

## ACKNOWLEDGEMENTS

This work has been financed by the National Authority for Research and Innovation in the frame of Nucleus programme-contract 4N/2016 and project PN-II-PT-PCCA-2011-3.1-1350.

## REFERENCES

[1] A. Bernthsen, "Zur Kenntniss des Methylenblau und verwandter Farbstoffe", *Berichte Dtsch. Chem. Ges,* vol. 16, no. 2, pp. 2896-2904, 1883. [http://dx.doi.org/10.1002/cber.188301602249]

[2] P. Charpentier, "Phenothiazine derivatives and process for their preparation", *Serial No. 219,519,* 1951.

[3] B. Zwanenburg, *Phenothiazines and 1,4-benzothiazines. Chemical and Biomedical aspects*, Elsevier: Amsterdam, 1988.

[4] J.M. Ford, W.C. Prozialeck, and W.N. Hait, "Structural features determining activity of phenothiazines and related drugs for inhibition of cell growth and reversal of multidrug resistance", *Mol. Pharmacol,* vol. 35, no. 1, pp. 105-115, 1989. [PMID: 2563302]

[5] R. Liu, and F.J. Sharom, "Site-Directed Fluorescence Labeling of P-Glycoprotein on Cysteine Residues in the Nucleotide Binding Domains", *Biochemistry (Mosc.),* vol. 35, no. 36, pp. 11865-11873, 1996.

[6] A. Jaszczyszyn, K. Gąsiorowski, P. Świątek, W. Malinka, K. Cieślik-Boczula, J. Petrus, and B. Czarnik-Matusewicz, "Chemical structure of phenothiazines and their biological activity", *Pharmacol. Rep,* vol. 64, no. 1, pp. 16-23, 2012. [http://dx.doi.org/10.1016/S1734-1140(12)70726-0] [PMID: 22580516]

[7] A.B. Hendrich, O. Wesołowska, N. Motohashi, J. Molnár, and K. Michalak, "New phenothiazine-type multidrug resistance modifiers: anti-MDR activity versus membrane perturbing potency", *Biochem. Biophys. Res. Commun,* vol. 304, no. 2, pp. 260-265, 2003. [http://dx.doi.org/10.1016/S0006-291X(03)00580-1] [PMID: 12711308]

[8] A. Ramu, and N. Ramu, "Reversal of multidrug resistance by phenothiazines and structurally related compounds", *Cancer Chemother. Pharmacol,* vol. 30, no. 3, pp. 165-173, 1992. [http://dx.doi.org/10.1007/BF00686306] [PMID: 1628364]

[9] M. Alcolea Palafox, M. Gil, J.L. Núñez, and G. Tardajos, "Study of phenothiazine and N-methyl phenothiazine by infrared, raman, 1H-, and 13C-NMR spectroscopies", *Int. J. Quantum Chem,* vol. 89, no. 3, pp. 147-171, 2002. [http://dx.doi.org/10.1002/qua.10314]

[10] M.L. Pascu, B. Danko, A. Martins, N. Jedlinszki, T. Alexandru, V. Nastasa, M. Boni, A. Militaru, I.R. Andrei, A. Staicu, A. Hunyadi, S. Fanning, and L. Amaral, "Exposure of chlorpromazine to 266 nm laser beam generates new species with antibacterial properties: contributions to development of a new process for drug discovery", *PLoS One,* vol. 8, no. 2, p. e55767, 2013. [http://dx.doi.org/10.1371/journal.pone.0055767] [PMID: 23405212]

[11] T. Alexandru, A. Staicu, A. Pascu, E. Radu, A. Stoicu, V. Nastasa, A. Dinache, M. Boni, L. Amaral, and M.L. Pascu, "Characterization of mixtures of compounds produced in chlorpromazine aqueous solutions by ultraviolet laser irradiation: their applications in antimicrobial assays", *J. Biomed. Opt,* vol. 20, no. 5, p. 051002, 2015. [http://dx.doi.org/10.1117/1.JBO.20.5.051002] [PMID: 25365041]

[12] A.M. Armada, T. Alexandru, D. Machado, B. Danko, A. Hunyadi, A. Dinache, V. Nastasa, M. Boni, J. Ramos, M. Viveiros, J. Molnar, M.L. Pascu, and L. Amaral, "The in vitro activity of products formed from exposure of chlorpromazine to a 266 nm laser beam against species of mycobacteria of human interest", *In Vivo,* vol. 27, no. 5, pp. 605-610, 2013. [PMID: 23988894]

[13] A. Simon, T. Alexandru, M. Boni, V. Damian, A. Stoicu, V. Dutschk, and M.L. Pascu, "Interaction of solutions containing phenothiazines exposed to laser radiation with materials surfaces, in view of biomedical applications", *Int. J. Pharm,* vol. 475, no. 1-2, pp. 270-281, 2014. [http://dx.doi.org/10.1016/j.ijpharm.2014.08.031] [PMID: 25148730]

[14] M.C. Morán, T. Tozar, A. Simon, A. Dinache, A. Smarandache, I.R. Andrei, M. Boni, M.L. Pascu, F. Cirisano, and M. Ferrari, "Toxicity study in blood and tumor cells of laser produced medicines for application in fabrics", *Colloids Surf. B Biointerfaces,* vol. 137, pp. 91-103, 2016. [http://dx.doi.org/10.1016/j.colsurfb.2015.06.041] [PMID: 26187648]

[15] R. Schwalbe, L. Steele-Moore, and A. C. Goodwin, *Antimicrobial Susceptibility Testing Protocols,* CRC Press, 2007.

[16] M.B. Coyle, *Manual of Antimicrobial Susceptibility Testing*, American Society For Microbiology, 2005.

[17] C.R. Mahon, D.C. Lehman, and G. Manuselis, *Textbook of Diagnostic Microbiology,* Saunders/Elsevier, 2011.

[18] K. Rogers, *Bacteria and Viruses*, Enciclopedia Britannica: New York, 2011.

[19] C. Ramos, T.R. Licht, C. Sternberg, K.A. Krogfelt, and S. Molin, "Monitoring bacterial growth activity in biofilms from laboratory flow chambers, plant rhizosphere, and animal intestine", *Methods Enzymol,* vol. 337, pp. 21-42, 2001. [http://dx.doi.org/10.1016/S0076-6879(01)37004-0] [PMID: 11398430]

[20] R.J. Doyle, *Microbial Growth in Biofilms: Special environments and physicochemical aspects,* Gulf Professional Publishing, 2001.

[21] N.M. Abdallah, S.B. Elsayed, M.M. Mostafa, and G.M. El-gohary, "Biofilm forming bacteria isolated from urinary tract infection, relation to catheterization and susceptibility to antibiotics", *Can. J. Vet. Res,* vol. 2, no. 10, pp. 172-178, 2011.

[22] U. Kohler, and F. Kreuter, *Data Analysis Using Stata,* Stata Press, 2005.

[23] W. Paulus, *Microbicides for the Protection of Materials: A Handbook*, Springer Science & Business Media, 2012.

[24] Y.S. McCarter, *Manual of Antimicrobial Susceptibility Testing*, American Society for Microbiology, 2005.

[25] A. De Meijere, *Science of Synthesis: Houben-Weyl Methods of Molecular Transformations*, vol. 17. Thieme, 2014.

[26] P. Guilfoile, and I.E. Alcamo, *Antibiotic-Resistant Bacteria*, Infobase Publishing, 2007.

[27] J.J. Aaron, M.D. Seye, S. Trajkovska, and N. Motohashi, "Bioactive Phenothiazines and Benzo[a]phenothiazines: Spectroscopic Studies, and Biological and Biomedical Properties and Applications", In: *Bioactive Heterocycles VII*, Springer: Berlin, 2008.

[28] N. Komatsu, N. Motohashi, M. Fujimaki, and J. Molnár, "Induction of a protective immunity in mice against Escherichia coli by phenothiazines, 10-[n-(phthalimido)alkyl]-2-substitute--10H-phenothiazines and 1-(2-chloroethyl)-3-(2-substituted-10H-phenothiazines-10-yl)alkyl-1 -ureas", *In Vivo,* vol. 11, no. 1, pp. 13-16, 1997. [PMID: 9067767]

[29] S. Sharma, H. Kaur, and G.K. Khuller, "Cell cycle effects of the phenothiazines: trifluoperazine and chlorpromazine in Candida albicans", *FEMS Microbiol. Lett,* vol. 199, no. 2, pp. 185-190, 2001. [http://dx.doi.org/10.1111/j.1574-6968.2001.tb10672.x] [PMID: 11377865]

[30] M. Martins, S.G. Dastidar, S. Fanning, J.E. Kristiansen, J. Molnar, J-M. Pagès, Z. Schelz, G. Spengler, M. Viveiros, and L. Amaral, "Potential role of non-antibiotics (helper compounds) in the treatment of multidrug-resistant Gram-negative infections: mechanisms for their direct and indirect activities", *Int. J. Antimicrob. Agents,* vol. 31, no. 3, pp. 198-208, 2008. [http://dx.doi.org/10.1016/j.ijantimicag.2007.10.025] [PMID: 18180147]

[31] J.E. Kristiansen, and K. Gaarslev, "The antibacterial effect of selected neuroleptics on Vibrio cholerae", *Acta Pathol. Microbiol. Immunol. Scand. [B],* vol. 93, no. 1, pp. 49-51, 1985. [PMID: 2858954]

[32] W. Paulus, *Microbicides for the Protection of Materials,* Springer Netherlands: Dordrecht, 1993. [http://dx.doi.org/10.1007/978-94-011-2118-7]

[33] G. Broasca, and D. Farima, "Antimicrobial treatment on textile surfaces", In: *Rev. Piel. Încălţăminte*, vol. 10, no. 4, pp. 61-66, 2010.

[34] L. Amaral, J. Kristiansen, and V. Lorian, "Synergic effect of chlorpromazine on the activity of some antibiotics", *J. Antimicrob. Chemother,* vol. 30, no. 4, pp. 556-558, 1992. [http://dx.doi.org/10.1093/jac/30.4.556] [PMID: 1490927]

[35] M. Viveiros, A. Jesus, M. Brito, C. Leandro, M. Martins, D. Ordway, A.M. Molnar, J. Molnar, and L. Amaral, "Inducement and reversal of tetracycline resistance in Escherichia coli K-12 and expression of proton gradient-dependent multidrug efflux pump genes", *Antimicrob. Agents Chemother,* vol. 49, no. 8, pp. 3578-3582, 2005. [http://dx.doi.org/10.1128/AAC.49.8.3578-3582.2005] [PMID: 16048990]

[36] J. Molnár, J. Ren, J.E. Kristiansen, and M.J. Nakamura, "Effects of some tricyclic psychopharmacons and structurally related compounds on motility ofProteus vulgaris", *Antonie van Leeuwenhoek,* vol. 62, no. 4, pp. 315-320, 1992. [http://dx.doi.org/10.1007/BF00572600]

[37] G. Spengler, A. Miczák, E. Hajdú, M. Kawase, L. Amaral, and J. Molnár, "Enhancement of plasmid curing by 9-aminoacridine and two phenothiazines in the presence of proton pump inhibitor 1-(--benzoxazolyl)-3,3,3-trifluoro-2-propanone", *Int. J. Antimicrob. Agents,* vol. 22, no. 3, pp. 223-227, 2003. [http://dx.doi.org/10.1016/S0924-8579(03)00207-3] [PMID: 13678825]

[38] S. Sabatini, G.W. Kaatz, G.M. Rossolini, D. Brandini, and A. Fravolini, "From phenothiazine to 3-phenyl-1,4-benzothiazine derivatives as inhibitors of the Staphylococcus aureus NorA multidrug efflux pump", *J. Med. Chem,* vol. 51, no. 14, pp. 4321-4330, 2008. [http://dx.doi.org/10.1021/jm701623q] [PMID: 18578473]

[39] J. Chevalier, A. Mahamoud, M. Baitiche, E. Adam, M. Viveiros, A. Smarandache, A. Militaru, M.L. Pascu, L. Amaral, and J-M. Pagès, "Quinazoline derivatives are efficient chemosensitizers of antibiotic activity in Enterobacter aerogenes, Klebsiella pneumoniae and Pseudomonas aeruginosa resistant strains", *Int. J. Antimicrob. Agents,* vol. 36, no. 2, pp. 164-168, 2010. [http://dx.doi.org/10.1016/j.ijantimicag.2010.03.027] [PMID: 20494558]

[40] H. Nikaido, and J-M. Pagès, "Broad-specificity efflux pumps and their role in multidrug resistance of Gram-negative bacteria", *FEMS Microbiol. Rev,* vol. 36, no. 2, pp. 340-363, 2012. [http://dx.doi.org/10.1111/j.1574-6976.2011.00290.x] [PMID: 21707670]

[41] K. Gulshan, and W.S. Moye-Rowley, "Multidrug resistance in fungi", *Eukaryot. Cell,* vol. 6, no. 11, pp. 1933-1942, 2007. [http://dx.doi.org/10.1128/EC.00254-07] [PMID: 17873085]

[42] "Cancer multidrug resistance", *Nat. Biotechnol,* vol. 18, suppl. , pp. IT18-IT20, 2000. [PMID: 11221708]

[43] M.M. Gottesman, T. Fojo, and S.E. Bates, "Multidrug resistance in cancer: role of ATP-dependent transporters", *Nat. Rev. Cancer,* vol. 2, no. 1, pp. 48-58, 2002. [http://dx.doi.org/10.1038/nrc706] [PMID: 11902585]

[44] J.E. Kristiansen, O. Hendricks, T. Delvin, T.S. Butterworth, L. Aagaard, J.B. Christensen, V.C. Flores, and H. Keyzer, "Reversal of resistance in microorganisms by help of non-antibiotics", *J. Antimicrob. Chemother,* vol. 59, no. 6, pp. 1271-1279, 2007. [http://dx.doi.org/10.1093/jac/dkm071] [PMID: 17403708]

[45] J.E. Kristiansen, and L. Amaral, "The potential management of resistant infections with non-antibiotics", *J. Antimicrob. Chemother,* vol. 40, no. 3, pp. 319-327, 1997. [http://dx.doi.org/10.1093/jac/40.3.319] [PMID: 9338482]

[46] L. Amaral, M. Martins, and M. Viveiros, "Enhanced killing of intracellular multidrug-resistant Mycobacterium tuberculosis by compounds that affect the activity of efflux pumps", *J. Antimicrob. Chemother,* vol. 59, no. 6, pp. 1237-1246, 2007. [http://dx.doi.org/10.1093/jac/dkl500] [PMID: 17218448]

[47] L. Amaral, M. Martins, and M. Viveiros, "Phenothiazines as anti-multi-drug resistant tubercular agents", *Infect. Disord. Drug Targets,* vol. 7, no. 3, pp. 257-265, 2007. [http://dx.doi.org/10.2174/187152607782110022] [PMID: 17897062]

[48] S. Sabatini, G.W. Kaatz, G.M. Rossolini, D. Brandini, and A. Fravolini, "From phenothiazine to 3-phenyl-1,4-benzothiazine derivatives as inhibitors of the Staphylococcus aureus NorA multidrug efflux pump", *J. Med. Chem,* vol. 51, no. 14, pp. 4321-4330, 2008. [http://dx.doi.org/10.1021/jm701623q] [PMID: 18578473]

[49] H.D. Coutinho, J.G. Costa, E.O. Lima, V.S. Falcão-Silva, and J.P. Siqueira-Júnior, "In vitro interference of Hyptis martiusii Benth. & chlorpromazine against an aminoglycoside-resistant Escherichia coli", *Indian J. Med. Res,* vol. 129, no. 5, pp. 566-568, 2009. [PMID: 19675386]

[50] J-M. Pagès, and L. Amaral, "Mechanisms of drug efflux and strategies to combat them: challenging the efflux pump of Gram-negative bacteria", *Biochim. Biophys. Acta,* vol. 1794, no. 5, pp. 826-833, 2009. [http://dx.doi.org/10.1016/j.bbapap.2008.12.011] [PMID: 19150515]

[51] W. Liang, C. Yang, J. Qi, H. Peng, J. Duan, H. Liu, and D. Zheng, "Modulating multidrug resistance through inhibiting of protein kinase C activity by phenothiazines", *Chin. Sci. Bull,* vol. 43, no. 14, pp. 1196-1200, 1998. [http://dx.doi.org/10.1007/BF02883224]

[52] J.H. Choi, Y.R. Yang, S.K. Lee, S-H. Kim, Y-H. Kim, J-Y. Cha, S-W. Oh, J-R. Ha, S.H. Ryu, and P-G. Suh, "Potential inhibition of PDK1/Akt signaling by phenothiazines suppresses cancer cell proliferation and survival", *Ann. N. Y. Acad. Sci,* vol. 1138, no. 1, pp. 393-403, 2008. [http://dx.doi.org/10.1196/annals.1414.041] [PMID: 18837915]

[53] S.B. Rho, B-R. Kim, and S. Kang, "A gene signature-based approach identifies thioridazine as an inhibitor of phosphatidylinositol-3-kinase (PI3K)/AKT pathway in ovarian cancer cells", *Gynecol. Oncol,* vol. 120, no. 1, pp. 121-127, 2011. [http://dx.doi.org/10.1016/j.ygyno.2010.10.003] [PMID: 21035837]

[54] M.K. Hwang, Y.K. Min, and S.H. Kim, "Calmodulin inhibition contributes to sensitize TRAIL-induced apoptosis in human lung cancer H1299 cells", *Biochem. Cell Biol,* vol. 87, no. 6, pp. 919-926, 2009. [http://dx.doi.org/10.1139/O09-058] [PMID: 19935877]

[55] B. Pajak, J. Molnar, H. Engi, and A. Orzechowski, "Preliminary studies on phenothiazine-mediated reversal of multidrug resistance in mouse lymphoma and COLO 320 cells", *In Vivo,* vol. 19, no. 6, pp. 1101-1104, 2005. [PMID: 16277030]

[56] K. Pluta, M. Jeleń, B. Morak-Młodawska, M. Zimecki, J. Artym, and M. Kocięba, "Anticancer activity of newly synthesized azaphenothiazines from NCIs anticancer screening bank", *Pharmacol. Rep,* vol. 62, no. 2, pp. 319-332, 2010. [http://dx.doi.org/10.1016/S1734-1140(10)70272-3] [PMID: 20508288]

[57] Z. Zhelev, H. Ohba, R. Bakalova, V. Hadjimitova, M. Ishikawa, Y. Shinohara, and Y. Baba, "Phenothiazines suppress proliferation and induce apoptosis in cultured leukemic cells without any influence on the viability of normal lymphocytes. Phenothiazines and leukemia", *Cancer Chemother. Pharmacol,* vol. 53, no. 3, pp. 267-275, 2004. [http://dx.doi.org/10.1007/s00280-003-0738-1] [PMID: 14663628]

[58] I. Gil-Ad, B. Shtaif, Y. Levkovitz, M. Dayag, E. Zeldich, and A. Weizman, "Characterization of phenothiazine-induced apoptosis in neuroblastoma and glioma cell lines: clinical relevance and possible application for brain-derived tumors", *J. Mol. Neurosci,* vol. 22, no. 3, pp. 189-198, 2004. [http://dx.doi.org/10.1385/JMN:22:3:189] [PMID: 14997012]

[59] M. Suwalsky, F. Villena, C.P. Sotomayor, S. Bolognin, and P. Zatta, "Human cells and cell membrane molecular models are affected in vitro by chlorpromazine", *Biophys. Chem,* vol. 135, no. 1-3, pp. 7-13, 2008. [http://dx.doi.org/10.1016/j.bpc.2008.02.014] [PMID: 18372093]

[60] M. Wainwright, "Photodynamic therapy: the development of new photosensitisers", *Anticancer. Agents Med. Chem,* vol. 8, no. 3, pp. 280-291, 2008. [http://dx.doi.org/10.2174/187152008783961888] [PMID: 18393787]

[61] M. Wainwright, and L. Amaral, "The phenothiazinium chromophore and the evolution of antimalarial drugs", *Trop. Med. Int. Health,* vol. 10, no. 6, pp. 501-511, 2005. [http://dx.doi.org/10.1111/j.1365-3156.2005.01417.x] [PMID: 15941412]

[62] F. Harris, L.K. Chatfield, and D.A. Phoenix, "Phenothiazinium based photosensitisersphotodynamic agents with a multiplicity of cellular targets and clinical applications", *Curr. Drug Targets,* vol. 6, no. 5, pp. 615-627, 2005. [http://dx.doi.org/10.2174/1389450054545962] [PMID: 16026282]

[63] D.A. Phoenix, and F. Harris, "Light activated compounds as antimicrobial agents - patently obvious?", *Recent Pat Antiinfect Drug Discov,* vol. 1, no. 2, pp. 181-199, 2006. [http://dx.doi.org/10.2174/157489106777452656] [PMID: 18221144]

[64] H. Ficheux, "Photodynamic therapy: principles and therapeutic indications", *Ann. Pharm. Fr,* vol. 67, no. 1, pp. 32-40, 2009. [http://dx.doi.org/10.1016/j.pharma.2008.10.013] [PMID: 19152849]

[65] L. Danaila, and M.L. Pascu, *Lasers in Neurosurgery,* Acad. Rom. Ed., Bucharest, , 2001.

[66] M.L. Pascu, V. Nastasa, A. Smarandache, A. Militaru, A. Martins, M. Viveiros, M. Boni, I.R. Andrei, A. Pascu, A. Staicu, J. Molnar, S. Fanning, and L. Amaral, "Direct modification of bioactive phenothiazines by exposure to laser radiation", *Recent Pat Antiinfect Drug Discov,* vol. 6, no. 2, pp. 147-157, 2011. [http://dx.doi.org/10.2174/157489111796064515] [PMID: 21517738]

[67] T. Alexandru, M.L. Pascu, B. Danko, V. Nastasa, M. Boni, A. Militaru, I.R. Andrei, A. Staicu, A. Hunyadi, A. Armada, M. Viveiros, and L. Amaral, "Generation and biological evaluation of the products formed from the exposure of Phenothiazine to a 266 nm laser beam",*Proc. SPIE,* vol. 8882, pp. 88820T-7, 2013. [http://dx.doi.org/10.1117/12.2032666]

[68] I.R. Andrei, T. Tozar, A. Dinache, M. Boni, V. Nastasa, and M.L. Pascu, "Chlorpromazine transformation by exposure to ultraviolet laser beams in droplet and bulk", *Eur. J. Pharm. Sci,* vol. 81, pp. 27-35, 2016. [http://dx.doi.org/10.1016/j.ejps.2015.09.017] [PMID: 26432595]

[69] V. Nastasa, V. Pradines, I.R. Andrei, M. Boni, M.L. Pascu, and R. Miller, "Studies about the generation and characterisation of microdroplets with a controlled content", *Optoelectron. Adv. Mater.-Rapid Commun,* vol. 4, no. 11, pp. 1916-1919, 2010.

[70] M.L. Pascu, G.V. Popescu, C.M. Ticos, and I.R. Andrei, "Unresonant interaction of laser beams with microdroplets", *J. Eur. Opt. Soc. Rapid Publ.,* vol. 7, p. 12001, 2012. [http://dx.doi.org/10.2971/jeos.2012.12001]

[71] M.L. Pascu, I.R. Andrei, M. Ferrari, A. Staicu, A. Smarandache, A. Mahamoud, V. Nastasa, and L. Liggieri, "Laser beams resonant interaction with micro-droplets which have a controlled content", *Colloid Surface A,* vol. 365, no. 1-3, pp. 83-88, 2010. [http://dx.doi.org/10.1016/j.colsurfa.2010.02.036]

## CHAPTER 14

# Application of Optically Modified Medicines in Fighting Pseudotumours

**Ruxandra Pirvulescu**[1,*], **Tatiana Tozar**[2], **Alexandru Stoicu**[2] and **Mihail Lucian Pascu**[2,3]

[1] *University of Medicine and Pharmacy "Carol Davila", Emergency University Hospital, Ophthalmology Clinic, Bucharest, Romania*

[2] *National Institute for Laser, Plasma and Radiation Physics, Magurele, Ilfov, Romania*

[3] *Faculty of Physics, University of Bucharest, Magurele, Ilfov, Romania*

**Abstract:** Results are shown about the applications of medicines modified by exposure to optical radiation on pseudotumours indusced in rabbit eyes. These consist, first, in characterising modifications produced on medicines by exposing them to optical/laser radiation. Selected medicines are (i) cytostatics: methotrexate, 5-fluorouracil, (ii) benzopyridine derivatives: BG 204, BG 1120, (iii) phenothiazines: chlopromazine. They were used as solutions either in water or in natural saline. The anti-tumour properties of resulted solutions are measured after applying them on pseudotumours induced in rabbit eyes by Schmidt-Erfurth method. The use of methotrexate on pseudotumours has shown that eye conjunctive and neovascularisation disappear after 1-2 treatments which consists in exposure of eye injected with unirradiated methotrexate solution to cw Hg lamp radiation. Even if the use of methotrexate fastens the recovery, one should avoid direct exposure of eye to UV-Vis radiation. 5-fluorouracil was exposed to nitrogen pulsed laser beam which was sent to the eye already impregnated with cytostatics. Eye conjunctive and neovascularisation disappeared after 1-2 treatments, recommending 5-fluorouracil to cure such pseudotumours by combining its action with that of 337.1 nm laser. BG 1120 was used only exposed to cw Xe lamp incoherent radiation, without irradiating the eyes impregnated with solutions of it. BG 1120 modified by exposure to optical radiation leads to faster decrease of inflammations associated with pseudotumours and of neovascularisation of conjunctive tissue. For chlorpromazine, the most efficient in the recovery of a pseudotumour tissue is the sample irradiated 20 min, at 10 mg/ml concentration in water.

**Keywords:** Absorption spectroscopy, Benzopyridine derivatives, Chlorpromazine, Cytostatics, 5-fluorouracil, FTIR, Hg-lamp, Laser inducedfluorescence, Laser spectroscopy , Malignant tumours, Methotrexate,

* **Corresponding author Ruxandra Pirvulescu:** University of Medicine and Pharmacy "Carol Davila", Emergency University Hospital, Ophthalmology Clinic, Bucharest, Romania; Tel: 0040 21 3180562; Fax: 0040 21 3180554; E-mail: ruxandrapascu78@gmail.com

Multiple drug resistance, Nitrogen pulsed laser, Ophthalmology, Phenothiazine derivatives, Photochemistry, Pseudotumours, Solid state laser, Xenon lamp.

## INTRODUCTION

Multiple drug resistance or multi-drug resistance (MDR) is acquired by bacteria, in a broader sense by microorganisms, but it is also signalled in the malignant tumours treatment where the cancer tissues may not be controlled after a particular time moment of their development in human body, if a given stage in cancer development was reached. The tumours are resistant to treatment, in particular to chemotherapy, and new methods to treat them and other drugs are requested to overpass or even cure this illness. In this chapter, results about the use of some drugs are reported which although, normally, do not have anti-tumour effects may generate through exposure to optical radiation photoproducts that by individual or combined action may prove to be efficient at least in the treatment of pseudotumours tissues produced on rabbit eye conjunctive.

The reported results are organised in two parts: the first, contains the description of drugs exposed to optical incoherent radiation emitted by continuous mode (cw) high pressure gas lamps (Hg and Xe) or by pulsed UV lasers. The second shows the effects that are exhibited by some exposed medicines on pseudotumour tissues either when drugs are injected in the eye conjunctive and exposed to optical UV and visible radiation *in situ*, or when drugs are exposed first to optical radiation, modified by it and then injected subconjunctivally in rabbit eyes. Of course, drugs in the two categories are different. Out of the utilised drugs, cytostatics played a core place, since they are already widely used in the treatment of proliferative processes. Optical beams were used to produce molecular structural changes of such drugs and to improve their effectiveness in tumour treatments.

Other classes of medicines were benzopyridine and phenothiazine derivatives that do not have necessarily anticancer effects, but may generate photoproducts during exposure to incoherent or coherent optical radiation which may be active compounds against tumours. Results in this respect obtained on pseudotumours produced by Schmidt-Erfurth method are shown in the chapter.

## MEDICINES

### Methotrexate (MTX)

MTX has the chemical formula $C_{20}H_{22}N_8O_5$ and molecular weight 454.44 g/mol; it is a yellow to orange-brown odourless crystalline powder that interferes with DNA and RNA synthesis [1, 2]. The MTX is sensitive to light, oxidation and hydrolysis [1, 2]. It is soluble in diluted solution of alkali hydroxides and carbonates, insoluble in chloroform, alcohol, ether and water and slightly soluble in diluted hydrochloric acid [3]. In therapeutic use, MTX acts as abortifacient [4], antirheumatic [5], nonsteroidal [6], antineoplastic [7], antimetabolites [8], dermatologic [9], immunosuppressive agents [10], and enzyme inhibitor [11].

MTX stops the DNA synthesis by inhibiting the action of dihydrofolate reductase (DHFR), thus reducing the tetrahydrofolate (THF) in the cell. THF has a major role in DNA synthesis [1, 12]. The difference between MTX and THF is the substitution with an amino group (-$NH_2$) of the oxygen in the pteridine ring. Photochemical studies on folic acid have shown the photo-dissociation effects of light on MTX [13] by measuring consecutive changes in absorption and fluorescence spectra with increasing exposure time to radiation; 6-formilpteridinic acid and p-aminobenzilglutamic derivatives are formed [1, 14]. Relatively recent studies and reports have proposed the combination MTX-radiation as a possible cancer treatment procedure [1, 14]. A detailed description and the molecular geometry of MTX can be found in Chapter 11.

### 5-fluorouracil (5-FU)

5-FU having chemical formula $C_4H_3FN_2O_2$ and molecular weight 130.077 g/mol, is an odourless, crystalline powder used for its antineoplastic activity [15]. It is soluble in water and methanol [16] and stable after exposure to environmental air [17]. 5-FU is a pyrimidine derivative belonging to the cytostatic group that includes the so called antimetabolites. It is primarily used to treat colon, breast and pancreas cancer and, not least, ocular tumours. Antimetabolites generally alter DNA or RNA synthesis by interfering with the synthesis of nucleic acids [18].

The utility of 5-FU is limited by the fact that tumour tissue acquires resistance to it relatively quick [18, 19]. Therefore, it became necessary to discover new techniques potentiating 5-FU action, the exploitation of its photo-physical properties being a recommended approach.

At the same time, the combination between cytostatics and laser therapy as an alternative to cancer treatment was tested [12, 14, 17, 18, 20, 21]. The administrated cytostatic is selectively accumulated in tumours and the tissue is then irradiated with an optical beam emitted at properly chosen wavelength [13]. The tumour is destroyed by the chemotherapy effect of cytostatics and by the resulting compounds produced *via* irradiation [22]. This therapy could also induce an immune response against the treated tumour. In addition, the reaction could even lead to disappearance of metastatic tissue. A detailed description and the molecular geometry of 5-FU can be found in Chapter 11.

## Benzopyridinic Compounds

Benzopyridines are compounds recently developed, which due to their azaheterocyclic component can exert antifungal, antibacterial and anticancer effects [23]. Out of this category, the molecular structures of the compounds labelled BG 204 and BG 1120, are shown in Figs. (**1** and **2**). They are experimentally produced and have been supplied by the Faculty of Pharmacy, Mediterranean University, Marseille, France.

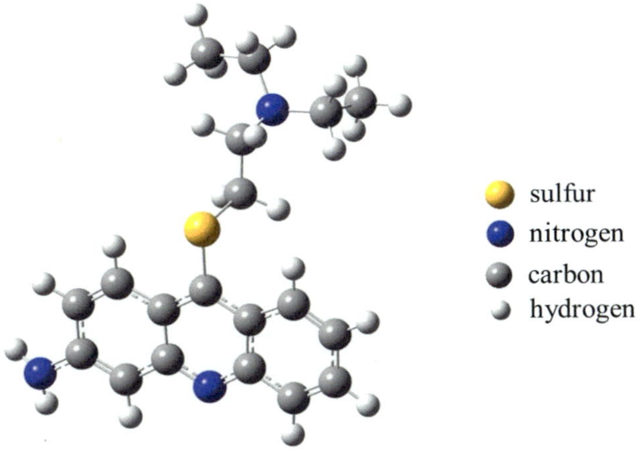


**Fig. (1).** 3D optimised structure of BG 204 for fundamental electronic state [obtained by utilising Gaussian09: DFT 6-31G (d,p)].

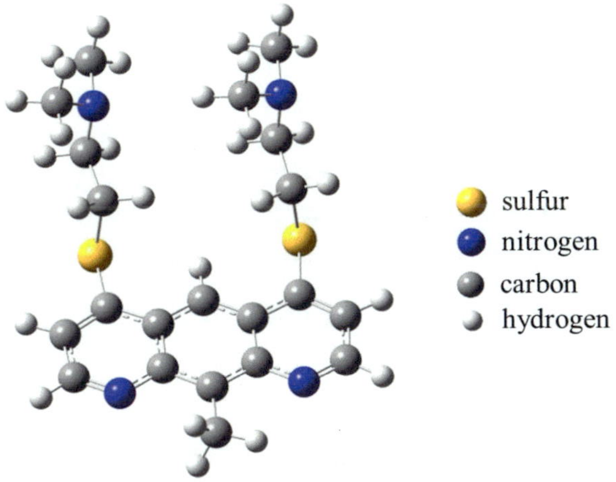


**Fig. (2).** 3D optimised structure of BG 1120 for fundamental electronic state [obtained by utilising Gaussian09: DFT 6-31G (d,p)].

Studies have shown the possible involvement of azaheterocyclic compounds in the treatment, prevention or amelioration of diseases involving angiogenesis. The compounds may be used in a variety of disorders such as abnormal tissue growth, neoplasia, hyperplasia and diabetic retinopathy [24]. In cancerous process, angiogenesis is of outmost importance in maintaining or increasing tumour size [25]; angiogenesis is a core component of the pathology of ocular neovascularisation diabetic retinopathy [24].

Thus, it has been demonstrated that the targeted treatment of angiogenesis may inhibit tumour growth [24, 25]. In addition, the new azaheterocyclic compounds were shown as being pain relievers and having anti-inflammatory properties [24].

### Phenothiazine Derivative

Chlorpromazine (CPZ) is a phenothiazine derivative primarily used as an antipsychotic drug. Phenothiazines were introduced first in clinical use in 1940 [25] and soon after, their antineoplastic properties were discovered. Certain phenothiazines derivatives possess, potentially, anti-tumour properties in which case target proteins associated to tumours have significant roles in cell signalling transduction cascades [27]. Studies regarding CPZ showed a potential use of it in cancer treatment. CPZ acts as a proliferation inhibitor or induces apoptosis of leukaemia cells [28, 29]. In these cases, CPZ induces a decrease of the viability of cells without affecting the normal lymphocyte [28] or melanoma cells [29]. Also, for tamoxifen-resistant and tamoxifen-sensitive human breast cancer cells it can improve the cytotoxic effect of tamoxifen [30].

Relatively recent studies reported that Gram-positive and Gram-negative bacteria are more susceptible to CPZ that is irradiated with a laser beam emitted at 266 nm by a Nd:YAG laser system, when compared to antimicrobial activity of unirradiated CPZ [31 - 34]. Such studies indicated that CPZ could have a great potential in clinical anti-tumour therapy and the results obtained on bacteria encourage further studies of the effects of irradiated CPZ on tumour cells.

## SPECTROSCOPIC STUDIES ON MEDICINES

### Methods

A first set of compounds that include MTX and 5-FU, has been studied and their effects on pseudotumour tissues were investigated before and after modifying them by exposure to optical radiation. A second class of studied substances are benzopyridinic compounds (generically called BG). For both cases, changes have been produced at molecular level of the parent compounds and their effectiveness in treating pseudotumours obtained on laboratory animals was highlighted.

A third class of investigated drugs includes the phenothiazines and their derivatives, out of which CPZ was the most explored. For these compounds, interaction with UV laser radiation generates photoproducts with anti-tumour activity.

In the following studies, for the irradiation of drug solutions and/or tissues impregnated with drugs were used four different optical radiation sources: Hg lamp, Xe lamp, $N_2$ laser and Nd:YAG laser. The lamps emit in continuous mode (cw) while each laser type emits a pulsed beam that may be used as a single pulse, a train of pulses, or a free (at operator's wish) number of pulses.

High pressure lamps emit radiation in UV and visible, the main common characteristics being: continuous spectra emission on which discrete spectral lines are overlapped, incoherent radiation emission and continuous emission in time. The emission spectral range of Hg lamp is distributed between 300-600 nm, having higher intensity lines at 365 nm, 406 nm and 435 nm and lower intensity lines at 337 nm, 549 nm and 580 nm. Xe lamp emits between 320 nm and 420 nm with a maximum at 365 nm and exhibits high intensity lines positioned at 325 nm, 380 nm and 415 nm.

The nitrogen pulsed laser (NPL) is a gas laser that emits UV radiation at 337.1 nm, in the UVA region. The radiation is coherent and the emission is pulsed, in the form of short pulses (full time width-FTW of a pulse is, typically, 10 ns) emitted with repetition rates between 1 and 100 pulses per second (pps). The peak power per pulse is 350 kW and the average energy of the laser beam is 3.5 μJ at 10 pps.

The pulsed Nd:YAG laser is a solid state laser that emits at 1,064 nm and combined with nonlinear crystals at 532 nm, 355 nm and 266 nm having a full time width at half maximum (FTWHM) of 6 ns for a pulse repetition rate of 10 pps. The average energy of the laser beam emitted at 266 nm and 10 pps is 6.5 mJ.

For the efficiency study of the samples against the pseudotumour tissue, MTX saline solutions (NaCl 0.9%) at $5\times10^{-5}$ M, pH 8.4 were irradiated using UV-Vis non-coherent beam emitted by a Hg lamp (power density 100 $W/m^2$); the irradiation time was 1, 3 and 5 min.

The same experiment was performed using 5-FU saline solutions (NaCl 0.9%) at $10^{-4}$ M. The same volume of solution, exposure time intervals and irradiated surface as for MTX application were used. The difference is that for this experiment 5-FU solutions were irradiated with a $N_2$ pulsed laser beam.

The third experiment used benzopyridine compounds, BG 204 and BG 1120, both dissolved in distilled water at $5\times10^{-5}$ M and exposed to UV-Vis radiation, emitted by a cw Xe lamp, up to 60 min and respectively, 30 min. The power density of the incoherent radiation was 11 mW/cm$^2$. The experimental set-up is shown in Fig. (**3**). Irradiated benzopyridine solutions were then injected in the rabbit eyes that contained an induced pseudotumour [22].

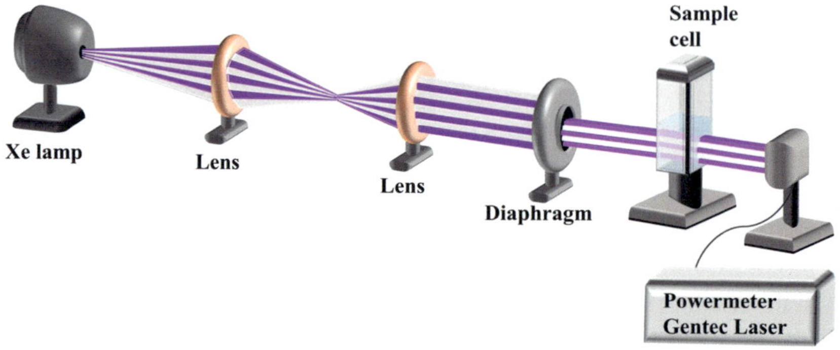


**Fig. (3).** Experimental set-up used to irradiate benzopyridine solutions; the pair of lenses adjusts the size of the optical beam, so that the solution is irradiated uniformly.

The fourth experiment included the study of CPZ effects on rabbit eyes at concentrations of 10 mg/ml and 20 mg/ml in ultrapure water irradiated 20 min and 240 min with a laser beam emitted at 266 nm having 6.5 mJ average energy. The solutions were applied on pseudotumour tissue after being irradiated.

The modifications produced in molecular structures of drugs by exposure to optical radiation (otherwise called irradiation) were investigated by UV-Vis and IR absorption spectroscopy and by fluorescence emission studies. The standard equipment utilised for UV-Vis absorption spectroscopy was a Perkin Elmer Lambda 950 spectrophotometer, for fluorescence emission an Aminco-Bowman spectrofluorimeter and for IR spectroscopy FTIR equipment Nicolet Magna-IR 550 and a FTIR Nicolet™ iS™50. For CPZ, the fluorescence emission was recorded in real time and the laser induced fluorescence (LIF) was collected using an optical fibre (400 μm core diameter), placed at 90° with respect to the incident beam, coupled to an Ocean Optics HR4000 spectrometer (optical resolution 0.64 nm). The LIF recorded spectra represent the average of the spectra recorded during an integration time of 10 s.

## RESULTS AND DISCUSSIONS

The spectroscopic studies regarding unirradiated and irradiated MTX (with an Hg lamp) and unirradiated and irradiated 5-FU (with a 337.1 nm laser beam emitted by a $N_2$ laser) are presented in Chapter 11. These experiments include UV-Vis and IR absorption as well as LIF studies. It is suggested that MTX and 5-FU molecules undergo structural changes upon UV-Vis irradiation and that the main photoproducts have a structure related to the parental compound.

## BG 204

The absorption spectra of unirradiated and 60 min irradiated (cw Xe lamp) solution of BG 204 dissolved in distilled water at $5\times10^{-5}$ M are given in Fig. (**4**). In each case, the absorption spectrum shows broad bands with peaks located at approximately 240 nm, 273 nm, 351 nm, 369 nm, 388 nm and 480 nm. One may observe that with increasing irradiation time the peak intensities of the absorption bands decrease. This indicates the photo-degradation of BG 204 molecules under the action of UV-Vis radiation.

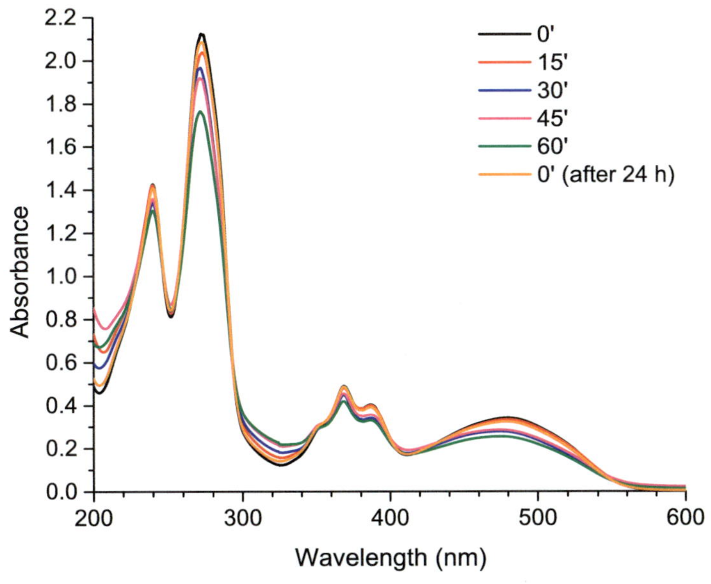


**Fig. (4).** Absorption spectra of BG 204 solutions ($5\times10^{-5}$ M in distilled water) for several irradiation time intervals with cw Xe lamp.

To assess the time stability of BG 204 solution, a sample of the same solution prepared at the beginning of the experiments was kept, without being irradiated, under the same conditions of temperature, pressure and ambient light as the sample subjected to irradiation. After 24 h the absorption spectrum was measured and the obtained values remained within the measurement errors, which reveals that the sample has not changed and that it is stable over time, at least for time intervals up to 24 h. This is of interest for experimental studies since allows to know the time interval after exposure to optical radiation within which drug samples remain stable and may be used.

The fluorescence excitation and emission spectra of BG 204 ($5\times10^{-5}$ M in distilled water) unirradiated and irradiated up to 60 min with a cw Xe lamp are presented in Figs. (**5-8**).

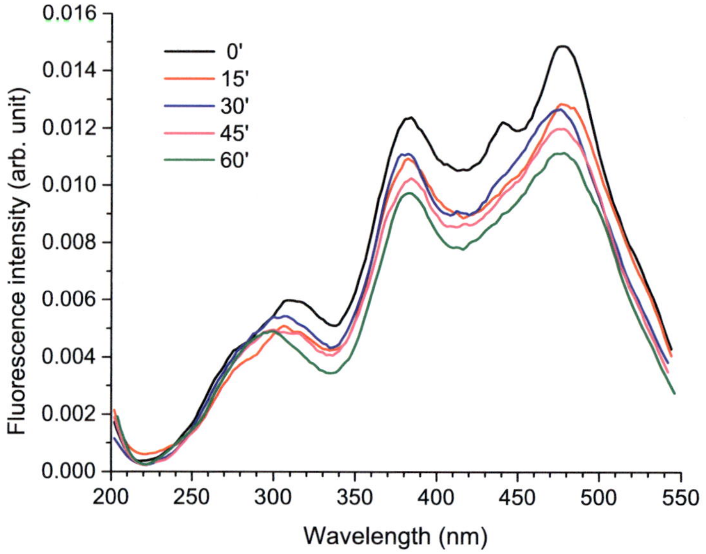


**Fig. (5).** Fluorescence excitation spectra of BG 204 solution ($5\times10^{-5}$ M in distilled water) for several irradiation times with cw Xe lamp; $\lambda_{em}$ = 570 nm.

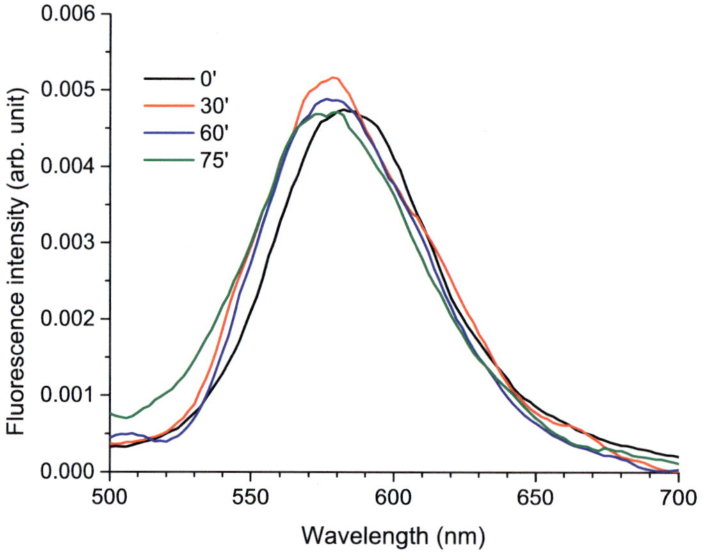


**Fig. (6).** Fluorescence emission spectra of BG 204 solutions ($5\times10^{-5}$ M in distilled water) for several irradiation times with cw Xe lamp; $\lambda_{ex}$ = 270 nm.

The fluorescence excitation spectra, Fig. (**5**), were measured in the 200-550 nm excitation spectral range when measured at 570 nm. The spectrum for the unirradiated solution shows four peaks at 310 nm, 382 nm, 440 nm and 478 nm, respectively. With increasing irradiation time several effects appear; first, there is a decrease in the intensity of emitted fluorescence radiation accompanied by a translation of fluorescence peaks to shorter wavelengths. This effect may be due to the production of fluorescence quenchers during irradiation, possible quenchers being radicals extracted from the structure of the BG 204 molecule or singlet oxygen molecules produced in solution by its exposure to UV-Vis optical radiation.

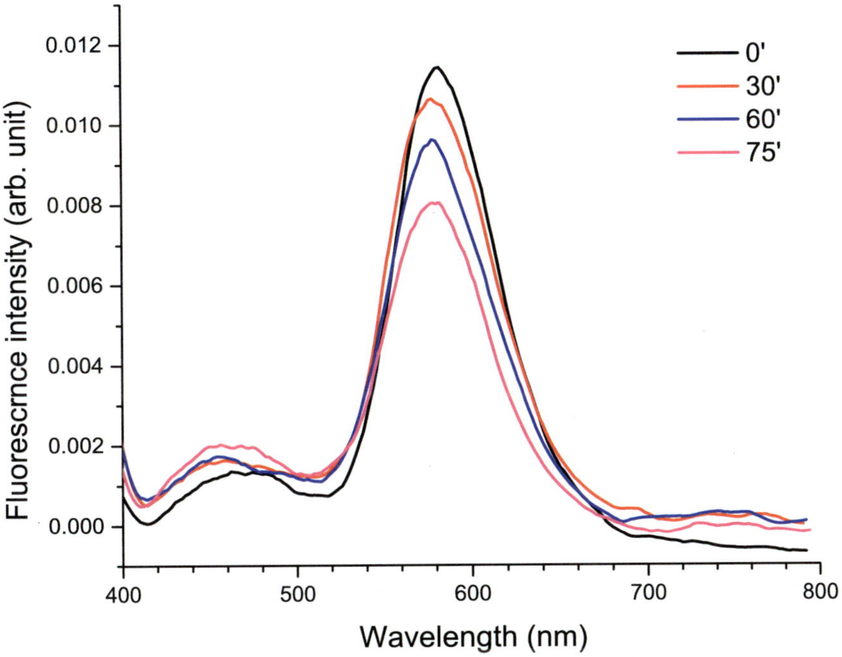


**Fig. (7).** Fluorescence emission spectra of BG 204 solutions (5×10$^{-5}$ M in distilled water) for several irradiation times with cw Xe lamp; $\lambda_{ex}$ = 370 nm.

A second kind of effect is the disappearance of the fluorescence band with prolonged irradiation time; the peak placed at λ = 440 nm, for instance, disappears after 45 min exposure time. A third type of effect is the blue shift of the fluorescence peak that can be observed for the band measured at λ = 310 nm where it appears to be shifted with 20 nm.

Fluorescence emission spectra for the same samples were also measured and are shown in Figs. (**6-8**).

In each case, BG 204 concentration was $5\times10^{-5}$ M and the irradiation time varied between 0 and 75 min; fluorescence excitation radiations had, each, the wavelength corresponding, respectively, to a peak of the absorption spectrum, namely: Fig. (**6**) excitation performed at 270 nm, Fig. (**7**) excitation at 370 nm and Fig. (**8**) excitation at 460 nm.

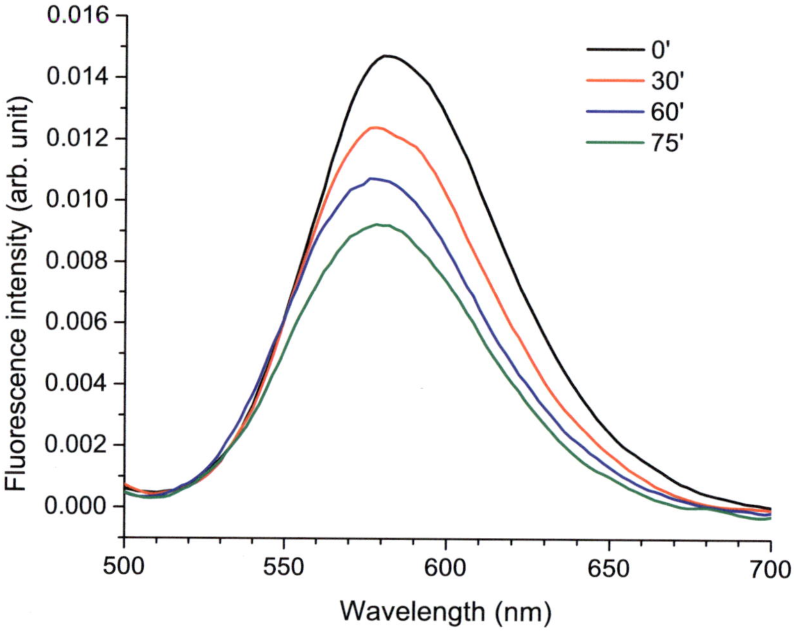


**Fig. (8).** Fluorescence emission spectra of BG 204 solutions ($5\times10^{-5}$ M in distilled water) for several irradiation times with cw Xe lamp; $\lambda_{ex}$ = 460 nm.

In all cases, the fluorescence emission spectrum looks like a wide band with its peak centred at 580 nm. The maximum fluorescence intensity decreases in all cases with increasing exposure time to incoherent UV-Vis radiation indicating that BG 204 suffers conformational changes, new photoproducts being formed.

To record the FTIR spectra of quinazoline solutions exposed to cw Xe lamp radiation up to 3 h, substances were poured onto a KBr crystal and allowed to dry for 24 h. The solid samples were then grinded and pressed at $1.575\times10^5$ kg/cm$^2$, to yield pellets which were used to measure FTIR spectra [35]. Three spectra were recorded, respectively, for unirradiated, 1 h irradiated and 3 h irradiated BG 204 solutions. As can be seen in Fig. (**9**), there are some observable changes between unirradiated and irradiated BG 204 samples.

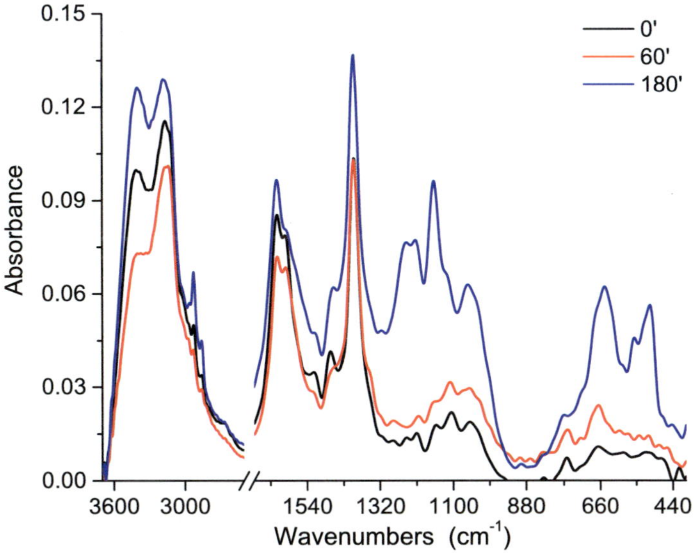


**Fig. (9).** FTIR spectra of unirradiated and irradiated BG 204 up to 180 min with a cw Xe lamp.

The sample irradiated 3 h shows noticeable changes of FTIR spectrum in the spectral domains 670-500 $cm^{-1}$ and 1,300-1,000 $cm^{-1}$. The peak between 520 and 500 $cm^{-1}$ may be attributed to S-S stretching vibration (aliphatic disulphides), while the one situated between 670-640 $cm^{-1}$ is attributable to stretching vibration of the C-S (aliphatic disulphides). These observations suggest that there was a cleavage of the S-C bond from -S-C-N- chain. The stretching vibration of C-S bond (1,230-1,300 $cm^{-1}$) may suggest that after irradiation, the chemical balance is moved towards a lactam form. These changes demonstrate the formation of thiol and disulphide compounds and the emergence of tautomers (Figs. **10, 11**) [36].

The bands at 1,630 $cm^{-1}$ that corresponds to the bending vibration of N-H bond, at 2,850 $cm^{-1}$ attributed to the alkyl groups $CH_2$ and at 2,925 $cm^{-1}$ corresponding to $CH_2$+$CH_3$ do not change too much.

Strong bands were observed at 1,400 $cm^{-1}$, 3,180 $cm^{-1}$ and 3,400 $cm^{-1}$ assigned to stretching vibration of N-H bond from ammonium ion which may be obtained from amines (equation 1).

$$R_3N + HOH \rightarrow R^3NH^+ + OH^- \quad \textbf{(1)}$$

All observations lead to the conclusion that irradiation produces important changes in the structure of molecules. Formation of ammonium ions suggests the existence of free radicals (-OH), which are hold responsible for the induction of tumour necrosis. Photo-transformation evidence suggests that used molecules are photo-reactive and their therapeutic effect can be improved by exposure to UV radiation.

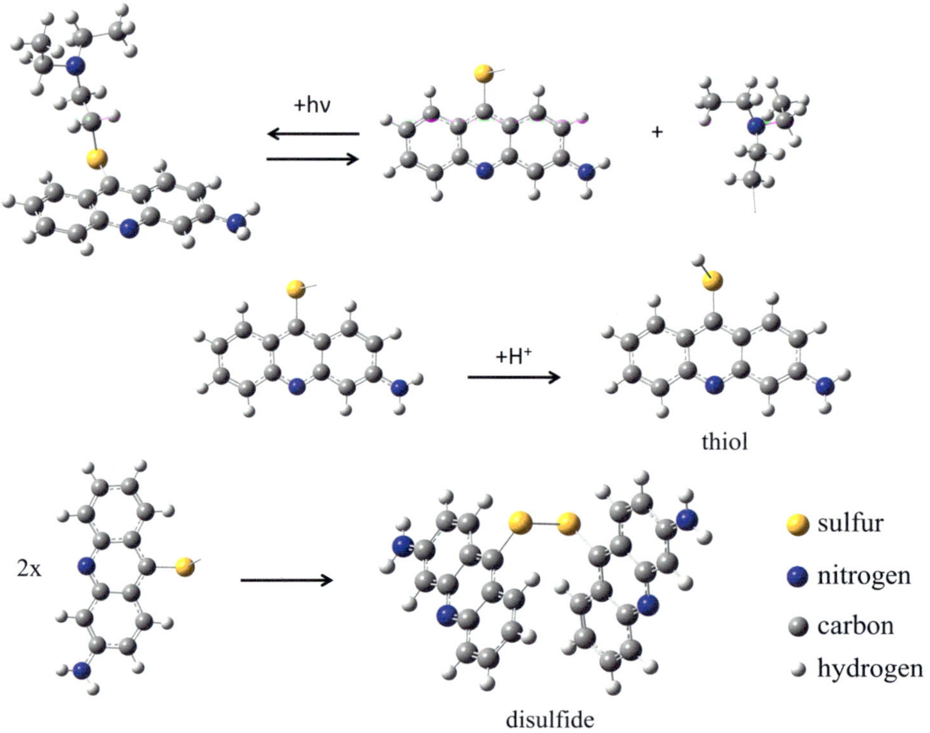


**Fig. (10).** Possible photo-chemical transformations resulted from the interaction of BG 204 distilled water solution with UV-Vis radiation resulted from FTIR analysis.

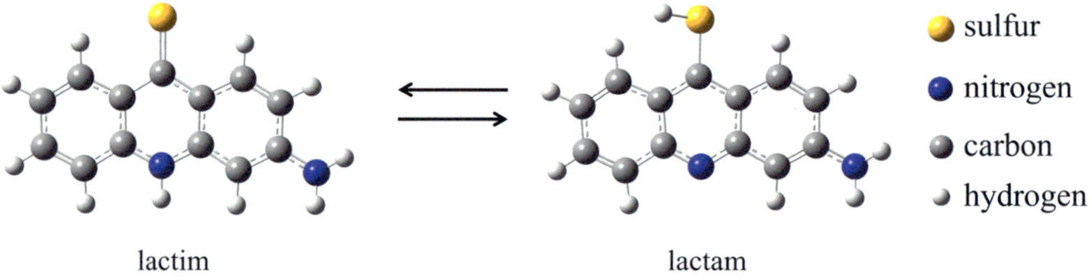


**Fig. (11).** Tautomer forms of the molecule containing the thiol group resulted from FTIR analysis of BG 204.

## BG 1120

BG 1120 dissolved in distilled water at $5\times10^{-5}$ M was exposed to UV-Vis radiation, emitted by a cw Xe lamp, up to 30 min. As shown in Fig. (**12**), modifications in the absorption spectrum of BG 1120 with increasing irradiation time are observed.

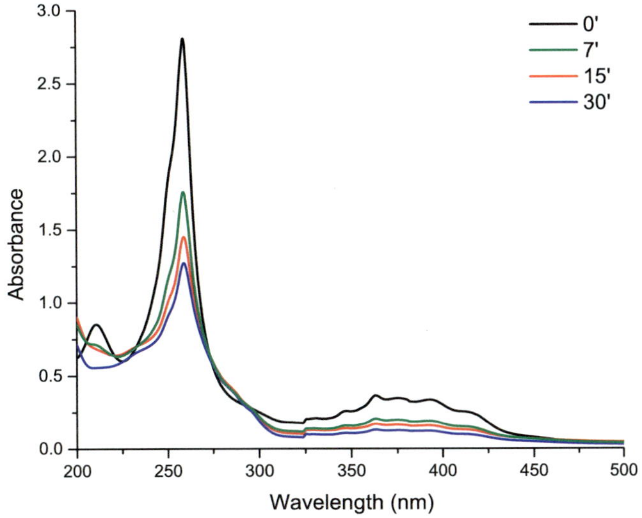


**Fig. (12).** Absorption spectra of BG 1120 solutions ($5\times10^{-5}$ M in distilled water) for several irradiation times with cw Xe lamp.

All absorption spectra have a narrow band peak at around 260 nm and a second broad band maximum at 375 nm. The absorption intensities of both bands decrease with increasing exposure time.

The evolution of absorption spectra suggests changes in the electronic structure of BG 1120 that occur after exposure to optical radiation, as reflected by lower absorption band at 260 nm and increased absorption band at 375 nm. Also, the band at 210 nm disappears after 15 min irradiation with light emitted by the same lamp.

The stability of unirradiated samples for different storage conditions was tested in order to determine the optimum time interval allowed to use BG 1120 solution in the treatment of pseudotumours. For BG 1120 solution stored in dark at 4°C during 10 days can be concluded that it remains stable, indicating that may be used, after preparation in this period of time, if kept in dark and at 4°C.

The same type of measurement was made and solution kept at room temperature (20°C) for 10 days. The absorption spectra of the sample showed a slight change in the absorption spectrum, which leads to reduction of the main absorption peak at 260 nm and to a slight increase of the secondary peak at 220 nm. These measurements recommend that BG 1120 solutions in distilled water should be preserved in dark at 4°C.

The degree of polymerisation of BG 1120 molecules in the solvent was also determined, using standard methods of measurement and calculation [1, 20, 37]. The range of concentrations was $10^{-3}$-$10^{-6}$ M and from the curve describing the degree of polymerisation resulted a value of 0.83 which indicates that in distilled water solutions are present only monomeric forms of BG 1120.

Measurements of fluorescence emission spectra excited at 300 nm for exposure times up to 30 min were carried out and the obtained data are shown in Fig. (**13**). The fluorescence spectra show changes regarding the wavelength and the fluorescence intensity evolutions with increasing exposure times.

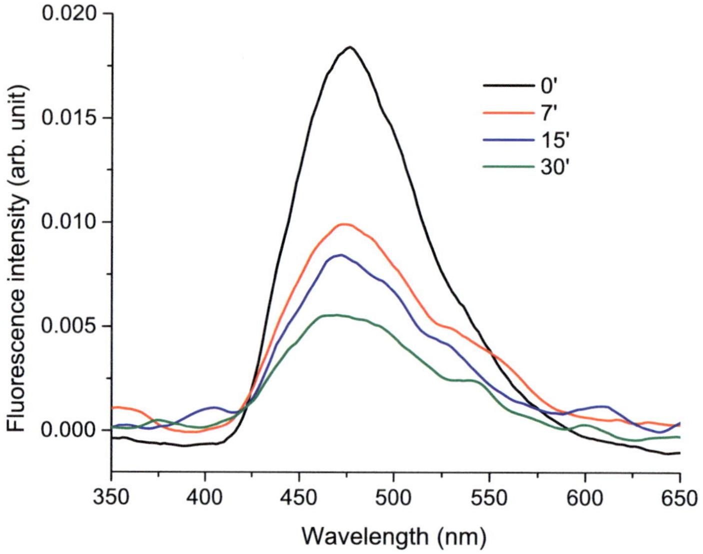


**Fig. (13).** Fluorescence emission spectra of BG 1120 solutions ($5\times10^{-5}$ M in distilled water) for several irradiation times with cw Xe lamp; $\lambda_{ex}$ = 300 nm.

The unirradiated solution presents a band with the main peak at 475 nm; with increasing irradiation time the peak suffers a hypsochromic shift of 10 nm and a hypochromic one as follows: the fluorescence intensity is decreasing with 46% after 7 min exposure, with 54% after 15 min and with 70% after 30 min, compared to the unirradiated sample spectrum.

The fluorescence excitation spectra measured at 460 nm for unirradiated and irradiated (up to 30 min) BG 1120 solutions in water are shown in Fig. (**14**). The unirradiated solution has three bands with peaks at 406 nm, 377 nm and 310 nm, respectively. The first peak suffers during exposure to cw Xe lamp radiation a bathochromic shift of 4 nm and a hypochromic one, the fluorescence intensity decreasing as follows: 39% for the solution irradiated 7 min, 57% for 15 min and 65% for 30 min, compared to evolution of the unirradiated solution.

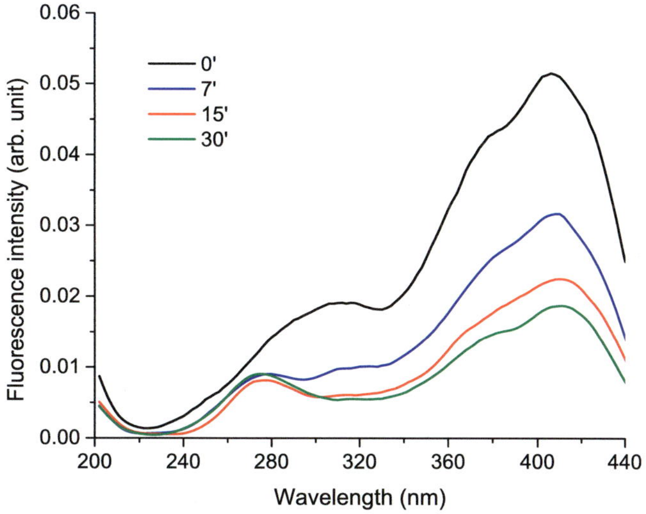


**Fig. (14).** Fluorescence excitation spectra of BG 1120 solutions ($5\times10^{-5}$ M in distilled water) for several irradiation times with cw Xe lamp; $\lambda_{em}$ = 460 nm.

The broader band having the peak at 310 nm shows a hypsochromic shift of 3 nm in the first 7 min of exposure to UV-Vis radiation and another band shows-up at 275 nm. After an irradiation of 15 min the band with the peak at 310 nm disappears and the one at 275 nm does not suffer any change, remaining unchanged for 30 min irradiation time interval.

The maximum fluorescence intensity decreased in all cases with increasing exposure time to incoherent UV-Vis and new bands are formed indicating, as in the case of the excitation and emission fluorescence spectra, that BG 1120 suffers conformational changes, and new photoproducts are generated.

The unirradiated and irradiated BG 1120 solutions in distilled water were analysed by FTIR and characteristic vibrations of chemical bonds of molecules were identified. The IR spectrum measured for unirradiated BG 1120 was compared to the FTIR spectrum of 180 min irradiated sample and its major modifications may be observed (Fig. **15**).

FTIR spectra of unirradiated and irradiated BG 1120 are characterised in the 3,500-3,100 $cm^{-1}$ spectral range by the stretching vibration of N-H in $NH_4^+$ and in the 1,750-1,500 $cm^{-1}$ range by the ring stretching vibration of -C=C-. In FTIR spectra measured on 180 min irradiated BG 1120 the following characteristic bands may be identified: 680 $cm^{-1}$ representing in-plane bending vibration of bonds -C-S-H, 1,400 $cm^{-1}$ for ring stretching vibration of thiol and 1,098 $cm^{-1}$, 1,638 $cm^{-1}$ and 1,579 $cm^{-1}$ attributed to the vibrations of -C=S in lactam form.

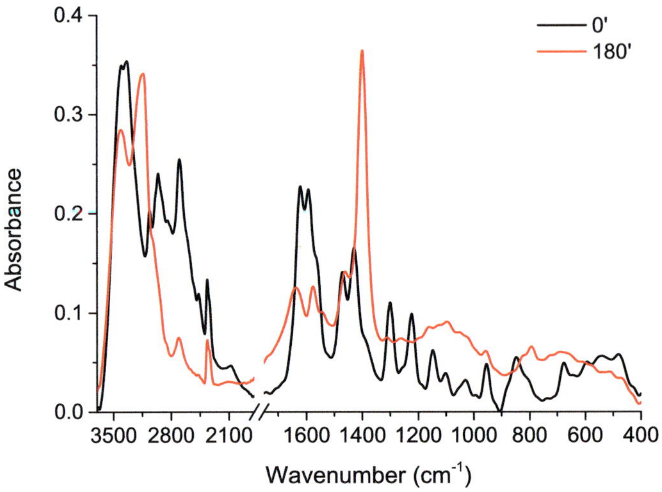


**Fig. (15).** FTIR spectra of unirradiated and 180 min irradiated BG 1120 solution in distilled water with a cw Xe lamp.

Considering the results obtained from absorption, fluorescence and FTIR spectra analysis, it can be concluded that BG 1120 molecule undergoes transformations that yield new photoproducts. The possible photo-chemical transformations resulted from the interaction of BG 1120 with UV-Vis radiation and are evidenced by FTIR analysis (Fig. **16**).

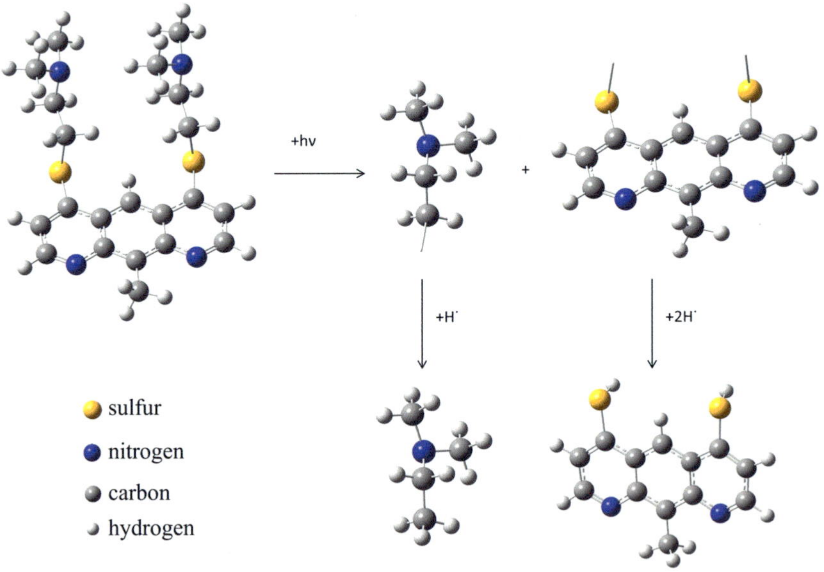


**Fig. (16).** Possible photo-chemical transformations resulted from the interaction of BG 1120 distilled water solution with UV-Vis radiation resulted from FTIR analysis.

The dithiol form (bis-lactim) obtained at the end of irradiation sequence is in dynamic equilibrium with the tautomer lactim-bis lactam (Fig. **17**) form that can be active in the treatment of tumour tissues [36].

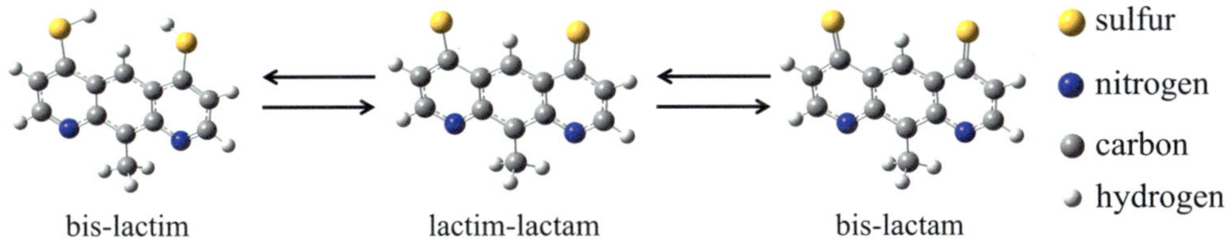


**Fig. (17).** Tautomer forms of molecule containing the thiol group resulted from FTIR analysis of BG 1120.

On the other hand, the changes that occur in BG 204/BG 1120 molecules are present within minutes after exposure; a convenient time for exposure applicable to the subsequent experiments appears to be about 15 min, according to absorption and fluorescence data.

In conclusion, the photo-transformations of BG 1120 show that molecules may be photo-reactive and their therapeutic effect may be improved by exposure to laser radiation in the visible and ultraviolet spectral ranges.

## Chlorpromazine

Absorption spectra were recorded using a Perkin Elmer Lambda 950 spectrophotometer, at a concentration of 0.2 mg/ml in ultrapure water to avoid saturation of the absorption signal between 180 nm and 280 nm, for an optical path of 1 mm, at a resolution of 0.05 nm. The absorption spectra of unirradiated CPZ and irradiated CPZ (at 266 nm) are presented in Fig. (**18**).

For unirradiated CPZ two bands are observed, with peaks at 254 nm and 306 nm. The irradiation of samples produces modifications of absorption spectra; the intensities of the peaks decrease with longer exposure time to UV laser radiation. In addition, irradiation does not change the wavelength of the 254 nm peak but the maximum at 306 nm is replaced by a broadened shoulder. With increase of irradiation time, changes become obvious in the absorption spectra, indicating transformation at molecular level.

The LIF spectra of irradiated, up to 240 min, CPZ solution (20 mg/ml) in ultrapure water are presented in Fig. (**19**) and are characterised by a band having the peak at 500 nm. The beam used for irradiation is emitted at 266 nm by a Nd:YAG laser system, this UV radiation being the fourth harmonic of the fundamental laser beam. The fluorescence intensity increases in the first 5 min of irradiation and then decreases. It is quenched by the end of 180 min exposure to UV laser radiation.

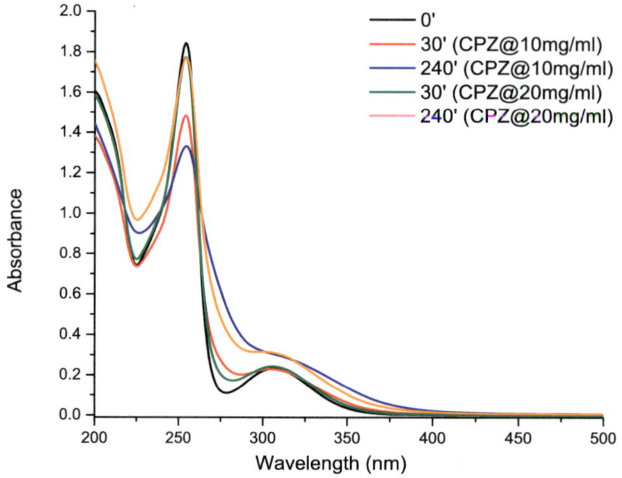


**Fig. (18).** Absorption spectra of CPZ solutions (10 mg/ml and 20 mg/ml in ultrapure water) for 20 min and 240 min irradiation time intervals with 266 nm emitted by a Nd:YAG laser system.

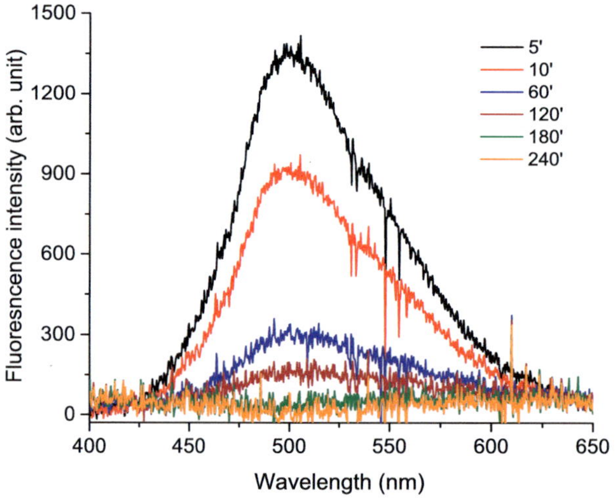


**Fig. (19).** LIF spectra of CPZ solution (20 mg/ml in ultrapure water) irradiated with a 266 nm laser beam emitted by a Nd:YAG pulsed FHG laser system up to 240 min.

Simultaneously, the intensity peak undergoes a hypsochromic shift of 2 nm by the end of 10 min irradiation time, suggesting the generation of new photoproducts. The changes in LIF spectra of CPZ solution irradiated with 266 nm laser beam indicate that the photoproducts that are formed are derived from the parental compound by substituent modification in the molecular ring structure [31, 38].

The IR spectra of CPZ samples were recorded using a FTIR Nicolet™ iS™50 spectrometer between 3,600 to 700 $cm^{-1}$ at 4 $cm^{-1}$ resolution; each shown spectrum is an average over 32 spectra. The samples were dried on KRS-5 (thallium bromo-iodide) support, and their IR spectra were subtracted from the spectra measured on solutions. Using FTIR, the characteristic vibrations of different groups belonging to CPZ were identified. As results from Fig. (**20**), the unirradiated CPZ FTIR spectra are characterised by the following molecular vibrations: (i) 755 $cm^{-1}$: out of plane bending vibration of 4-adiacent H of the ring system, (ii) 804 $cm^{-1}$: stretching vibration of C-Cl bond, (iii) 1,590 $cm^{-1}$ and 1,566 $cm^{-1}$: aromatic ring stretching vibrations (C=C-C), (iv) 865-1,180 $cm^{-1}$: C-N stretching vibrations, (v) 950-1,150 $cm^{-1}$: in plane C-H bending vibration, (vi) 3,420 $cm^{-1}$: N-H stretching vibration, (vii) 3,152 $cm^{-1}$: =C-H sp hybridised, (viii) 3,053 and 3,020 $cm^{-1}$: =C-H $sp^2$ hybridised, (ix) 2,962 $cm^{-1}$: =C-H $sp^3$ hybridised and (x) 2,672-2,472 $cm^{-1}$: $N^+H$ hydrate vibrations.

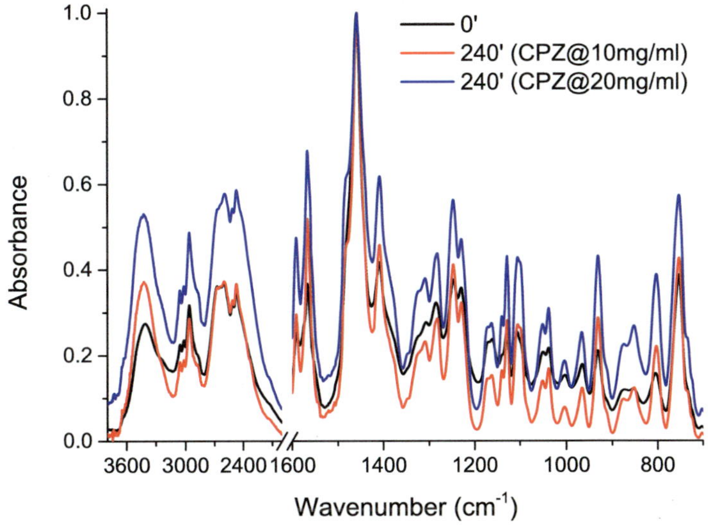


**Fig. (20).** FTIR spectra of unirradiated and 240 min irradiated CPZ with 266 nm laser beam for 10 mg/ml and 20 mg/ml concentration in ultrapure water solutions.

In the spectral range 3,600-700 $cm^{-1}$ there is no significant modification of IR spectra of irradiated 10 mg/ml and 20 mg/ml CPZ ultrapure water solutions compared to unirradiated solution. Only a new band appears at 1,348 $cm^{-1}$ resulting from the combination of O-H deformation vibration and C-O stretching vibration (phenol) which indicates the formation of hydroxyl photoproducts. The major contribution of the IR spectra for 10 mg/ml and 20 mg/ml CPZ irradiated 240 min is given by the parental compound, demonstrated in [32] using TLC analysis. So, no major modification can be observed for these concentrations.

# LABORATORY ANIMAL STUDIES

## Methotrexate

### *Materials and Methods*

Given the anatomical and physiological similarities between human and rabbit eyes, rabbits were used for experimental animal studies, treating and handling them as requested by existing EU laws regarding laboratory animals. Each step of the experiments including rabbits was made under anaesthesia, with Xylazine 4 mg/kg i.m., Ketamine 10-20 mg/kg i.v. and Atropine 1%, 0.5 ml i.v. Furthermore, all rabbits were monitored regarding their appetite, general condition, and body temperature. The experimental model proposed was that introduced in Schmidt-Erfurth *et al.* [22]. The experiment included 6 rabbits that were treated on both eyes. On each eye a pseudotumour was produced by sewing a 6.0 chromic catgut stitch at the sclero-corneal limbus. After several days pseudotumours were obtained and then each one was treated by injecting MTX (diluted in natural saline at $5\times10^{-5}$ M, pH=8.4) subconjunctivally, in pseudotumour area. Injected volume was 0.2 $cm^3$. Three rabbit eyes were injected with MTX solution unexposed to radiation and were kept as control. Other three eyes were irradiated 1 min after being injected, with an Hg lamp optical beam, emitted in cw regime in UV-Vis with 100 $W/m^2$ power density. The lamp was placed at approximately 30 cm from the eye, thus allowing for a 10 mm corneal-conjunctival circular area to be irradiated. Other three eyes were irradiated with the same lamp during three min and the last three eyes were irradiated 5 min. All eyes (besides the control) were irradiated three times a week for a time interval of 4 weeks. At the end of the experiment, eyes were submitted to anatomo-pathological examination, using a microscope set at 40X magnification factor.

### *Results and Discussions*

The reason the tissues were irradiated after being injected with MTX was to point out *in situ* molecular changes of MTX molecules, so that, even if this changes happen very fast (in nano- or pico-seconds), the modified molecules could interact with the pseudotumour tissue. This hypothesis is even more applicable for longer lifetime MTX molecules (seconds) or for MTX molecules irreversibly modified/ generated photoproducts.

Since neovascularisation plays a major role in malignant tissue proliferation or in tissue inflammation, destroying neovessels may have a key-role in anti-tumour therapy. The anatomopathological examinations showed the evolution of each pseudotumour according to the received treatment.

Fig. (**21**) show the histological aspect of the pseudotumours treated with unirradiated MTX (diluted in natural saline at $5\times10^{-5}$ M, pH 8.4, control-eyes). The picture shows a diffuse area of the inflammatory tissue.

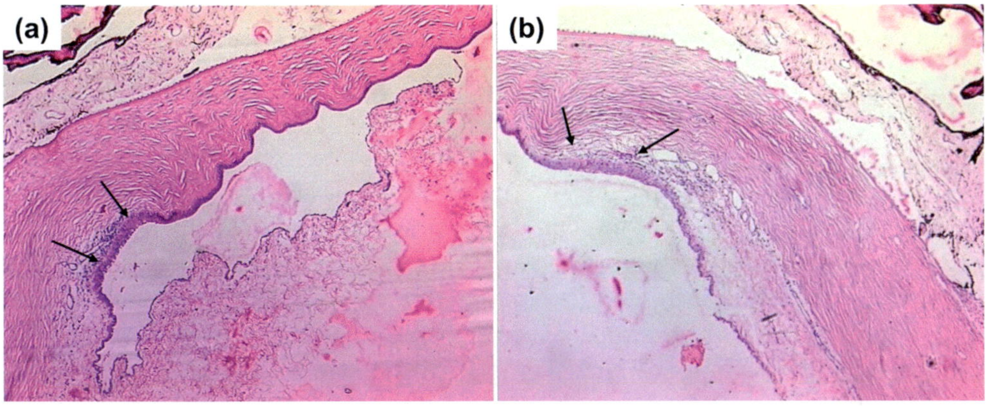


**Fig. (21).** Histological cross-section of a pseudotumour treated with unirradiated MTX, at 40X magnification factor (10×4), tissue coloured by hematoxiline-eosine: (**a**) arrows point out the inflammatory tissue; (**b**) arrows point out the areas of inflammatory tissue and oedema.

Fig. (**22a** and **b**) show the histological aspect of the pseudotumour tissue treated with MTX and exposed to the UV-Vis radiation, compared to control eye. There can be noticed partial fibrosis of tissue, thrombosis in neovessels and lower inflammation areas than in control eyes.

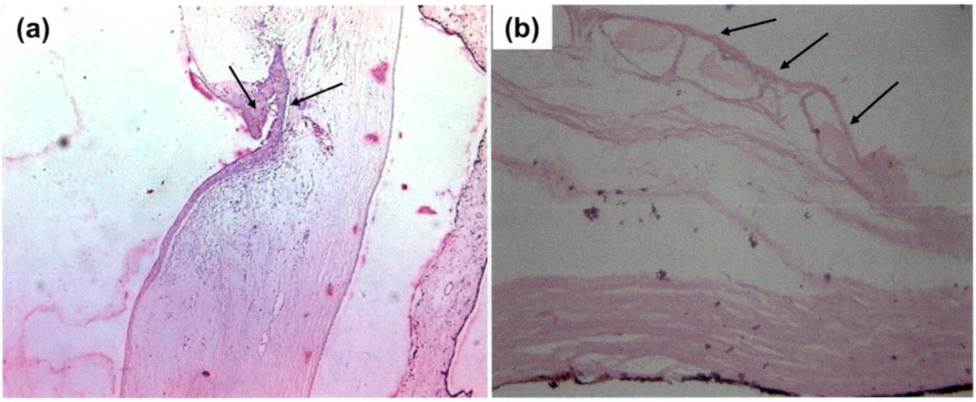


**Fig. (22).** Cross-section on the pseudotumour tissue injected with MTX at 40X magnification factor (10×4), tissue coloured by hematoxiline-eosine: (**a**) irradiated after injection 3 min: arrows show the neovessels and the less inflammatory tissue, (**b**) irradiated after injection 5 min: arrows show the massive thrombosis of neovessels.

## *Conclusions*

This model featuring eye conjunctive tissue inflammation and neovascularisation simulate, although quite approximately, real the tumour situation. This preliminary experiment yielded encouraging results in the sense that the eye conjunctive and neovascularisations disappeared after 1-2 treatments. Such results point out the photo-sensitising properties of MTX and its possible use in the treatment of conjunctive neovascularisation and cancer tumour.

## 5-Fluorouracil

### *Materials and Methods*

The experimental model regarding the animal eye reproduced the similar model applied by Schmidt-Erfurth and co-workers [22]. It refers to neovascular inflammatory corneal-conjunctival tissue. An experiment, made to identify possible application on rabbit eyes was performed using a nitrogen pulsed laser. Pseudotumours were produced by sewing a catgut stitch (diameter=0.88 mm) at rabbit eye sclera-corneal limbus. The experiment was performed on six rabbits (12 eyes). The rabbit eyes were previously injected with $10^{-4}$ M, 5-FU solutions in saline water (0.9% NaCl). Three rabbit eyes were injected with 5-FU (0.2 $cm^3$) at the pseudotumour site; they were not irradiated and were used as controls. The next three eyes were irradiated with a nitrogen pulsed laser beam for 1 min. Other three eyes were exposed 3 min and the last three eyes were irradiated 5 min. Eyes were irradiated using an optical fibre whose transversal section was 4 mm.

### *Results and Discussions*

Fig. (**23**) presents the eye pseudotumour before injection and irradiation.

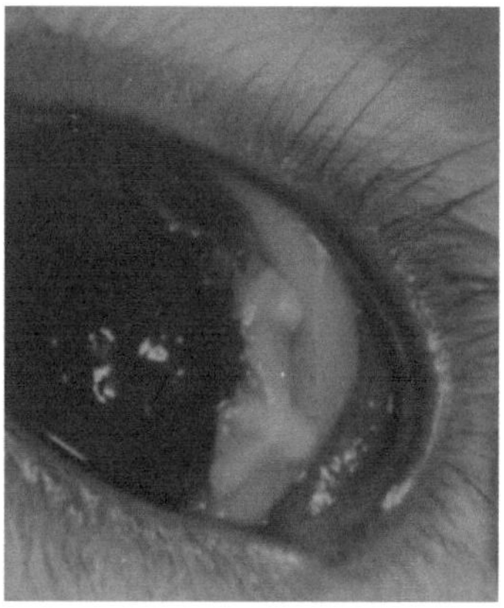

**Fig. (23).** Rabbit eye pseudotumour prior to 5-FU injection and exposure to radiation.

All the eyes exposed to laser radiation were irradiated 3 times a week. The treatment duration was 4 weeks after which the pathological examination of conjunctive tissue was made using a microscope Nikon 6 set at a 10X magnification factor. Images concerning conjunctive tissue in this experiment are shown in Figs. (**24-26**).

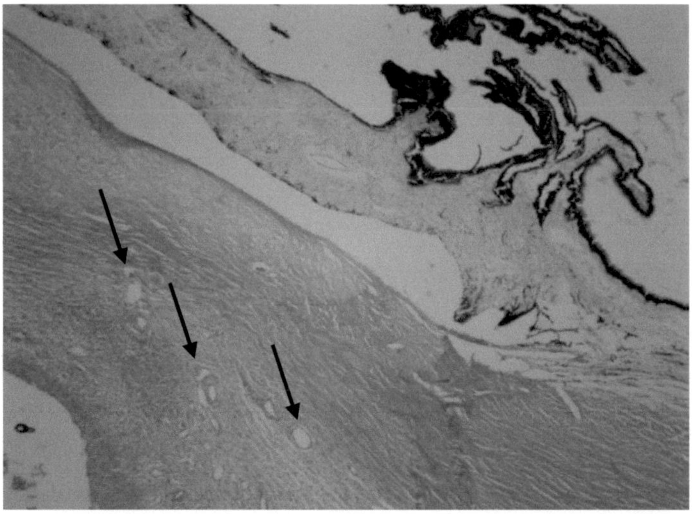

**Fig. (24).** Conjunctive tissue images at treatment with 5-FU ($10^{-4}$ M solutions in saline water, 0.9% NaCl) and exposure to UV nitrogen pulsed laser beam, at 100X magnification factor (10×10). The tissue was coloured by van Gieson dye. The image shows homogeneous dye blemishes which represent neoformation vessels without their own walls (see arrows). Eye exposed to nitrogen pulsed laser 1 min after 5-FU injection does not show inflammation regress.

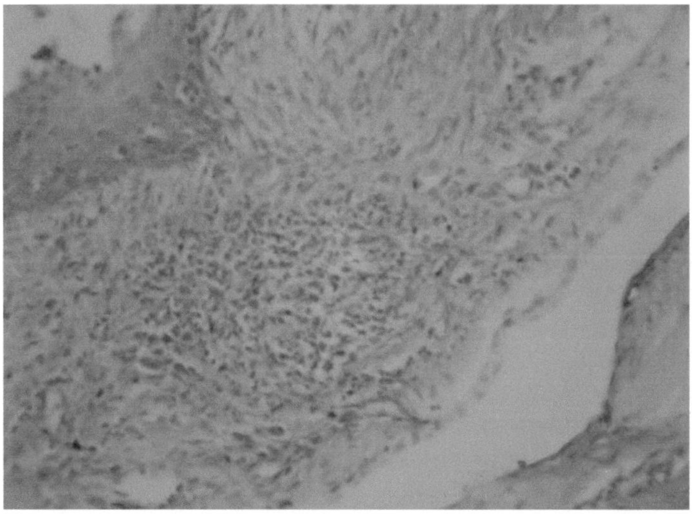

**Fig. (25).** Conjunctive tissue images at treatment with 5–FU ($10^{-4}$ M solutions in saline water, 0.9% NaCl) and exposure to UV NPL beam, at 200X magnification factor (10×20). The tissue was coloured by van Gieson dye. The image is obtained for a tissue irradiated with nitrogen pulsed laser during 3 min after injection with 5-FU; no inflammation regress may be noticed

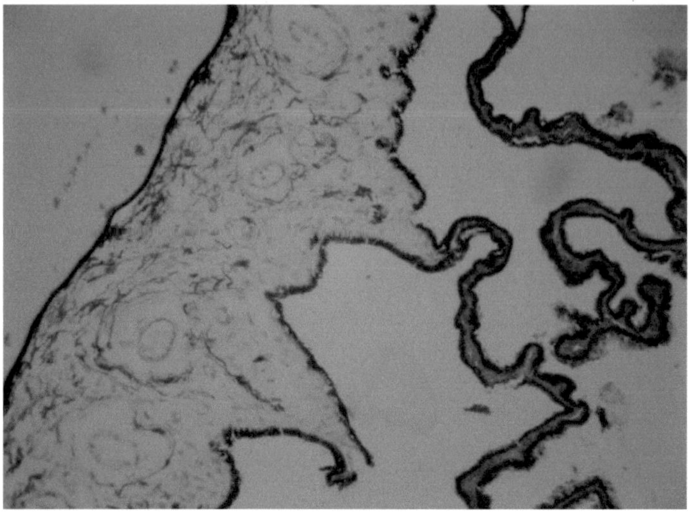

**Fig. (26).** Conjunctive tissue images at treatment with 5-FU ($10^{-4}$ M solutions in saline water, 0.9% NaCl) and exposure to UV NPL beam, at 100X magnification factor (10×10). The tissue was coloured by van Gieson dye. The image presents the rabbit eye injected with 5-FU and non-irradiated, used as control. It shows a small area of inflammatory tissue.

5-FU molecule is not fluorescent when excited at 337.1 nm, but the solution presents an intense fluorescence after exposure to UV-Vis light. As shown when the substance was studied from the physical/optical point of view, the absorption spectra of 5-FU did not exhibit any structural modification, but the FTIR spectra have shown that 5-FU changed his molecular geometry after irradiation. Exposure to UV-Vis radiation transformed 5-FU molecule into a tautomeric form.

These spectroscopic studies were coupled with preliminary tests on 5-FU photosensitising properties measured on rabbit eyes conjunctive. Comparative data were obtained about neovascularisations evolution of conjunctive between an untreated rabbit eye and an impregnated eye with cytostatic at the concentration of $10^{-4}$ M in natural saline and exposed to UV light beam emitted by a $N_2$ laser in near UV.

The energy of the $N_2$ pulsed laser beam used to irradiate 5-FU was concentrated at $\lambda$=337.1 nm. The pseudotumour was exposed to laser radiation within the limits of 10 mm spot diameter. The energy used for 1 min irradiation was 0.18 J. For 3 min exposure, the energy was 0.54 J, and for 5 min 0.9 J. The doses of radiation (in terms of energy reported to the irradiated surface) were 0.21 J/cm$^2$, 0.63 J/cm$^2$ and 1.05 J /cm$^2$, respectively.

On the other hand, if the doses of radiation used for exposure of 5-FU are compared to those used for MTX, there are two main differences: 5-FU was exposed to $N_2$ pulsed laser beam, whereas for MTX a cw Hg lamp was used. The amount of radiation was approximately $4\times10^3$ times higher for 5-FU than for MTX, for each irradiation time. The second difference is that the higher amount of radiation used for 5-FU was concentrated on just one wavelength (337.1 nm), whereas for MTX, the radiation was distributed on UV-Vis spectrum. These two differences could explain the fact that, when stimulated with a higher energy monochromatic UV fascicle, the molecular modification of the cytostatic could be more efficient.

### *Conclusions*

This model featuring eye conjunctive tissue inflammation and neovascularisation simulates the real tumour case. This preliminary experiment yields encouraging results: eye conjunctive and neovascularisations disappeared after 1-2 treatments. The obtained data point out photo-sensitising properties of 5-FU and recommend its possible use in the treatment of conjunctive neovascularisation and cancer.

## Benzopyridinic Compounds

It was demonstrated since the early 90s that azaheterocycles are efficient chemosensitisers. In this respect, thioacridines and pyridoquinoline derivatives have been successfully tested. In addition, some of these compounds carry on anticancer properties, due to their capability to intercalate into DNA. To compare with, other reports show that cytostatics such as MTX and 5-FU may become more efficient in acting on pseudotumour tissues when exposed to UV-Vis optical radiation.

### *Materials and Methods*

There have been studied four test sample compounds with a common benzo[b]pyridine nucleus, the results being reported in the following. The compounds belong to acridine and pyridoquinoline series (Fig. **1** and **2**). Chemical data of BG 204 and BG 1120 compounds were shown above in this chapter. The studies reported in this section contain only data about the use in the context of these experiments of BG 1120 solutions in distilled water.

For *in vivo* studies it was used the experimental model of the rabbit eye (Schmidt-Erfurth [22]) featuring conjunctive tissue inflammation and neovascularisation. Pseudotumours with new vascularisation were induced by sewing a catgut stitch 2.0 at the sclero-corneal limbus and by injecting a prostaglandinic analogue substance (TRAVOPROST 0.1 ml-which has a slightly proinflammatory action) nearby, under the conjunctiva. Four rabbits were used, for each of them keeping one unaltered eye as reference.

Fig. (**27**) shows one of the pseudotumours tissues which was produced (the obtained injuries looked approximately the same in all four eyes). 12 days after this procedure the eyes were injected with BG 1120 solution in distilled water at the sclera-corneal limbus, under the conjunctiva, near the affected area.

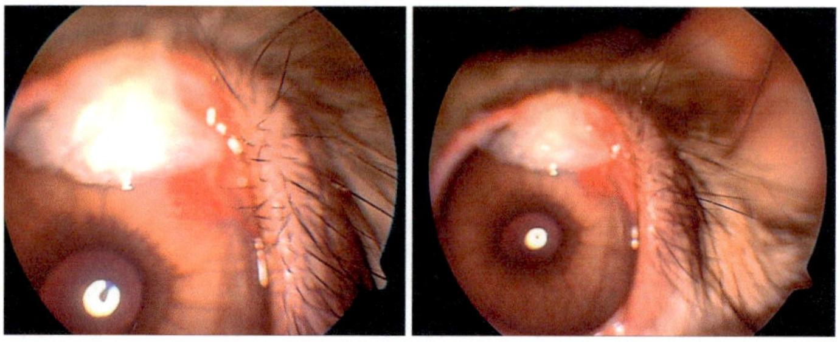

**Fig. (27).** Pseudotumour induced in rabbit eye before BG 1120 injection and exposure to radiation shown at two random magnifications.

The 1st eye was not treated. The 2nd eye was injected with 0.1 ml BG 1120 at the concentration $5\times10^{-5}$ M in distilled water, the solution being not irradiated. The 3rd eye was injected with the same quantity of irradiated substance with cw Xe lamp during 15 min. In the 4th eye was injected the unirradiated substance and then the eye was irradiated *in vivo* with the cw Xe lamp 15 min.

After two days, the eyes were submitted to the anatomopathological examination with a Nikon 6 microscope. The tissues were coloured by hematoxiline-eosine.

### *Results and Discussions*

As Fig. (**28**) shows-1st eye (not treated) - surrounding the necrotic stitch (see arrow) a large quantity of inflammatory cells may be observed, as well as neovascularisations and a small amount of fibrosis.

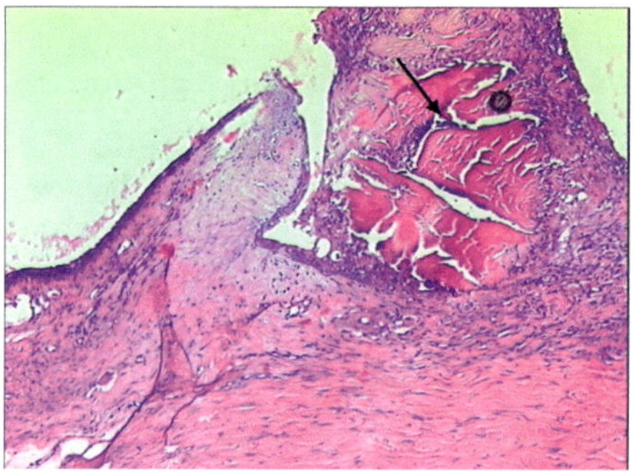

**Fig. (28).** Rabbit eye untreated tissue; image recorded for the 1st eye, at 40X magnification factor (10×4). The tissue was coloured by hematoxiline-eosine.

The 2nd eye (in which un-irradiated substance was introduced) has the same tissue modifications, however much reduced (Fig. **29**).

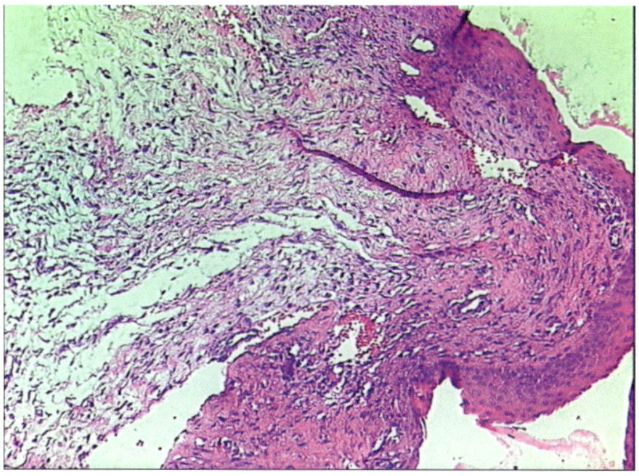

**Fig. (29).** Image of tissue of 2nd rabbit eye impregnated with unirradiated BG 1120 solution in distilled water, at 100X magnification factor (10×10). The tissue was coloured by hematoxiline-eosine.

In the 3rd eye (impregnated with irradiated substance) the quantity of the inflammatory infiltrate and oedema is decreased comparing to the 2nd eye (Fig. **30**).

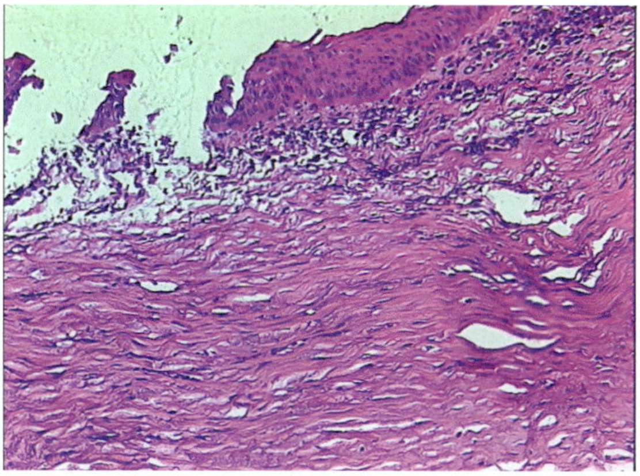

**Fig. (30).** Image of tissue belonging to the 3rd eye (impregnated with irradiated BG 1120) solution, at 100X magnification factor (10×10). The tissue was coloured by hematoxiline-eosine.

In the 4th eye (Fig. **31**) while the neovascularisations and the inflammatory cells are diminished, there is a large quantity of fibrosis which might be due to optical irradiation of the eye used in treatment procedure.

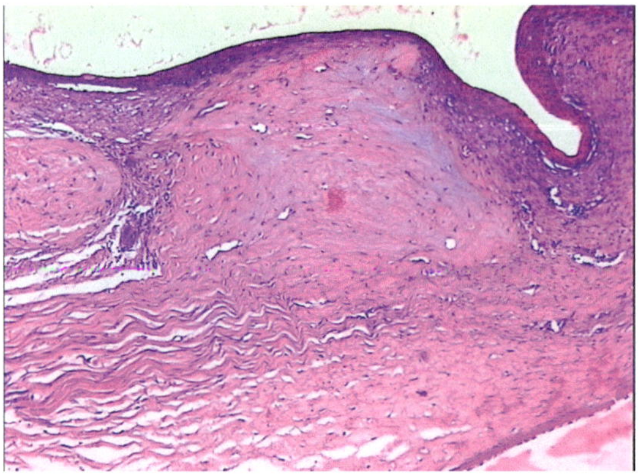

**Fig. (31).** Image of tissue in the 4th eye (impregnated with unirradiated BG 1120 solution and exposed to cw light emitted by Xe lamp), at 40X magnification factor (10×4). The tissue was coloured by hematoxiline-eosine.

This kind of image leads to the conclusion that optical radiation should be focalised on the treated area only, to avoid any damage of the healthy tissue.

## *Conclusions*

While some of the analysed chemical compounds present stability at the UV irradiation, others (BG 1120) present a decrease of the absorption, excitation and emission band intensities with the increase of irradiation time, which fact suggests their molecular transformations. The photo-transformation of the analysed samples shows that these molecules are photo-reactive and their therapeutic effect could be improved by exposure to UV radiation, as it is suggested by the data shown here.

BG 1120 was able to decrease the inflammation and the neovascularisations in the conjunctival tissue, especially after being irradiated with cw Xe lamp.

Two mechanisms could be involved in the process: (i) the cytostatic mechanism which implies the interference with the synthesis of the DNA and the cellular replication in the tissues with active proliferation, and (ii) the photo-sensitising mechanism that leads to the singlet oxygen and free radicals formation and finally to necrosis of the tissue involved.

This experiment points out the photo-sensitising properties of some, yet to be studied, substances and their possible use in the treatment of some important affections such as conjunctive neovascularisation and cancer.

## Chlorpromazine

MDR became and remains a worldwide health issue, which requires a flexible approach to find medicines able to overcome it. One possible method could be the exposure of existing medicines to UV laser beams to generate photoproducts that may be efficient against bacteria and/or malignant tumours. This is particularly applicable for substances which are not antibiotics (non-antibiotics) but may generate, by exposure to laser beams, isomers and/or photoproducts that might have antibiotic properties.

A particular class of medicines for these studies is represented by phenothiazines which are used as neuroleptics (CPZ, TZ are two compounds in this class) but which may work as agents to treat malaria or tuberculosis (particularly TZ) as well [33, 34]. Recent studies performed on larger volumes (bulk=1.5 ml) of CPZ solutions in ultrapure water at several concentrations showed that hundreds of photo-reaction products are obtained by exposure to UV laser radiation at 266 nm wavelength or 337.1 nm or even 355 nm and that application of solutions containing mixtures of the obtained photoproducts on Gram-positive bacteria evidences a higher antibacterial effect of the mixture than that of the CPZ parent compound [34, 36]. On the other hand, there is MDR acquired by malignant tumours; in some cases, infections of MDR tumour tissues with MDR bacteria are mentioned [39].

### *Materials and Methods*

The study was performed on rabbit eyes; rabbits were treated according to animal ethics regulations of the University of Medicine and Pharmacy Carol Davila in Bucharest. The protocol was approved by the Ethics Committee of Scientific Research of the University of Medicine and Pharmacy Carol Davila (Code PO-3--F-03, No. 34). All treatment injections were performed under anaesthesia with Xylazine 4 mg/kg i.m., Ketamine, 10-20 mg/kg i.v. and Atropine 0.5 ml of solution 1% to minimise suffering. Pseudotumour tissues on rabbit eyes were produced as above using the Schmidt-Erfurth method which consists in the insertion of a propylene suture wire (5.0) at the sclerodermal limbus [22].

In Fig. (**32**) is shown the overall image of a pseudotumour produced on one rabbit eye. CPZ solutions at 20 mg/ml and 10 mg/ml in ultrapure water were exposed, previous to treatment, to 266 nm laser beam having 6.5 mJ average energy, time intervals between 5 and 240 min. The laser beam is emitted by a Nd:YAG system which has the frequency of the fundamental emitted beam quadrupled [40].

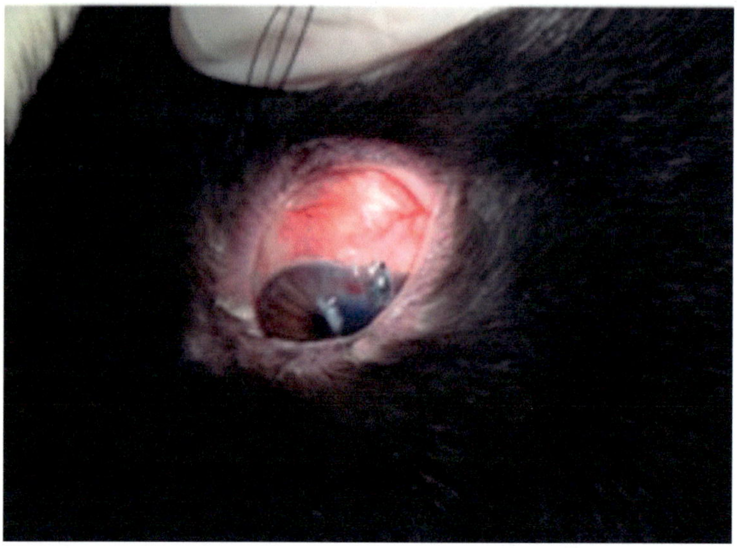

**Fig. (32).** Image of a pseudotumour obtained on rabbit eye using Schmidt-Erfurth [22] method.

The treatment was applied on 5 rabbits which were injected with CPZ subconjunctivally, at the limbus, right near the pseudotumour area; rabbit no. 1 was kept as control I and had pseudotumours on both eyes that were not treated; rabbit no. 2 was control II and had pseudotumours treated with unexposed CPZ at two different concentrations (one eye 0.1 ml of solution at 20 mg/ml in ultrapure water and the second 0.1 ml of solution at 10 mg/ml). Rabbits no. 3, 4 and 5 were treated with CPZ solutions at two concentrations irradiated at 266 nm different time intervals as follows: for rabbits no. 3 and 4, one eye was injected with 0.1 ml CPZ 20 mg/ml solution and the second eye was injected with 0.1 ml CPZ 10 mg/ml solution, both irradiated 20 min; these two rabbits were studied for observing the reproducibility of the results; as for rabbit no. 5, one eye was injected with 0.1 ml solution at 20 mg/ml and the other with 0.1 ml solution at 10 mg/ml the irradiation time of solutions being in both cases 4 h [40]. At the end of treatment the eyes were extracted and anatomopathological examination was performed.

### *Results and Discussions*

Results are organised on pairs of extracted eyes and are shown as images of parts of tissues selected in accordance with the dimensions and qualities of the effects produced on them. In Fig. (**33**) is shown the histopathologic image of a sample of pseudotumour tissue extracted from control rabbit I (rabbit no. 1). Areas A and B (Fig. **33**) include cells of inflammatory type; rare, normal vascular elements may be seen in zone C (Fig. **33**) at the end of the three days of observation.

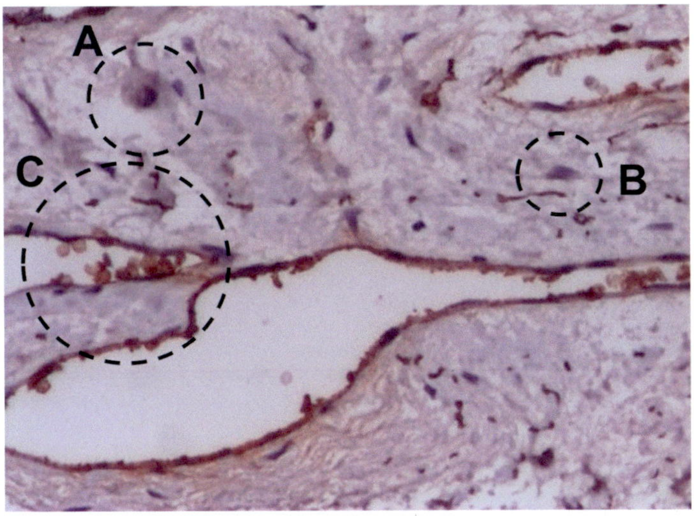


**Fig. (33).** Histological image of a sample of pseudotumour tissue of control rabbit I; A), B) inflammatory cells, C) vascular elements. Magnification factor 400X (10×40). The tissue was coloured by van Gieson dye.

In Fig. (**34**), histological images are shown from the eye of control rabbit II, where 0.1 ml of unirradiated CPZ solution at 20 mg/ml in ultrapure water was injected; a granular inflammatory tissue (A) may be noticed as well as a moderate chronic infiltrate with frequent eosinophils (at treatment with 20 mg/ml solution).

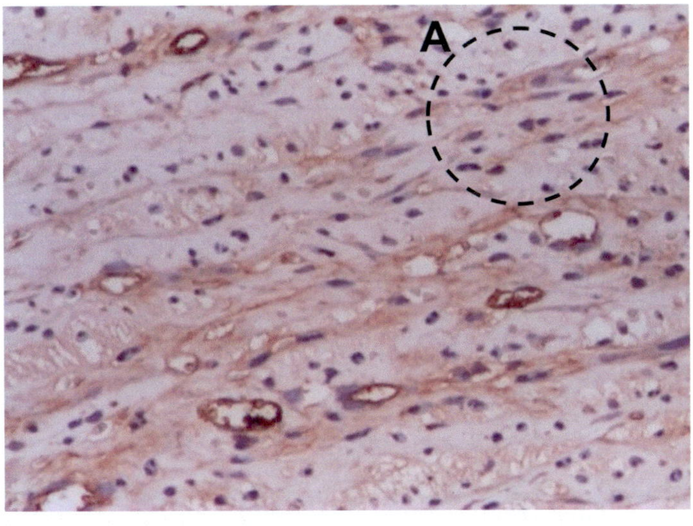


**Fig. (34).** Histological findings on eye in control rabbit II; area A shows granular inflammatory tissue (CPZ 20 mg/ml in ultrapure water). Magnification factor 20x (10×2).

In Fig. (**35**) (where the treatment is made with 10 mg/ml solution of CPZ in ultrapure water), necrotic masses of tissues (area A) rich in eosinophilic infiltrates (area B) are shown. There are not significant differences between control I and control II rabbit eyes, *i.e.* the unirradiated CPZ solutions do not contribute to faster destruction of the pseudotumours.

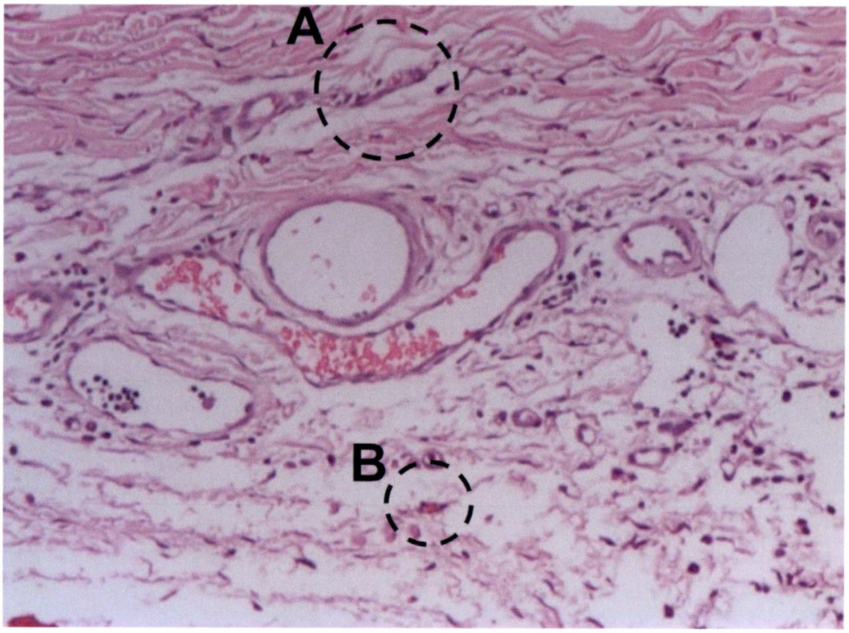


**Fig. (35).** Histological findings in control rabbit II (CPZ 10 mg/ml in ultrapure water) at 200X magnification (10×20); tissue coloured by hematoxiline-eosine: (A) necrotic tissue and (B) eosinophilic infiltrates.

For treated rabbits no. 3 and 4, CPZ solutions at the two concentrations were irradiated 20 min. This time interval lags between 5 min and 37.5 min for which the modifications of CPZ solutions were measured. The generated photoproducts at 20 min exposure to UV laser beam are the same as those reported in the above mentioned measurements the differences being their relative concentrations.

Fig. (**36**) shows the results after the treatment with CPZ solutions of rabbit no. 3. In Fig. (**36a**) is presented the image of a sample of pseudotumour tissue after treatment with CPZ at 20 mg/ml in ultrapure water. One may observe rich chronic inflammations (A), frequent eosinophils (B) and necrotic nodular tissue masses (C), as well as frequent blood vessels (D). In Fig. (**36b**) the image shown is that of the eye tissues treated with irradiated CPZ at 10 mg/ml in ultrapure water and exhibits the same effects. However, extension of damages is lower at treatment with 10 mg/ml than with 20 mg/ml, which recommends the use of CPZ solutions at 10 mg/ml for treatment, after their exposure to UV laser beam.

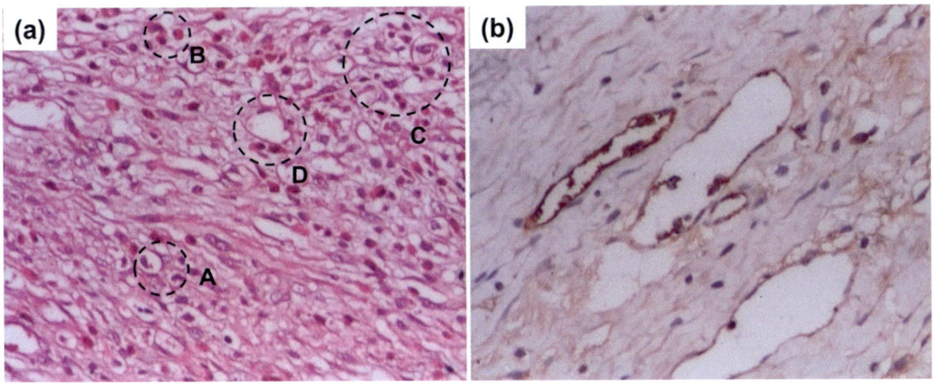


**Fig. (36).** Effect of CPZ solution at $5\times10^{-5}$ M in distilled water exposed 20 min at 266 nm pulsed laser beam; (**a**) chronic inflammations (A), frequent eosinophils (B), necrotic nodular tissue masses (C), and frequent blood vessels (D); at 200X magnification (10×20), tissue coloured by hematoxiline-eosine; (**b**) the same area at 200X magnification factor (10×20), tissue coloured by van Gieson.

In Fig. (**37**) is shown the histopathologic image of a sample of pseudotumour tissue for an eye extracted from rabbit no. 4. The eye was treated with 10 mg/ml CPZ solution in ultrapure water exposed 20 min to 266 nm laser beam. The results in this case are comparable with those obtained on rabbit no. 3; one may also observe a nodular necrotic mass produced in the pseudotumour tissue.

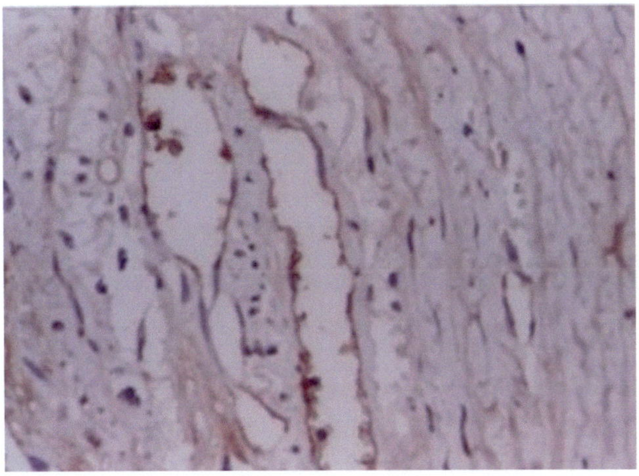

**Fig. (37).** Histological image of a part (sample) of pseudotumour tissue extracted from rabbit eye no. 4. Similar aspect to rabbit eye no. 3; in addition, necrotic tissue is observed. Magnification factor at 400X magnification factor (10×40). The tissue coloured by hematoxiline-eosine.

In Figs. (**38a** and **b**) are shown histopathologic images of samples of eye pseudotumour tissues obtained from rabbit no. 5. In Fig. (**38a**) the eye was injected with 0.1 ml CPZ solution at 20 mg/ml in ultrapure water, exposed 4 h at 266 nm and in Fig. (**38b**) the tissue was injected with 0.1 ml CPZ solution at 10 mg/ml irradiated in the same conditions.

In both images one may observe rich, chronically inflammated infiltrates which contain frequent eosinophils (areas A) as well as nodular necrotic tissular masses (areas B) inserted in the striate muscle tissue. Relatively frequent blood vessels exhibiting trombosis (Fig. **38b**, area C) are noticed.

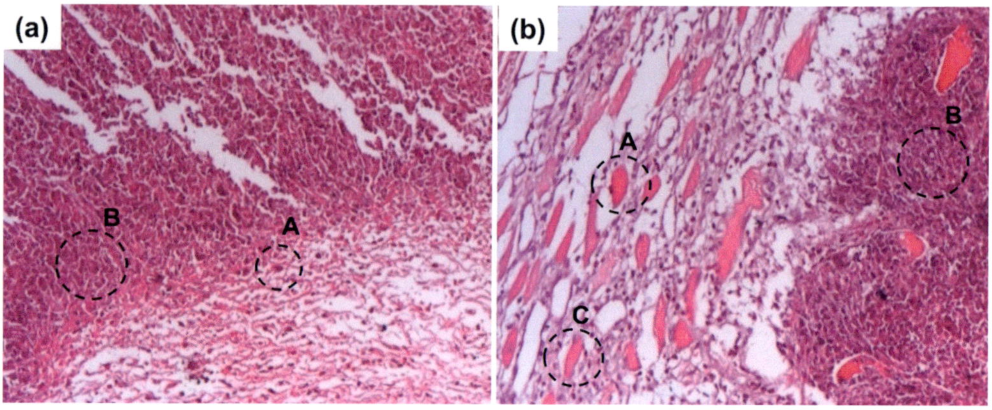


**Fig. (38).** Histopathologic images of samples of eye pseudotumour tissues obtained from rabbit eye no.5. (**a**) tissue injected with CPZ 20 mg/ml in ultrapure water exposed 4h at 266 nm; eosinophils (A), nodular necrotic tissue (B); (**b**) eosinophils (A), nodular necrotic tissue (B), frequent thrombosis of vessels (C); Magnification factor at 40X magnification factor (10×4). The tissue coloured by hematoxiline-eosine.

The main conclusion that results out of Figs. (**36-38**) is that the treatment with irradiated CPZ solution is most efficient in recuperating the qualities of corneal tissue when the concentration 10 mg/ml is utilised. Although exposure time for the solution used to treat rabbit no. 5 is 4 h (18 times longer than in the case of solutions used to treat rabbit no. 3) the effects are not spectacularly better than in the case of solutions exposed 20 min.

## *Conclusions*

The use of CPZ water solution exposed to 266 nm laser beam to treat pseudotumour tissues produced on rabbit eyes shows that treatment results, in terms of recuperating the qualities of healthy corneal tissues, depend on CPZ concentration at the beginning of laser irradiation and on exposure time length. At the same time, one knows that the exposed solutions contain hundreds of new photoproducts. In the experiments on rabbits, solutions exposed to UV laser beam were applied on pseudotumour tissues produced on eyes, *i.e.* the mixtures in water of the photoproducts obtained by irradiation. At this stage, it is not possible to specify which, out of generated photoproducts, individual or as a group with synergistic effects, is efficient in pseudotumour cure but the overall effects may be clearly outlined [34]. Since exposed CPZ solutions are stable in time, they may be applied on eye at any time after being processed by light. Application of CPZ irradiated solutions on rabbit eyes pseudotumours seems to produce a faster recovery of tissues with respect to control eyes, but it remains to quantify the differences. The principle and methods described here could serve for fighting MDR or tumour processes.

## GENERAL CONCLUSIONS

MDR is a concept that describes not only the behaviour of microorganisms at interaction with antibiotics or other drugs, but the answer of some tumours, in particular malignant tumours, at treatment using chemotherapy related methods, as well.

In this chapter are shown, based on literature reports made mainly by the authors, data that demonstrate that a method to fight MDR developed by tumours may be the transformation of some photo-sensitive medicines in new photoproducts by exposure to ultraviolet and/or visible optical radiation. The photoproducts prove to have, at least in mixtures, tumouricidal properties demonstrated at their interaction with pseudotumours induced in rabbit eyes conjunctive that simulate real malignant tumours. At the same time, in some cases the parent compounds may change their forms and/or symmetry and may become more efficient in fighting MDR tumours.

The efficiency in generating new photoproducts may differ from one drug to another function of drugs concentrations and characteristics of the optical beams used to irradiate them. The resulted solutions after exposure to optical beams would contain remnants of initial parent compound, but also the generated photoproducts having different lifetimes in solutions.

The checked drugs reported here were first two cytostatics, MTX and 5-FU. MTX at $5\times10^{-5}$ M in natural saline (pH 8.4) was exposed at cw Hg lamp radiation while injected in pseudotumour tissues produced in rabbit eye conjunctive. The experiments have demonstrated that the eye conjunctive and neovascularisations disappear after 1-2 treatments pointing out the photo-sensitising properties of MTX and its possible use to treat conjunctive neovascularisation and even cancer tumours. At the same time, it was demonstrated that exposure of rabbit eye impregnated with MTX solution leads to fibroses of the conjunctive that recommend to use irradiated MTX previous to application in eye and not to expose it in the eye. The same conclusions may be synthetised about 5-FU when it is irradiated in UV at 337.1 nm using a laser beam emitted by a nitrogen pulsed laser.

The second class of drugs used in experiments contained benzopyridine derivatives, in particular BG 1120 at $5\times10^{-5}$ M in distilled water exposed to cw Xe lamp emitted radiation. The photo-transformation of the analysed BG 1120 samples shows that these molecules are photo-reactive and their therapeutic effect could be improved by exposure to UV radiation. BG 1120 diminished the inflammation and the neovascularisations in the conjunctival tissue, especially after being irradiated with cw Xe lamp. Two mechanisms could be involved in the process: (i) the cytostatic mechanism which implies the interference with the synthesis of DNA and cellular replication in tissues with active proliferation, and (ii) the photo-sensitising mechanism that leads to singlet oxygen and free radicals formation and finally to necrosis of the treated tissue.

The third class of medicines investigated here is the phenothiazines derivatives class, out of which CPZ action exposed to 266 nm pulsed laser beam is described in more detail. The main conclusion is that the treatment with irradiated CPZ solution is most efficient in recuperating the qualities of corneal tissue when the concentration of 10 mg/ml is utilised and the irradiation time, previous to injection in the eye, is 20 min. The CPZ exposed solutions contain hundreds of new photoproducts and, consequently, mixtures in water of the photoproducts obtained by irradiation were applied on pseudotumour tissues produced on eyes.

In general, it was not possible to specify which, out of generated photoproducts, individually or as a group with synergistic effects, are efficient in pseudotumour cure but the overall effects may be evidenced. The performed studies have shown that the exposed solutions of medicines are stable in time and may be applied on eye at any time after being processed by light. This allows better experimental conditions for applications. As a trend, the application of irradiated solutions on rabbit eyes pseudotumours seems to produce a faster recovery of tissues with respect to control eyes and it remains to quantify the differences.

Another conclusion with significant consequences in applications is that laser beams are more efficient in modifying drug solutions in view of further application in MDR fighting. This is due to the fact that the beams energies reported to unit spectral width (J/nm for instance) are orders of magnitude larger with respect to the incoherent light sources (such as Hg or Xe lamps emitting even in cw regime). So, the selectivity of the laser beam action on drug molecules as well as the efficiency of generating new photoproducts are much higher.

## CONFLICT OF INTEREST

The authors confirm that this chapter content has no conflict of interest.

## ACKNOWLEDGEMENTS

This work has been financed by the National Authority for Research and Innovation in the frame of Nucleus programme-contract 4N/2016 and project PN-II-PT-PCCA-2011-3.1-1350.

## REFERENCES

[1] M.L. Pascu, A. Staicu, L. Voicu, M. Brezeanu, B. Carstocea, R. Pascu, and D. Gazdaru, "Methotrexate as a photosensitiser", *Anticancer Res,* vol. 24, no. 5A, pp. 2925-2930, 2004. [PMID: 15517898]

[2] *METHOTREXATE,CAMEO Chemicals, NOAA*. Available at: https://cameochemicals.noaa.gov/chemical/20602

[3] Pubchem, *methotrexate, C20H22N8O5 - PubChem.* Available at: https://pubchem.ncbi.nlm.nih.gov/compound/methotrexate#section=Melting-Point

[4] E.A. Schaff, U. Penmetsa, S.H. Eisinger, and P. Franks, "Methotrexate. A single agent for early abortion", *J. Reprod. Med,* vol. 42, no. 1, pp. 56-60, 1997. [PMID: 9018647]

[5] T. Pincus, Y. Yazici, T. Sokka, D. Aletaha, and J.S. Smolen, "Methotrexate as the anchor drug for the treatment of early rheumatoid arthritis", *Clin. Exp. Rheumatol,* vol. 21, no. 5, suppl. Suppl. 31, pp. S179-S185, 2003. [PMID: 14969073]

[6] P. Bertin, N. Carpentier, P. Vergne, C. Bonnet, B. Bannwarth, J. Dehais, and R. Treves, "Methotrexate and non-steroidal anti-inflammatory agent combination in rheumatoid arthritis", *Therapie,* vol. 52, no. 2, pp. 133-137, 1997. [PMID: 9231508]

[7] O.S. Nilsson, F.C. Bauer, L-Å. Broström, and U. Nilsonne, "Effect of the antineoplastic agent methotrexate on experimental heterotopic new bone formation in rats", *Cancer Res,* vol. 44, no. 4, pp. 1653-1656, 1984. [PMID: 6367970]

[8] S. B. Kaye, "New antimetabolites in cancer chemotherapy and their clinical impact", *Br. J. Cancer,* vol. 78, no. Suppl 3, pp. 1-7, 1998. [http://dx.doi.org/10.1038/bjc.1998.747]

[9] A.C. Holliday, M.N. Moody, and A. Berlingeri-Ramos, "Methotrexate: role of treatment in skin disease", *Skin Therapy Lett,* vol. 18, no. 3, pp. 4-9, 2013. [PMID: 23674145]

[10] S. Herman, N. Zurgil, P. Langevitz, M. Ehrenfeld, and M. Deutsch, "The immunosuppressive effect of methotrexate in active rheumatoid arthritis patients vs. its stimulatory effect in nonactive patients, as indicated by cytometric measurements of CD4+ T cell subpopulations", *Immunol. Invest,* vol. 33, no. 3, pp. 351-362, 2004. [http://dx.doi.org/10.1081/IMM-120039865] [PMID: 15495793]

[11] "Methotrexate Pathway (Cancer Cell), Pharmacodynamics", *PharmGKB*. Available at: https://www.pharmgkb.org/pathway/PA2039

[12] I. Mogos, "A new therapeutic concept with applications in bone malignant tumours which means chemotherapy with high doses of Methotrexate in association with dynamic photostimulation monitored by thermograph", *Romanian J. Biophys,* vol. 3-4, pp. 165-172, 2000.

[13] A. Thomas, F.G. Einschlag, M.R. Féliz, and A.L. Capparelli, "First steps in the photochemistry of folate in alkaline medium", *J. Photochem. Photobiol. Chem,* vol. 116, no. 3, pp. 187-190, 1998. [http://dx.doi.org/10.1016/S1010-6030(98)00304-9]

[14] K. Robinson, T.F. Pilot, and J.E. Meany, "Time dependent inhibition of xanthine oxidase in irradiated solutions of folic acid, aminopterin and methotrexate", *Physiol. Chem. Phys. Med. NMR,* vol. 22, no. 2, pp. 95-103, 1990.
[PMID: 2100009]

[15] *FLUOROURACIL, CAMEO Chemicals, NOAA.*04-Jul-2016 Available at: https://cameochemicals.noaa.gov/chemical/5005

[16] G. G. Hawley, *Hawley's condensed chemical dictionary,* Wiley, 2001.

[17] S. Hayes, "Remington: The Science and Practice of Pharmacy, Vol. 1-2, 22nd edition", In: *J. Med. Libr. Assoc,* vol. 102, no. 3, pp. 220-221, 2014.

[18] M.L. Pascu, M. Brezeanu, L. Voicu, A. Staicu, B. Carstocea, and R.A. Pascu, "5-fluorouracil as a phosensitiser", *In Vivo,* vol. 19, no. 1, pp. 215-220, 2005.
[PMID: 15796177]

[19] J. Jin, M. Huang, H-L. Wei, and G-T. Liu, "Mechanism of 5-fluorouracil required resistance in human hepatocellular carcinoma cell line Bel(7402)", *World J. Gastroenterol,* vol. 8, no. 6, pp. 1029-1034, 2002.
[http://dx.doi.org/10.3748/wjg.v8.i6.1029] [PMID: 12439919]

[20] A. Staicu, M.L. Pascu, I. Mogos, M. Enescu, S. Truica, L. Voicu, D.M. Gazdaru, A. Radu, and S. Gazdaru, "Contribution to the spectroscopic study of cytostatics molecules", *Proc. SPIE,* vol. 4430, pp. 680-683, 2001.
[http://dx.doi.org/10.1117/12.432908]

[21] A.H. Thomas, G. Suárez, F.M. Cabrerizo, R. Martino, and A.L. Capparelli, "Study of the photolysis of folic acid and 6-formylpterin in acid aqueous solutions", *J. Photochem. Photobiol. Chem,* vol. 135, no. 2-3, pp. 147-154, 2000.
[http://dx.doi.org/10.1016/S1010-6030(00)00304-X]

[22] U. Schmidt-Erfurth, R. Birngruber, and T. Hasan, "Photodynamic therapy in ocular vascular disease", *IEEE J. Sel. Top. Quantum Electron,* vol. 2, no. 4, pp. 988-996, 1996.
[http://dx.doi.org/10.1109/2944.577328]

[23] J. Chevalier, A. Mahamoud, M. Baitiche, E. Adam, M. Viveiros, A. Smarandache, A. Militaru, M.L. Pascu, L. Amaral, and J-M. Pagès, "Quinazoline derivatives are efficient chemosensitizers of antibiotic activity in Enterobacter aerogenes, Klebsiella pneumoniae and Pseudomonas aeruginosa resistant strains", *Int. J. Antimicrob. Agents,* vol. 36, no. 2, pp. 164-168, 2010.
[http://dx.doi.org/10.1016/j.ijantimicag.2010.03.027] [PMID: 20494558]

[24] A. Hansen, T. Jørgensen, and U. Olsen, *Use of N-substituted azaheterocyclic compounds for the manufacture of a pharmaceutical composition for the treatment of indications related to angiogenesis,* December 1st, 1999. EP1135129A1

[25] T. Tanaka, Y. Cao, J. Folkman, and H.A. Fine, "Viral vector-targeted antiangiogenic gene therapy utilizing an angiostatin complementary DNA", *Cancer Res,* vol. 58, no. 15, pp. 3362-3369, 1998.
[PMID: 9699667]

[26] A.N. Chakrabarty, *Non Antibiotics: A New Class of Unrecognised Antimicrobics,* National Institute of Science Communication, 1998.

[27] L. Qi, and Y. Ding, "Potential antitumor mechanisms of phenothiazine drugs", *Sci. China Life Sci,* vol. 56, no. 11, pp. 1020-1027, 2013.
[http://dx.doi.org/10.1007/s11427-013-4561-6] [PMID: 24132550]

[28] Z. Zhelev, H. Ohba, R. Bakalova, V. Hadjimitova, M. Ishikawa, Y. Shinohara, and Y. Baba, "Phenothiazines suppress proliferation and induce apoptosis in cultured leukemic cells without any influence on the viability of normal lymphocytes. Phenothiazines and leukemia", *Cancer Chemother. Pharmacol,* vol. 53, no. 3, pp. 267-275, 2004.
[http://dx.doi.org/10.1007/s00280-003-0738-1] [PMID: 14663628]

[29] I. Gil-Ad, B. Shtaif, Y. Levkovitz, J. Nordenberg, M. Taler, I. Korov, and A. Weizman, "Phenothiazines induce apoptosis in a B16 mouse melanoma cell line and attenuate in vivo melanoma tumor growth", *Oncol. Rep,* vol. 15, no. 1, pp. 107-112, 2006.
[PMID: 16328041]

[30] C.W. Yde, M.P. Clausen, M.V. Bennetzen, A.E. Lykkesfeldt, O.G. Mouritsen, and B. Guerra, "The antipsychotic drug chlorpromazine enhances the cytotoxic effect of tamoxifen in tamoxifen-sensitive and tamoxifen-resistant human breast cancer cells", *Anticancer Drugs,* vol. 20, no. 8, pp. 723-735, 2009.
[http://dx.doi.org/10.1097/CAD.0b013e32832ec041] [PMID: 19584708]

[31] T. Alexandru, A. Staicu, A. Pascu, E. Radu, A. Stoicu, V. Nastasa, A. Dinache, M. Boni, L. Amaral, and M.L. Pascu, "Characterization of mixtures of compounds produced in chlorpromazine aqueous solutions by ultraviolet laser irradiation: their applications in antimicrobial assays", *J. Biomed. Opt,* vol. 20, no. 5, p. 051002, 2015.
[http://dx.doi.org/10.1117/1.JBO.20.5.051002] [PMID: 25365041]

[32] T. Alexandru, M.L. Pascu, B. Danko, V. Nastasa, M. Boni, A. Militaru, I.R. Andrei, A. Staicu, A. Hunyadi, A. Armada, M. Viveiros, and L. Amaral, "Generation and biological evaluation of the products formed from the exposure of Phenothiazine to a 266nm laser beam", *Proc. SPIE,* vol. 8882, p. 88820T, 2013.
[http://dx.doi.org/10.1117/12.2032666]

[33] A.M. Armada, T. Alexandru, D. Machado, B. Danko, A. Hunyadi, A. Dinache, V. Nastasa, M. Boni, J. Ramos, M. Viveiros, J. Molnar, M.L. Pascu, and L. Amaral, "The in vitro activity of products formed from exposure of chlorpromazine to a 266 nm laser beam against species of mycobacteria of human interest", *In Vivo,* vol. 27, no. 5, pp. 605-610, 2013.
[PMID: 23988894]

[34] M.L. Pascu, B. Danko, A. Martins, N. Jedlinszki, T. Alexandru, V. Nastasa, M. Boni, A. Militaru, I.R. Andrei, A. Staicu, A. Hunyadi, S. Fanning, and L. Amaral, "Exposure of chlorpromazine to 266 nm laser beam generates new species with antibacterial properties: contributions to development of a new process for drug discovery", *PLoS One,* vol. 8, no. 2, p. e55767, 2013.
[http://dx.doi.org/10.1371/journal.pone.0055767] [PMID: 23405212]

[35] M.L. Pascu, B.D. Carstocea, A. Staicu, M.A. Ionita, S. Truica, and R. Pascu, "Spectroscopic studies of drugs used in the treatment of malignant tumors in ophthalmology", *Proc. SPIE,* vol. 4606, pp. 52-60, 2001.
[http://dx.doi.org/10.1117/12.446717]

[36] R.A. Pascu, M. Trifu, M. Dumitrescu, A. Mahamoud, A. Staicu, B. Carstocea, and M.L. Pascu, "In vivo studies of the effects of alkyl substituted benzo[b]pyridinium ompounds exposed to optical radiation", *Rom. Rep. Phys,* vol. 60, no. 3, pp. 899-908, 2008.

[37] M.L. Pascu, I. Mogos, M. Enescu, A. Staicu, S. Truica, L. Voicu, D.M. Gazdaru, M.O. Pascu, and A. Radu, "Optical properties of cytostatic drugs used in cancer treatment", *Proc. SPIE,* vol. 4606, pp. 66-71, 2001.
[http://dx.doi.org/10.1117/12.446699]

[38] J.J. Aaron, M.D. Seye, S. Trajkovska, and N. Motohashi, "Bioactive Phenothiazines and Benzo[a]phenothiazines: Spectroscopic Studies, and Biological and Biomedical Properties and Applications", In: *Bioactive Heterocycles VII,* Springer, 2008.
[http://dx.doi.org/10.1007/7081_2008_125]

[39] M.L. Pascu, A. Staicu, L. Voicu, M. Brezeanu, B. Carstocea, R. Pascu, and D. Gazdaru, "Methotrexate as a photosensitiser", *Anticancer Res,* vol. 24, no. 5A, pp. 2925-2930, 2004. [PMID: 15517898]

[40] M.L. Pascu, R.A. Pirvulescu, M.O. Romanitan, and A. Smarandache, "Laser induced fluorescence in biomedicine", In: *Laser Manual,* Officina Editoriale Oltrarno, 2014.

# CHAPTER 15

# Interaction of Medicines Exposed to Laser Beams with Fabrics of Interest for Biomedical Applications

**Ágota Simon**[1,2] and **Mihail Lucian Pascu**[1,2,*]

[1] *National Institute for Laser, Plasma and Radiation Physics, Magurele, Ilfov, Romania*

[2] *Faculty of Physics, University of Bucharest, Magurele, Ilfov, Romania*

**Abstract:** Photosensitive phenothiazine drugs subjected to UV laser radiation exhibit alteration of the initial parental compounds and provide newly generated photoproducts with possible bacteriostatic, bactericidal and antitumor effects. These new compounds with enhanced properties may also have outstanding wetting abilities when applied on fabrics of interest for applications, in order to develop new tools for drug delivery to patients, biological targets, or even surfaces of manmade equipment, constructions, or space ships. However, prior to biomedical applications, surface properties including contact angle measurements, require more thorough studies since such properties may control the delivery process of medicines to target tissues. This chapter shows the results obtained on the wettability of cotton, polyester and Parafilm M surfaces by laser irradiated phenothiazine aqueous solutions. In the present study, pendant chlorpromazine, promazine and promethazine droplets have been generated, then by simply bringing them into contact with the respective surfaces, detachment from a vertical capillary took place thus achieving the formation of sessile droplets. Results have evidenced the fact that some of the implemented drugs, containing the photoproducts obtained by prolonged exposure to laser radiation of the parent compounds, indicated better wetting abilities compared to their unexposed control, hence providing new promising perspectives.

**Keywords:** Absorption, Adsorption, Advancing contact angle, Chlorpromazine, Contact angle hysteresis, Cotton, Non-antibiotics, Parafilm M, PES, Polyester, Promazine, Promethazine, Receding contact angle, Sessile droplet, Surface tension, Surface wettability, UV laser, Wetting properties.

## INTRODUCTION

Wetting processes may be regarded as a key player in many applications, such as biology [1, 2], medicine [3, 4], cosmetics [5], industry [6, 7] as well as space

* **Corresponding author Mihail Lucian Pascu:** Laser Department, National Institute for Laser, Plasma and Radiation Physics, Magurele, Ilfov, Romania; Tel/Fax: 0040 21 4575739; E-mail: mihai.pascu@inflpr.ro

**Mihail Lucian Pascu (Ed.)**

science [8 - 11]. Concerning biomedical applications, studies of biological fluids in interaction with biological/alien surfaces have a significant role in numerous treatments [12]. Bearing this in mind, wetting processes have been under investigations in (i) dentistry (*e.g.* gingival surface wettability in contact with intra-oral water [13], wetting behaviour of dental implants [14], denture base materials wettability with natural and artificial saliva [15]), (ii) ophthalmology (*e.g.* tear film wettability with or without contact lens [16], tear film breakup depending on corneal surface wettability [17]) and (iii) vascular biology (*e.g.* inhibiting the adhesion of blood components by increasing wettability of implanted materials [18]). Another interest resides in studying the interaction of medicine solutions with different surfaces implemented in medical treatments (*e.g.* medicinal bandages, transdermal patches), as well as with human skin.

One possible approach to determine the degree of wetting consists in contact angle (CA) measurements.

As shown in Chapters 2 and 10, the CA formed by a liquid drop on an ideal solid surface, *i.e.* flat, rigid, smooth, homogeneous and with no contact angle hysteresis (CAH), is governed by the thermodynamic equilibrium of the drop under the action of three interfacial tensions, described by Young's equation:

$$\gamma_{LV} \cos\theta = \gamma_{SV} - \gamma_{SL} \quad \textbf{(1)}$$

where $\gamma_{LV}$, $\gamma_{SV}$ and $\gamma_{SL}$ represent the liquid-vapor, solid-vapor and solid-liquid interfacial tensions, respectively, and $\theta$ is the CA arisen from the balance of the intermolecular interactions between solid-liquid [19], illustrated in Fig. (**1**).

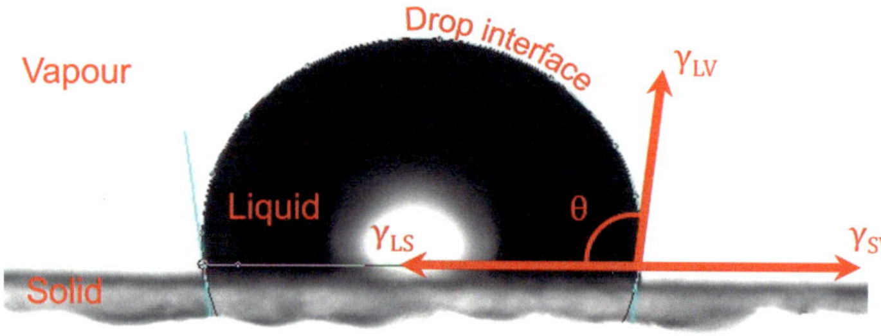


**Fig. (1).** Schematic representation of a liquid drop and CA formed by the intersection of the liquid-solid and the liquid-vapor interface.

A liquid drop tends to avoid contact with surface when cohesive forces between liquid's molecules are stronger than adhesive forces between solid'-liquid's molecules. Otherwise, the drop will spread onto surface [19], as depicted in Fig. (**2**).

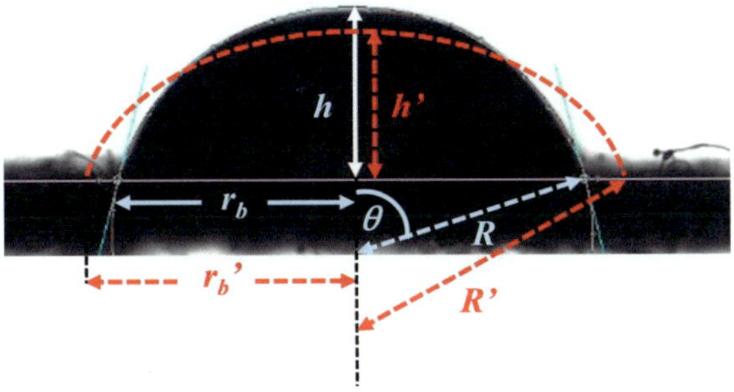


**Fig. (2).** Evolution of a sessile drop dimensions when spreading of a liquid drop onto a solid surface occurs.

The majority of surface and colloid scientists regard Thomas Young as the father of CA and wetting fundamentals, although most probably the earliest direct recognition, concerning wetting phenomena may be attributed to Galileo Galilei, who might be perceived as the grandfather of the field [20 - 22]. His observations of a dense, flat and thin solid floating on water allowed him to advance the idea of spreading and wetting in 1612 [19, 21]. The respective early findings were followed by the introduction of CA equation on ideal solid surfaces in 1805 by Young [23], while in 1869 Dupré correlated the Young equation with the work of adhesion [24]. The studies which followed included investigations on the shape of pendant drops, carried out by Worthington in 1881, whereas in order to analyse drop profiles he implemented the Laplace curvature theory instead of measuring the formed tangents of drop profiles [25]. However, the first experimentally measured CA results, in the time interval 1877-1898, can be associated with the work of Quincke [19]. Further research regarding CA in terms of surface forces and hysteresis can be assigned to Lord Rayleigh, who examined in 1890 the advancing and receding of a liquid drop onto a solid surface [19, 26]. Later on, Bartell, Hatch and Mack [27, 28] have carried out trigonometric calculations by considering a spherical geometry of droplets, obtaining CA from the height and diameter of small drops utilising the following equation:

$$tan\frac{\theta}{2} = \frac{h}{r_b} \quad \textbf{(2)}$$

where $h$ represents the height and $r_b$ is the base radius of the drop. The first correlation between the CA and the roughness of a specific surface, by taking into consideration the area of liquid/solid interface, was defined by Wenzel in 1936 [29]. His investigations were further extended by Cassie and Baxter, who performed studies on porous surfaces by including the interfacial area fraction approach, in 1944 [30]. A review on CA history is detailed in [19].

The wettability of solid target surfaces by medicine solutions involves CA measurements as primary data, thus indicating the degree of wetting during the interaction. CA lower than 90° correspond to a good wettability, in that case the liquid spreads over an extensive surface area, while those greater than 90° are correlated with a poor wettability, the liquid minimizing its contact with the solid surface and forming a compact droplet [31], as explained in Table **1**.

**Table 1. Wetting, adhesiveness and solid-liquid interactions characteristics, function of CA.**

| Contact angle | Wetting | Adhesiveness | Solid-liquid interactions |
|---|---|---|---|
| $\theta = 0°$ | Complete | Good | Strong |
| $\theta \ll 90°$ | Good | Good | Strong |
| $\theta \gg 90°$ | Poor | Poor | Weak |
| $\theta = 180°$ | Non | Poor | Weak |

Within the present chapter, cotton (CO) and polyester (PES) textiles have been implemented as hydrophilic target surfaces [32], taking into consideration their use in medical treatments. In the case of such surfaces, equilibrium conditions are seldom achieved during the actual wetting processes [31] since droplets are continuously evolving as spreading takes place. Fig. (**3**) shows the time evolution of left and right CA of a phenothiazine droplet placed on hydrophilic fabric.

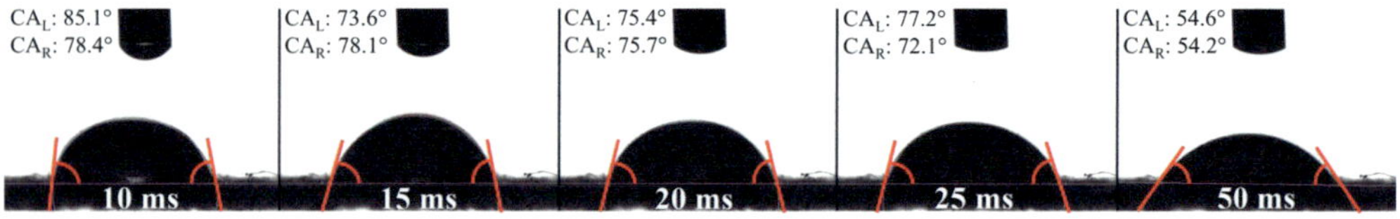


**Fig. (3).** Images depicting contact angle evolution of a spreading promethazine drop on PES target surface.

Regarding hydrophobic surfaces, advancing (ACA) and receding contact angles (RCA) have been determined alongside their respective hysteresis. CAH, appearing possibly as a result of surface roughness and chemical heterogeneity, has been treated in [22], according to which CA reaches its maximum value when liquid front advances and the triple line is pinned at the edge, while in the case of receding liquid, the triple line is pinned down until a minimum value of CA is reached.

The current chapter treats ACA and RCA measurements performed on Parafilm M surfaces [32], consisting in the addition of drops at every stage followed by their step by step removal (Fig. **4**).

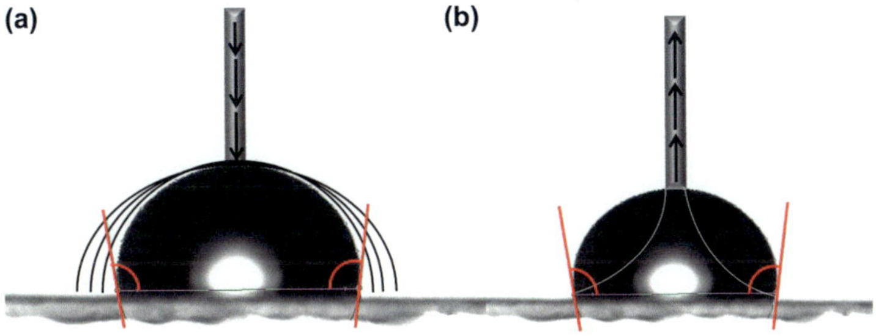


**Fig. (4).** Schematic representation of (**a**) ACA and (**b**) RCA on a hydrophobic surface*.

The present study has been carried out under terrestrial conditions. Results of a similar study, under hypergravity conditions are introduced in Chapter 16.

## MATERIALS AND METHODS

### Medicine Solutions

Chlorpromazine (CPZ), promazine (PZ) and promethazine (PMZ) phenothiazine derivatives, acquired from Sigma-Aldrich as hydrochloride powders at a purity higher than 98% and described in detail in Chapters 9, 11 and 14, have been implemented in wetting studies on different target surfaces.

Medicine solutions preparation consisted in dissolving the solutes into ultrapure water, prior to laser exposure, thus attaining a 20 mg/ml concentration. After applying the irradiation procedure, but before performing measurements, the solutions have been stored at 2-5°C and in a dark environment, in order to prevent possible further modifications in their chemical structures due to their photosensitivity.

### Laser Exposure

The 266 nm laser beam, emitted as fourth harmonic of the fundamental radiation of a pulsed Nd:YAG laser, has been employed in the irradiation procedure of phenothiazine solutions. Drug solutions were subjected to a 6.5 mJ average beam energy, monitored in real time by a Gentec-EO powermeter, the auxiliary irradiation conditions being detailed in Chapter 11. The exposure of solutions ranged between 1, 15, 30, 60, 120 and 240 min, respectively, an unexposed sample being always kept as control. In order to avoid precipitates formation, an AREC-X, VELP Scientifica magnetic stirrer was utilised during irradiation at 600 rpm to agitate the 2 ml drug solution placed inside a sealed quartz cuvette.

Real time laser induced fluorescence (LIF) signals emitted by samples were collected by an optical fiber coupled with an Ocean Optics HR4000CG-UV-NIR high resolution spectrometer. After exposure, solutions were characterised by absorption UV-Vis-NIR spectroscopy, pH and surface tension measurements, thin layer chromatography (TLC) analysis as well as epifluorescence microscopy, such results are discussed in Chapters 9 and 14. Moreover, phenothiazine solutions have been employed to impregnate target surfaces. The present chapter shows a new insight into the wetting ability of laser modified non-antibiotic solutions in terms of CA and wetted diameter (WD) measurements with respect to solid state surfaces.

## Target Surfaces

Three different target surfaces, hydrophilic CO and PES test materials as well as hydrophobic Parafilm M paraffin wax surface, purchased from wfk Testgewebe, and Bemis, respectively, have been explored in order to gain a better understanding of their interaction with phenothiazine solutions.

Prior to measurements, CO and PES surfaces underwent a severe washing procedure, by utilising warm tap water. A comparative surface tension measurement was performed after each washing cycle between the used and the clean tap water. Regarding CO, 10 cycles were needed to reach the value of control water; as for PES, 5 cycles were enough to obtain unblemished material, due to the specific property of PES to be easily washed. Thus, obtained hydrophilic surfaces were subjected to a cutting procedure, then placed on an adjustable FSH 30 DataPhysics sample holder made of polyoxymethylene, in order to be stretched, while the hydrophobic Parafilm M surface was positioned onto a microscope glass slide.

## Contact Angle Measurements

CA formed by phenothiazine drops on CO, PES and Parafilm M surfaces represented the primary data of the study alongside determination of their WD. An OCA 40Micro system connected to a SCA 20 DataPhysics software has been used as a measurement instrument.

First, medicine pendant droplet generation was performed by the electronic single direct dosing subsystem SD-DM mounted directly on the illumination housing of the OCA instrument as a stand-alone system. The SD-DM, equipped with a disposable syringe as well as a 0.5 mm outer diameter (OD) needle, controlled the dosing volume and rate, respectively. Concerning measurements made on the hydrophilic surfaces, a 0.5 μl dosing volume with a 1.18 μl/s rate has been implemented, generating 2 μl droplets.

Secondly, the produced droplet was gently brought into contact with the solid surface. Upon contact with the target, detachment of droplet took place, thus sessile droplets being formed. Their visualisation was made by employing a high-speed image processing system UpHSC 1000, able to record up to 1600 frames/s. In order to acquire the desired high quality images, droplets located on the surface have been lighted by a cold LED system and their images were magnified by a zoom lens having software controlled motorized focus and adjustment of the observation angle. Taking into consideration the high absorption rate of phenothiazine aqueous solutions into the hydrophilic target surfaces, a 200 frames/s recording speed has been selected, whereas for the hydrophobic surface, images were captured with 5 frames/s, as a result of the slow evaporation process which replaced the absorption in the fabric.

It has to be mentioned that on Parafilm M, ACA and RCA measurements have been undertaken by adding a volume of 1 µl at each stage until the generation of a 10 µl drop was obtained, followed immediately by the removal of the accumulated drops by the same 1 µl dosing quantity, using a 10.6 µl/s rate.

Finally, the obtained images have been analysed, by establishing the time of the impact of droplets as a first step, followed by the definition of the contact line or otherwise called baseline between medicine drops and target surfaces and as a final step by determining the droplet's contour.

## RESULTS AND DISCUSSIONS

In the present study, results on the interaction of unexposed and laser exposed CPZ, PZ and PMZ solutions with CO, PES and Parafilm M surfaces, respectively, are shown.

Since phenothiazines are amphiphilic molecules, they possess similar behaviour as surfactants. The importance of surfactants has been detailed comprehensively in Chapter 12 of the book. This can explain why all studied phenothiazine solutions exhibited reduced surface tensions in contrast to ultrapure water, thus providing enhanced wetting ability compared to the solvent used for their preparation. These outstanding wetting properties have been manifested for the hydrophilic surfaces, phenothiazine solutions being immediately spread onto, fast absorbed and easily penetrated into the fibres of permeable CO and PES textiles. Due to the very rapid absorption rate of medicine solutions into such surfaces, the wetting process was characterized by a very low rate of evaporation.

Regarding the CO target surface, Fig. (**5**) highlights the better wetting ability of both unexposed and laser exposed CPZ solutions in contrast to ultrapure water; thus, solutions at 100 times greater concentration, *i.e.* 20 mg/ml, have presented much better wetting properties than at 0.2 mg/ml. One has to mention that the initially prepared CPZ solutions at 20 mg/ml as well as those exposed to laser radiation at the respective concentration have been diluted down to 0.2 mg/ml, having as objective the study regarding the influence of concentration on surface wettability.

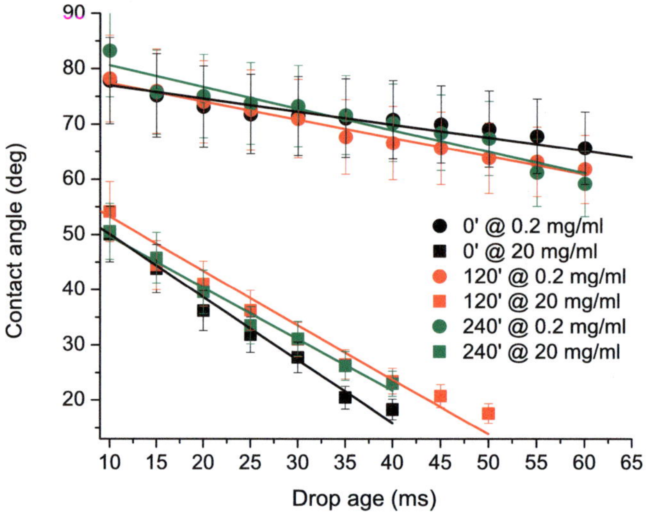


**Fig. (5).** Time evolution of CA on CO as a function of the concentration of unirradiated as well as 120 and 240 min laser irradiated CPZ solutions in ultrapure water*.

Drop absorbency can be determined by observing the evolution of spreading droplet profiles by looking at influencing factors such as WD and drop height. Collected data indicated a continuous decrease in WD and height of droplets, as long as hydrophilic surfaces have been implemented in the experiment, causing deviation from the hemi-spherical shape of droplets after their impact with the target surface.

At the same time, one must take into consideration that in the case of enhanced wetting properties, decreased CA corresponds to increased WD, while the opposite situation validates the case of poor wetting characteristics.

Fig. (**6**) illustrates the evolution of WD on CO surface of the same CPZ solutions and their concentrations as presented in Fig. (**5**). One may notice that laser irradiated 120 and 240 min samples exhibited inflated WD, corresponding to the plotted CA data.

According to experimental data, the time evolution of WD follows the distribution of a quadratic polynomial function.

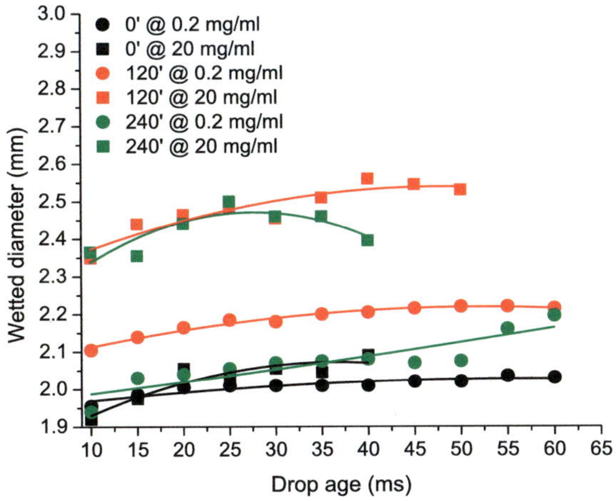


**Fig. (6).** WD evolution in time on CO of unexposed as well as 120 and 240 min laser exposed CPZ aqueous solutions at 0.2 and 20 mg/ml concentrations, respectively.

As far as PES target surface is concerned, Fig. (**7**) emphasises a faster absorption alongside better wetting ability of both unexposed and laser exposed CPZ samples at 20 mg/ml, in correlation to the 100 times fold lower concentration. It has been noticed that at 0.2 mg/ml, the absorption of drug solutions presented a slow evolution over a prolonged period of time.

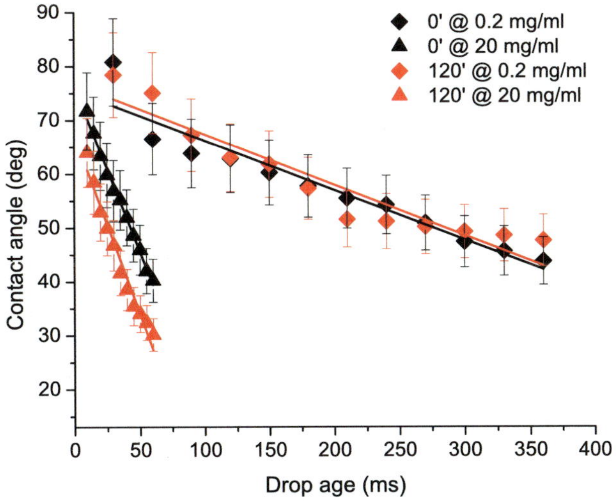


**Fig. (7).** CA evolution of control and 120 min irradiated CPZ solutions at 0.2 and 20 mg/ml concentrations in ultrapure water on PES*.

Fig. (**8**) depicts the correlation of low and high concentrations of CPZ droplets generated onto PES surface from the point of view of their WD.

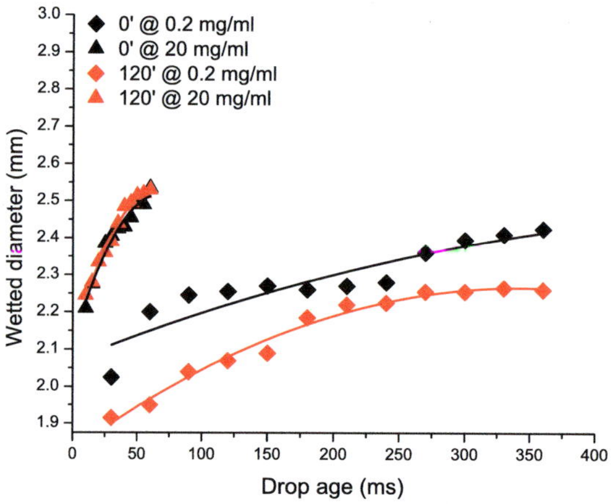


**Fig. (8).** WD evolution on PES of unexposed and 120 min laser exposed aqueous CPZ at 0.2 and 20 mg/ml.

The outcome of the time evolution of WD fitted, as it was expected, with the obtained CA characteristics, the largest diameters being obtained in the case of both control and laser exposed 20 mg/ml CPZ solutions.

An analogy between the time evolution of CA and WD on CO and PES target surfaces of the three different phenothiazines has been performed, Figs. (**9-14**) displaying the unexposed controls alongside the 120 and 240 min laser exposed samples. It can be noticed that all three medicine solutions have formed smaller CA with CO, except the 120 and 240 min irradiated PZ samples, indicating its better wetting ability in contrast to PES. Regarding WD, all laser exposed samples showed increased values on CO compared to unexposed controls.

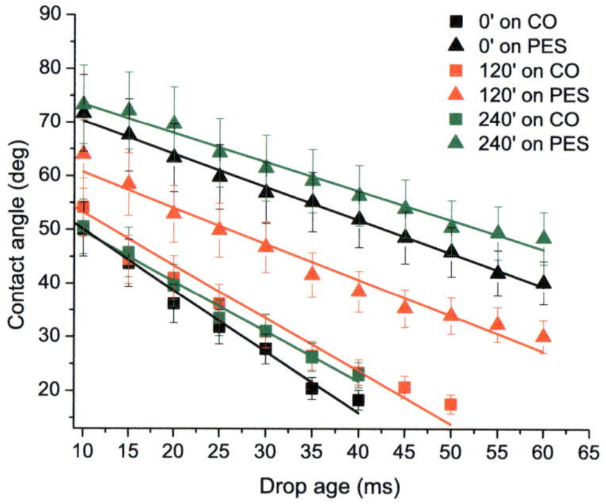


**Fig. (9).** Comparison of CA evolution on CO and PES surfaces of unexposed, 120 and 240 min exposed CPZ solutions at 20 mg/ml concentration in ultrapure water*.

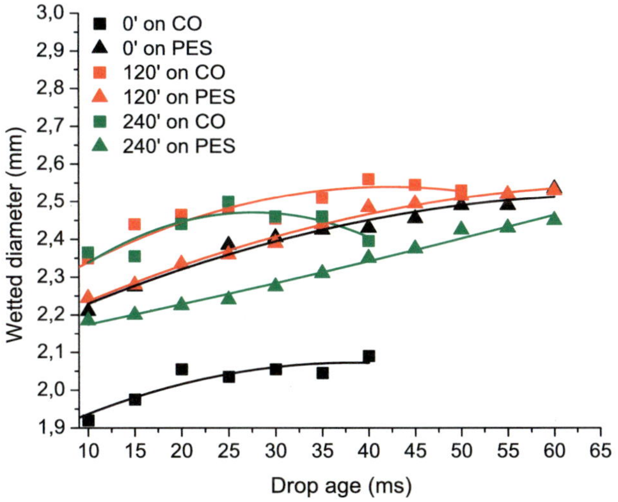


**Fig. (10).** WD evolution on CO and PES of control, 120 and 240 min irradiated 20 mg/ml aqueous CPZ.

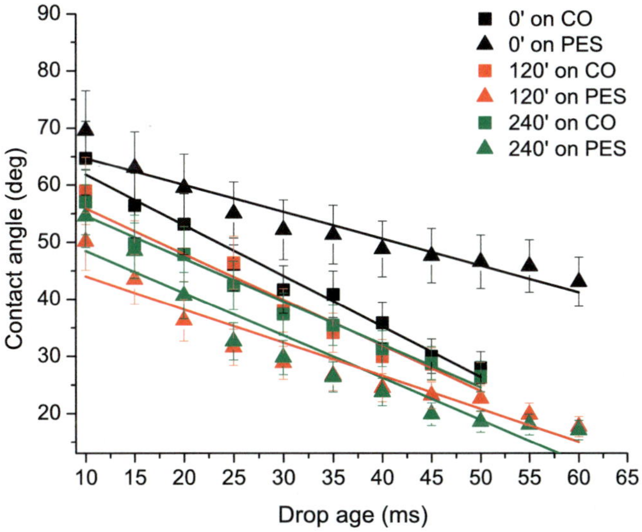


**Fig. (11).** CA evolution on CO and PES by unexposed, 120 and 240 min exposed aqueous 20 mg/ml PZ*.

On one hand, the lowest CA values have been obtained for 120 min exposed CPZ solution out of the entire set applied on PES textile, while the best wetting properties have been attributed to the control CPZ on CO, as presented in Fig. (**9**). On the other hand, overall, the highest CA were achieved for the 240 min irradiated CPZ solutions on PES.

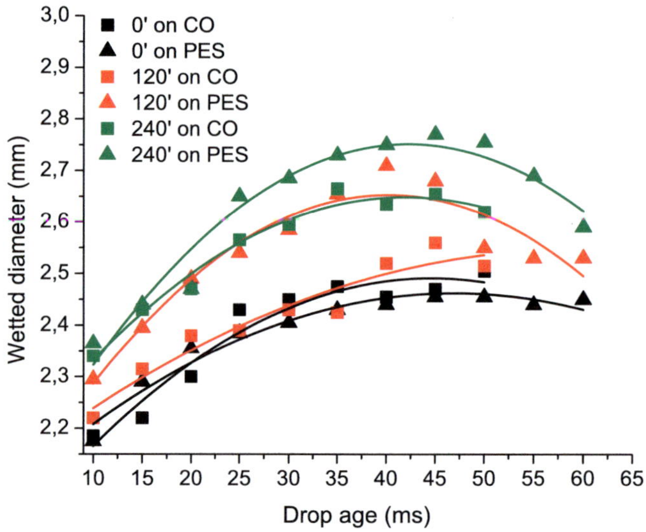


**Fig. (12).** Time evolution of WD on CO correlated to PES of unexposed as well as 120 and 240 min laser exposed PZ samples at 20 mg/ml concentration in ultrapure water.

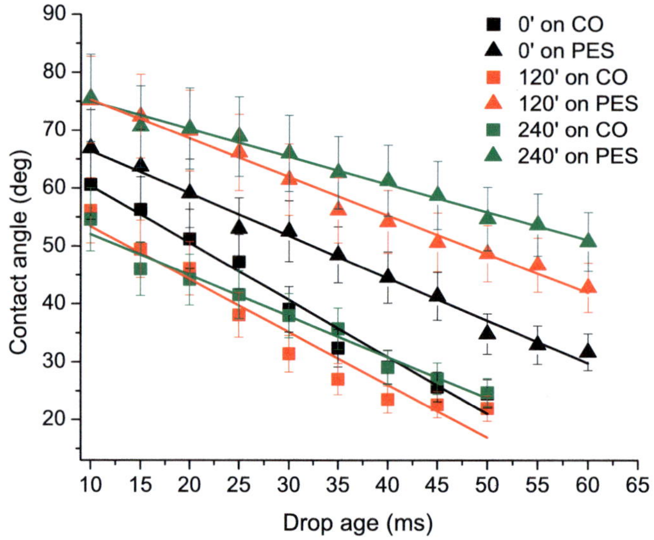


**Fig. (13).** CA comparison of unexposed, 120 and 240 min exposed 20 mg/ml PMZ on CO and PES surfaces*.

The largest WD, as illustrated in Fig. (**10**), have been obtained for CO surface, except the control CPZ. A similar evolution can be observed for all CPZ samples, when spreading onto and absorption by PES target surface occurred.

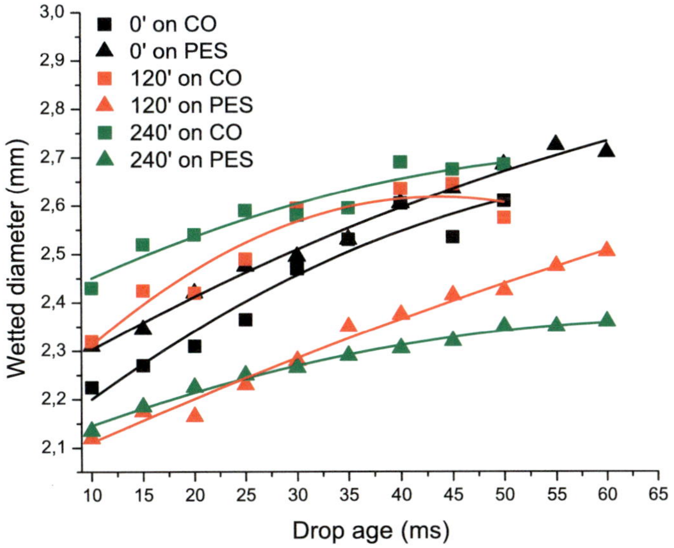


**Fig. (14).** WD for control, 120 and 240 min irradiated 20 mg/ml aqueous PMZ on CO compared to PES.

Regarding PZ, solutions subjected to 120 and 240 min laser exposure have shown preeminent wetting properties on PES (Fig. **11**), although the respective feature alters from its expected surface wettability behaviour.

The highest values of CA pertain to the unexposed PZ solution, notably to the sample placed on PES surface, as illustrated in Fig. (**11**).

Fig. (**12**) exhibits increased wetting diameters for both irradiation time intervals in the case of PES, while the lowest diameter values belong to the control PZ sample on PES and CO, respectively, being in good agreement with data shown in Fig. (**11**).

Also, one may observe that diameters pursued a quadratic polynomial function growth until a maximum value is reached, followed by their decay.

Fig. (**13**) shows the CA evolution of PMZ containing droplets on CO and PES surfaces. The best wetting properties have been discernible in the case of laser exposed PMZ containing droplets on CO, however the respective samples presented poor wetting ability on PES surface compared to the control PMZ.

WD produced by PMZ droplets are depicted in Fig. (**14**). The obtained results have confirmed data presented in Fig. (**13**), since the 120 and 240 min irradiated solutions exhibited the highest values of WD on CO, while the lowest on PES.

In the following, Figs. (**15-18**) present a comparison between unexposed and 240 min laser exposed 20 mg/ml CPZ, PZ and PMZ samples on CO and PES, respectively.

Fig. (**15**) reveals a better impregnation of CO target surface by the prolonged irradiated phenothiazines in contrast to their unirradiated control samples, an exception being the case of CPZ.

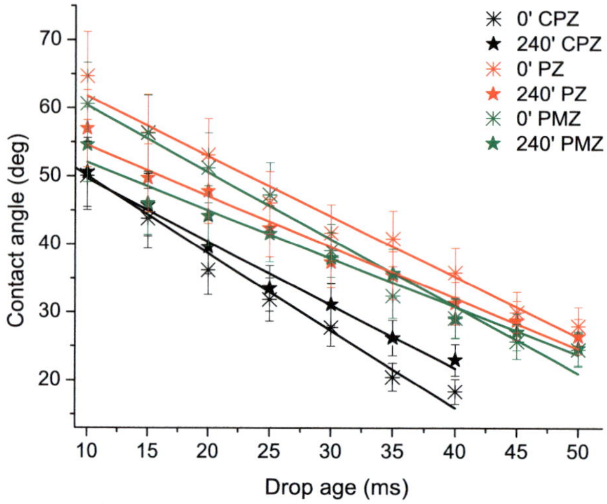


**Fig. (15).** Comparison of CA evolution of unexposed and prolonged laser exposed aqueous phenothiazine droplets at 20 mg/ml concentration on CO target surface*.

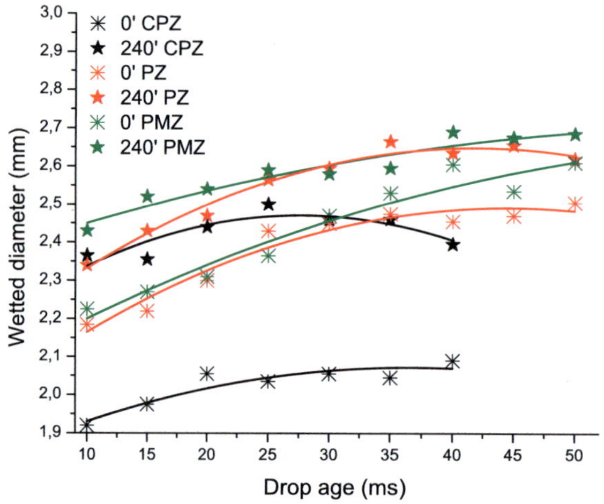


**Fig. (16).** Analogy in terms of WD on CO surface between unirradiated and longtime irradiated 20 mg/ml CPZ, PZ and PMZ ultrapure water solutions.

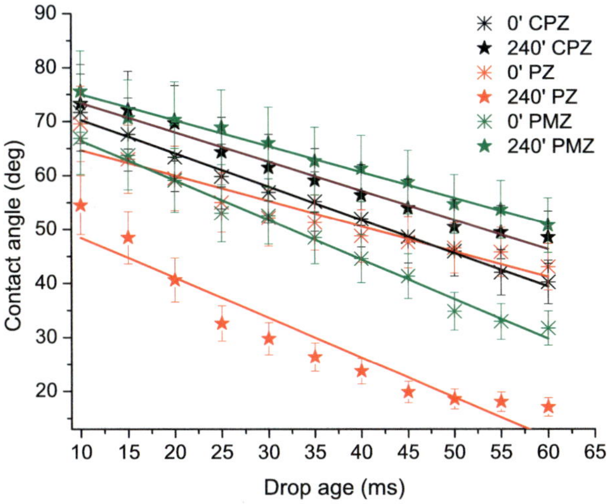


**Fig. (17).** CA evolution of unexposed and 240 min exposed drug solutions on PES target surface.

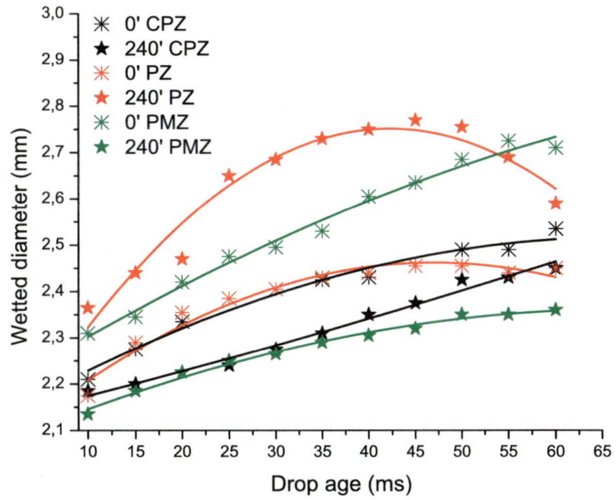


**Fig. (18).** Comparison of WD on PES surface of unirradiated as well as prolonged irradiated CPZ, PZ and PMZ solutions in ultrapure water at 20 mg/ml concentration.

However, both unirradiated and irradiated CPZ display the best wetting properties amongst the three drug solutions. An explanation may reside in the fact that laser exposed solutions contain photoproducts generated *via* irradiation process, acting as dominating factors and constituents when their microfluidic properties are measured.

As for WD, Fig. (**16**) shows that 240 min exposed PZ and PMZ have the highest values, meaning better wetting properties compared to their control.

Although the lowest CA values have been obtained for unexposed and laser exposed CPZ samples (Fig. **15**), the highest WD do not correspond to the respective medicine solutions as it was expected, the control CPZ exhibiting the lowest WD, for instance.

Concerning the time evolution of CA on PES target surface, Fig. (**17**) illustrates the three studied aqueous phenothiazines at 20 mg/ml concentration, comparing the control with the prolonged irradiated samples from the point of view of their wetting ability. It has been observed that on PES surface, PZ solution subjected to extended irradiation indicated improved wetting properties in contrast to its control sample, presenting also the most outstanding wetting ability amongst all medicine solutions.

However, one has to state the fact that irradiated CPZ and PMZ solutions did not exhibit better properties in comparison to that of the unirradiated controls, the poorest wetting ability corresponding to the 240 min exposed PMZ and CPZ on PES target surface.

Data regarding WD generated by 20 mg/ml phenothiazine aqueous drops on PES, shown in Fig. (**18**), present a good agreement with the determined CA and thus PZ exposed to prolonged irradiation indicates the highest values of WD. One may observe that the diameter increases until reaches a maximum value and then starts decreasing once the droplet penetration into the PES fibres advances, the time evolution of WD being described by a quadratic polynomial function.

The lowest diameters have been found for exposed CPZ and PMZ, confirming CA data presented in Fig. (**17**).

ACA and RCA have been determined for the highly hydrophobic Parafilm M target surface, curves corresponding to the 240 min irradiated samples at 0.2 mg/ml concentration being shown in Fig. (**19**).

One may notice that by increasing the generated medicine droplet from 1 to 10 µl, CA values slightly decrease from 104.05° to 100.77°, while by withdrawing 1 µl by 1 µl, angles dramatically decreases down to 49.9°, thus presenting hysteresis. The decline between ACA and RCA is given in Fig. (**20**), in some cases differing by as much as 55°.

Moreover, Fig. (**20**) indicates an increment in CAH in the case of small drop volumes as the irradiation time interval of CPZ solutions increases.

The significance of CAH has been explored in detail in [33 - 38]. The explanation for the hysteresis behaviour may be due to the existence of hydrophobic domains, which are specific for this kind of surface. These regions pin the motion of the solutions as they advance, leading to the increase in measured CA and moreover they hold back the contracting movements of medicine solutions when they recede, causing decrease in CA [38].

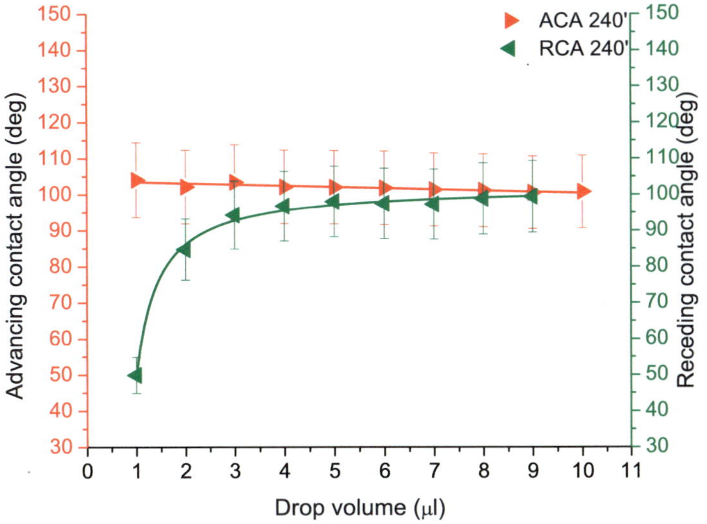


**Fig. (19).** ACA and RCA dependence on volume of the 240 min exposed 0.2 mg/ml CPZ water droplet*.

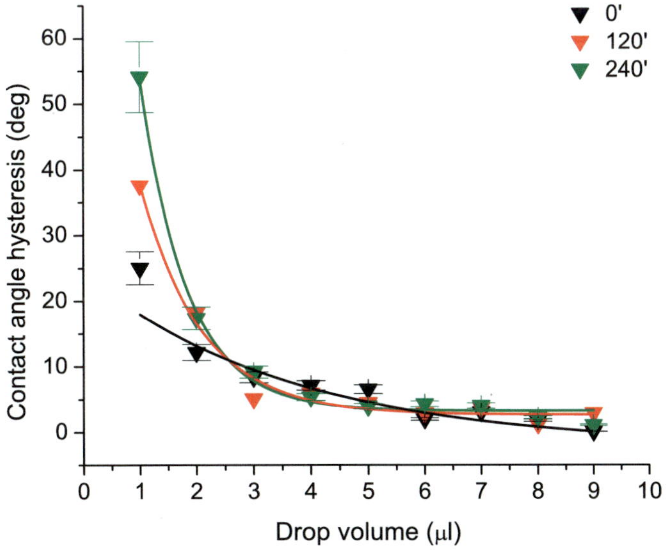


**Fig. (20).** CAH *versus* drop volume of unirradiated control as well as 120 and 240 min laser irradiated 0.2 mg/ml CPZ in ultrapure water*.

## CONCLUSIONS

Within the current chapter observations regarding the wettability of hydrophilic and hydrophobic target surfaces by UV laser modified medicine solutions have been presented. According to the obtained results, both unirradiated as well as irradiated CPZ, PZ and PMZ aqueous solutions exhibit improved wetting properties on CO and PES compared to the ultrapure water used as solvent. All CA on the respective hydrophilic surfaces indicated values lower than 90°, thus obtaining favourable surface wettability. It has been noticed that both unexposed and exposed CPZ have great wetting abilities on CO, while on PES the irradiated PZ has better wettability. Regarding WD, all three phenothiazines exposed to prolonged laser radiation offered outstanding wetting properties on CO, whereas on PES the laser exposed PZ proved to have the highest WD.

Being aware of the particular properties of phenothiazine medicines subjected to UV laser radiation, one may better understand their interaction with different target surfaces in view of biomedical applications. Since in many cases the standard drug delivery methods have failed, impregnated fabrics present promising tools to deliver lower drug quantities slowly, in order to obtain a more efficient treatment in an optimal time interval.

## NOTES

* Reprinted from Int. J. Pharm., vol. 475, no. 1-2, A. Simon, T. Alexandru, M. Boni, V. Damian, A. Stoicu, V. Dutschk, M. L. Pascu, Interaction of solutions containing phenothiazines exposed to laser radiation with materials surfaces, in view of biomedical applications, 270-281, (2014), with the permission from Elsevier.

## CONFLICT OF INTEREST

The authors confirm that this chapter content has no conflict of interest.

## ACKNOWLEDGEMENTS

This work has been financed by the National Authority for Research and Innovation in the frame of Nucleus programme-contract 4N/2016. The studies have been performed under the umbrella of COST Action MP1106.

## REFERENCES

[1] L. Xu, H. Zhu, H.E. Ozkan, and H.W. Thistle, "Evaporation rate and development of wetted area of water droplets with and without surfactant at different locations on waxy leaf surfaces", *Biosystems Eng.*, vol. 106, no. 1, pp. 58-67, 2010. [http://dx.doi.org/10.1016/j.biosystemseng.2010.02.004]

[2] A.Y. Stark, T.W. Sullivan, and P.H. Niewiarowski, "The effect of surface water and wetting on gecko adhesion", *J. Exp. Biol.*, vol. 215, no. Pt 17, pp. 3080-3086, 2012. [http://dx.doi.org/10.1242/jeb.070912] [PMID: 22875772]

[3] P.G. Campbell, and L.E. Weiss, "Tissue engineering with the aid of inkjet printers", *Expert Opin. Biol. Ther.,* vol. 7, no. 8, pp. 1123-1127, 2007.
[http://dx.doi.org/10.1517/14712598.7.8.1123] [PMID: 17696812]

[4] H. Schott, "Contact angles and wettability of human skin", *J. Pharm. Sci.,* vol. 60, no. 12, pp. 1893-1895, 1971.
[http://dx.doi.org/10.1002/jps.2600601233] [PMID: 5158017]

[5] A. Trybala, A. Bureiko, N. Kovalchuk, O. Arjmandi-Tash, Z. Liu, and V. Starov, "Wetting properties of cosmetic polymeric solutions on hair tresses", *Colloids Interface Sci. Commun.,* vol. 9, pp. 12-15, 2015.
[http://dx.doi.org/10.1016/j.colcom.2016.02.001]

[6] S. Karlsson, A. Rasmuson, I.N. Björn, and S. Schantz, "Characterization and mathematical modelling of single fluidised particle coating", *Powder Technol.,* vol. 207, no. 1-3, pp. 245-256, 2011.
[http://dx.doi.org/10.1016/j.powtec.2010.11.006]

[7] R. Sliz, Y. Suzuki, T. Fabritius, and R. Myllyla, "Influence of temperature on wetting properties of thin films in organic solar cells applications", *Colloids Surf. Physicochem. Eng. Asp.,* vol. 443, pp. 182-187, 2014.
[http://dx.doi.org/10.1016/j.colsurfa.2013.11.008]

[8] A. Diana, M. Castillo, D. Brutin, and T. Steinberg, "Sessile drop wettability in normal and reduced gravity", *Microgravity Sci. Technol.,* vol. 24, no. 3, pp. 195-202, 2012.
[http://dx.doi.org/10.1007/s12217-011-9295-0]

[9] Z-Q. Zhu, D. Brutin, Q-S. Liu, Y. Wang, A. Mourembles, J-C. Xie, and L. Tadrist, "Experimental investigation of pendant and sessile drops in microgravity", *Microgravity Sci. Technol.,* vol. 22, no. 3, pp. 339-345, 2010.
[http://dx.doi.org/10.1007/s12217-010-9224-7]

[10] O.A. Kabov, and D.V. Zaitsev, "The effect of wetting hysteresis on drop spreading under gravity", *Dokl. Phys.,* vol. 58, no. 7, pp. 292-295, 2013.
[http://dx.doi.org/10.1134/S1028335813070045]

[11] D. Brutin, Z-Q. Zhu, O. Rahli, J-C. Xie, Q-S. Liu, and L. Tadrist, "Evaporation of ethanol drops on a heated substrate under microgravity conditions", *Microgravity Sci. Technol.,* vol. 22, no. 3, pp. 387-395, 2010.
[http://dx.doi.org/10.1007/s12217-010-9200-2]

[12] D.M. Eckmann, D.P. Cavanagh, and A.B. Branger, "Wetting characteristics of aqueous surfactant-laden drops", *J. Colloid Interface Sci.,* vol. 242, no. 2, pp. 386-394, 2001.
[http://dx.doi.org/10.1006/jcis.2001.7719]

[13] H.C. van der Mei, D.J. White, and H.J. Busscher, "On the wettability of soft tissues in the human oral cavity", *Arch. Oral Biol.,* vol. 49, no. 8, pp. 671-673, 2004.
[http://dx.doi.org/10.1016/j.archoralbio.2004.03.002] [PMID: 15196985]

[14] F. Rupp, L. Scheideler, M. Eichler, and J. Geis-Gerstorfer, "Wetting behavior of dental implants", *Int. J. Oral Maxillofac. Implants,* vol. 26, no. 6, pp. 1256-1266, 2011.
[PMID: 22167431]

[15] A.-T. Farcaşiu, and M. Păuna, "Denture base materials wettability with natural and artificial saliva", *Romanian J. Oral Rehabil.,* vol. 3, no. 3, 2011.

[16] L.C. Thai, A. Tomlinson, and M.G. Doane, "Effect of contact lens materials on tear physiology", *Optom. Vis. Sci.,* vol. 81, no. 3, pp. 194-204, 2004.
[http://dx.doi.org/10.1097/00006324-200403000-00012] [PMID: 15017179]

[17] A. Sharma, "Acid-base interactions in the cornea-tear film system: surface chemistry of corneal wetting, cleaning, lubrication, hydration and defense", *J. Dispers. Sci. Technol.,* vol. 19, no. 6-7, pp. 1031-1068, 1998.
[http://dx.doi.org/10.1080/01932699808913229]

[18] M. Ombelli, D.M. Eckmann, and R.J. Composto, "Biomimetic dextran coatings on silicon wafers: Thin film properties and wetting", *MRS Proc,* vol. 734, 2002
[http://dx.doi.org/10.1557/PROC-734-B10.7]

[19] H.Y. Erbil, "The debate on the dependence of apparent contact angles on drop contact area or three-phase contact line: A review", *Surf. Sci. Rep.,* vol. 69, no. 4, pp. 325-365, 2014.
[http://dx.doi.org/10.1016/j.surfrep.2014.09.001]

[20] J. MacLachlan, *Galileo Galilei: First physicist,* Oxford University Press, 1999.

[21] R.J. Good, "Contact angle, wetting, and adhesion: a critical review", *J. Adhes. Sci. Technol.,* vol. 6, no. 12, pp. 1269-1302, 1992.
[http://dx.doi.org/10.1163/156856192X00629]

[22] K.L. Mittal, *Advances in contact angle, wettability and adhesion*vol. 1. John Wiley & Sons, 2013.
[http://dx.doi.org/10.1002/9781118795620]

[23] T. Young, *An essay on the cohesion of fluids*, Royal Society of London, 1805.

[24] P. D. Athanase M. Dupré, *Théorie mécanique de la chaleur,* Gauthier-Villars, 1869.

[25] A. Worthington, *On pendent drops*, Royal Society of London, 1881.

[26] L. Rayleigh, *On the theory of surface forces*, 1890.

[27] G.L. Mack, "The determination of contact angles from measurements of the dimensions of small bubbles and drops. I. The spheroidal segment method for acute angles", *J. Phys. Chem.,* vol. 40, no. 2, pp. 159-167, 1935.
[http://dx.doi.org/10.1021/j150371a001]

[28] G.L. Mack, and D.A. Lee, "The determination of contact angles from measurements of the dimensions of small bubbles and drops. II. The sessile drop method for obtuse angles", *J. Phys. Chem.,* vol. 40, no. 2, pp. 169-176, 1935.
[http://dx.doi.org/10.1021/j150371a002]

[29] R.N. Wenzel, "Resistance of solid surfaces to wetting by water", *Ind. Eng. Chem.,* vol. 28, no. 8, pp. 988-994, 1936.
[http://dx.doi.org/10.1021/ie50320a024]

[30] A.B. Cassie, and S. Baxter, "Wettability of porous surfaces", *Trans. Faraday Soc.,* vol. 40, pp. 546-551, 1944.
[http://dx.doi.org/10.1039/tf9444000546]

[31] M.J. Rosen, and J.T. Kunjappu, *Surfactants and interfacial phenomena,* John Wiley & Sons, 2012.
[http://dx.doi.org/10.1002/9781118228920]

[32] A. Simon, T. Alexandru, M. Boni, V. Damian, A. Stoicu, V. Dutschk, and M.L. Pascu, "Interaction of solutions containing phenothiazines exposed to laser radiation with materials surfaces, in view of biomedical applications", *Int. J. Pharm.,* vol. 475, no. 1-2, pp. 270-281, 2014.
[http://dx.doi.org/10.1016/j.ijpharm.2014.08.031] [PMID: 25148730]

[33] L. Gao, and T.J. McCarthy, "Contact angle hysteresis explained", *Langmuir,* vol. 22, no. 14, pp. 6234-6237, 2006.
[http://dx.doi.org/10.1021/la060254j] [PMID: 16800680]

[34] J.F. Joanny, and P.G. de Gennes, "A model for contact angle hysteresis", *J. Chem. Phys.,* vol. 81, no. 1, pp. 552-562, 1984.
[http://dx.doi.org/10.1063/1.447337]

[35] R.E. Johnson, and R.H. Dettre, "Contact angle hysteresis. III. Study of an idealized heterogeneous surface", *J. Phys. Chem.,* vol. 68, no. 7, pp. 1744-1750, 1964. [http://dx.doi.org/10.1021/j100789a012]

[36] J.W. Krumpfer, and T.J. McCarthy, "Contact angle hysteresis: a different view and a trivial recipe for low hysteresis hydrophobic surfaces", *Faraday Discuss.,* vol. 146, pp. 103-111, 2010. [http://dx.doi.org/10.1039/b925045j] [PMID: 21043416]

[37] L.W. Schwartz, and S. Garoff, "Contact angle hysteresis on heterogeneous surfaces", *Langmuir,* vol. 1, no. 2, pp. 219-230, 1985. [http://dx.doi.org/10.1021/la00062a007]

[38] G. Bracco, and B. Holst, *Surface science techniques*, vol. 51. Springer, 2013. [http://dx.doi.org/10.1007/978-3-642-34243-1]

**CHAPTER 16**

# Microvolumetric Droplets in Air in Hypergravity Conditions

**Ágota Simon[1,2], Alexandru Stoicu[1], Tatiana Tozar[1], Ionuţ Relu Andrei[1], Săndel Simion[1], Jack J. W. A. van Loon[3,4], Alan Dowson[4] and Mihail Lucian Pascu[1,2,*]**

[1] *National Institute for Laser, Plasma and Radiation Physics, Magurele, Ilfov, Romania*

[2] *Faculty of Physics, University of Bucharest, Magurele, Ilfov, Romania*

[3] *Dutch Experiment Support Center (DESC), Department of Oral and Maxillofacial Surgery / Oral Pathology, VU University Medical Center (VUmc) & Academic Centre for Dentistry Amsterdam (ACTA), Amsterdam, The Netherlands*

[4] *European Space Agency (ESA), European Space Research and Technology Centre (ESTEC), TEC-MMG, Noordwijk, The Netherlands*

**Abstract:** The interaction of laser modified medicine solutions with hydrophilic and hydrophobic target surfaces has been investigated under the effect of simulated hypergravity conditions, employing the Large Diameter Centrifuge (LDC) facility, developed by the European Space Agency (ESA). Experiments have been performed within the HyperMed project under the aegis of the ESA "Spin Your Thesis!" 2015 programme. During centrifugation, real-time video files have been recorded regarding generation of ultrapure water, unexposed and laser exposed chlorpromazine aqueous pendant droplets, followed by their detachment due to the exerted high gravitational accelerations and finally by the formation of sessile droplets on target surfaces. In this way, information about the volume of the generated droplet, the degree of wetting and its time evolution at different hypergravity levels has been obtained. Phenothiazine solutions irradiated with UV laser radiation indicate reduced surface tension, thus presenting better wetting properties. Target surfaces impregnated with medicine solutions may constitute an unconventional tool and even vector in developing new drug delivery systems. Such a wetting process under high g-level conditions may be useful in space medicine applications. Microorganisms can survive, grow and even proliferate under the effect of increased gravity. Therefore, upon launching of a spacecraft, during a long-term mission in microgravity conditions, astronauts and spacecraft surfaces may require treatment and decontamination, respectively, against onboard infectious microbes. Since non-terrestrial gravity may alter drug properties, medicine droplets behaviour in interaction with target surfaces under hypergravity conditions is the aim of the present study.

* **Corresponding author Mihail Lucian Pascu:** Laser Department, National Institute for Laser, Plasma and Radiation Physics, Magurele, Ilfov, Romania; Tel/Fax: 0040 21 4575739; E-mail: mihai.pascu@inflpr.ro

**Keywords:** Activated charcoal, Aluminium, Contact angle, Cotton, Hydrophilic surface, Hydrophobic surface, Hypergravity conditions, HyperMed project, Large Diameter Centrifuge, Long-Term space mission, Microbial infections, Multiple drug resistance, Nd:YAG laser, Parafilm M, Phenothiazines, Space medicine, "Spin Your Thesis!".

## INTRODUCTION

Since the access to spaceflight research is limited, different ground-based facilities are employed in order to induce outer space conditions. Gravity influences at a large extent countless processes, thus emphasising the impact of weight on life and physical systems. Ground-based simulators of both, microgravity and hypergravity, allow scientists to perform preliminary studies and test hardware as well as to determine experiments' parameters and protocols. Such an approach provides cost-effective platforms that can be implemented in gravitational research related to space missions [1, 2].

The term microgravity refers to a micro-g environment, which is also referred to as weightlessness or zero-g. However, such denomination should be avoided since neither space vehicles nor ground-based simulators can achieve and sustain real zero gravity. In microgravity conditions the net sum of all forces is not actually zero, but just very small compared to terrestrial conditions [3]. For instance, onboard space platforms like the International Space Station (ISS) or Shuttles the g-level may range between $\approx 10^{-3}$ and $10^{-6}$ g, depending on the location within the spacecraft and the frequency of vibration induced by aboard machinery, thruster operation for positional orientation or reboost, docking/berthing, astronauts' activity and micrometeorite impacts [1, 4 - 9]. On the other hand, the hypergravity terminology defines the condition wherein gravity surpasses that on the surface of Earth, *i.e.* exceeds 1 g. By taking into consideration that all organisms on Earth have evolved under 1 g condition, their exposure to high g-levels might lead to significant alterations of their structures and/or behavior [2].

One cannot rule out the realistic possibility of transporting living microorganisms through space. The first mention of the respective hypothesis was made as early as the 5$^{th}$ century BC by Anaxagoras [10], which later gained a more scientific approach through prominent scientists such as Berzelius (1834), Richter (1865), Kelvin (1871) and Helmholtz (1879) [11 - 14]. A more comprehensive study of the hypothesis was proposed by Arrhenius (1903) [15], and was discussed by Hoyle and Wickramasinghe (1974) [16, 17]. According to Arrhenius' suggestion, the transport of living bacteria from one solar system to another may be possible by the radiation pressure of its home star [15]. Although this kind of transfer might be difficult, the transfer within our solar system should not be neglected

[18]. Since in extraterrestrial environment microbes may better survive shielded, given the strong UV space radiation [19], they might be able to be transported inside meteorites, for example [20, 21]. The hypothesized process of microorganisms journey from one planet to another, consists in the ejection phase of bacteria-bearing rocks from the planet of origin as the result of an impact, where organisms would be, besides other factors, subjected to extreme accelerations. The following phase corresponds to the space travel alongside the associated environmental hazards such as vacuum, radiation and desiccation. Finally, the last phase involves the atmospheric entry, where the microbes would be exposed to heating and deceleration upon landing on Earth, for instance [20]. According to predictions made by Mastrapa and colleagues [20], a $3\times10^5$ g maximum acceleration would be possible due to an impact ejection from Mars. In order to fight microorganisms that were submitted to conditions during space fligths, one should have available medicines and transport vectors that contain them which have been submitted to the same conditions.

Another importance of hypergravity conditions arises during launches and reentries, the exposure being up to 3.2 g and around 1.4 g (as reported by NASA Ames Research Center), respectively (this could be much more in Soyuz systems where up to 20-40 g for a very short time can be produced). The fastest man-made object re-entry on record with successful landing was constituted by the Stardust Sample Return Capsule in 2006, with a hypervelocity of $\approx$ 46000 km/h. A drastic profile characterised the capsule reentry, ranging from Mach 36 to subsonic speed within 110 s, the peak deceleration lasting 40 s at 34 g at an altitude of 55 km [22].

Thus, investigations performed at high g conditions become worthwhile when one has to identify micro-organism survival and growth rate in outer space rocks, *e.g.* meteorites, asteroids, comets and planetoids, and onboard spacecrafts as well.

In order to tackle the issue of microbes endurance at high acceleration and jerk, ballistic experiments have been made by Benardini *et al.* [23], resulting in $\approx$75-100% survival of spores of *Bacillus* species basalt isolates after their exposure to $1.5\times10^6$ m/s$^2$ and $1.5\times10^{10}$ m/s$^3$. Survival of *B. subtilis* spores and *Deinococcus radiodurans* cells during a hypothetical ejection from Mars have been determined and reported in [20]. *B. subtilis* spores subjected to extended periods of time of ultracentrifugation at $4.36\times10^5$ g peak acceleration, presented endurance at the respective order of magnitude of acceleration [20]. According to [24], it was observed that studied Gram-positive and Gram-negative bacteria not only survived, but also experienced robust growth at hyperaccelerations up to$\approx2\times10^4$ g, *Escherichia coli* and *Paracoccus denitrificans* proliferating even at 403,627 g.

Pathogens that withstand increased gravitational conditions may constitute a threat to the health of crew members and may have impact on hardware reliability, thus influencing flight safety, especially during space missions. However, not only hypergravity constitutes a potential risk. Microgravity, being associated with defective immune functions recognized in the case of the returning astronauts after the Apollo and Skylab mission, also may be regarded as a threatening factor [25 - 27]. It is well-known that *Drosophila* fly is considered as an established model of the human innate immune function sharing common elements in cellular and humoral immunity systems, clotting and wound healing, as well as signaling pathways [25, 28]. The fly was recently used to model immune responses in both microgravity and hypergravity conditions, after being infected with *Beauveria bassiana* fungus. In microgravity environment, the infected flies failed to activate the Toll pathway. Under the effects of hypergravity, post-infection survival of the wild type as well as the immune response mutant flies, have increased [25].

Another risk may reside in the development and evolution of microbes onboard spacecraft, where the recycling of air, water, food and waste takes place [26]. As evidenced in [26, 27, 29] under microgravity conditions, pathogens are able to turn into more virulent forms. The bacterial pathogen *Salmonella typhimurium* grown aboard Space Shuttle Atlantis Mission STS-115 (2006) exhibited alterations in bacterial gene expression and enhanced virulence compared to the identical ground controls [29]. A space research project, undertaken in 2008 has revealed that under low-shear environments the ability of *Streptococcus pneumoniae* to adhere to and infect respiratory cells showed an increase [29]. Changes in drug sensitivity of bacteria, such as *E. coli*, *Klebsiella pneumoniae*, *Enterococcus faecium* and *Bacillus cereus*, arose after being subjected to the space environment and their resistance to several antibiotics increased significantly, as uncovered by the mission of the ShenZhou VIII spacecraft (2011) [29]. During the mission of the orbital station Mir, an entire "flora" of bacteria and/or fungi, consisting of a total number of 234 species and 65 genes, has been discovered in the closed/confined medium of the space station [30]. Also, throughout the operation of Mir, 20 bacterial isolates have been catalogued by [31] from drinking and recycled water.

As space voyages became more frequent with the advancement of explorations, space medicine requires alternative treatment solutions by developing new drugs or improving the action of existing ones. An unconventional method employed to fight multiple drug resistance (MDR) onboard spacecrafts, consists in the exposure of photosensitive phenothiazine aqueous solutions to UV laser radiation in order to generate photoproducts with enhanced antimicrobial activity against bacterial and fungus infections.

The interaction of chlorpromazine (CPZ), belonging to the phenothiazine class of drugs, with various target surfaces plays an important role in biomedical applications, in particular in designing new delivering procedures of medicines to target tissues [32, 33]. In Chapter 15 is discussed the wettability of target surfaces by phenothiazine solutions at Earth's gravity conditions, while the present chapter tackles the problem of surface wettability by means of contact angles (CA) under the effect of increased gravity.

## MATERIALS AND METHODS

### Medicine Solution

Among phenothiazines, CPZ introduced and presented in Chapter 9 and Chapter 13, was employed in hypergravity experiments, being purchased from Sigma-Aldrich in hydrochloride form at a purity > 98%. The non-antibiotic powder has been dissolved in ultrapure water, at 20 mg/ml concentration. Solution preparation was undertaken just before exposure to UV laser beam and subsequently stored at 2-5°C and in a dark environment.

### Laser Exposure

CPZ solutions, prior to be utilised in hypergravity conditions, have been exposed for 4 h to a 6.5 mJ average energy of the laser beam emitted at 266 nm *via* fourth harmonic generation effect by a Nd:YAG laser, with the same characteristics as in the case of the phenothiazines irradiation for terrestrial wetting properties studies, described in terms of irradiation conditions in Chapter 15.

### Target Surfaces

Wettability of cotton (CO) and activated charcoal, as hydrophilic target surfaces, and Parafilm M and aluminium, considered as hydrophobic surfaces, have been studied under hypergravity conditions. Cotton test material was purchased from wfk Testgewebe, pharmacological charcoal from Biofarm and Parafilm M from Bemis. Raw Al material was used to create the target surface by a polishing process in which it was used emery powder and sand paper. In order to obtain an unblemished CO surface, consecutive washing cycle procedures by warm tap water have been employed as described in Chapter 15. CO and Parafilm M materials were cut into 100 mm diameter circular pieces, according to the placing requirements on an Al made disk-shaped target surface support, and afterwards stretched by an Al ring, as represented in Figs. (**1a** and **c**). Concerning activated charcoal surfaces, the 12.3 mm diameter and 4 mm thick pills have been placed inside milled indentations of the disk, as depicted in Fig. (**1b**).

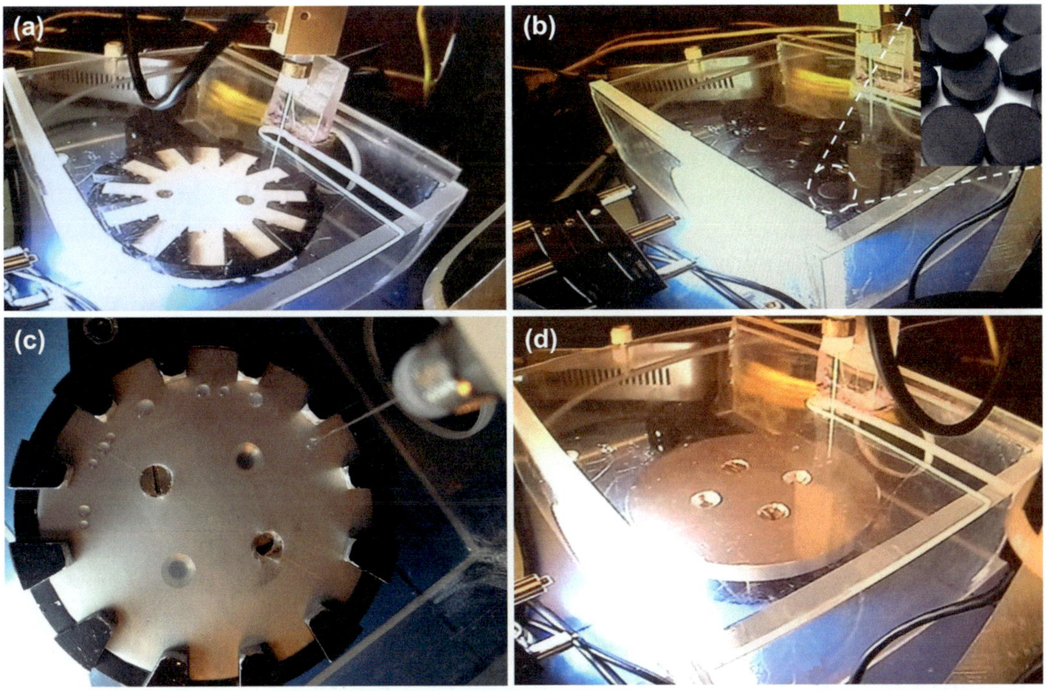


**Fig. (1).** Materials used in CA experiments. (**a**) CO, (**b**) activated charcoal, (**c**) Parafilm M and (**d**) Al surfaces visualised during the hypergravity experiment at the LDC facility.

Fig. (**1**) presents the studied target surfaces under the effect of high g, Fig. (**1d**) showing the Al target surface.

## Simulated Hypergravity - Large Diameter Centrifuge

In order to attain the wished hypergravity levels, the LDC ground-based high g simulator, developed by ESA at the European Space Research and Technology Centre (ESTEC), was employed. The LDC is equipped with four rotating arms, on which a total number of six swinging gondolas can be attached alongside a gondola placed in the centre of the rotation stage, utilised for control samples, as illustrated in Fig. (**2**) and Movie **1** [34]. Each arm has been designed to lock the gondolas in eight standard locations. When in full swing-out, the LDC has an 8 m diameter. As for the HyperMed experiment, the gondolas have been secured at the extremity of each arm in order to put to use the maximum capabilities offered by the facility [35].

The effect of inertial shear is diminished in a centrifuge with large radius and therefore it is limited in the LDC. However, it is recommended that interactions occur at locations tangential with respect to the center of rotation of the gondola

[34 - 36]. One should also consider that various g levels are present at different locations of the gondola and for that reason the experimental set-up has been placed on the gondola's floor, thus achieving the best use of the LDC geometry. The respective facility is able to reach and sustain up to a period of 6 months a 20 g maximum gravitational level; each gondola supports an 80 kg maximum payload with an inside working space of 500×500×720 mm (W×D×H). In order to achieve the maximum high g-level, a time of ≈ 60 s is necessary to speed up the system from rest, utilising a 22 kW motor as a driven engine. A 220 VAC power supply is provided inside each gondola. A programmable standard Windows-LabView based user interface is available at LDC, allowing programming of rotational profiles as well as the monitoring of the experimental set-up during centrifugation by using real time video connection [34 - 36]. Since vibrations constitute an accompanying, not negligible component of rotating devices, attention was invested in their limitation within the system, considering an upper limit of 1.2 $V_{rms}$ for industrial machinery according to ISO-10816-3/Class II. A maximum vibration level of 0.0712 $V_{rms}$ within a gondola as well as a 0.12 $V_{rms}$ within the LDC main structure operating with the entire set of gondolas fully loaded, respectively, has been established at the highest g-level, thus complying with the stated standards [34]. In addition, the final vibration levels at sample might also have been caused by attenuation or amplification of the experiment hardware itself.

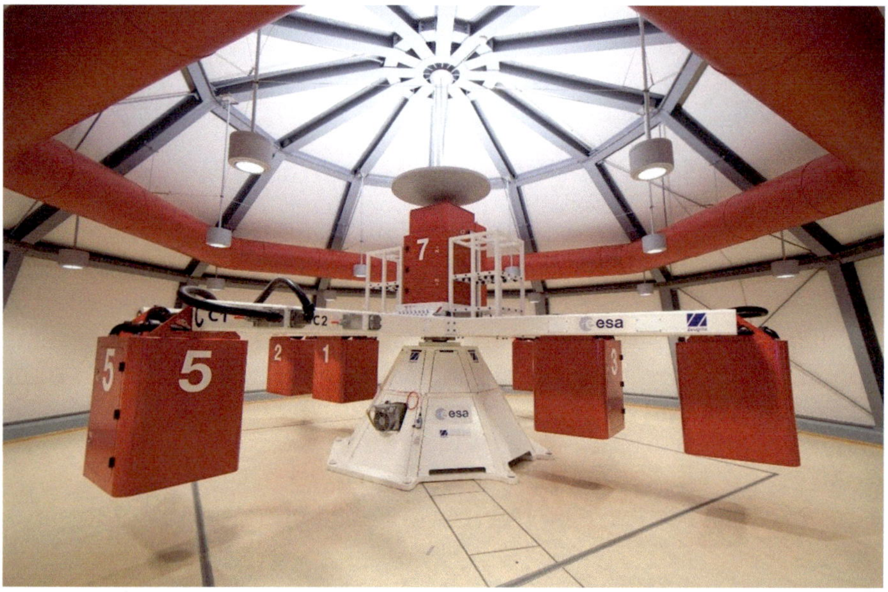


**Fig. (2).** The LDC facility at ESTEC [35]*.

## The HyperMed Project Experimental Set-up

The HyperMed experiment allowed generation of microdroplets of drug solution by utilising a remotely controlled Cole-Parmer Compact Syringe Pump 75900-00 by a native dispersing program. The studied solutions have been stored in a 100 µl glass Hamilton Gastight Syringe with Luer Tip. In order to avoid leakage of the subsystem or reabsorption of the solutions, induced by high g conditions, a Bürkert 6650-A00 solenoid valve with a DN size of 0.4 mm and 2/2-way function coupled with a single manifold was intercalated between the syringe and the 0.72 mm OD/ 0.41 mm ID Hamilton Kel-F Hub blunt, electropolished capillary with the help of proper Hamilton Luer Lock adapters. A National Instruments USB digital I/O device, coupled with an in-house special designed electronic circuit, has been employed to remotely control the valve. The droplet generation subsystem, designed of easily interchangeable components, facilitated the implementation of different medicine solutions.

The experiment was performed at 2, 5, 10 and 20 g.

Once pendant droplets have been generated, their detachment took place due to the subjected high g conditions inside the gondola, thus reaching the target surfaces as sessile droplets. The interaction occurred in a sealed poly(methyl methacrylate) enclosure, termed the test cell (see Fig. **3**). The hydrophilic and hydrophobic surfaces, placed onto the target surface support, have been mounted onto a PRM1/MZ8 Thorlabs Motorized Precision Rotation Stage and rotated by a TDC001 T-Cube DC Servo Controller *via* a DC servo motor, yielding a velocity of 25 deg/s and a resolution of 1 arcsec over the entire 360° rotation plane. A smooth and continuous motion has been achieved by the controller, which allowed an automated control of the stage *via* the APT System Software (Thorlabs). The rotation of target surfaces was necessary in order to acquire data at required g-levels using the same working solution on the same surface without stopping the LDC.

A rugged Phantom Miro 3 high-speed and high g supporting camera (Vision Research) has been employed to record pendant microdroplets generation and behaviour, as well as their impact and interaction with target surfaces. Within the current experiment, a 600×320 pixels resolution was selected, thereby obtaining an 1800.18 frames/s resulting in 555.5 µs time between 2 consecutive frames. In order to provide high quality images for motion analysis by eliminating the blur, the speed of electronic shutter could be decreased as low as 2 µs. The CMOS sensor 17.6×13.2 mm, is characterised by 22 µm pixel size and 12-bit depth. Regarding video data acquisition and processing, the Phantom Camera Control Application 2.6 software was implemented. The obtained images have been used to determine surface wettability by means of CA formed by medicine solutions on target surfaces, absorption/evaporation time, drop height as well as wetted diameter.

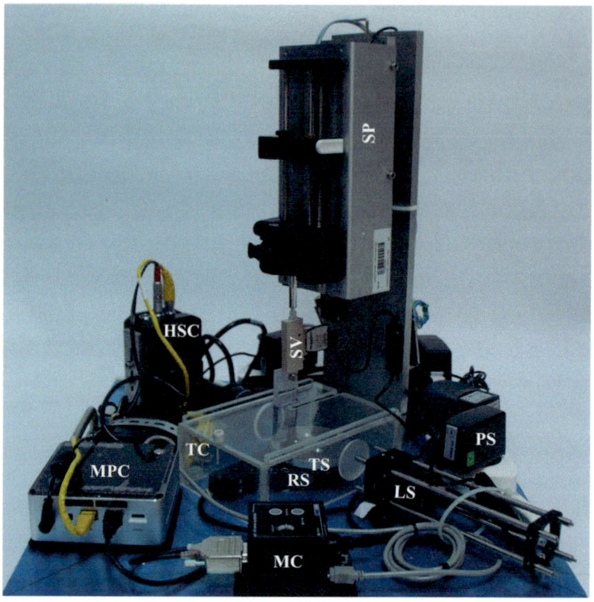


**Fig. (3).** The HyperMed experimental set-up. Configuration: SP-syringe pump, SV-solenoid valve, TS-target surface, TC-test cell, RS-rotational stage, MC-motor controller, HSC-high-speed camera, LS-light source, MPC-mini PC and PS-power supply.

In order to capture high quality images, medicine droplets required proper visualisation through the backlit illumination technique. As depicted in Fig. (**3**), the respective illumination scenario consists in positioning opposite of the high-speed camera a Thorlabs LIU 004 white LED Array Light Source with 1700 $\mu W/cm^2$ intensity, 100 mA maximum current and 24 V maximum voltage, thus producing a silhouette of the studied droplets. With this technique, a dark and sharp image of the drop on a white background is acquired, hence facilitating the detection of the contact line alongside the contour of the sessile drop by utilizing a CA software. In-house polished optical glasses, placed in Ø35 mm aperture Thorlabs CP03/M cage plates, were used to obtain diffused light. The respective plates were mounted on an X-axis movement structure in order to achieve an easily adaptable set-up for different environmental conditions.

A Remote Desktop Protocol provided connection between the mini PC installed inside the gondola and the Experimental Control Equipment located in the Operation Control Station, in order to have control of the experiment while rotating [36].

By taking into account the fact that acceleration always points downwards inside the gondola, thus pressing the set-up towards its floor [36], no fixations were needed, however each component required a firm fixation and a balanced mass distribution onto a 450×500 mm (W×D) aluminium platform, in order to resist the highest g conditions.

## RESULTS AND DISCUSSIONS

From the data collected during the "Spin Your Thesis!" experimental campaign, dynamic processes dominating the interaction of microdroplets with target surfaces have been observed. The origin of the respective dynamics might be due to both: (i) different parameters, such as gravitational acceleration, surface tension and solution density, mass of the droplets as well as their volume, and (ii) the inherent vibrations of the LDC, generated at the selected high g levels.

Vibrations of liquid drops have been described in previous papers [37 - 40], mentioning on one hand the polygonal vibrations, observed for gravity-flattened drops, and on the other, the axisymmetric vibrations, examined for incompressible liquid drop in partial contact with a spherical bowl [39]. Several theoretical studies have been carried out by neglecting the effect of gravity [39, 41 - 43]; however in [44, 45] some approximations were presented concerning the vibrations of gravity flattened drops, thus being altered from the spherical and the cylindrical cases. Also, various vibration modes and the process of their transmission for superheated drops of liquid nitrogen and liquid oxygen placed on a plate at room temperature have been shown [44].

During the experiment conducted in hypergravity conditions, images captured inside the LDC gondola exhibited similar features as those obtained for falling drops from a vibrating ceiling [46]. Also, it has to be mentioned that vibrations occurred not only in the case of sessile drops but for pendant droplets as well.

Regarding pendant droplets, within the HyperMed project the dependency of the generated droplet volume by hypergravity levels was determined. The volume at which the drop detached from the capillary decreased as the gravitational acceleration increased, is highlighted in Fig. (**4**).

A sigmoidal decreasing tendency is characterising droplet volumes, indicating a drop from as high as 4.9 μl to as low as 0.1 μl, when the LDC induced hypergravity rose from 2 to 20 g.

Concerning the dynamics, the hydrophilic target surfaces have presented a continuous evolution, since working solutions containing droplets have been rapidly absorbed into such surfaces, as shown in the case of CO in Fig. (**5**). CA formed by irradiated medicine droplets on CO surface underwent a decline as a function of time (data not shown), exhibiting values lower than 90°, thus indicating favourable surface wettability.

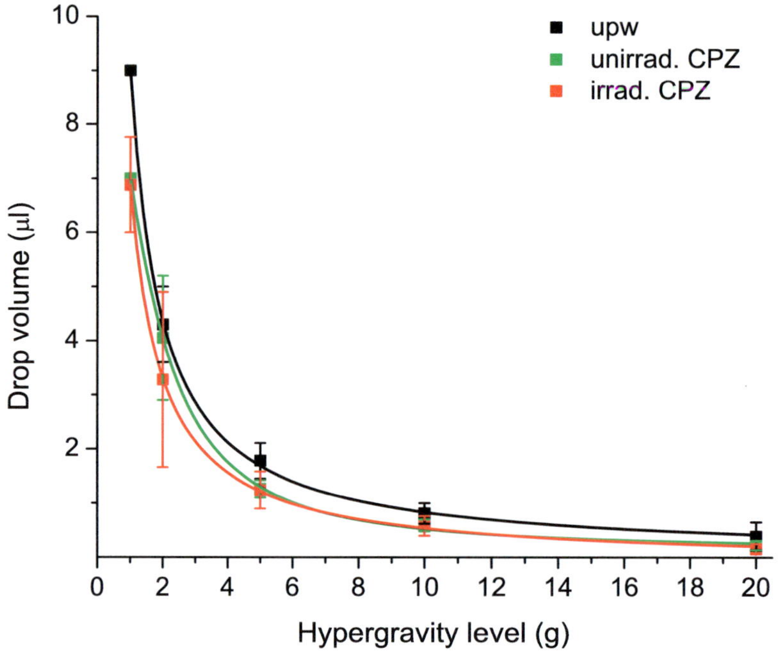


**Fig. (4).** Drop volume dependence on g level. Volumes corresponding to average values of twelve drops generated at same gravity level.

The first 50 ms as illustrated in Fig. (**5**), starting from the impact of the drop, correspond to oscillations in CA values due to the above mentioned droplet vibrations. After a certain time, depending on the generated volume and the simulated high g level, it was followed by a continuous three-parameter exponential decay of CA (data not shown) alongside with a quasi-stabilisation process in terms of vibrations. It can be observed that these factors highly influence the droplets absorption time into CO fibres. For higher volumes, a longer time interval is required in order to obtain total absorption compared to that of small droplet volumes; as for high g levels the absorption takes place faster than in the case of terrestrial gravitational conditions, considering that droplets are pushed downwards into the surface by the applied gravitational force. Evolution of the absorption of irradiated CPZ solution on CO surface is shown in Movie **2** for 2 g hypergravity level and in Movie **3** for 20 g.

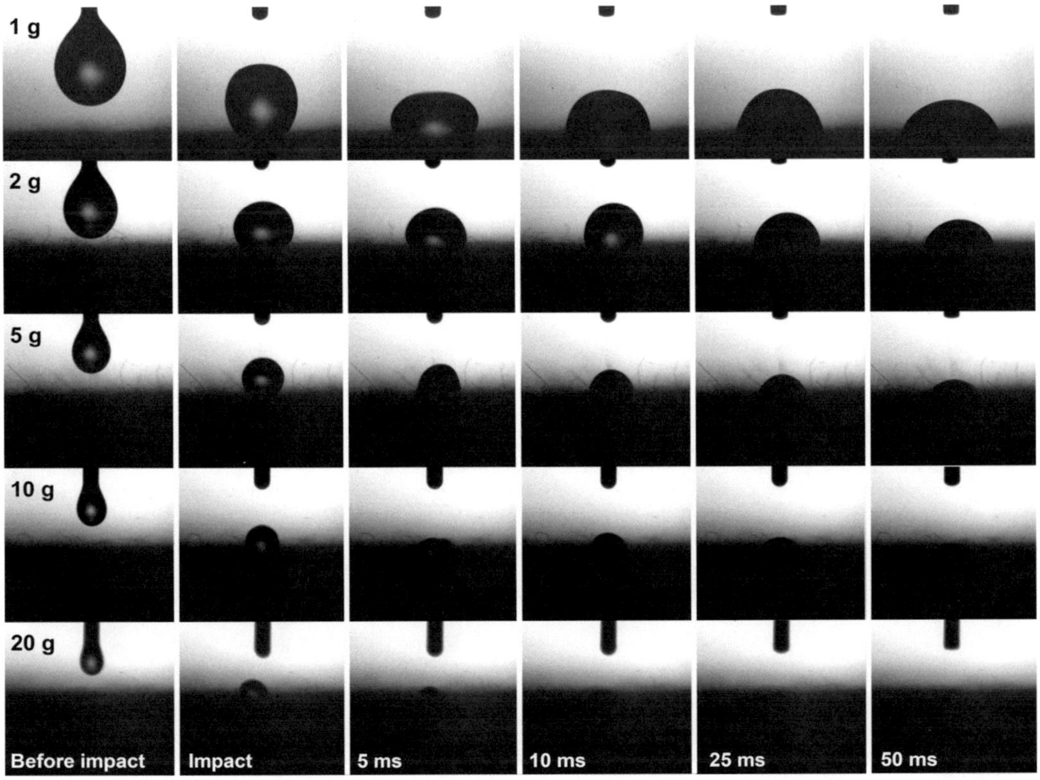


**Fig. (5).** Time evolution of laser exposed CPZ droplets on CO under the effect of various g conditions.

The present study aimed to compare the wetting ability of laser exposed CPZ droplets with that of unexposed (the same concentration) drug solution and ultrapure water, respectively. Experiments on ultrapure water were necessary, since this is the solvent utilized in the unirradiated and irradiated phenothiazine solutions. It was also utilised due to its biocompatibility as well as for preventing solvent dependent reactions that may be crucial in biomedical applications. Fig. (**6**) emphasises the better wetting properties of the long time exposed CPZ solution compared to water at different gravity levels.

It has been discussed in Chapter 15 that phenothiazine solutions exhibit reduced surface tension with respect to water, thus providing very good wetting properties. As can be observed in Fig. (**6**), irradiated CPZ droplets presented smaller CA than those of ultrapure water even at 30 ms after impact. More than that, the obtained results revealed that unexposed/exposed medicine droplets had better wetting ability not just in the case of CO, but for all studied target surfaces. In some cases, laser irradiated solution proved to have even better properties.

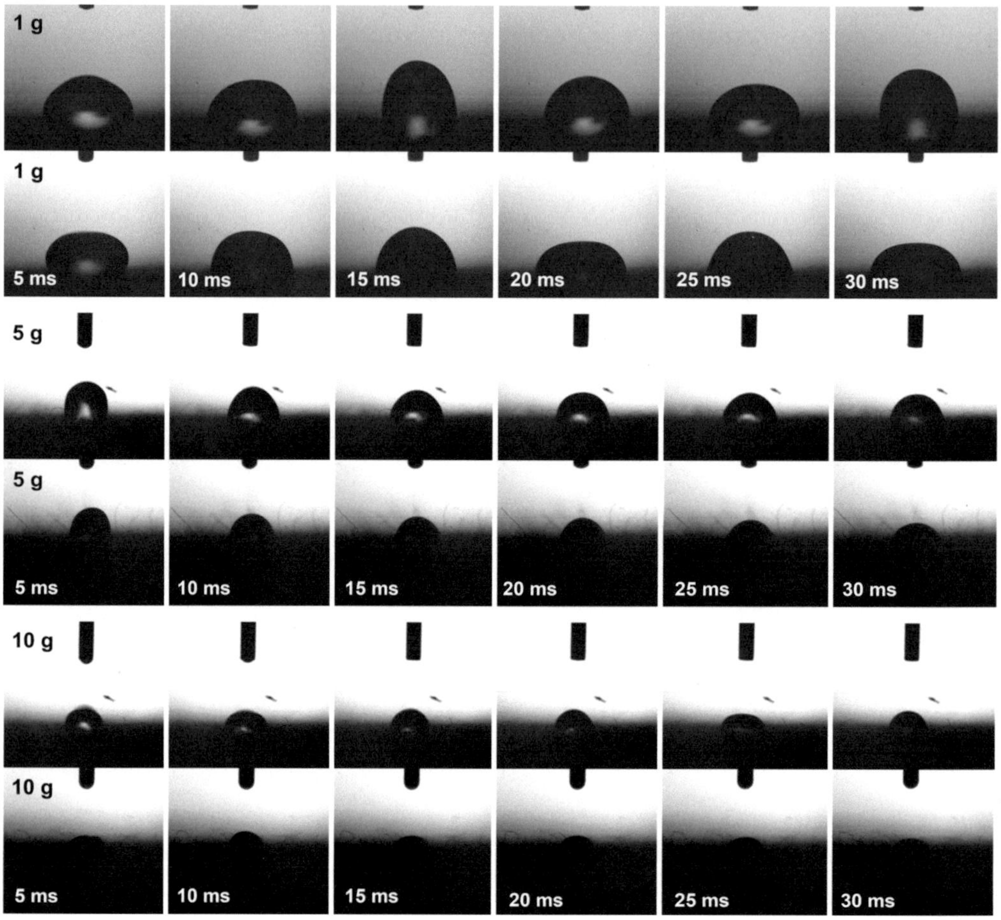


**Fig. (6).** Time evolution of ultrapure water and laser irradiated CPZ droplets after impact with CO surface at 1, 5 and 10 g, respectively.

As for hydrophobic surfaces, droplets oscillations were more pronounced in the first 50 ms from impact comparative to hydrophilic target surfaces, followed by their stabilisation, *ergo* CA exhibited relatively constant evolution in time, as shown for Parafilm M and Al in Fig. (**7**).

Fig. (**7**) emphasises alterations in droplet behaviour on different studied surfaces under hypergravity condition.

Activated charcoal presented outstanding wettability in the first 100 ms after drop impact. It can be observed that in the respective time interval, irradiated CPZ droplet exhibited enhanced wetting properties on charcoal with respect to CO target surface. However, complete wetting took seconds, even minutes after drops were applied on charcoal, while in the case of CO it has been achieved in milliseconds. Thus, CA showed constant evolution over quite long periods of time for charcoal (Movie **4**), in contrast to the drastic decline in the case of CO (see Movie **2**).

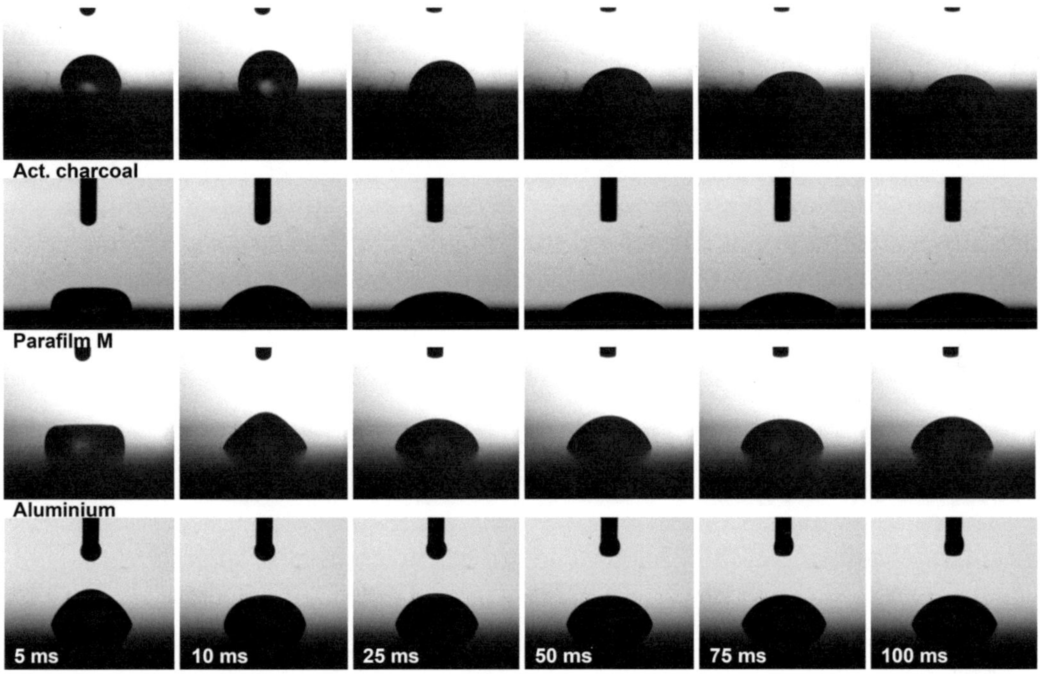


**Fig. (7).** Time evolution of laser exposed CPZ droplets on CO (1st row), activated charcoal (2nd row), Parafilm M (3rd row) and Al (4th row) target surfaces under the effect of 2 g.

As for Parafilm M (Movie **5**) and Al (Movie **6**), the absorption process has been replaced by the evaporation of droplets, providing constant CA evolution in time.

The results indicate that parameters, such as drop volume and g level, must be taken into account [47, 48], since they might influence CA measurements. In the scientific community, opinions are divided regarding the significance of the effect of droplet volume on CA, some considering that CA remains constant even as drop volumes vary [49], while others believe that CA may be modified below a critical diameter [50]. In order to gain a better understanding of the effect of hypergravity on wettability of the binom surface-solution, further investigations are required keeping constant one of the parameters while varying the other, thus being able to exclude their cumulative effects.

## CONCLUSIONS

Within this chapter a new insight into the wettability of hydrophilic and hydrophobic surface, under hypergravity conditions has been introduced.

As time spent in space by humans is rapidly increasing with the expansion of the scope of space explorations, novel space medicine procedures in view of biomedical applications are required. Regarding medical risks, it has to be taken into account that mutated microbes, transported onboard spacecraft, may pollute the space environment, and after returning to Earth, organisms with enhanced virulence and resistance may threaten human health on the ground [29]. Medicine solutions interacting with spacecraft surfaces may provide a solution for decontamination purposes. Therefore, studying the effect of space environment on such interactions one may obtain crucial data in protecting the health of astronauts during long-lasting space voyages as well as humans on Earth.

## NOTES

* Reprinted from ESA Large Diameter Centrifuge Summary.

## CONFLICT OF INTEREST

The authors confirm that this chapter content has no conflict of interest.

## ACKNOWLEDGEMENTS

This work has been financed by the National Authority for Research and Innovation in the frame of Nucleus programme-contract 4N/2016 and the projects PN-II-ID-PCE-2011-3-0922 as well as by the European Space Agency Education Office. The HyperMed team express their appreciation for the support of the ESA Education Office through Lily Ha, Lucas Pfeiffer and Nigel Savage during the sixth "Spin Your Thesis!" campaign.

## REFERENCES

[1] R. Herranz, R. Anken, J. Boonstra, M. Braun, P.C. Christianen, M. de Geest, J. Hauslage, R. Hilbig, R.J. Hill, M. Lebert, F.J. Medina, N. Vagt, O. Ullrich, J.J. van Loon, and R. Hemmersbach, "Ground-based facilities for simulation of microgravity: organism-specific recommendations for their use, and recommended terminology", *Astrobiology,* vol. 13, no. 1, pp. 1-17, 2013. [http://dx.doi.org/10.1089/ast.2012.0876] [PMID: 23252378]

[2] T. Frett, G. Petrat, J.J. van Loon, R. Hemmersbach, and R. Anken, "Hypergravity facilities in the ESA ground-based facility program -Current research activities and future tasks", *Microgravity Sci. Technol,* vol. 28, no. 3, pp. 205-214, 2016. [http://dx.doi.org/10.1007/s12217-015-9462-9]

[3] J.J. van Loon, "The gravity environment in space experiments", In: *Biology in space and life on Earth,* E. Brinckmann, Ed., Wiley, 2007, pp. 17-32. [http://dx.doi.org/10.1002/9783527617005.ch1]

[4] E.S. Nelson, "An examination of anticipated g-jitter on Space Station and its effects on materials processes", *NASA Technical Memorandum 103775,* 1994.

[5] A. Khoshnevis, A. Ahadi, and M.Z. Saghir, "On the influence of g-jitter and prevailing residual accelerations onboard International Space Station on a thermodiffusion experiment", *Appl. Therm. Eng,* vol. 68, no. 1-2, pp. 36-44, 2014. [http://dx.doi.org/10.1016/j.applthermaleng.2014.04.001]

[6] K. Jules, K. McPherson, K. Hrovat, E. Kelly, and T. Reckart, "A status report on the characterization of the microgravity environment of the International Space Station", *Acta Astronaut,* vol. 55, no. 3-9, pp. 335-364, 2004.

[7] N.J. Penley, C.P. Schafer, and J-D. Bartoe, "The International Space Station as a microgravity research platform", *Acta Astronaut,* vol. 50, no. 11, pp. 691-696, 2002. [http://dx.doi.org/10.1016/S0094-5765(02)00003-6] [PMID: 12035811]

[8] B.V. Tryggvason, W.M. Duval, R.W. Smith, K.S. Rezkallah, S. Varma, R.F. Redden, and R.A. Herring, "The vibration environment on the international space station: its significance to fluid-based experiments", *Acta Astronaut,* vol. 48, no. 2-3, pp. 59-70, 2001. [http://dx.doi.org/10.1016/S0094-5765(00)00140-5]

[9] R.A. Herring, and B. Tryggvason, "Recent measurements of experiment sensitivity to g-jitter and their significance to ISS facility development", *Symp. JJ -Mater. Space-Sci. Technol. Explor,* vol. 551, 1998.

[10] M. O'Leary, "Anaxagoras and the origin of panspermia theory", *iUniverse,* 2008.

[11] D.A. Vakoch, *Astrobiology, history, and society: Life beyond Earth and the impact of discovery,* Springer Science & Business Media, 2013. [http://dx.doi.org/10.1007/978-3-642-35983-5]

[12] L.J. Rothschild, and A. Lister, *Evolution on planet Earth: The Impact of the physical environment,* Academic Press, 2003.

[13] S.P. Thompson, *The life of William Thomson, Baron Kelvin of Largs*, Cambridge University Press, 2011.

[14] "The word: Panspermia", *New Sci,* 2006.

[15] S. Arrhenius, and H. Borns, *Worlds in the making; the evolution of the universe*, Harper: New York, London, 1908.

[16] F. Hoyle, and N.C. Wickramasinghe, "Bacterial life in space", *Nature,* vol. 306, no. 5942, pp. 420-420, 1983. [http://dx.doi.org/10.1038/306420a0] [PMID: 6646221]

[17] F. Hoyle, and N.C. Wickramasinghe, "The case for life as a cosmic phenomenon", *Nature,* vol. 322, no. 6079, pp. 509-511, 1986. [http://dx.doi.org/10.1038/322509a0]

[18] C. Mileikowsky, F.A. Cucinotta, J.W. Wilson, B. Gladman, G. Horneck, L. Lindegren, J. Melosh, H. Rickman, M. Valtonen, and J.Q. Zheng, "Natural transfer of viable microbes in space", *Icarus,* vol. 145, no. 2, pp. 391-427, 2000. [http://dx.doi.org/10.1006/icar.1999.6317] [PMID: 11543506]

[19] G. Horneck, "Survival of microorganisms in space: a review", *Adv. Space Res,* vol. 1, no. 14, pp. 39-48, 1981. [http://dx.doi.org/10.1016/0273-1177(81)90241-6] [PMID: 11541716]

[20] R.M. Mastrapa, H. Glanzberg, J.N. Head, H.J. Melosh, and W.L. Nicholson, "Survival of bacteria exposed to extreme acceleration: Implications for panspermia", *Earth Planet. Sci. Lett,* vol. 189, no. 1-2, pp. 1-8, 2001. [http://dx.doi.org/10.1016/S0012-821X(01)00342-9]

[21] G. Horneck, D.M. Klaus, and R.L. Mancinelli, "Space microbiology", *Microbiol. Mol. Biol. Rev,* vol. 74, no. 1, pp. 121-156, 2010.
[http://dx.doi.org/10.1128/MMBR.00016-09] [PMID: 20197502]

[22] NASA, *Stardust sample return,* Press Kit, 2006.

[23] J.N. Benardini, J. Sawyer, K. Venkateswaran, and W.L. Nicholson, "Spore UV and acceleration resistance of endolithic Bacillus pumilus and Bacillus subtilis isolates obtained from Sonoran desert basalt: implications for lithopanspermia", *Astrobiology,* vol. 3, no. 4, pp. 709-717, 2003.
[http://dx.doi.org/10.1089/153110703322736033] [PMID: 14987476]

[24] S. Deguchi, H. Shimoshige, M. Tsudome, S.A. Mukai, R.W. Corkery, S. Ito, and K. Horikoshi, "Microbial growth at hyperaccelerations up to 403,627 x g", *Proc. Natl. Acad. Sci. USA,* vol. 108, no. 19, pp. 7997-8002, 2011.
[http://dx.doi.org/10.1073/pnas.1018027108] [PMID: 21518884]

[25] K. Taylor, K. Kleinhesselink, M.D. George, R. Morgan, T. Smallwood, A.S. Hammonds, P.M. Fuller, P. Saelao, J. Alley, A.G. Gibbs, D.K. Hoshizaki, L. von Kalm, C.A. Fuller, K.M. Beckingham, and D.A. Kimbrell, "Toll mediated infection response is altered by gravity and spaceflight in Drosophila", *PLoS One,* vol. 9, no. 1, p. e86485, 2014.
[http://dx.doi.org/10.1371/journal.pone.0086485] [PMID: 24475130]

[26] N. Guéguinou, C. Huin-Schohn, M. Bascove, J-L. Bueb, E. Tschirhart, C. Legrand-Frossi, and J-P. Frippiat, "Could spaceflight-associated immune system weakening preclude the expansion of human presence beyond Earths orbit?", *J. Leukoc. Biol,* vol. 86, no. 5, pp. 1027-1038, 2009.
[http://dx.doi.org/10.1189/jlb.0309167] [PMID: 19690292]

[27] J.W. Wilson, C.M. Ott, K. Höner zu Bentrup, R. Ramamurthy, L. Quick, S. Porwollik, P. Cheng, M. McClelland, G. Tsaprailis, T. Radabaugh, A. Hunt, D. Fernandez, E. Richter, M. Shah, M. Kilcoyne, L. Joshi, M. Nelman-Gonzalez, S. Hing, M. Parra, P. Dumars, K. Norwood, R. Bober, J. Devich, A. Ruggles, C. Goulart, M. Rupert, L. Stodieck, P. Stafford, L. Catella, M.J. Schurr, K. Buchanan, L. Morici, J. McCracken, P. Allen, C. Baker-Coleman, T. Hammond, J. Vogel, R. Nelson, D.L. Pierson, H.M. Stefanyshyn-Piper, and C.A. Nickerson, "Space flight alters bacterial gene expression and virulence and reveals a role for global regulator Hfq", *Proc. Natl. Acad. Sci. USA,* vol. 104, no. 41, pp. 16299-16304, 2007.
[http://dx.doi.org/10.1073/pnas.0707155104] [PMID: 17901201]

[28] B. Lemaitre, and J. Hoffmann, "The host defense of Drosophila melanogaster", *Annu. Rev. Immunol,* vol. 25, pp. 697-743, 2007.
[http://dx.doi.org/10.1146/annurev.immunol.25.022106.141615] [PMID: 17201680]

[29] L. Su, D. Chang, and C. Liu, "The development of space microbiology in the future: the value and significance of space microbiology research", *Future Microbiol,* vol. 8, no. 1, pp. 5-8, 2013.
[http://dx.doi.org/10.2217/fmb.12.127] [PMID: 23252487]

[30] N.D. Novikova, "Review of the knowledge of microbial contamination of the Russian manned spacecraft", *Microb. Ecol,* vol. 47, no. 2, pp. 127-132, 2004.
[http://dx.doi.org/10.1007/s00248-003-1055-2] [PMID: 14994178]

[31] B. Song, and L.G. Leff, "Identification and characterization of bacterial isolates from the Mir space station", *Microbiol. Res,* vol. 160, no. 2, pp. 111-117, 2005.
[http://dx.doi.org/10.1016/j.micres.2004.10.005] [PMID: 15881827]

[32] G. Tiwari, R. Tiwari, B. Sriwastawa, L. Bhati, S. Pandey, P. Pandey, and S.K. Bannerjee, "Drug delivery systems: An updated review", *Int. J. Pharm. Investig,* vol. 2, no. 1, pp. 2-11, 2012.
[http://dx.doi.org/10.4103/2230-973X.96920] [PMID: 23071954]

[33] P.V. Devarajan, S.M. Dawre, and R. Dutta, "Infectious diseases: Need for targeted drug delivery", In: *Targeted drug delivery : Concepts and design,* P.V. Devarajan, S. Jain, Eds., Springer International Publishing, 2015, pp. 113-148.
[http://dx.doi.org/10.1007/978-3-319-11355-5_3]

[34] J.J. van Loon, J. Krause, H. Cunha, J. Goncalves, H. Almeida, and P. Schiller, "The Large Diameter Centrifuge, LDC, for life and physical sciences and technology", *Proc. of the Life in Space for Life on Earth Symposium,* 2008 Angers, France

[35] "ESA Large Diameter Centrifuge summary", *European Space Agency.*

[36] J. J. W. A. van Loon, and A. Dowson, "Large Diameter Centrifuge (LDC) experimenter users manual", *European Space Agency, v. 2, r. 3,* 2016.

[37] K. Adachi, and R. Takaki, "Vibration of a flattened drop. I. Observation", *J. Phys. Soc. Jpn,* vol. 53, no. 12, pp. 4184-4191, 1984. [http://dx.doi.org/10.1143/JPSJ.53.4184]

[38] R. Takaki, and K. Adachi, "Vibration of a flattened drop. II. Normal mode analysis", *J. Phys. Soc. Jpn,* vol. 54, no. 7, pp. 2462-2469, 1985. [http://dx.doi.org/10.1143/JPSJ.54.2462]

[39] M. Strani, and F. Sabetta, "Free vibrations of a drop in partial contact with a solid support", *J. Fluid Mech,* vol. 141, pp. 233-247, 1984. [http://dx.doi.org/10.1017/S0022112084000811]

[40] R.W. Smithwick, and J.A. Boulet, "Vibrations of microscopic mercury droplets on glass", *J. Colloid Interface Sci,* vol. 130, no. 2, pp. 588-596, 1989. [http://dx.doi.org/10.1016/0021-9797(89)90134-3]

[41] M. Strani, and F. Sabetta, "Viscous oscillations of a supported drop in an immiscible fluid", *J. Fluid Mech,* vol. 189, pp. 397-421, 1988. [http://dx.doi.org/10.1017/S0022112088001077]

[42] U. Olgac, D. Izbassarov, and M. Muradoglu, "Direct numerical simulation of an oscillating droplet in partial contact with a substrate", *Comput. Fluids,* vol. 77, pp. 152-158, 2013. [http://dx.doi.org/10.1016/j.compfluid.2013.03.007]

[43] S. Ramalingam, D. Ramkrishna, and O. A. Basaran, "Free vibrations of a spherical drop constrained at an azimuth", *Phys. Fluids 1994-Present,* vol. 24, no. 8, p. 82102, 2012. [http://dx.doi.org/10.1063/1.4742339]

[44] R. Takaki, *Self-induced vibration of a drop*, Forma, 2015.

[45] M. Perez, Y. Brechet, L. Salvo, M. Papoular, and M. Suery, "Oscillation of liquid drops under gravity: Influence of shape on the resonance frequency", *Europhys. Lett. EPL,* vol. 47, no. 2, pp. 189-195, 1999. [http://dx.doi.org/10.1209/epl/i1999-00371-6]

[46] H-Y. Kim, "Drop fall-off from the vibrating ceiling", *Phys. Fluids,* vol. 16, no. 2, pp. 474-477, 2004. [http://dx.doi.org/10.1063/1.1637352]

[47] C.W. Extrand, and S.I. Moon, "When sessile drops are no longer small: transitions from spherical to fully flattened", *Langmuir,* vol. 26, no. 14, pp. 11815-11822, 2010. [http://dx.doi.org/10.1021/la1005133] [PMID: 20553001]

[48] C.W. Extrand, and S.I. Moon, "Contact angles of liquid drops on super hydrophobic surfaces: understanding the role of flattening of drops by gravity", *Langmuir,* vol. 26, no. 22, pp. 17090-17099, 2010. [http://dx.doi.org/10.1021/la102566c] [PMID: 20964303]

[49] C.E. Cansoy, "The effect of drop size on contact angle measurements of superhydrophobic surfaces", *RSC Advances,* vol. 4, no. 3, pp. 1197-1203, 2013. [http://dx.doi.org/10.1039/C3RA45947K]

[50] R.J. Good, and M.N. Koo, "The effect of drop size on contact angle", *J. Colloid Interface Sci,* vol. 71, no. 2, pp. 283-292, 1979. [http://dx.doi.org/10.1016/0021-9797(79)90239-X]

**CHAPTER 17**

# Lasing by Optically Pumped Pendant Droplets

**Mihai Boni**[1], **Ionut Relu Andrei**[1], **Angela Staicu**[1], **Viorel Nastasa**[1] and **Mihail Lucian Pascu**[1,2,*]

[1] *National Institute for Laser, Plasma and Radiation Physics, Magurele, Ilfov, Romania*

[2] *Faculty of Physics, University of Bucharest, Magurele, Ilfov, Romania*

**Abstract:** Pendant droplets have optical properties that reproduce the characteristics of corresponding bulk materials. A particular case is the droplet hanging in air, made of a laser dye solution and optically side-pumped. Here are shown results about laser induced fluorescence (LIF) emitted by a single pendant droplet excited with 532 nm pumping pulsed laser beam when its content is solution of rhodamine 6G in water kept as such, or doped with $TiO_2$. By changing the geometry of fluorescent medium from bulk to pendant droplet and its volume, significant LIF changes occur. The main difference is amplification of emitted fluorescence in pendant droplets, explained by the confinement of light *via* total internal reflection in droplet. A new technique is shown, developed to distinguish temporal changes of a droplet emission in air. It was observed that the presence in the droplet of rhodamine 6G solution in ultrapure water of $TiO_2$ nanoparticles induces LIF spectra modifications depending on nanoparticles number density and laser beam pumping energy. A concentration of $10^{11}$ part/cm$^3$ nanoparticles favours formation of two new emission bands shifted towards blue with respect to main band. The increase with one order of magnitude of $TiO_2$ number density, produces the disappearance of these bands regardless the pumping energy. Data suggest that addition of $TiO_2$ nanoparticles to rhodamine droplet solutions influences emission spectra which can be modulated by varying nanoparticles concentration and pumping beam energy.

**Keywords:** Fluorescence, Laser active medium, Laser dye, Laser induced fluorescence, LIF, Lasing, Microdroplet, Nd:YAG laser, Spherical cavity, Rhodamine 6G, Rh6G, $TiO_2$ nanoparticles.

## INTRODUCTION

As it was already presented in previous chapters particularly in Chapter 9, the resonant interaction(s) of laser beam(s) with pendant droplet(s) triggers an increased interest for further possible applications. This is correlated with the fact that pendant droplets are suitable for different applications such as lasing [1 - 4],

* **Corresponding author Mihail Lucian Pascu:** Laser Department, National Institute for Laser, Plasma and Radiation Physics, Magurele, Ilfov, Romania; Tel/Fax: 0040 214575739; E-mail: mihai.pascu@inflpr.ro

sensing, molecular modifications by exposure to laser beams and transport of substances/medicines to targets [5 - 7].

A pendant droplet that contains a laser dye, may work as a particular type of dye laser, when optically pumped, if it has suitable dimensions (liquid volume and resulting diameter) and a chosen content (laser dye, solvent, concentration in the solute). Functioning of a microvolumetric droplet as a dye laser is conditioned by its proper placing with respect to the pumping laser beam in a given pumping geometry and by the pumping laser beam energy correlated with beam waist and focus dimensions and position. Given the characteristics of the pendant droplet, this kind of laser may be considered as a droplet-, or an optofluidic- dye laser. In this case, the liquid droplet represents laser active medium and absorbs the pumping beam energy emitting induced fluorescence which is further amplified. The shape of the droplet correlated with the shape of the laser pumping beam may produce multiple total internal reflections leading to a confinement of pumping and fluorescence light similar to that produced in a closed quasi-spherical optical resonator. The liquid-air surface of the pendant droplet provides the optical feedback at selected wavelengths corresponding to the morphology-dependent resonances of a spherical droplet [1] or whispering gallery modes functioning [8, 9]. The properties of the laser light such as wavelength, spatial mode structure, emission direction(s) and tunability depend on the properties of the liquid medium and on pumping conditions.

At the same time, several reports on the effect of nanoparticles on dye laser emission were made. Droplets doped with nanoparticles contain scattering formations constituted by these nanoparticles. When losses in the droplet are lower than gain due to longer optical path obtained by light scattering produced by nanoparticles conglomerates, one may obtain a random droplet laser [10]. After the first report about random laser [10], studies on the effect of different nano-scatterers present in the dye laser medium such as $TiO_2$ [11 - 13] and ZnO [12] nanoparticles, polymers [14], intralipids [2], emulsions [4], and CdSe nanostructures [15] were reported.

In the following some results about the studies on laser induced fluorescence (LIF) are shown and its amplification by laser pumped pendant droplets that contain rhodamine 6G (Rh6G) solutions in water and are hanging in air. A comparison is made between bulk solutions and pendant droplets of the same materials as bulk, in terms of pumping geometries of active media underlining specific phenomena obtained in droplet. In the case of pendant droplet, the emission signal is studied function of different experimental parameters such as: droplet's content, laser dye concentration, pumping geometry and fluorescence collection geometry, emitted light collection time [16].

Measurements of temporal structure/changes of droplet emission are presented, conducted by varying the fluorescence signal acquisition gate width and gate delay of the analysing spectrograph in correlation to laser pulse - droplet interaction time. By using this sampling technique it was obtained the temporal evolution of emission spectra during the interaction of the laser pumping pulse with a droplet.

Also, studies on the emission spectra of microdroplets containing solutions in water doped with $TiO_2$ nanoparticles are presented. The excitation is made by the second harmonic of a pulsed Nd:YAG laser in green at 532 nm, pulse duration at half maximum 6 ns, pulse repetition rate 10 pulses per second, energy varied between 6 and 10 mJ. LIF emitted spectra are analysed function of $TiO_2$ concentration and laser pumping energy. A comparison between emission spectra for pendant droplets generated in air and containing $TiO_2$ nanoparticles, and water solution droplets pumped in the same conditions is made [17]. A combined action of droplet spherical cavity/resonator that produces morphology dependent resonances (or WGM) and of $TiO_2$ nano-scatterers on the emission of the dye takes place, and its effect is evaluated function of droplet size, nanoparticles density and pumping energy.

**Emission Spectra Function of Geometry of Fluorescent Medium and/or Geometry of Collection System**

The experimental set-up presented in Chapter 9 is used for droplet's irradiation. Laser beam passes through a lens with focal length of 150 mm (Fig. **1a**). Keeping constant the droplet's position with respect to the collecting optical fibre, the emission signal is measured for different positions of $L_1$ with respect to the droplet.

LIF intensity depends on the distance lens-droplet (Fig. **1b**). One may observe that collected fluorescence decreases close to the focal point. In this region, the power density is increased and is localised on a small volume along the droplet diameter (on laser beam axis) and as a consequence the number of excited dye molecules is reduced. Therefore, around the focus point, the unresonant effects [18] are dominant and the droplet is destroyed. Going beyond the focal point, the intensity of droplet emission starts to increase, but with smaller rates compared to that of droplet placed diametrically opposed with respect to the focal point position.

Based on this result in the experiments performed further the distance between the lens $L_1$ and the droplet was set at 100 mm. In this case the beam waist is 2.5 mm and almost the entire droplet with diameter 2.67 mm (volume 10 μl) is covered. The area of the irradiated surface on droplet is a spherical cap of 0.3 $cm^2$.

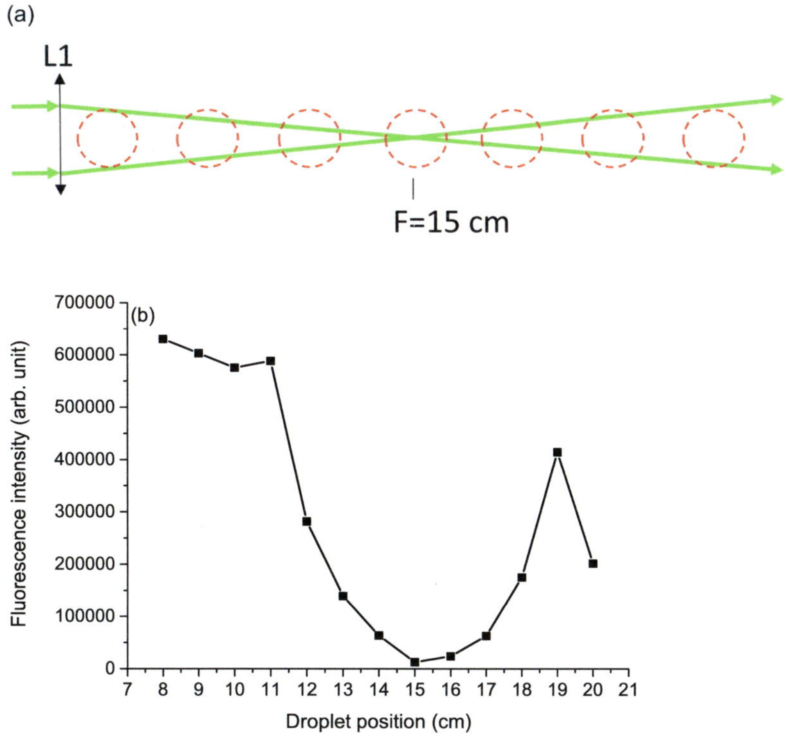


**Fig. (1).** Pumping geometry. (**a**) Droplet (red dashed circle) position function of focus point position fixed at f = 15 cm, (**b**) intensity of LIF spectra of Rh6G at $10^{-5}$ M solutions in ultrapure water obtained at different distances between the lens $L_1$ and the droplet.

As shown in Chapter 9 the same experimental set-up was used to characterise the emission spectra for pendant droplet and bulk [19]. Results are presented in Fig. (**2**) for a solution of Rh6G at $10^{-5}$ M in water.

It can be observed that for the pendant droplet in air the peak intensity of the fluorescence spectra and in fact the overall fluorescence are amplified, and the fluorescence peak reaches about an order of magnitude higher value with respect to bulk when excited with a beam having the same pumping parameters. This amplification of fluorescence signal occurs although the droplet's volume is much lower than the volume of the cuvette. In droplet case, the shape of the spectrum is changed and fluorescence peak has a bathochromic shift with respect to bulk solution emission.

According to literature reports, this occurs when active medium is placed in a spherical micro-resonator [20 - 22]. In the present case, the sphericity of the droplet makes possible multiple total internal reflections of the pumping beam and fluorescence radiation and ensures in this way a resonant cavity effect for the light emitted by the active medium.

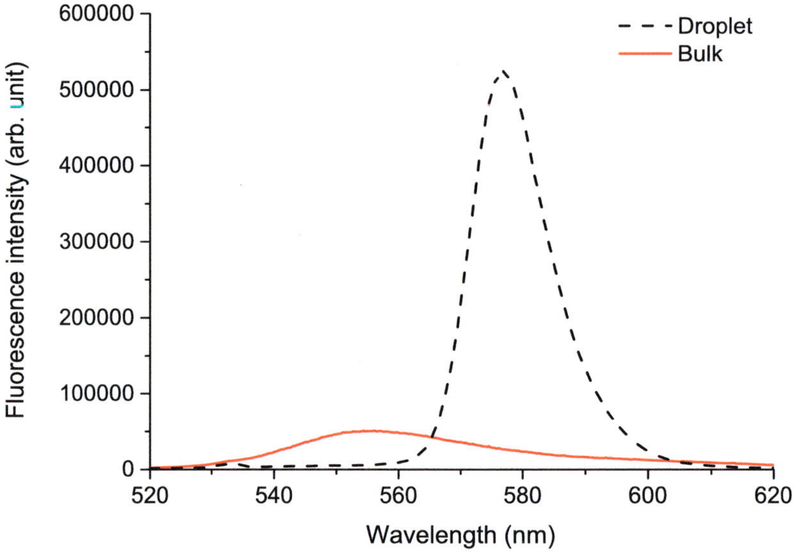


**Fig. (2).** Comparison between fluorescence spectra emitted by a 10 µl droplet and a bulk sample (placed in a spectrophotometric cuvette) containing solution ($10^{-5}$ M) in water. The pumping energy/pulse was 2.2 mJ. Each spectrum is an average of 10 pulses.

Thereby, that part of the radiation emitted by molecules that falls at the interface of droplet with surrounding air at angles larger than 48.59° (the total reflection angle) will suffer a total reflection. This radiation could interact with other excited molecules and can lead to stimulated emission. The obtained effect is an amplification of the signal in droplet compared to bulk.

Studies also followed-up the influence of Rh6G concentration in ultrapure water on LIF; concentrations from $10^{-5}$ M to $10^{-3}$ M were used and LIF results are shown in Fig. (**3**) for bulk and for pendant droplet. For LIF measured on bulk at $10^{-5}$ M samples, a typical broadband fluorescence is obtained. When concentration is increased, LIF intensity increases due to the higher number of molecules that are excited. When concentration exceeds $10^{-4}$ M, LIF intensity starts to decrease because reabsorption processes and formation of non-fluorescent dimers take place in solution. At the same time, by increasing the concentration the fluorescence bands are red-shifted which indicates reabsorption, as well. In droplet, by increasing dye concentration the intensity of the emission peak is slowly decreasing and the position of it has a strong bathochromic shift.

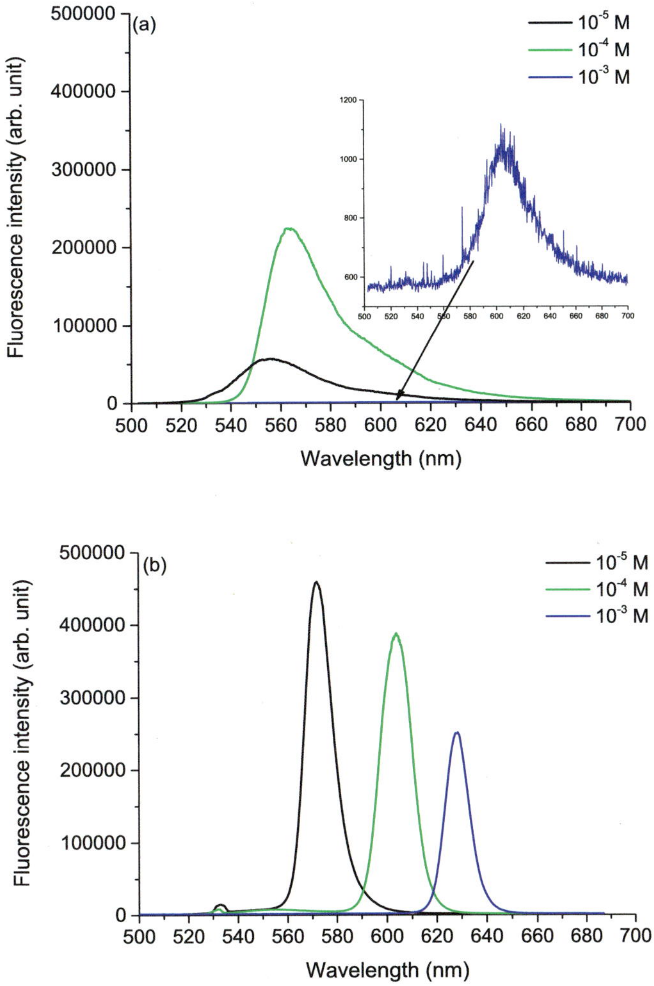


**Fig. (3).** Comparison between fluorescence spectra of solution emitted by bulk (sample in spectrophotometric cuvette) and droplet in pendant position in air at different concentrations. (**a**) the fluorescence broadband of Rh6G at several concentrations measured in 1 ml cuvette, (**b**) LIF of Rh6G at different concentrations measured in 10 µl pendant droplets in air. Pumping energy per pulse: 2.2 mJ. Each spectrum is an average on 10 pulses.

The droplet emission spectra may be measured when the signal collection geometry is changed, *i.e.* the position of the collection optical fibre is moved along a line that is parallel with the incident beam axis in the equatorial plane of the droplet. Results are shown in Fig. (**4**) that shows LIF spectra measured for three different positions of the optical fibre (front, centre and back) when dye concentration is $10^{-5}$ M in water. For optical fibre positioned in front of the droplet measured spectra have three different peaks. The first, at 532 nm represents the excitation laser beam that is reflected by the droplet's surface. The following peak belongs to a broad band emission and is positioned at 560 nm; it represents fluorescence obtained also in bulk samples measurements.

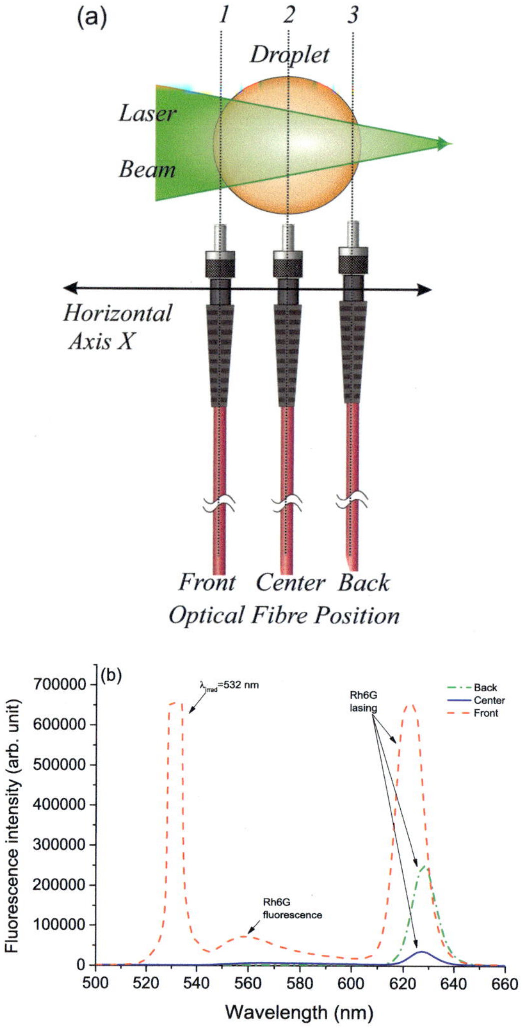


**Fig. (4).** LIF measured on droplet at several positions of collecting optical fibre. (**a**) LIF signal collection scheme, (**b**) LIF spectral distribution emitted by solution of Rh6G in water at $10^{-5}$ M measured for different position of the collecting fibre in the equatorial plane of the droplet.

The third peak at 625 nm represents the main difference between LIF spectra measured in droplet compared to bulk at the same concentration of the fluorophore. This peak originates from the amplification of the fluorescence signal inside the droplet since this acts as a spherical, closed resonant cavity.

Moving the optical fibre towards the back of the droplet (positions 2 and 3, Fig. **4a**), the reflection of the incident beam decreases as well as fluorescence intensity and only the third peak is observed. The intensity changes of the third peak with respect to the collecting position are presented in Fig. (**4b**). The peak has a maximum value for the front position of the collecting optical fibre with respect to the droplet and decreases significantly when the optical fibre reaches the position that corresponds to the centre of the droplet. When optical fibre is moved towards the back of the droplet, emission intensity increases again.

These data suggest that the geometry of irradiation leaves a number of molecules unexcited which can reabsorb the emitted fluorescence leading to a lower intensity of the collected signal measured in position two. Another observation is that lasing peak is shifted towards longer wavelengths when the optical fibre is moved from point 1 to 3. This can be explained by a gradient of concentration produced in droplet's volume or mass, due to exposure to the pumping laser beam. First layers of the droplet that interact with the laser beam could contain a higher number of destroyed molecules due to the relatively high energy of pumping laser beam, resulting in a decreasing of unexcited molecules concentration in this part of the droplet compared with its back.

In other experiment, LIF was collected at different angles around the droplet, by moving the optical fibre in its equatorial (horizontal) plane (Fig. **5a**). In Fig. (**5b**) is represented the dependence of LIF intensity function of collecting angle at a concentration of $10^{-3}$ M in water. The signal collection at the 0° angle corresponds to a signal acquisitioned at the back face of the droplet with respect to the propagation direction of pumping laser beam and along the normal at droplet surface. A signal collection at the 180° angle corresponds to the diametrically opposed position of that at 0° angle.

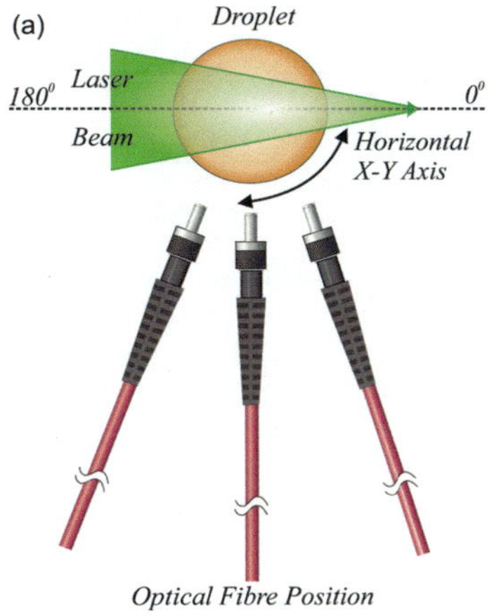


*Fig. 5 contd.....*

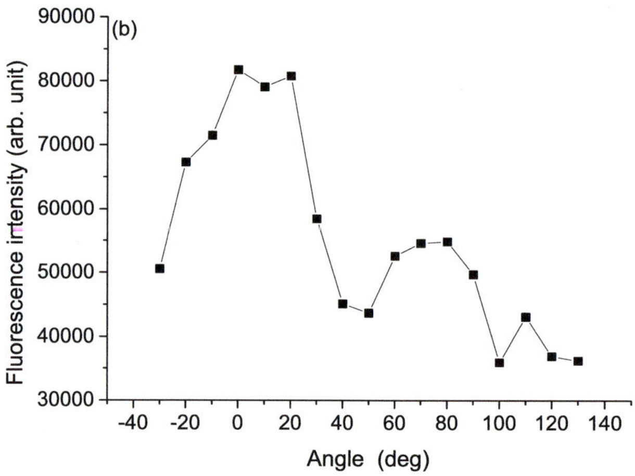


**Fig. (5).** Angular dependence of LIF intensity around the droplet (half volume only). (**a**) Sketch of set-up (top view) for LIF collection in horizontal plane; (**b**) LIF signal intensity measured at different collection angles.

## Measurements of Temporal Structure of Droplet emission

To study LIF amplification (lasing) that takes place inside a pendant droplet in air, the volume was reduced from 7 µl to 1µl. For these experiments, the laser beam was sent on the droplet in a quasi-parallel geometry, slightly convergent, through a lens with the focal length f=150 mm (Fig. **6**). Droplet was positioned at 122 mm with respect to lens in such a way that laser beam spot covers, at the limit, the entire droplet. The beam waist in the interaction plane had the linear dimension around droplet's diameter that is 1.24 mm (for 1 µl volume). Laser beam divergence was around 4.83°.

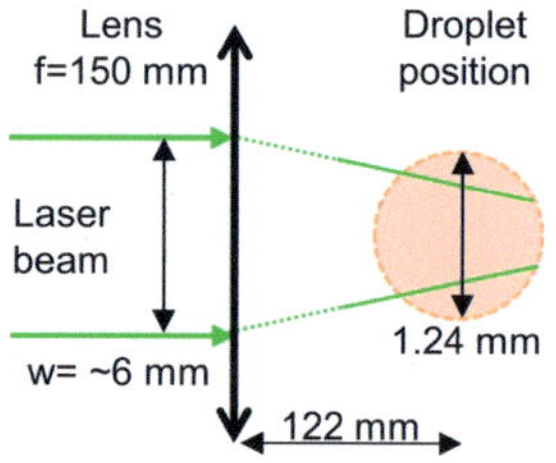


**Fig. (6).** Scheme of the irradiation geometry in amplified LIF measurements in droplets; w: laser beam waist.

Due to the specificity of the Nd:YAG laser construction, the emitted pulse width increases and the energy decreases when the Q-switch is decreasing. For laser operation at an optimum Q-switch delay (150 µs) the pulse full width at half maximum (FTWHM) is 6 ns and the beam energy is of the order of hundreds of mJ. Since the experiments were carried out at low energies, up to 20 mJ, the laser energy has been fixed at a value over, but close, to the maximum energy necessary in experiments. During experiments, the fixed laser energy was adjusted using a set of absorptive neutral density filters. In such conditions the laser pulse width was 9.14 ns large (Fig. 7) obtained for a laser Q-switch delay of 107 μs.

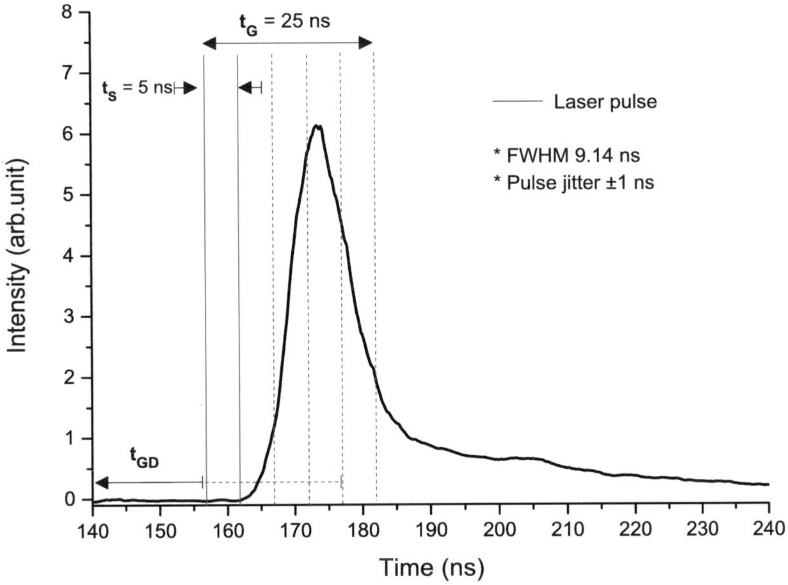


**Fig. (7).** Temporal shape of laser pulse at low laser Q-switch delay. Arrows indicate, in order, the overall gate width $t_G$ = 25 ns, the sampling gate width $t_S$ = 5 ns, and the gate delay $t_{GD}$, respectively. The vertical red lines mark the current sampling gate width and the dashed red lines mark the next sampling gates. The horizontal dashed red line marks the gate delay progressively increased in steps of 5 ns during sampling experiments.

During experiments, the spectrograph was synchronised by a TTL signal from laser through a digital delay generator. A time delay $t_{GD}$ for spectrograph gate opening was introduced with respect to TTL synchronisation signal to achieve overlapping of the acquisition gate width with the laser pulse (Fig. 7). LIF signal was integrated for full time laser pulse duration or sampled by 5 ns time window [23].

In the first case, to frame the laser pulse, an acquisition gate width $t_G$=25 ns was used. For signal acquisition by sampling, a gate width $t_S$=5 ns was set and the gate delay $t_{GD}$ was progressively increased from 5 ns to 25 ns in steps of 5 ns from one measurement to the next (Fig. 7). Thus, through five successive measurements it was scanned the entire time domain covered by LIF emission. Acquisition through the sampling method allows the temporal solving of LIF spectrum so that one can monitor and know the behaviour of LIF emission during pumping laser beam-droplet interaction.

Further, are presented LIF measurements made on pendant droplets in air with concentrations of laser dye between $10^{-5}$ M and $10^{-3}$ M. By using the technique presented above is obtained a measured deconvolution of the droplet emission for each concentration. Each spectrum is measured for a single laser pulse and one fresh droplet, with a window of 5 ns and with a delay between 0 ns and 25 ns.

In Fig. (**8**) are presented the emission spectra obtained on a pendant droplet containing laser dye at $10^{-5}$ M in water. It can be observed how the fluorescence spectra are formed by recording LIF measurements at different delays. At this concentration the fluorescence spectra show one peak at 575 nm and their shape is not changed when the delay time is modified. On the other hand, the peak intensity increases when delay time is increased, reaching a maximum for 15 ns and showing then a faster decrease that follows the intensity distribution in time of the pumping pulse. By changing the gate delay, the peak's position is moving towards higher wavelengths if compared with the case of 5 ns delay; for delays higher than 15 ns the peak is blue-shifted.

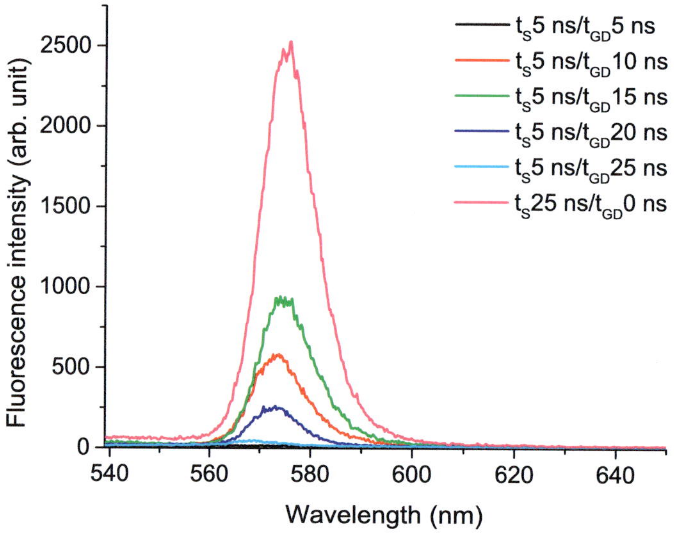


**Fig. (8).** LIF spectra measured on 1 µl pendant droplet in air, containing Rh6G solution in water at $10^{-5}$ M. Droplets were irradiated at 10 mJ pumping energy. Spectra are obtained keeping the gate width fixed at 5 ns (25 ns in last case) and modifying the gate delay in steps of 5 ns. Each spectrum is measured on first laser pulse.

By increasing the solution concentration to $5\times10^{-5}$ M, the same measurements have been performed and results are shown in Fig. (**9**). Measurements of the total emitted signal for temporal length of the pumping pulse (time gate delay 0 s and measuring time gate width 25 ns) show two distinct and well-shaped peaks: one, at 583 nm and the other at 565 nm, with 1.5 the ratio of intensities between their peaks. It was shown above that for this kind of solution measured in pendant droplets, by increasing the concentration, the emission peak is shifted to longer wavelengths. As a consequence, one may assume that the peak obtained at 583 nm for $5\times10^{-5}$ M concentration in water is analogue with the LIF peak at 575 nm for $10^{-5}$ M. The measured deconvolution of droplet emission at $5\times10^{-5}$ M concentration, highlights that the emission signal in the first 20 ns ($t_{GD}$ 10 ns and 15 ns, and $t_G$ = 5 ns) contains only the signal of the peak obtained at 583 nm. Thus, this presents an evolution in time similar with that at the previous concentration. Using the same time gate width of 5 ns, but a time gate delay of 20 ns this first peak vanishes, and the second peak arises at 565 nm.

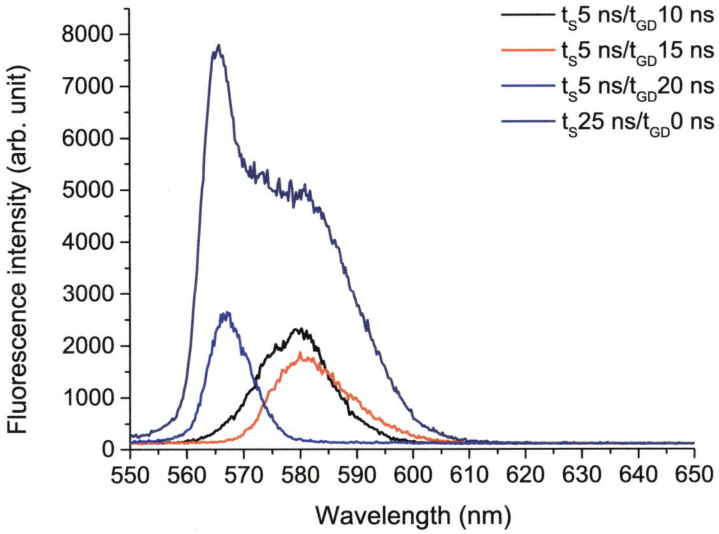


**Fig. (9).** LIF spectra measured on 1 µl pendant droplet in air containing Rh6G solution in water at $5\times10^{-5}$ M. Droplets were irradiated at 18 mJ pumping energy. Spectra are obtained keeping the gate width fixed at 5 ns and modifying the gate delay in steps of 5 ns; the last is obtained integrating the measured signal by using $t_G$ = 25 ns and $t_{GD}$ = 0 ns. Each spectrum is measured on the first laser pumping pulse.

Similar results are obtained for $10^{-4}$ M, $5\times10^{-4}$ M or $10^{-3}$ M concentrations as presented in Figs. (**10-12**), respectively.

It can be observed that by increasing the concentration to $10^{-4}$ M, the vanishing of the first peak and the rise of the second take place faster (after 15 ns). By increasing the concentration, both peaks have a red-shift, *e.g.* the first moving towards 615 nm and the second at 602 nm for $5\times10^{-4}$ M concentration in water (Fig. **11**).

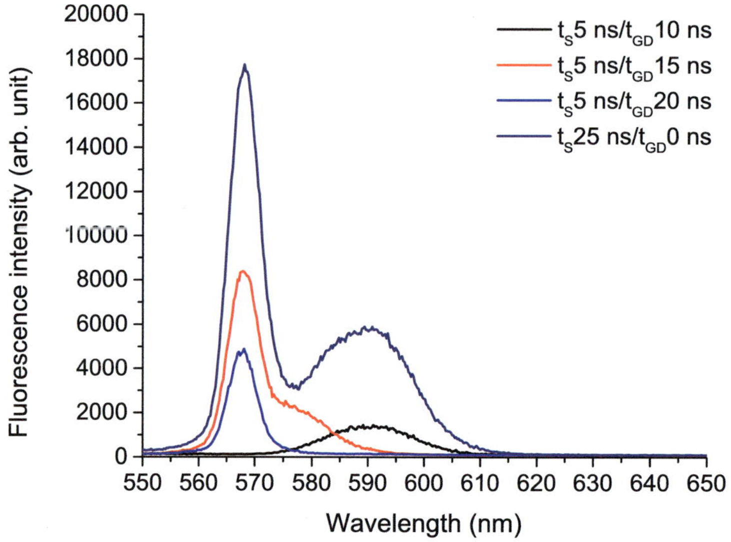


**Fig. (10).** LIF spectra measured on a 1 µl pendant droplet containing Rh6G solution in water at $10^{-4}$ M. Droplets were irradiated at 19 mJ pumping beam energy. Spectra are obtained keeping gate width fixed at 5 ns and modifying the time gate delay in steps of 5 ns. Each spectrum is measured on the first laser pumping pulse.

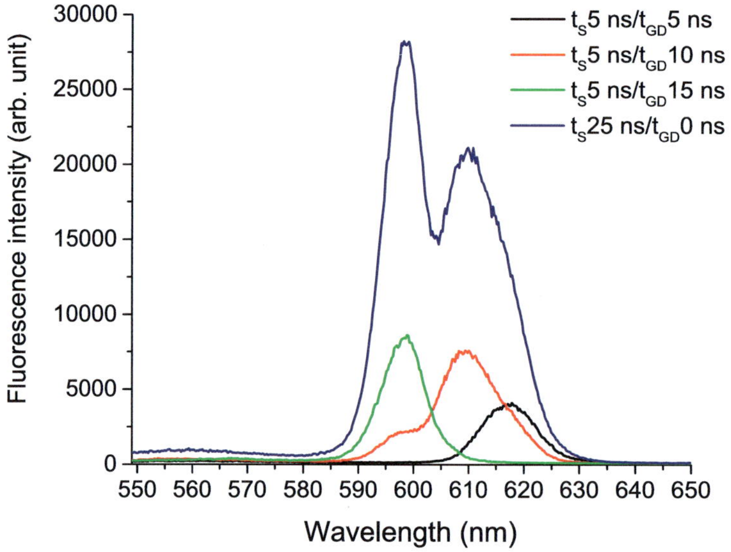


**Fig. (11).** LIF spectra measured on a 1 µl pendant droplet in air containing solution in water at $5\times10^{-4}$ M. Droplets were irradiated at 18 mJ pumping energy. Spectra are obtained keeping the gate width fixed at 5 ns and modifying the time gate delay in steps of 5 ns. Each spectrum is measured on the first laser pumping pulse.

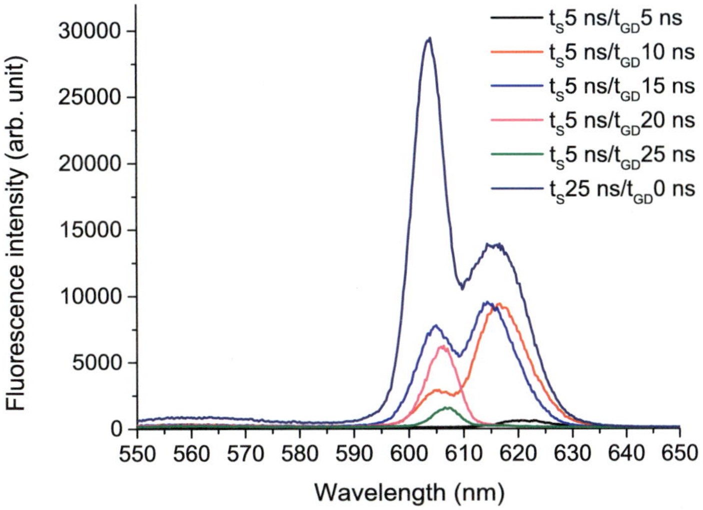


**Fig. (12).** LIF spectra measured on a pendant droplet (V = 1 µl) in air containing Rh6G in water at $10^{-3}$ M. Droplets were irradiated at 20 mJ pumping energy. Spectra are obtained keeping the gate width fixed at 5 ns and modifying the time gate delay in steps of 5 ns. Each spectrum is measured on the first laser pumping pulse.

In Fig. (**13**) are presented measurements for delay 0 ns and time gate delay 25 ns for $10^{-3}$ M Rh6G concentration by varying the energy of the pumping laser from 5 to 26 mJ. For the lowest energy at 5 mJ, the emission spectrum contains only one peak at 616 nm, and FWHM is 15 nm. By increasing the pumping laser beam energy to 10 mJ, the second peak is obtained at 604 nm with the FWHM of 7 nm. Further increasing the pumping energy, leads to the fact that the intensities of the two peaks are insignificantly changed, but their peaks are moved towards shorter wavelengths (Fig. **13a**).

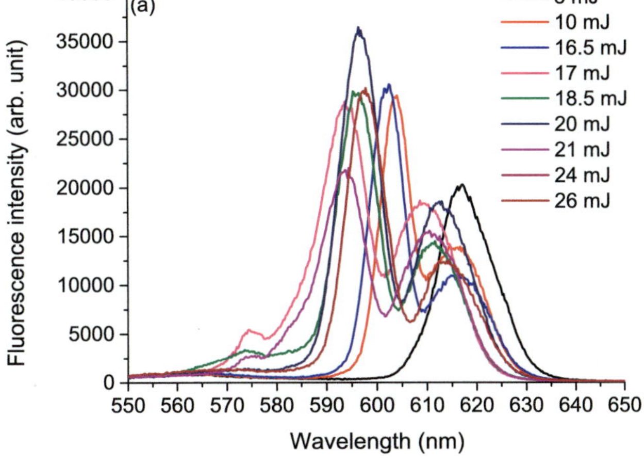


*Fig. 13 contd.....*

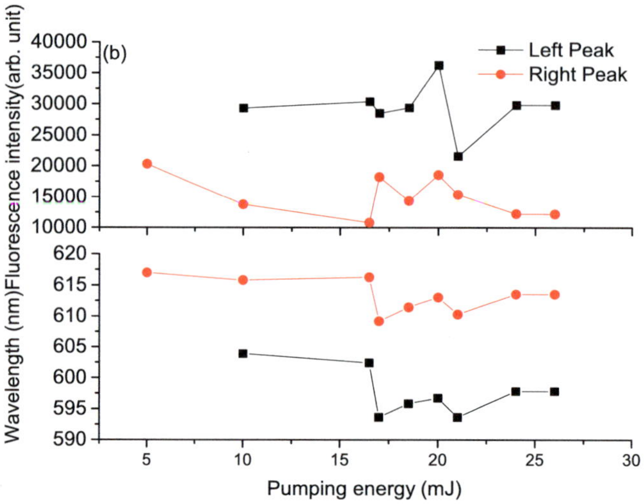


**Fig. (13).** (**a**) Laser induced fluorescence spectra measured on a 1 μl pendant droplet in air containing Rh6G solution in water at $10^{-3}$ M; droplets were irradiated at different pumping energies; spectra are obtained keeping the gate width fixed at 25 ns and the delay 0 ns; each spectrum is measured for the first laser pulse. (**b**) The dependence of the intensity and of the wavelength for both peaks function of pumping energy.

Considering the dependence of the shape of the spectra on pumping laser beam energy and on time delay, one may conclude that the second peak arises for increased energies and that its appearance is a process that supposes a threshold energy at about 10 mJ (Fig. **13b**).

In Fig. (**14**) are plotted the peaks of the amplified LIF measured for different concentrations with the window fixed at 25 ns and 0 ns delay. The energy used for irradiation, for each spectrum, varied and was chosen as the maximum energy for which the unresonant effects are negligible with respect to the resonant effects. One may observe that by increasing the dye concentration, the peaks show a bathochromic shift.

## Spectra of Rhodamine 6G Water Solutions, Doped with $TiO_2$ Nanoparticles

Solutions in pure water of Rh6G at $5\times10^{-4}$ M were doped with $TiO_2$ nanoparticles from a stock colloidal solution at the concentration 0.24 M, pH 2.5 and containing about $10^{17}$ particle/ml. The 1 μl droplets are generated in air with a computer controlled programmable system. The studied solutions were Rh6G at $5\times10^{-4}$ M in water only and Rh6G at $5\times10^{-4}$ M in water containing $TiO_2$ nanoparticles at two different concentrations, $10^{11}$ particles/cm$^3$ and $10^{12}$ particles/cm$^3$ (pH 5). The pH of the solutions was adjusted using concentrated solutions of NaOH or HCl and it was controlled with a Lab 860 pH meter.

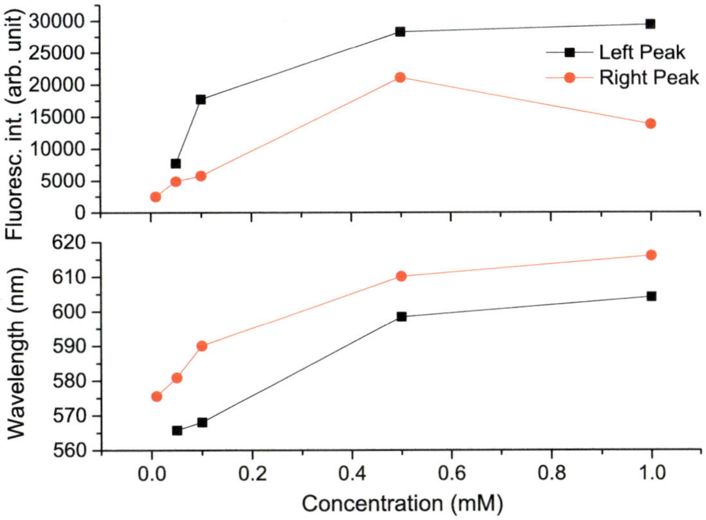


**Fig. (14).** Intensity and wavelength of LIF peak measured on a 1 μl pendant droplet in air containing Rh6G solution in water at different concentrations of laser dye. Droplets were irradiated at different pumping energies, where the resonant effect is still higher than the unresonant effects. Spectra are obtained keeping gate width of 25 ns and the delay 0 ns. Each spectrum is measured for the first pumping laser pulse on different droplets.

The experimental set-up used for LIF studies on pendant droplets is detailed in [4, 16, 24]. The laser excitation is made by the second harmonic of a pulsed Nd:YAG laser at 532 nm, with pulse FTWHM 6 ns, pulse repetition rate 10 pps, and the average laser beam energy varied between 6 mJ and 10 mJ. Fluorescence signal is collected at 90° with respect to the incident beam by an optical fibre with 400 μm core diameter. The analysis and detection of the signal is made by a spectrograph coupled with an ICCD. LIF spectra are registered pulse by pulse for sequences of maximum 200 laser pulses with a spectral resolution of 0.8 nm or 0.03 nm. The detection chain is synchronised by a Digital Delay Generator DG 535 (SRS-Stanford Research Systems).

The effect of $TiO_2$ nanoparticles, added to Rh6G water solutions in pendant droplets, on emitted LIF spectra when optically pumped was studied. LIF spectra excited at 532 nm and beam energy 6 mJ per pulse are shown in Fig. (**15a**) for pendant droplets in air that contain solutions of Rh6G at $5\times10^{-4}$ M in pure water (pH=5) without containing $TiO_2$. The spectra emitted by a droplet doped with nanoparticles at $10^{11}$ part/cm$^3$ are shown in Fig. (**15b**) and with $10^{12}$ part/cm$^3$ in Fig. (**15c**). Spectra are registered in the 560-630 nm spectral range and are exhibited for the first 10 pumping laser pulses.

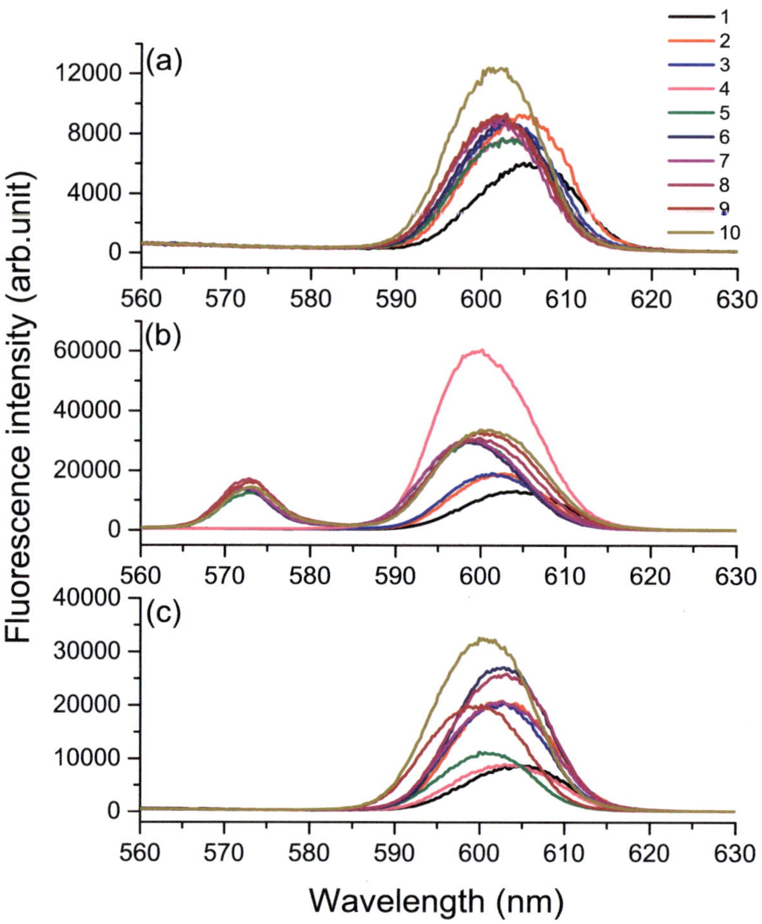


**Fig. (15).** LIF spectra of pendant droplets in air of Rh6G water solutions at $5\times10^{-4}$ M containing $TiO_2$ nanoparticles at (**a**) 0, (**b**) $10^{11}$ part/cm$^3$ and (**c**) $10^{12}$ part/cm$^3$ number density, for the first 10 pumping laser pulses. Excitation at 532 nm, laser pulse energy 6 mJ.

At 6 mJ excitation energy, one may notice small influences of $TiO_2$ addition. The dispersed fluorescence spectra exhibit a main emission band with about 15 nm FWHM and the peak in the range 601.5-605.5 nm when no $TiO_2$ is introduced in samples, or in a range shifted to the blue 598.5-604 nm with $TiO_2$ at $10^{11}$ part/cm$^3$, and 599.2-605.2 nm with $TiO_2$ at $10^{12}$ part/cm$^3$.

Addition of $TiO_2$ at $10^{11}$ part/cm$^3$ increases LIF peak intensity about five times. The concentration of $TiO_2$ that is higher with one order of magnitude, induces an about two times decrease of fluorescence intensity with respect to the case of droplet doped with $TiO_2$ at number density of $10^{11}$ part/cm$^3$.

It can be noticed that for nanoparticles number density $10^{11}$ part/cm$^3$, another emission peak appears at shorter wavelength placed at 572.6 nm and without variations of peak wavelength or intensity with laser pumping shot.

All these observations are summarised in Fig. (**16**) where the emission peak wavelength range (a) and the intensity of the emission maximum (b) are plotted *versus* laser pulse shot for all the samples excited with 6 mJ.

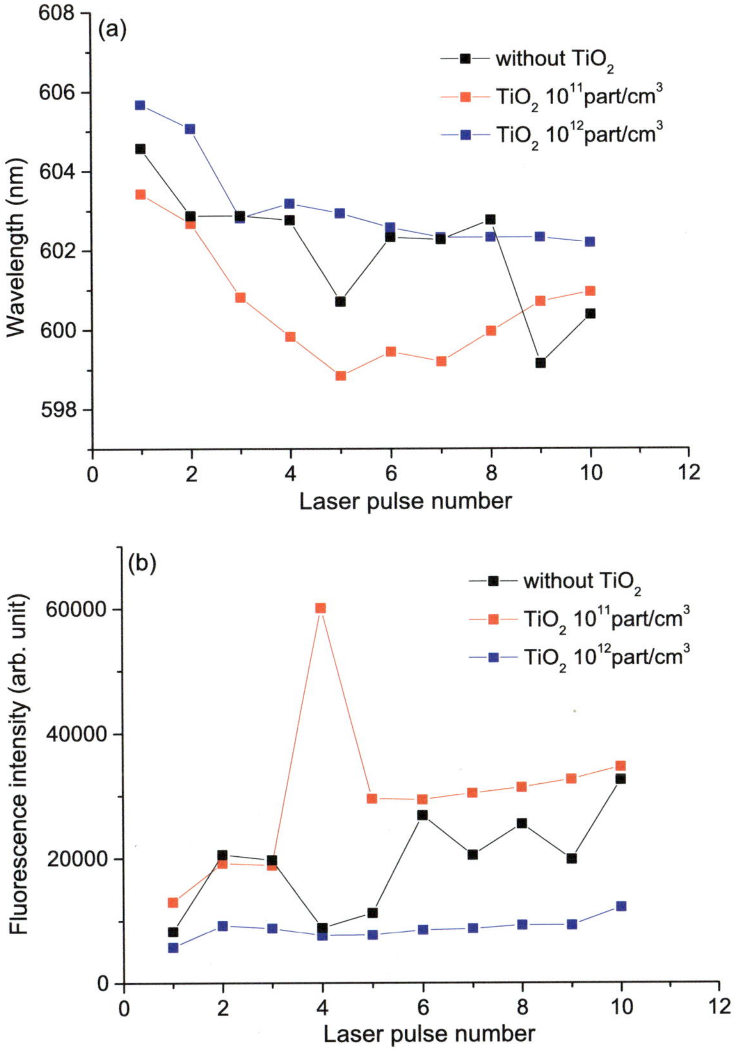


**Fig. (16).** The behaviour of the emission peak wavelength and intensity *versus* laser pulse number: (**a**) the peak wavelength *versus* laser pulse number, (**b**) the intensity of the maximum plotted *versus* laser pulse number. Excitation laser pulse energy 6 mJ.

In Fig. (**17**) the emission spectra for excitation energy of 8 mJ are shown for the same type of solutions. A very interesting behaviour of the spectra can be noticed when the $TiO_2$ number density is $10^{11}$ part/cm$^3$: besides the fluorescence peak placed at longer wavelength (the spectral range 595.3-599.7 nm), two other bands appear at shorter wavelengths that have a quite strong intensity variation from pulse to pulse.

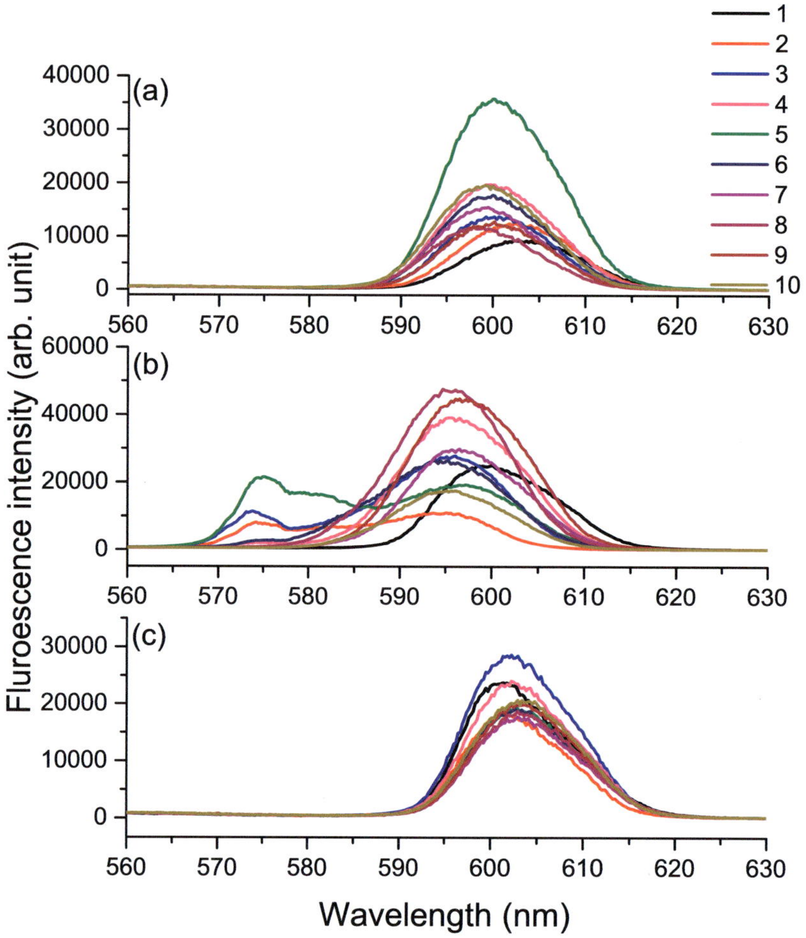


**Fig. (17).** LIF spectra of pendant droplets of Rh6G water solutions at $5\times10^{-4}$ M containing $TiO_2$ nanoparticles at (**a**) 0, (**b**) $10^{11}$ part/cm$^3$ and (**c**) $10^{12}$ part/cm$^3$ number density. For the first 10 pumping laser pulses, excitation at 532 nm, laser pulse energy was 8 mJ.

The other 2 bands that are formed are not observed for all laser shots. This is illustrated in Fig. (**18**), where the third peak's (placed towards blue) intensity and wavelength *versus* laser shot numbers are plotted.

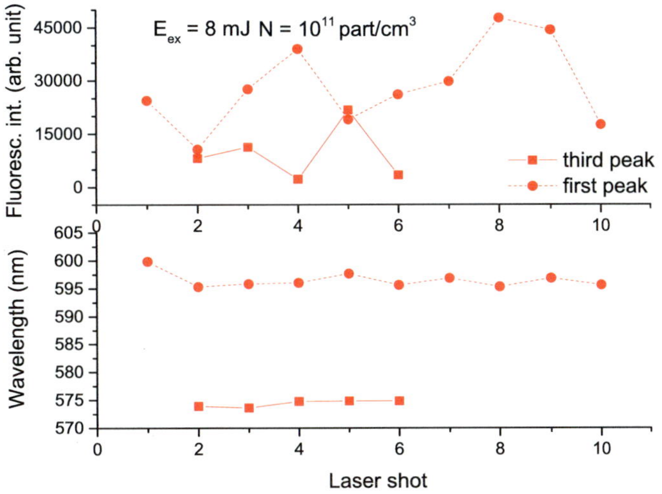


**Fig. (18).** The first and the third peaks wavelength (up) and intensity (bottom) variation with laser shot. Excitation energy was 8 mJ. Number density of $TiO_2$: $10^{11}$ part/cm$^3$.

The first peak intensity variation with laser shot for all the studied samples at 8 mJ excitation energy is shown in Fig. (**19**). The intensity of the fluorescence signal is stronger for the $10^{11}$ part/cm$^3$ concentration of $TiO_2$.

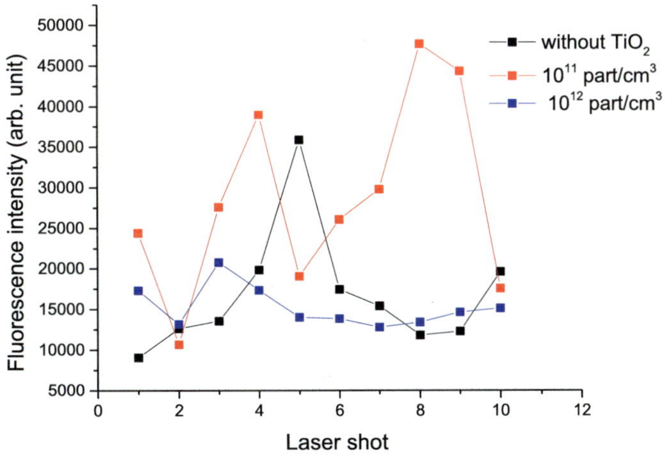


**Fig. (19).** The emission peak intensity plotted *versus* laser pulse number. Excitation energy: 8mJ. Droplet volume: 1 μl.

At this value of excitation energy (8 mJ), for the solutions without $TiO_2$ or $TiO_2$ at $10^{12}$ part/cm$^3$ concentration, the emission spectra changes are not as spectacular as in the previous case. Only one emission band is observed and fluorescence peaks are placed at wavelengths in the following ranges: 599-603 nm (for the samples without nanoparticles) and respectively 601.3-603.2 nm (for the samples with $TiO_2$ at $10^{12}$ part/$cm^3$.

Comparing these intervals, one may conclude that addition of $TiO_2$ at the higher density ($10^{12}$ part/$cm^3$) induces a shift to longer wavelengths of the LIF peak with respect to pure dye sample in contrast to the case of $TiO_2$ number density of one order of magnitude lower where the shift is towards blue. This can be observed in Fig. (**20**), where the wavelength for the first peak is shown for all the droplet samples excited at 8 mJ energy *versus* laser pulse number.

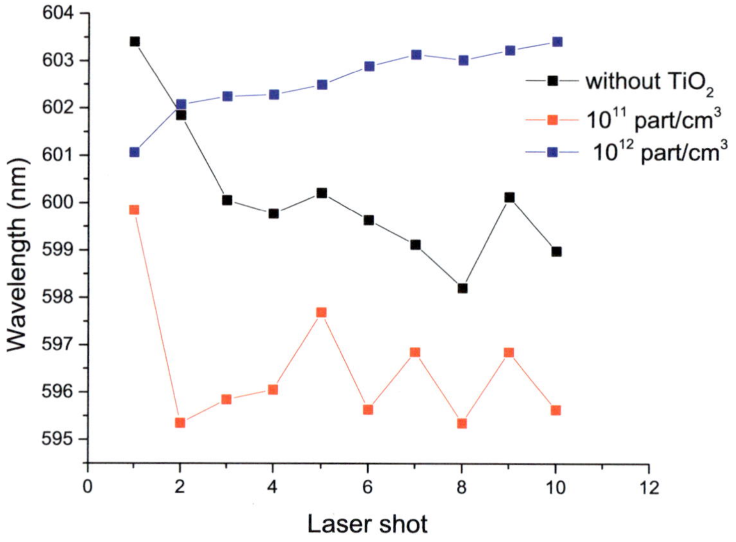


**Fig. (20).** The emission peak wavelength *versus* laser pulse number function of $TiO_2$ number density. Laser pulse energy: 8 mJ.

In Fig. (**21**), the emission spectrum for an excitation energy of 10 mJ is shown for the same type of samples: (a) without $TiO_2$; $TiO_2$ at two particular number densities: (b) $10^{11}$ part/$cm^3$ and (c) $10^{12}$ part/$cm^3$. The main LIF peak placed at longer wavelength and evidenced in all spectra is accompanied by one or two other bands placed to the blue spectral range. Therefore, by increasing the excitation energy to 10 mJ the two maxima are exhibited in the spectra even when no $TiO_2$ is added to droplet dye solution.

When nanoparticles are present at smaller concentration, the intermediate fluorescence peak vanishes and a hypsochromic shift of the other two takes place.

The $TiO_2$ number density of $10^{12}$ part/cm$^3$ in droplet sample has as consequence the disappearance of the peaks located at shorter wavelengths. Only the main peak (placed towards red) remains and its intensity and wavelength have no relevant variations from pulse to pulse.

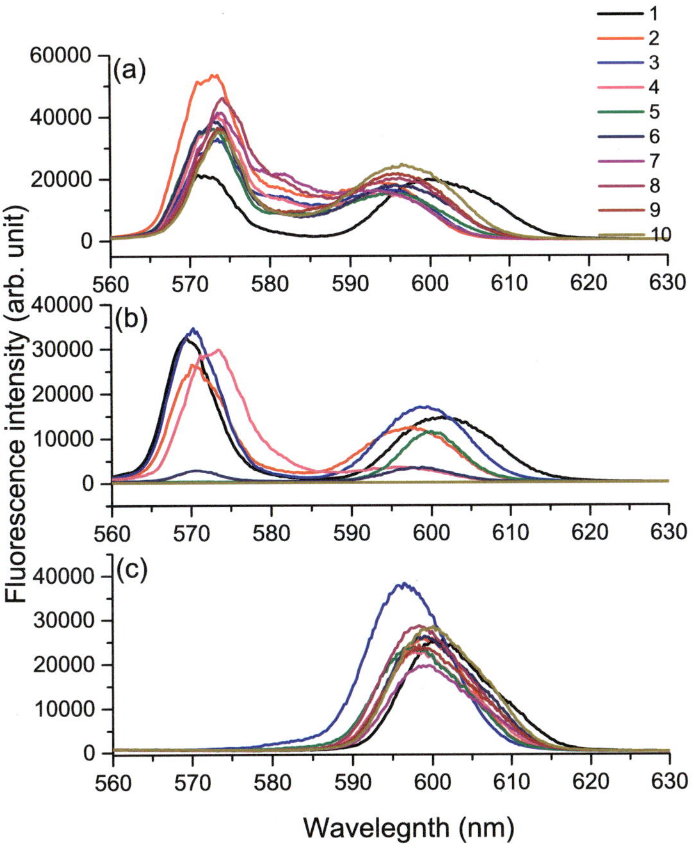


**Fig. (21).** Emission spectra of pendant droplets in air containing Rh6G solutions in ultrapure water doped with $TiO_2$ nanoparticles at: (**a**) 0; (**b**) $10^{11}$ part/cm$^3$; (**c**) $10^{12}$ part/cm$^3$ number density. For the first 10 pumping laser pulses, excitation at 532 nm, pumping beam energy was 10 mJ.

## CONCLUSIONS

In this chapter are shown measurements of laser induced fluorescence using as fluorescent medium different solutions of Rh6G in water kept as such, or doped with $TiO_2$. It was demonstrated that by changing the geometry of the fluorescent medium from bulk to pendant droplet and its volume, significant changes occur in the information obtained about LIF spectra. The main difference is an amplification of emitted fluorescence measured for pendant droplets and explained by the confinement of light by total internal reflection in the droplet that acts as a quasispherical closed optical resonator. The signals emitted by pendant droplets have been analysed function of solution's concentration and the LIF collecting geometry. There has been found that signal collection geometry has a strong influence on the shape, intensity and overall characteristics of LIF spectra.

A new technique was developed to distinguish the temporal changes of droplet emission by varying the signal acquisition time gate and time delay with respect to excitation/pumping laser pulse. The main result consists in evidencing two peaks in LIF spectra for certain conditions such as fluorophore concentration, pumping laser beam energy or droplet's volume in air. Highlighting how these two peaks evolve during irradiation, it was found that the peak obtained at longer wavelengths is formed and then vanishes in the first 15 ns from the beginning of the irradiation and, in contrast, the peak obtained at shorter wavelengths rises in the last 10 ns of the pumping pulse.

The emission spectra of microdroplets containing solutions in water doped with $TiO_2$ nanoparticles are shown. It was observed that the presence, in the droplet which contains laser dye solution, of $TiO_2$ nanoparticles induces emission spectra modifications depending on nanoparticles number density and laser beam pumping energy. A concentration of $10^{11}$ part/$cm^3$ nanoparticles favours the formation of two other emission bands that are shifted to the blue with respect to the main band. The increase with one order of magnitude of the $TiO_2$ number density produces the disappearance of these bands, regardless the pumping energy. The main emission band, *i.e.* the one placed at longer wavelength, has first a shift to the blue and then to the red with the increase of nanoparticles' concentration. At pumping energies of 6 mJ and 8 mJ, $TiO_2$ at lower concentration induces an enhancement of the emission bands with respect to pure dye solutions samples. When pumping energy is higher, *i.e.* 10 mJ, this enhancement effect is vanished.

These data suggest that $TiO_2$ nanoparticles addition to pure dye droplet solutions has an influence on the emission spectra which can be modulated by varying nanoparticles concentration and pumping laser energy.

## CONFLICT OF INTEREST

The authors confirm that this chapter content has no conflict of interest.

## ACKNOWLEDGEMENTS

This work has been financed by the National Authority for Research and Innovation in the frame of Nucleus programme-contract 4N/2016 and COST networks MP1106 and MP1205.

# REFERENCES

[1] H-M. Tzeng, K.F. Wall, M.B. Long, and R.K. Chang, "Laser emission from individual droplets at wavelengths corresponding to morphology-dependent resonances", *Opt. Lett,* vol. 9, no. 11, pp. 499-501, 1984.
[http://dx.doi.org/10.1364/OL.9.000499] [PMID: 19721646]

[2] S. Tanosaki, H. Taniguchi, K. Tsujita, and H. Inaba, "Microdroplet dye laser enhancing effects in dye-highly scattering intralipid mixture", *Appl. Phys. Lett,* vol. 69, no. 6, pp. 719-721, 1996.
[http://dx.doi.org/10.1063/1.117869]

[3] M. Aas, A. Jonáš, A. Kiraz, O. Brzobohatý, J. Ježek, Z. Pilát, and P. Zemánek, "Spectral tuning of lasing emission from optofluidic droplet microlasers using optical stretching", *Opt. Express,* vol. 21, no. 18, pp. 21380-21394, 2013.
[http://dx.doi.org/10.1364/OE.21.021380] [PMID: 24104013]

[4] M. Boni, V. Nastasa, I.R. Andrei, A. Staicu, and M.L. Pascu, "Enhanced fluorescence emitted by microdroplets containing organic dye emulsions", *Biomicrofluidics,* vol. 9, no. 1, p. 014126, 2015.
[http://dx.doi.org/10.1063/1.4913648] [PMID: 25784965]

[5] T. Alexandru, A. Staicu, A. Pascu, E. Radu, A. Stoicu, V. Nastasa, A. Dinache, M. Boni, L. Amaral, and M.L. Pascu, "Characterization of mixtures of compounds produced in chlorpromazine aqueous solutions by ultraviolet laser irradiation: their applications in antimicrobial assays", *J. Biomed. Opt,* vol. 20, no. 5, p. 051002, 2015.
[http://dx.doi.org/10.1117/1.JBO.20.5.051002] [PMID: 25365041]

[6] M.L. Pascu, B. Danko, A. Martins, N. Jedlinszki, T. Alexandru, V. Nastasa, M. Boni, A. Militaru, I.R. Andrei, A. Staicu, A. Hunyadi, S. Fanning, and L. Amaral, "Exposure of chlorpromazine to 266 nm laser beam generates new species with antibacterial properties: contributions to development of a new process for drug discovery", *PLoS One,* vol. 8, no. 2, p. e55767, 2013.
[http://dx.doi.org/10.1371/journal.pone.0055767] [PMID: 23405212]

[7] I.R. Andrei, T. Tozar, A. Dinache, M. Boni, V. Nastasa, and M.L. Pascu, "Chlorpromazine transformation by exposure to ultraviolet laser beams in droplet and bulk", *Eur. J. Pharm. Sci,* vol. 81, pp. 27-35, 2016.
[http://dx.doi.org/10.1016/j.ejps.2015.09.017] [PMID: 26432595]

[8] Y. Wu, and P.T. Leung, "Lasing threshold for whispering-gallery-mode microsphere lasers", *Phys. Rev. A,* vol. 60, no. 1, pp. 630-633, 1999.
[http://dx.doi.org/10.1103/PhysRevA.60.630]

[9] D. Melnikau, D. Savateeva, A. Chuvilin, R. Hillenbrand, and Y.P. Rakovich, "Whispering gallery mode resonators with J-aggregates", *Opt. Express,* vol. 19, no. 22, pp. 22280-22291, 2011.
[http://dx.doi.org/10.1364/OE.19.022280] [PMID: 22109070]

[10] N.M. Lawandy, R.M. Balachandran, A.S. Gomes, and E. Sauvain, "Laser action in strongly scattering media", *Nature,* vol. 368, no. 6470, pp. 436-438, 1994.
[http://dx.doi.org/10.1038/368436a0]

[11] F. Shuzhen, Z. Xingyu, W. Qingpu, Z. Chen, W. Zhengping, and L. Ruijun, "Inflection point of the spectral shifts of the random lasing in dye solution with TiO2 nanoscatterers", *J. Phys. Appl. Phys,* vol. 42, no. 1, p. 15105, 2009.
[http://dx.doi.org/10.1088/0022-3727/42/1/015105]

[12] A. Bavali, P. Parvin, S.Z. Mortazavi, M. Mohammadian, and M.R. Mousavi Pour, "Red/blue spectral shifts of laser-induced fluorescence emission due to different nanoparticle suspensions in various dye solutions", *Appl. Opt,* vol. 53, no. 24, pp. 5398-5409, 2014.
[http://dx.doi.org/10.1364/AO.53.005398] [PMID: 25321111]

[13] X. Wu, W. Fang, A. Yamilov, A.A. Chabanov, A.A. Asatryan, L.C. Botten, and H. Cao, "Random lasing in weakly scattering systems", *Phys. Rev. A,* vol. 74, no. 5, p. 53812, 2006.
[http://dx.doi.org/10.1103/PhysRevA.74.053812]

[14] B.R. Prasad, H. Ramachandran, A.K. Sood, C.K. Subramanian, and N. Kumar, "Lasing in active, sub-mean-free path-sized systems with dense, random, weak scatterers", *Appl. Opt,* vol. 36, no. 30, pp. 7718-7724, 1997.
[http://dx.doi.org/10.1364/AO.36.007718] [PMID: 18264291]

[15] A.K. Augustine, P. Radhakrishnan, V.P. Nampoori, and M. Kailasnath, "Enhanced random lasing from a colloidal CdSe quantum dot-Rh6G system", *Laser Phys. Lett,* vol. 12, no. 2, p. 25006, 2015.
[http://dx.doi.org/10.1088/1612-2011/12/2/025006]

[16] M. Boni, V. Nastasa, A. Staicu, I.R. Andrei, and M.L. Pascu, "Characterisation of Fluorescent Pendant Droplets", *Rom. Rep. Phys,* vol. 67, no. 4, pp. 1278-1287, 2015.

[17] M. Boni, A. Staicu, I. R. Andrei, A. Smarandache, V. Nastasa, Z. Saponjic, and M. L. Pascu, "Studies on the laser induced emission of microdroplets containing Rhodamine 6G solutions in water doped with TiO2 nanoparticles", *Colloid. Surface. A,* vol. 519, pp. 238-244, 2017.
[http://dx.doi.org/10.1016/j.colsurfa.2016.07.033]

[18] M.L. Pascu, G.V. Popescu, C.M. Ticos, and I.R. Andrei, "Unresonant interaction of laser beams with microdroplets", *J. Eur. Opt. Soc,* vol. 7, p. 12001, 2012.
[http://dx.doi.org/10.2971/jeos.2012.12001]

[19] M. Boni, V. Nastasa, A. Militaru, A. Smarandache, I.R. Andrei, A. Staicu, and M.L. Pascu, "Laser beams interaction with liquids in optofluidic experiments", *Rom. Rep. Phys,* vol. 64, pp. 1179-1194, 2012.

[20] R. Symes, R.M. Sayer, and J.P. Reid, "Cavity enhanced droplet spectroscopy: Principles, perspectives and prospects", *Phys. Chem. Chem. Phys,* vol. 6, no. 3, pp. 474-487, 2004.
[http://dx.doi.org/10.1039/b313370b]

[21] H. Azzouz, L. Alkhafadiji, S. Balslev, J. Johansson, N.A. Mortensen, S. Nilsson, and A. Kristensen, "Levitated droplet dye laser", *Opt. Express,* vol. 14, no. 10, pp. 4374-4379, 2006.
[http://dx.doi.org/10.1364/OE.14.004374] [PMID: 19516590]

[22] J. Ward, and O. Benson, "WGM microresonators: sensing, lasing and fundamental optics with microspheres", *Laser Photonics Rev,* vol. 5, no. 4, pp. 553-570, 2011.
[http://dx.doi.org/10.1002/lpor.201000025]

[23] M. Boni, I.R. Andrei, A. Staicu, and M.L. Pascu, "Time resolved fluorescence emission by laser pumped microliter droplets", *submitted for publication in ACS Photonics,* 2017.

[24] M.L. Pascu, I.R. Andrei, M. Ferrari, A. Staicu, A. Smarandache, A. Mahamoud, V. Nastasa, and L. Liggieri, "Laser beams resonant interaction with micro-droplets which have a controlled content", *Colloid. Surface. A,* vol. 365, no. 1-3, pp. 83-88, 2010.
[http://dx.doi.org/10.1016/j.colsurfa.2010.02.036]

CHAPTER 18

# Spectroscopy of Microdroplets: An Alternative to the Spectroscopy of Bulky Materials

**Mihail Lucian Pascu**[1,2,*], **Adriana Smarandache**[1], **Tatiana Tozar**[1] and **Ionut Relu Andrei**[1]

[1] *National Institute for Laser, Plasma and Radiation Physics, Magurele, Ilfov, Romania*

[2] *Faculty of Physics, University of Bucharest, Magurele, Ilfov, Romania*

**Abstract:** In this chapter data about the laser induced fluorescence and Raman spectra obtained from microvolumetric (some microliters) droplets are introduced when they are exposed to laser radiation of suitable characteristics in terms of wavelength, power, and focusing conditions. A comparison is made with the same spectra obtained for the same solutions and using the same laser beams, except focus characteristics, from bulky volumes (some milliliters) showing that, basically, the information obtained from a single droplet after interaction with a single laser beam pulse is the same with that obtained from the bulky sample. In some cases, function of optimisation degree of the excitation geometry, the accuracy of spectral data collected from a microdroplet is higher than for bulk. The equivalence of the spectral data obtained from droplets and bulk is due to the fact that the interaction takes place in small volumes of the samples in both cases. In droplet, the remaining solution components that have not interacted with the pumping laser beam are very few, whereas in bulk their numbers are higher and so, the radiation emitted at the interaction volume is perturbed by them. This experimental evidence leads to the recommendation to use in many cases interaction of laser beams with a single droplet to obtain reliable spectral data, rather than bulky samples.

**Keywords:** Chlorpromazine, Dimethyl sulfoxide, Droplet, Hydantoin derivatives, Laser beam, Laser induced fluorescence, Laser induced phosphorescence, Light scattering, Luminescence, Microdroplet, Microspectroscopy, Raman, Resonant interaction, Rhodamine 6G, Spectroscopy, Unresonant interaction.

## GENERAL CONSIDERATIONS

Studies about interaction of laser beams with microvolumetric droplets (further called microdroplets), require understanding of the behavior of samples having

* **Corresponding author Mihail Lucian Pascu:** Laser Department, National Institute for Laser, Plasma and Radiation Physics, Magurele, Ilfov, Romania; Tel/Fax: 0040 21 4575739; E-mail: mihai.pascu@inflpr.ro

very small volumes, of some microliters and even smaller towards nanoscale domain, at interaction with laser radiation field. In doing this, it is necessary to control, quite precisely, the mode structure of the laser beam, its waist, energy, time duration and if focused, the shape, dimensions and intensity distribution in the focus. A more accurate description of these properties is made in Chapter 7. The shape of the laser beam at input on a droplet is, in any case, critical.

Micro-spectroscopy as a term refers to the spectroscopy of samples with volumes smaller than 1 $mm^3$ (1 μl). The samples may be gaseous as is the case of optogalvanic spectroscopy in which atomic and/or molecular phenomena of relatively high temperature metal vapors are, typically, studied. More general, cold plasma devices use or may use very small volumes of plasma samples which are constituted by neutral atoms or molecules and of charged particles as well (electrons, positive and/or negative ions, possibly clusters, *etc.*). But very small liquid droplets may be also target samples in micro-spectroscopy, having volumes lower than 1 μl of liquid. The smallest sample volume which can be utilized is determined by experimenter's ability to generate and keep constant an as tiny as possible liquid volume, down to nanometric scale which is normally considered below 0.1 μl.

In micro-spectroscopy, samples may be groups of droplets in stationary position and jets or aerosols, but of particular interest are single droplets that interact with optical beams. In this case, as shown in this book, a droplet of microliter volume may hang on a capillary in air and the interaction may be produced with single laser pulses, repetitive pulses of laser radiation or even one or more cw beams. The target droplet may also be kept in levitated position using ultrasound sources or may be fixed in a certain position in a liquid with which its material is immiscible [1]. The use of sessile droplets for spectroscopic studies are used as an alternative [2 - 4]. The samples may be also microspheres of amorphous or crystalline materials that may provide spectral data about their components at interaction with optical beams [5, 6].

From spectroscopic point of view, droplets are optical "components" with the same properties as bulky objects made of the same material. Their small dimensions comparable with laser beam waist or focus dimensions are in most cases an advantage for performing accurate spectroscopic measurements. Up-to-date, the spectroscopic data that may be collected from a microdroplet are, mainly, related to laser induced luminescence (fluorescence or phosphorescence) and Raman scattering which may be obtained regardless of the sample volume from the origin point where the interaction of laser beam with the sample takes place. So, if a beam is focused, with focus diameter of some micrometers, the information about the sample is obtained from the volume described by focus, independently of how large the total volume of the sample actually is. In droplets case, the linear dimensions of laser beam are very close to the dimensions of microdroplets, so that the interaction beam-droplet implies practically all droplet's volume. A negligible number of molecules in the liquid droplet remains "untouched" by laser beam, so that the radiation that is coming out from droplet contains information only from the molecules that interacted with the beam. In other words, the excess molecules in the sample may not influence the quality of recorded signal if the interaction takes place in a droplet, comparatively with a bulky sample where signal may be jammed by the molecules that surround the interaction place.

In utilising microdroplets for spectroscopic purposes, an important requirement is to conserve their volume and shape and to keep fixed their positions during measurements. As shown in Chapter 8, spectroscopic measurements based on absorption of pumping laser radiation or on its scattering by droplet's material are accompanied by mechanical vibrations of droplets as entities, their deformations or even destruction, so that it is necessary to identify and chose the optimal laser beam parameters for such applications. Given the fact that the light forces at the utilised beam energies are quite close to the weight of the droplet, unresonant and resonant effects may interfere and this process must be minimized in order to obtain reproducible spectral data from a microdroplet. Another experimental parameter is the droplet temperature during measurements on the interaction with a pulsed laser beam. This should be kept constant so that the microdroplet does not have inner modifications due to temperature gradients and is not subject of other temperature related phenomena.

## EXPERIMENTAL DATA

### Fluorescence Spectra

In the following, experimental data about laser induced fluorescence (LIF) emitted by droplets in pendant position in air are first shown. Several variants of experimental set-ups used to perform this kind of measurements on microdroplets are introduced in Chapters 8, 9, 12 and 17. In Chapter 8, detailed comments are made about the ratio between the unresonant and resonant interaction with laser beams; then, Chapter 9 is dedicated exclusively to resonant interaction outlining the laser induced fluorescence effects obtained in microdroplets that contain different fluorophores. An example about laser induced fluorescence emitted by chlorpromazine (CPZ) in ultrapure water solution at 10 mg/ml concentration measured in bulk and droplet is shown in Fig. (**1**). In droplet experiments, the irradiation energy was 3 mJ in order to fast accumulate resonant effects and at the same time to avoid mechanical effects specific to unresonant interaction. The droplet is positioned so that the beam fully covers its droplet cross section. The beam waist has 2.5 mm diameter in the interaction plane with the droplet, *i.e.* close to droplet's diameter (considering it spherical) that is around 2.67 mm at 10 µl volume.

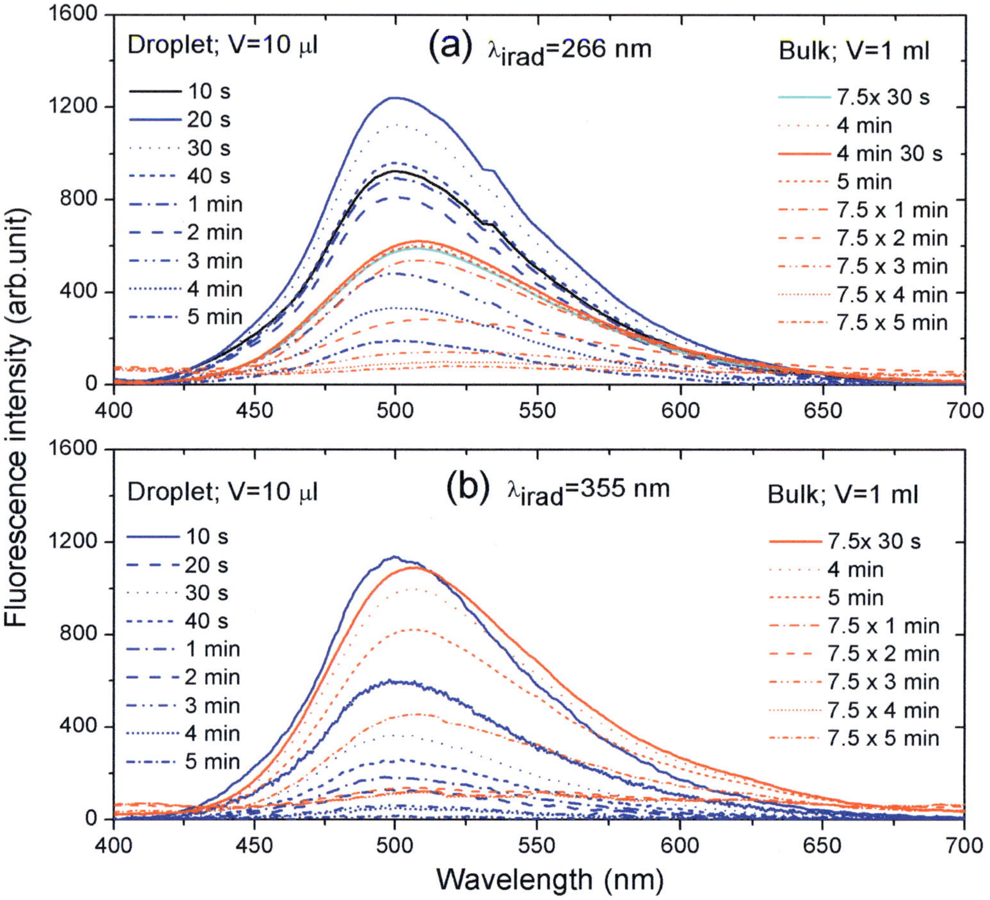


**Fig. (1).** Laser induced fluorescence spectra emitted by CPZ solution in ultrapure water at 10 mg/ml concentration, excited at (a) 266 nm and (b) 355 nm different time intervals. Irradiation times are applied to expose samples at the same irradiation doses.

The laser beams are provided by a pulsed Nd:YAG laser (Continuum, Surelite II) which emits the third (355 nm) and fourth (266 nm) harmonics of the fundamental beam (1.064 µm). The laser emits in pulsed regime 6 ns full time width at half-maximum pulses at 10 pps repetition rate. Laser beam energy used for irradiation in bulk is 7 mJ. In droplet as well as bulk irradiation, the respective laser beams are processed so that on samples they are not focused, but quasi-parallel.

In bulk experiments, CPZ solution is poured in a standard spectrophotometer cell that is positioned at 425 mm from the focusing lens; the laser beam waist is at the input on the spectrophotometer cell around 11 mm.

The fluorescence excitation is made at two wavelength 266 nm and 355 nm using the same exposure doses. The irradiation duration varies from 10 s to 5 min for droplet and from 3.75 min to 37.5 min for bulk. The irradiation time intervals were chosen so that exposure doses were the same for droplet and bulk at the irradiation geometries described above. The doses were calculated considering the average laser power incident on exposed area (average power density) during the irradiation time. The average power density calculated for droplet was 524 mW/cm$^2$ whereas for bulk it was 70 mW/cm$^2$. This means that the irradiation dose was about 7.5 times lower in bulk than in droplet for the same irradiation time. Consequently, the bulk samples are irradiated 7.5 times longer to obtain the same irradiation dose as in droplet [7].

LIF evolution for the two types of samples is the same. At 266 nm excitation, LIF increases at shorter exposure times and then a constant decrease with the irradiation time is measured, showing the decomposition of the fluorophores and the reorganization of the molecular content of the samples. The processes are easier to observe in droplet because the variations of LIF intensity from one exposure time to another are larger. At excitation with 355 nm, LIF spectra show a constant decrease in intensity for both kinds of samples.

Experiments reported in Chapter 17 go further, showing data about lasing emission by optically pumped pendant droplets that contain laser dye (rhodamine 6G) and $TiO_2$ nanoparticles.

Another example about LIF emitted by rhodamine 6G (Rh6G) in water at $5\times10^{-4}$ M concentration, when excited at 532 nm in bulk (1 ml volume) and microdroplet (1μl volume) is given in Fig. (**2**). Each spectrum is recorded for pumping with only one laser pulse for both targets and the pumping laser beam energy was 20 mJ for droplet and 100 mJ for bulk. Comparing the two spectra, in the same exposure conditions, it results that the microdroplet emits an overall LIF signal that is more intense than in bulk. A fluorescence broad band centered on 610 nm is emitted only by droplet and the lasing peak formation at about 570 nm is more important in droplet than in bulk.

LIF may be obtained not only from liquid droplets but from foam droplets as well; in Chapter 12 are shown data about this kind of spectra concluding that foam microdroplets are recommended as target samples to obtain spectral data on their constituents.

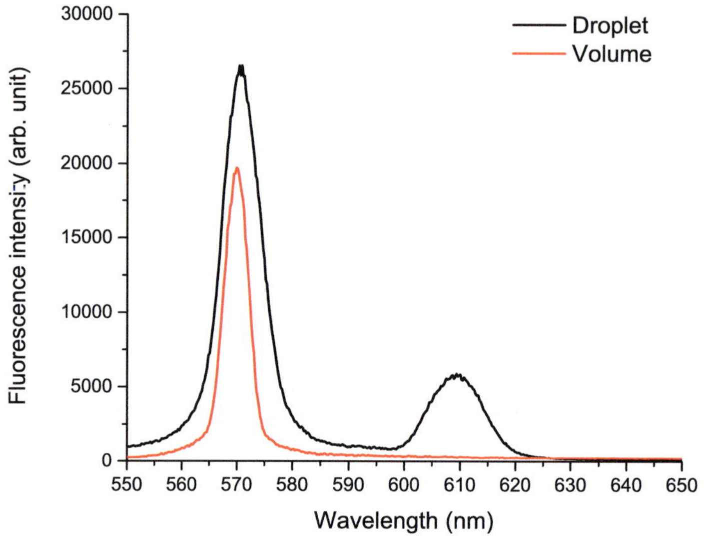


**Fig. (2).** LIF and lasing spectra emitted by Rh6G in water at $5x10^{-4}$ M excited with 532 nm by pumping a microdroplet or a bulky sample with a pulsed laser beam of 20 mJ, respectively 100 mJ energy and 10 pps pulse repetition rate. Recordings for one pumping pulse in both cases. Experimental set-up as in Chapter 17.

Another set of LIF spectra is obtained for hydantoins described in Chapter 11 and in [8]. For SZ-2 (that is non-toxic) in DMSO at $10^{-3}$ M, the spectra are shown in Fig. (**3**) when pumped at 266 nm in pulsed regime using an experimental set-up described in Chapter 17. The spectra are recorded, both for bulk and microdroplets, for 10 pulses (1 s) after exposing the target a number of seconds (up to 60 min) at 10 pulses per second repetition rate. This means that LIF spectra recorded for 1 s originate from samples previously modified by exposing them different time intervals to pumping laser pulses. This holds for microdroplet of 5.5 μl and for bulky sample of 1.5 ml with an optical path of 1 cm. The pumping laser beam energy was 6 mJ for bulk and 1 mJ for microdroplet and the exposure time was increased up to 60 min. The average 1 mJ beam energy was selected as an upper limit value in order to evaluate the effect of laser radiation on droplet's content and to prevent any modifications of the bead's shape due to unresonant laser beam action. Each fluorescence curve represents an average of 10 fluorescence curves, each curve being obtained for one pumping pulse on the same droplet. If the droplet is changed with a new one and the experiment is repeated, the reproduction of LIF curves remains within ±1% error range. The same procedure, with the same overall errors, is used for measurements at 6 mJ pumping energy.

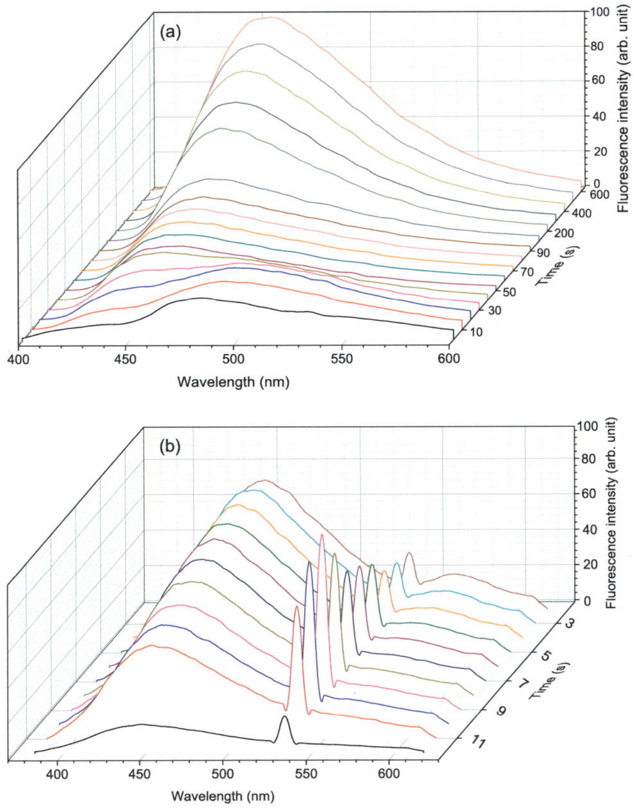


**Fig. (3).** LIF and lasing spectra emitted by hydantoin SZ-2 in DMSO at $10^{-3}$ M excited with 266 nm by pumping (a) a bulky sample of 1.5 ml volume and (b) 5.5 µl microdroplet. Pulsed laser beam energy: 6 mJ for bulk and 1 mJ for microdroplet at 10 pps pulse repetition rate. Recordings for one pumping pulse in both cases, after exposing the target a number of seconds, *i.e.* a number of pumping laser pulses.

The recordings show a different behavior of LIF spectra in bulk and microdroplet within the same spectral range. LIF modifications evidence structural changes of molecules produced after tens of minutes of bulk exposure to optical radiation, but after only a few seconds in droplet presentation.

In LIF spectra of the bulk sample of SZ-2 $10^{-3}$ M solution in DMSO is shown the evolution of two peaks (Fig. **3a**): * First, arises at approximately 435 nm and undergoes a red shift of 20 nm function of exposure time; the intensity increases with exposure time; * Second, shows-up at 485 nm, increases in intensity in the first stages of development exhibiting a 15 nm red shift in the first 20 s of laser exposure, and then gradually vanishes.

The microdroplet emits laser induced fluorescence with different spectral properties (Fig. **3b**). First, a much more rapid change of the fluorescence bands takes place when the droplet is irradiated, in terms of intensity and spectral distribution. A broad fluorescence band is measured between 400 nm and 510 nm noticing that the overall fluorescence spectral range is 400-600 nm. This band is vanished in approximately 15 s exposure to 266 nm, but during suppression, its peak walks off about 10 nm towards shorter wavelengths suggesting a destruction of the parent SZ-2 molecule. A second band is measured around 520 nm which is much narrower, evidencing a lasing effect produced in the quasi-spherical microdroplet. This peak shows also an evolution of its intensity accompanied by a walk off towards longer wavelengths after seconds of droplet's exposure to 266 nm pumping beam, until complete vanishing after 15 s.

## Raman Spectra

Other type of experiments centered on droplets as target samples refers to Raman scattering measurements. There are reports about Raman spectra obtained from water droplets when pumped at 532 nm [9]. In Fig. (**4**) is shown a Raman spectrum (obtained with a system as described in Chapter 17) of dimethyl sulfoxide (DMSO) for a bulk sample and a droplet of 7 µl volume excited at 532 nm. One may observe that the two Raman spectra contain basically the same information, although the intensity of the signal is lower in droplet than in bulk. For application on droplet, the beam was tailored to avoid droplet destruction at contact with laser pulse.

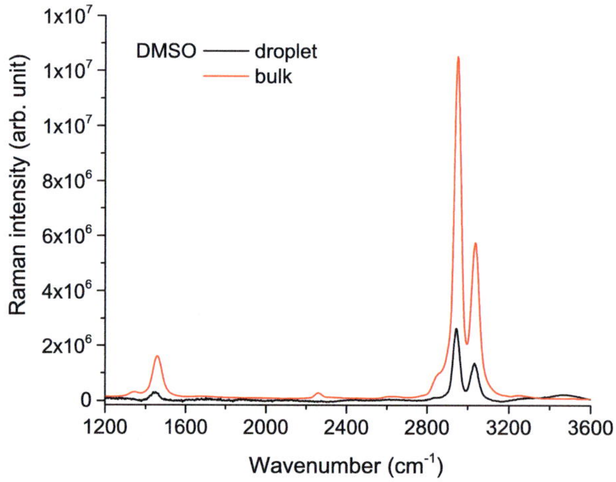


**Fig. (4).** Raman spectra of DMSO bulk and droplet when excited at 532 nm by pumping with a pulsed laser beam of 1 mJ energy at 10 pps pulse repetition rate. The used experimental set-up as in Chapter 17. Droplet volume 7 µl.

There is a decrease in intensity and a loss of several spectral lines in droplet by comparison with the spectrum registered for DMSO in bulk sample. This may be due to the fact that the laser beam energy used for Raman excitation in microdroplet is limited to, at most, a few mJ in order to avoid mechanical phenomena or breakdown that can take place at higher beam energies (see Chapters 3 and 9). Though, obtained Raman signals on beads strengthens the idea that, although it is very weak, Raman signal can be collected even from small samples such as microdroplets and this can be helpful to investigate various small samples of interest in biomolecular and biomedical studies.

## CONCLUSIONS

Comparing LIF and Raman data obtained from bulk and microdroplet samples one may conclude that the use of a single droplet for such studies is recommended since it provides the same quantity of accurate spectral data as bulk.

More, microdroplets may be used as a new kind of laser sources that emit lasing or laser radiation in the free space around them. This opens interesting perspectives for applications, the more so if one takes into account the very small volume of these type of lasers which makes them suitable for use in miniature systems.

Another advantage of droplets as samples for spectral measurements is that they are compatible with other kind of non-spectroscopic measurements such as thin layer chromatography where drops of solutions may be used to evidence compounds produced in the droplets, by exposure to laser beams, for instance. This is shown in more detail in Chapter 9.

More, beside LIF, phosphorescence studies may be performed on droplets with immediate applications such as singlet oxygen monitoring by using small sample volumes.

On the other hand, the use of microdroplets in spectroscopy experiments may have disadvantages which should be considered when tailoring measurements. One, is microdroplets sensitivity to environment conditions (temperature, humidity, air flow) due to their very small volume, which requires particular protection of samples during measurements. The second, is the possible fast evaporation of droplet material during measurements, which leads to the need to keep the droplet in stationary and protected conditions in confined environment during measurements. The third, is the sensitivity of the droplet with respect to the light beam with which it is interacting during measurements. This means mechanical stability of the droplet which imposes a recommended laser beam energy interval within which measurements should be performed. This could also mean internal temperature variations of the droplet during measurements, as well as modification of the molecular structures of its components under interaction with the laser beam.

## CONFLICT OF INTEREST

The authors confirm that this chapter content has no conflict of interest.

## ACKNOWLEDGEMENTS

Declared none.

## REFERENCES

[1] D. Foresti, M. Nabavi, M. Klingauf, A. Ferrari, and D. Poulikakos, "Acoustophoretic contactless transport and handling of matter in air", *Proc. Natl. Acad. Sci. USA,* vol. 110, no. 31, pp. 12549-12554, 2013.
[http://dx.doi.org/10.1073/pnas.1301860110] [PMID: 23858454]

[2] P. Müller, A. Kloss, P. Liebetraut, W. Mönch, and H. Zappe, "A fully integrated optofluidic attenuator", *J. Micromech. Microeng.,* vol. 21, no. 12, p. 125027, 2011.
[http://dx.doi.org/10.1088/0960-1317/21/12/125027]

[3] V. Vespini, S. Coppola, S. Grilli, M. Paturzo, and P. Ferraro, "Milking liquid nano-droplets by an IR laser: a new modality for the visualization of electric field lines", *Meas. Sci. Technol.,* vol. 24, no. 4, p. 45203, 2013.
[http://dx.doi.org/10.1088/0957-0233/24/4/045203]

[4] Y. Karadag, M. Aas, A. Jonáš, S. Anand, D. McGloin, and A. Kiraz, "Dye lasing in optically manipulated liquid aerosols", *Opt. Lett.,* vol. 38, no. 10, pp. 1669-1671, 2013.
[http://dx.doi.org/10.1364/OL.38.001669] [PMID: 23938905]

[5] M. Cai, O. Painter, K.J. Vahala, and P.C. Sercel, "Fiber-coupled microsphere laser", *Opt. Lett.,* vol. 25, no. 19, pp. 1430-1432, 2000.
[http://dx.doi.org/10.1364/OL.25.001430] [PMID: 18066238]

[6] T.T. Tam, D.Q. Trung, T.A. Vu, L.H. Minh, and D.N. Chung, "Wavelength shift in microsphere lasers", *J. Sci. Math. Phys.,* vol. 24, pp. 89-96, 2008.

[7] I.R. Andrei, T. Tozar, A. Dinache, M. Boni, V. Nastasa, and M.L. Pascu, "Chlorpromazine transformation by exposure to ultraviolet laser beams in droplet and bulk", *Eur. J. Pharm. Sci.,* vol. 81, pp. 27-35, 2016.
[http://dx.doi.org/10.1016/j.ejps.2015.09.017] [PMID: 26432595]

[8] A. Smarandache, V. Nastasa, M. Boni, A. Staicu, J. Handzlik, K. Kiec-Kononowicz, L. Amaral, and M.L. Pascu, "Laser beam resonant interaction of new hydantoin derivatives droplets for possible biomedical applications", *Colloid Surface A,* vol. 505, pp. 37-46, 2016.
[http://dx.doi.org/10.1016/j.colsurfa.2015.12.002]

[9] M. Boni, and V. Nastasa, "Andra Militaru, Adriana Smarandache, I.R. Andrei, Angela Staicu, M.L. Pascu, “Laser beams interaction with liquids in optofluidic experiments", *Rom. Rep. Phys.,* vol. 64, no. 1, pp. 1179-1194, 2014.

# SUBJECT INDEX

## A



## B

## C



## D

# E



# F

## G



## H



## I

## J



## K



## L

## M



## N

## O



## P

## Q



## R



## S

## T

## U



## V



## W



## X



## Z

Made in the USA
Las Vegas, NV
03 November 2026

57424491R00302